HVACR

Level One

SIXTH EDITION

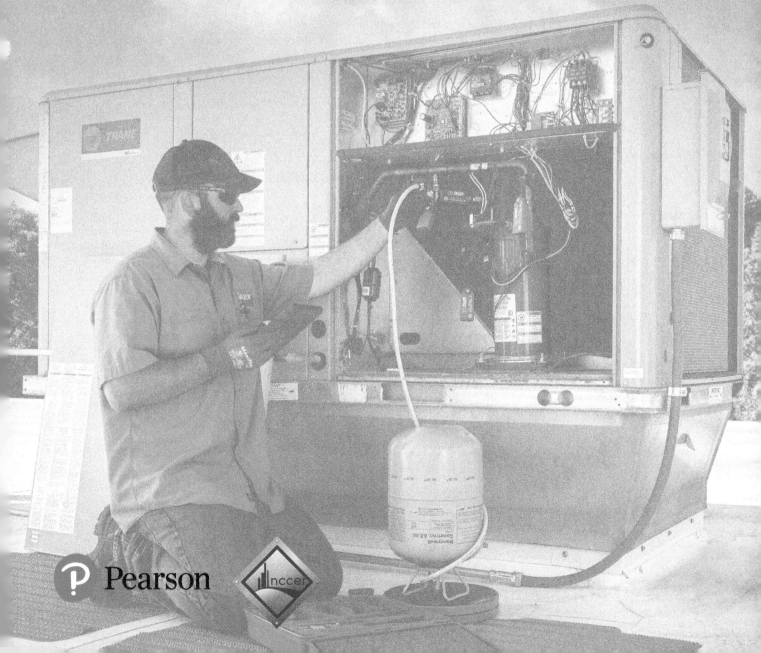

Pearson

nccer

NCCER

President and Chief Executive Officer: Boyd Worsham
Vice President of Innovation and Advancement: Jennifer Wilkerson
Chief Learning Officer: Lisa Strite
HVACR Project Manager: Lauren Corley
Senior Manager of Projects: Chris Wilson
Senior Manager of Production: Erin O'Nora
Testing/Assessment Project Manager: Elizabeth Schlaupitz
Testing and Assessment Project Coordinator: Chelsi Santana
Lead Technical Writer: Troy Staton
Art Manager: Carrie Pelzer
Production Artist: Judd Ivines
Permissions Specialist: Amanda Smith
Desktop Publishing Manager: James McKay
Production Specialists: Gene Page, Eric Caraballoso, Daphney Milian
Digital Content Coordinators: Rachael Downs, Yesenia Tejas
Managing Editor: Graham Hack
Editors: Jordan Hutchinson, Karina Kuchta, Hannah Murray, Zi Meng

Pearson

Manager of Project Management: Vanessa Price
Senior Content Producer: Alexandrina B. Wolf
Employability Solutions Coordinator: Monica Perez
Content Producers: Alma Dabral

Composition: NCCER
Content Technologies: Gnostyx
Printer/Binder: Lakeside Book Company
Cover Printer: Lakeside Book Company
Text Fonts: Palatino LT Pro and Helvetica Neue

Cover Image

Cover photo provided by: YELLOW JACKET®

6 2023

Scout Automated Print Code

Paperback
ISBN-10: 0-13-794984-7
ISBN-13: 978-0-13-794984-7

Hardcover
ISBN-10: 0-13-794983-9
ISBN-13: 978-0-13-794983-0

PREFACE

To the Trainee

The heating, ventilating, air-conditioning, and refrigeration (HVACR) industry doesn't just greatly impact the comfort of our indoor spaces. It also preserves the quality of our food supply and supports manufacturing at many levels. The spread of COVID-19 in recent years elevated the importance of indoor air quality and filtration. HVACR technicians install and maintain our refrigeration systems, ensuring our medical institutions, restaurants, and food processing facilities are maintained to the required standards. HVACR technicians serve a vital role in keeping many industries operational and contributing to the business sectors they serve.

Competent technicians are required to install, maintain, and troubleshoot HVACR systems. They must be able to read and interpret drawings or other specifications to maintain these systems. Technicians understand air movement, temperature and pressure relationships, and heat transfer applications. They often need basic knowledge in other crafts such as sheet metal, welding, pipefitting, and electrical practices to succeed in their chosen field.

The increasing advancement of technology prompts employers to recognize the importance of continuous education and keeping up with the latest equipment and techniques. Technical training and apprenticeship programs provide employers with skilled craftworkers who are well prepared to lead the way and confront new challenges.

NCCER's HVACR program has been designed by highly qualified subject matter experts. The four levels present an apprentice approach to the HVACR field, including both theoretical and practical skills essential to your success as an HVACR technician.

We wish you the best as you begin an exciting and promising career. This newly revised curriculum will help you enter the workforce with the knowledge and skills needed to perform productively in any facet of the HVACR craft.

New with *HVACR Level One*

NCCER is proud to release the newest edition of *HVACR Level One*. This book includes updates to the curriculum that will help engage you and provide the best training possible. These updates include video resources that depict relevant tasks that HVACR technicians must routinely complete.

NCCER Module 03106, *Basic Electricity*, now includes enhanced coverage of electrical safety and appropriate PPE. NCCER Module 03103, *Basic Copper and Plastic Piping Practices*, now offers information about both press-to-connect and push-to-connect fittings suitable for refrigerant piping. The Performance Tasks throughout this level were revised to reflect the most practical job tasks relevant to the trainee.

We wish you success as you progress through this training program. If you have any comments on how NCCER might improve upon this textbook, please complete the User Update form using the QR code on this page. NCCER appreciates and welcomes customer feedback. You may submit yours by emailing **support@nccer.org**. When doing so, please identify feedback on this title by listing *#HVACRL1* in the subject line.

Our website, **www.nccer.org**, has information on the latest product releases and training. Our website, **www.nccer.org**, has information on the latest product releases and training.

Your feedback is welcome. You may email your comments to **curriculum@nccer.org** or send general comments and inquiries to **info@nccer.org**.

SCAN ME

NCCER Standardized Curricula

NCCER is a not-for-profit 501(c)(3) education foundation established in 1996 by the world's largest and most progressive construction companies and national construction associations. It was founded to address the severe workforce shortage facing the industry and to develop a standardized training process and curricula. Today, NCCER is supported by hundreds of leading construction and maintenance companies, manufacturers, and national associations. The NCCER Standardized Curricula was developed by NCCER in partnership with Pearson, the world's largest educational publisher.

Some features of the NCCER Standardized Curricula are as follows:

- An industry-proven record of success
- Curricula developed by the industry, for the industry
- National standardization providing portability of learned job skills and educational credits
- Compliance with the Office of Apprenticeship requirements for related classroom training (*CFR 29:29*)
- Well-illustrated, up-to-date, and practical information

NCCER also maintains the NCCER Registry, which provides transcripts, certificates, and wallet cards to individuals who have successfully completed a level of training within a craft in NCCER's Curricula. *Training programs must be delivered by an NCCER Accredited Training Sponsor in order to receive these credentials.*

Online Badges

Show off your industry-recognized credentials online with NCCER's digital badges!

NCCER is now providing online credentials. Transform your knowledge, skills, and achievements into badges which you can share across social media platforms, send to your network, and add to your resume. For more information, visit **www.nccer.org**.

Cover Image Provider

The YELLOW JACKET® brand name is synonymous with quality HVAC/R and automotive service tools. Ritchie Engineering Company, Inc., based in Bloomington, Minnesota, engineers, manufactures, and continuously improves YELLOW JACKET® products. With an emphasis on quality and service, YELLOW JACKET® products are sold worldwide through a network of authorized HVAC&R and automotive wholesalers.

More than seventy years ago, Jack Ritchie set a goal for Ritchie Engineering to become the standard by which all other HVAC hoses are measured. In the last seven decades, the YELLOW JACKET® brand has not only become the standard in hoses, but also for tools, refrigeration gauges, manifolds, vacuum pumps and refrigerant recovery machines.

This has been accomplished through an unwavering commitment to excellence. While other companies sold out, sent work overseas or lessened quality, the relentless commitment of Ritchie Engineering employees has led to over 70 years of best-in-class products, first-class customer service and innovations that make the contractor's job easier, worldwide.

DESIGN FEATURES

Content is organized and presented in a functional structure that allows trainees to acces the information where they need it.

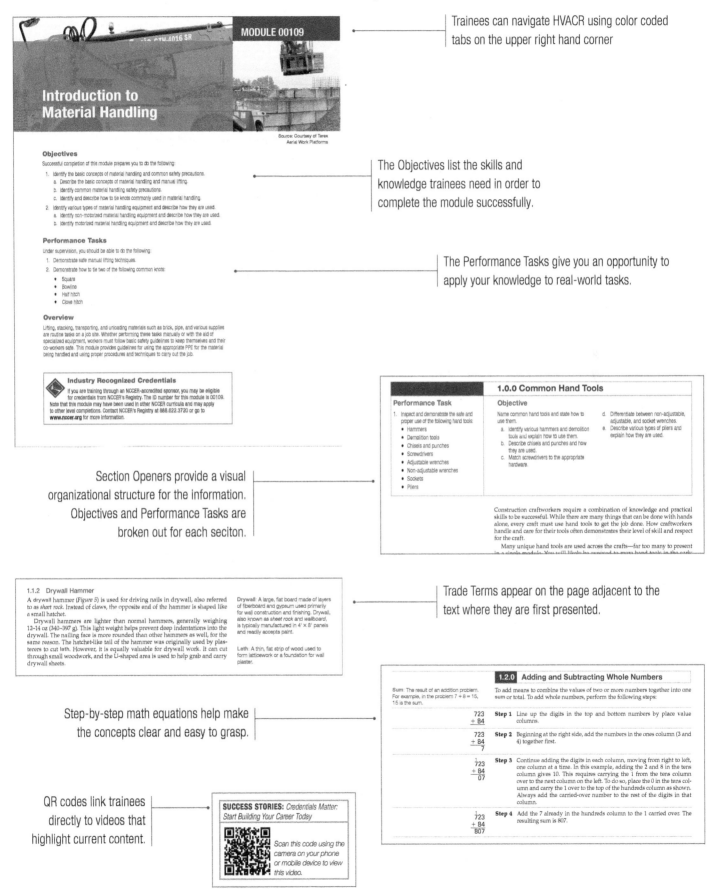

Trainees can navigate HVACR using color coded tabs on the upper right hand corner

The Objectives list the skills and knowledge trainees need in order to complete the module successfully.

The Performance Tasks give you an opportunity to apply your knowledge to real-world tasks.

Section Openers provide a visual organizational structure for the information. Objectives and Performance Tasks are broken out for each seciton.

Trade Terms appear on the page adjacent to the text where they are first presented.

Step-by-step math equations help make the concepts clear and easy to grasp.

QR codes link trainees directly to videos that highlight current content.

Important information is highlighted, illustrated, and presented to facilitate learning.

Placement of images near the text description and details such as callouts and labels help trainees absorb information.

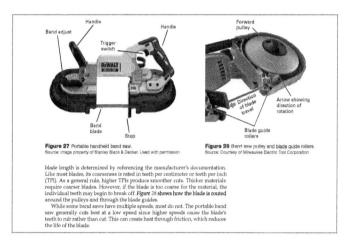

Figure 27 Portable handheld band saw.
Source: Image property of Stanley Black & Decker. Used with permission

Figure 28 Band saw pulley and blade guide rollers.
Source: Courtesy of Milwaukee Electric Tool Corporation

blade length is determined by referencing the manufacturer's documentation. Like most blades, its coarseness is rated in teeth per centimeter or teeth per inch (TPI). As a general rule, higher TPIs produce smoother cuts. Thicker materials require coarser blades. However, if the blade is too coarse for the material, the individual teeth may begin to break off. *Figure 28* shows how the blade is routed around the pulleys and through the blade guides.

While some band saws have multiple speeds, most do not. The portable band saw generally cuts best at a low speed since higher speeds cause the blade's teeth to rub rather than cut. This can create heat through friction, which reduces the life of the blade.

Preparing Drills with Keyless Chucks

Most cordless drills use a keyless chuck. While the steps for preparing a cordless drill are similar, there are some small differences. Follow the steps below when preparing to use drills with keyless chucks:

Step 1 Disconnect the drill from its power source by removing the battery pack before loading a bit.

Step 2 As shown in (*Figure 7A*), open the chuck by turning it counterclockwise until the jaws are wide enough to insert the bit shank.

Step 3 Insert the bit shank into the chuck opening (*Figure 7B*). Keeping the bit centered in the opening, turn the chuck by hand until the jaws grip the bit shank.

Step 4 Tighten the chuck securely with your hand so that the bit does not move (*Figure 7C*). You are now ready to use the cordless drill.

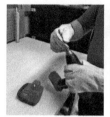

(A) Insert the Bit Shank.   (B) Keep Bit Straight and Partially   (C) Tighten the Chuck Securely.
Tighten the Chuck.

Figure 7 Loading the bit on a keyless chuck.
Source: Cianbro Corporation

New boxes highlight safety and other important information for trainees. Warning boxes stress potentially dangerous situations, while Caution boxes alert trainees to dangers that may cause damage to equipment. Note boxes provide additional information on a topic.

WARNING!

A portable band saw always cuts in the direction of the user. For that reason, workers must be especially careful to avoid injury when using this type of saw. Always wear appropriate PPE and stay focused on the work.

CAUTION

Never assume anything. It never hurts to ask questions, but disaster can result if you don't ask. For example, do not assume that an electrical power source is turned off. First ask whether the power is turned off, then check it yourself to be completely safe.

NOTE

This training alone does not provide any level of certification in the use of fall arrest or fall restraint equipment. Trainees should not assume that the knowledge gained in this module is sufficient to certify them to use fall arrest equipment in the field.

Did You Know?

Louis Henry Sullivan, an American architect in the late 19th century, created a new style of architecture that resulted in buildings that were tall but still considered beautiful, a unique concept at the time. Called the "Father of Skyscrapers," he is most known for his design of the Wainwright Building in St. Louis.

These boxed features provide additional information that enhances the text.

Orthographic Drawings

Orthographic drawings are used for elevation drawings. They show straight-on views of the different sides of an object with dimensions that are proportional to the actual physical dimensions. In orthographic drawings, the designer draws lines that are scaled-down representations of real dimensions. Every 12 inches, for example, may be represented by ¼ inch on the drawing. Similarly, in an example using metric measurements with a ratio of 1:2, every 30 millimeters may be represented by 15 millimeters on the drawing.

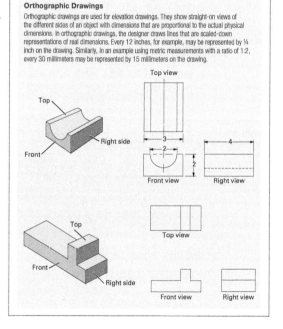

Around the World
GOST

While OSHA serves to protect workers by setting safety standards in the United States, other systems are used internationally. One such set of technical standards used on a regional basis is known as GOST. GOST standards are more far-reaching than OSHA standards, as they cover a much broader range of topics than worker safety alone. The first set of GOST standards were published in 1968 as state standards for the former Soviet Union. After the Soviet Union was dismantled, GOST became a regional standard used by many previous members of the Soviet Union. Although countries may also have some standards of their own, countries such as Belarus, Moldova, Armenia, and Ukraine continue to use GOST standards as well. The standards are no longer administered by Russia, however. Today, the standards are administered by the Euro-Asian Council for Standardization, Metrology and Certification (EASC).

Going Green
Biodiesel

Cranes and other equipment used in rigging operations consume lots of fuel—just like all the other pieces of equipment at a typical job site. Most large trucks and construction equipment run on diesel fuel. These vehicles and machines could go green and use biodiesel instead. Biodiesel is a plant oil based fuel made from soybeans, canola, and other waste vegetable oils. It is even possible to make biodiesel from recycled frying oil from restaurants. Biodiesel is considered a green fuel since it is made using renewable resources and waste products. Biodiesel can be combined with regular diesel at any ratio or be run completely on its own. This means any combination of biodiesel and regular petroleum diesel can be used or switched back and forth as needed.

But what benefits does biodiesel have over traditional fuels?

- It's environmentally friendly. Biodiesel is sustainable and a much more efficient use of our resources than diesel.
- It's non-toxic. Biodiesel reduces health risks such as asthma and water pollution linked with petroleum diesel.
- It produces lower greenhouse gas emissions. Biodiesel is almost carbon-neutral, contributing very little to global warming.
- It can improve engine life. Biodiesel provides excellent lubricity and can significantly reduce wear and tear on your engine.

Think about the environmental impact that would occur if every vehicle and piece of equipment at every job site were converted to biodiesel. The use of biodiesel also continues to increase in Europe, where Germany produces the majority of these fuels. However, even tiny countries such as Malta and Cyprus have some level of production.

Going Green looks at ways to preserve the environment, save energy, and make good choices regarding the health of the planet.

Review questions at the end of each section and module allow trainees to measure their progress.

1.0.0 Section Review

1. The person primarily responsible for your safety is _____.
   a. your foreman
   b. your instructor
   c. yourself
   d. your employer

2. The color commonly used for informational signs is _____.
   a. green
   b. red
   c. yellow
   d. blue

3. The SDS for any chemical used at a job site must be available _____.
   a. at the job site
   b. online
   c. at the contractor's office
   d. at the nearest hospital

Module 00101 Review Questions

1. The four leading causes of death in the construction industry include electrical hazards, struck-by hazards, caught-in or caught-between hazards, and _____.
   a. vehicular incidents
   b. falls
   c. radiation exposure
   d. chemical burns

2. A sign that has a white background with a green panel with white lettering is a _____.
   a. general information sign
   b. safety instruction sign
   c. caution sign
   d. danger sign

3. To properly dispose of oily rags, they must be _____.
   a. stored in a container designed for the purpose
   b. washed thoroughly and returned to use
   c. taken outdoors and thrown into a dumpster
   d. burned at the end of the shift

4. Keeping your work area clean and free of scraps or spills is referred to as _____.
   a. managing
   b. organizing
   c. housekeeping
   d. stacking and storing

NCCERCONNECT

This interactive online course is a unique web-based supplement that provides a range of visual, auditory, and interactive elements to enhance training. Also included is a full eText.

Visit **www.nccerconnect.com** for more information!

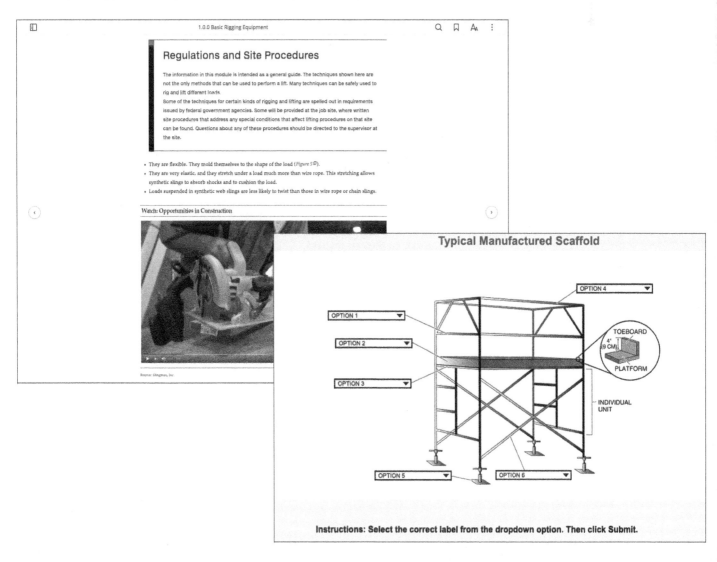

ACKNOWLEDGMENTS

This curriculum was revised as a result of the vision and leadership of the following sponsors:

Brown & Root	Epting Distributing
Central Oregon Community College	Haynes Mechanical
CMS Mechanical	Standard Heating & Air Conditioning
Epic Controls LLC	Virginia Technical Academy

This curriculum would not exist without the dedication and unselfish energy of those volunteers who served on the Authoring Team. A sincere thanks is extended to the following:

Kathleen Egan	Dave Jensen
Thomas Egan	Jim Koehn
Todd Ferrara	Joseph Moravek
David Gillespie	David G. Perkins
Allen Heath	Arthur Wilson

A final note: This book is the result of a collabortive effort involving the production, editorial, and development staff at Pearson Education, Inc., and NCCER. Thanks to all of the dedicated people involved in the many stages of this project.

NCCER PARTNERS

To see a full list of NCCER Partners, please visit:

www.nccer.org/about-us/partners.

 You can also scan this code using the camera on your phone or mobile device to view these partnering organizations.

SCAN ME

CONTENTS

Module 03106 Basic Electricity

Module 03108 Introduction to Heating

Module 03107 Introduction to Cooling

Module 03109 Air Distribution Systems

Module 03103 Basic Copper and Plastic Piping Practices

Module 03104 Soldering and Brazing

Module 03105 Basic Carbon Steel Piping Practices

HVACR   Appendixes

Introduction to HVACR

Source: iStock@Maudib

Objectives

Successful completion of this module prepares you to do the following:

1. State the basic principles of heating, ventilation, air conditioning, and refrigeration.
    a. State the basic principles of heating.
    b. State the basic principles of ventilation.
    c. State the basic principles of air conditioning.
    d. State the basic principles of refrigeration.
2. Describe the principles that guide HVACR installation and service techniques.
    a. Identify some common HVACR safety concerns and guidelines.
    b. Describe the role of HVACR technicians in energy-efficient system operation and state how LEED principles affect green-building design and construction.
    c. Describe common HVACR contractor and technician licensure requirements.
    d. Identify important codes and permits.
3. Identify HVACR career paths available and describe common training processes.
    a. Identify the personal characteristics needed to be a successful craft professional.
    b. Describe career opportunities in the various business segments of the HVACR industry.
    c. Describe how HVACR training is delivered and the general requirements and structure of an apprenticeship program.

Performance Tasks

This is a knowledge-based module. There are no Performance Tasks.

Overview

Many millions of homes and businesses across the world have a heating system. A large percentage of them have comfort cooling systems, and the majority of them also contain a refrigeration system in one form or another. Millions of stores and restaurants also house a great deal of refrigeration equipment.

Craftworkers trained in the HVACR industry have the opportunity to install systems in new construction and service equipment in existing facilities. The HVACR craft is constantly growing and changing, offering endless opportunities for those willing to learn and develop the necessary skills. This module represents the starting point of your journey into an exciting craft.

Digital Resources for HVACR

Scan this code using the camera on your phone or mobile device to view the digital resources related to this craft.

SCAN ME

Industry Recognized Credentials

If you are training through an NCCER-accredited sponsor, you may be eligible for credentials from NCCER's Registry. The ID number for this module is 03101. Note that this module may have been used in other NCCER curricula and may apply to other level completions. Contact NCCER's Registry at 1.888.622.3720 or go to **www.nccer.org** for more information.

You can also show off your industry-recognized credentials online with NCCER's digital badges. Transform your knowledge, skills, and achievements into badges that you can share across social media platforms, send to your network, and add to your resume. For more information, visit **www.nccer.org**.

1.0.0 Basic HVACR Principles

Performance Tasks

There are no Performance Tasks in this section.

Objective

State the basic principles of heating, ventilation, air conditioning, and refrigeration.

a. State the basic principles of heating.
b. State the basic principles of ventilation.
c. State the basic principles of air conditioning.
d. State the basic principles of refrigeration.

Today, the heating, ventilating, air conditioning, and refrigeration (HVACR) industry provides the means to control the temperature, humidity, and even the cleanliness of the air in our homes, schools, offices, and factories. Comfort air conditioning and refrigeration are based on the same principles, so there are many common elements in the training required for these two craft areas.

A good way to begin your journey into the craft is to learn a little about each of the terms that make up the HVACR acronym.

1.1.0 Heating

Early humans learned to burn wood as a source of heat for both comfort and cooking. That hasn't changed much in some cases. Many of us look forward to going camping and "living off the land," using the same techniques. Even in our homes, we continue to enjoy the warmth and glow of a fire, but more often than not, we do it only because we want to do it.

Thankfully, it is no longer necessary to huddle around a fireplace to stay warm. Instead, a central heating source such as a furnace or boiler does the job using the principles of **heat transfer**. We can create heat in a convenient, controlled manner and transfer that heat to where it is needed. Air and water are commonly used to help us move heat from one place to another.

For example, in a common household furnace, fuel oil, natural gas, or propane is burned to create heat, which warms metal plates called *heat exchangers* (*Figure 1*). Air from the living spaces is circulated over the heat exchangers using a fan. Heat from the heat exchangers is transferred to the air, and warmer air is then sent to the living spaces. This type of system is known as a *forced-air system*. The forced-air system is one of the most common approaches to centralized heating currently used in US homes because of its installation flexibility, performance, and installation cost.

Water is also a common heat-exchange medium for a heating system. Water is heated in a boiler (*Figure 2*), then pumped through pipes to heat exchangers. The heat it contains is then transferred to the air. The heat exchangers may be baseboard heating elements located in the space, or they may be coils with air blown across them. Systems that use water or steam as the primary heat-exchange medium are known as **hydronic** systems. In the United States, hydronic

Heat transfer: The movement of heat from a warmer substance to a cooler substance.

Hydronic: Describes a system that uses water as a heat transfer medium.

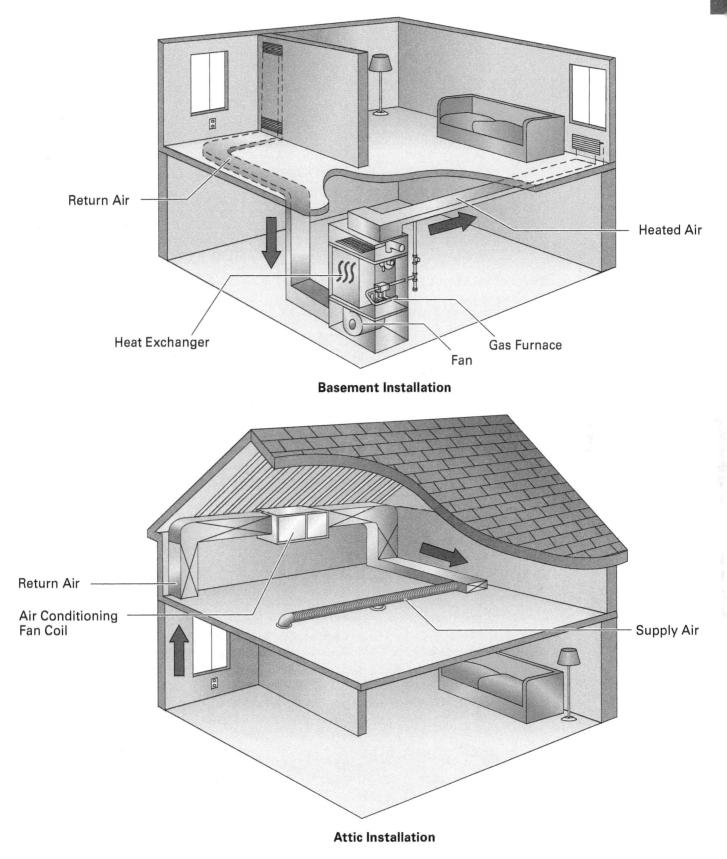

Return Air

Heated Air

Heat Exchanger

Fan

Gas Furnace

Basement Installation

Return Air

Air Conditioning
Fan Coil

Supply Air

Attic Installation

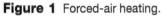 **Figure 1** Forced-air heating.

systems are common in homes built in the North and Midwest. However, they are also very common in large commercial applications and industrial facilities nationwide.

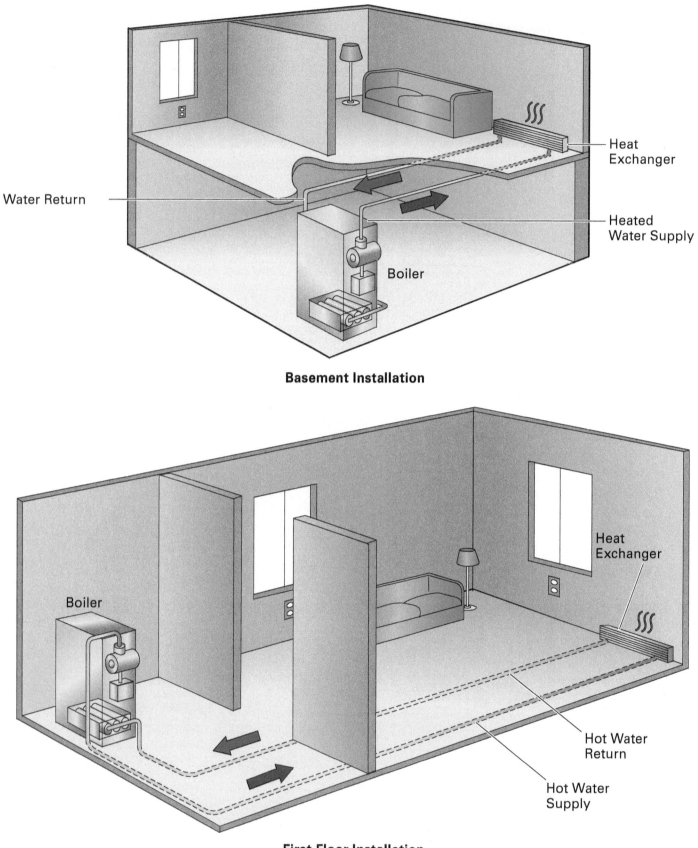

Basement Installation

First-Floor Installation

Figure 2 Hydronic heating.

Natural gas, propane, and fuel oil are the most widely used fossil fuels for heating. In urban areas where natural gas is readily available, it is the fuel of choice. Propane and fuel oil heat are common in rural areas where natural gas pipelines are not available. Both propane and fuel oil must be stored in tanks on the property where the equipment is located and must be refilled periodically from a truck.

Electricity is also a common heat source. In an electric-resistance heating system, electricity flows through coils of heavy wire, causing the wires to become hot. Air from the conditioned space is blown over the coiled wires and the heat is transferred to the air (*Figure 3*). Because electricity can be expensive, electric-resistance heating alone is no longer common in cold climates. It is more likely to be used in warmer climates where heat is seldom required.

Electric **heat pumps** are used across the United States. They are often preferred over electric-resistance heat alone, due to their superior energy efficiency and the fact that they also provide cooling. However, heat pumps are often supplemented by electric-resistance heating elements for those instances where the heat pump cannot provide sufficient heat. Although heat pumps are also a form of electric heating, they can produce heat more efficiently by using the **mechanical refrigeration cycle**.

In rural areas, a heating system may still be fueled by wood or coal. Due to increases in the cost of fossil fuels such as oil and gas over the last several decades, and the fact that wood is a relatively cheap and renewable resource, wood-burning stoves and furnaces remain popular. This is especially true for homes and businesses that are completely "off the grid," meaning that there is no access to electric utilities.

Heat pumps: Comfort systems or water heaters that can produce heat by reversing the standard mechanical refrigeration cycle.

Mechanical refrigeration cycle: A process that depends on machinery and phase-changing, heat-transfer mediums (refrigerants) to provide a cooling effect, typically using the refrigerant to absorb heat from one location and transfer it to another. May also be referred to as the *vapor compression cycle*.

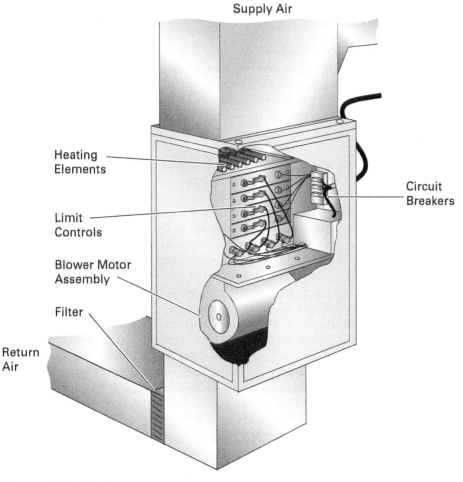

Supply Air

Heating Elements

Circuit Breakers

Limit Controls

Blower Motor Assembly

Filter

Return Air

Figure 3 Electric furnace.

Did You Know?

The Laws of Thermodynamics

There are laws that govern thermodynamic processes. However, these aren't laws established by a government. They belong to Mother Nature, but they are referred to as the *laws of thermodynamics*. The science behind each of them is quite deep, and a thorough discussion of them would require another textbook. They are presented here for your consideration, but only with the simplest possible explanations. Each of them is typically referred to by number:

- *First Law of Thermodynamics* — This one is also referred to as the *Law of Conservation of Energy*, and you may recall it from your days in high school. It states that energy can neither be created nor destroyed in an isolated system. However, it can be transferred or changed from one form of energy to another.

- *Second Law of Thermodynamics* — The second law is a little more complicated. It states that the entropy of an isolated system is always increasing. The term *entropy* refers to disorder, and disorder is always increasing. Heat seeks disorder, or chaos, by dispersing itself. If you boil water in a teapot, then set the pot on a surface, the heat will seek to move into the countertop, cabinets, and anything else connected to it, including the air around the pot. Thus, the heat energy has become more chaotic.

- *Third Law of Thermodynamics* — A perfectly crystalline substance at a temperature of absolute zero has zero entropy. In other words, there is no disorder. It is the most ordered state that matter can attain.

- *Zeroth Law of Thermodynamics* — A fourth law was developed after the first three, but it is fundamental to the others. For that reason, it isn't called the Fourth Law. It is instead referred to as the *Zeroth Law*. It states that, when two systems are in thermodynamic equilibrium with a third system, the two original systems must be in thermal equilibrium with each other. In other words, if two boxes of apples each weigh precisely the same as a third box of apples, then the first two boxes of apples must be precisely the same weight.

Again, these are concepts that require a great deal of study to fully understand and apply. As an HVACR service technician, you won't need to refer to them to do your job well, but you may find the concepts interesting and worthy of further study.

1.2.0 Ventilation

One way to describe ventilation is to say it is the introduction of fresh air into an enclosed space in order to improve air quality. Fresh air entering a building provides needed oxygen and dilutes airborne contaminants. The air in our homes, schools, and offices contains dust, pollen, and mold spores, as well as gases and odors from a variety of sources. The introduction of fresh outside air and air filtration are two common tools used to help keep the air in built environments clean and fresh.

Humidity control and exposing the conditioned airstream to **ultraviolet (UV) light** are two additional methods used to improve indoor air quality in some buildings. UV light is used in medical facilities and similar buildings to reduce or eliminate bacterial growth inside equipment and air ducts.

Many industrial environments require special ventilation and air cleaning systems. Such systems are needed to eliminate **noxious** particles and fumes that may be created by the processes and materials used at the facility.

Energy-efficient building construction that allows very little **infiltration** can contribute to indoor air quality (IAQ) problems. Building construction today aims to minimize infiltration around windows, doors, and small openings in the structure. In the 1990s, indoor air quality became a major concern, and the term *sick building syndrome* was coined to represent buildings in which the air or surfaces held excessive amounts of dust, gases, mold, viruses, or other contaminants that affected the occupants. The recognition of this problem has led to the much wider use of mechanical ventilation, electronic air cleaners, and other measures aimed at improving IAQ. The problem has also created opportunities in the HVACR industry to manufacture, sell, install, and service equipment designed to improve IAQ.

Various organizations have developed strict standards governing the quality of air in both commercial and industrial environments. Ventilation standards are an important part of the effort. In commercial buildings, indoor air must be consistently replaced with fresh air. Although various equipment and devices are used to condition and clean air, providing the occupants with a consistent supply of fresh air and the oxygen it contains is essential.

Ultraviolet (UV) light: A form of electromagnetic radiation naturally generated by the sun that has a shorter wavelength than visible light. UV light can also be generated artificially and used to destroy bacteria, viruses, and similar contaminants.

Noxious: Harmful to health; potentially poisonous or fatal.

Infiltration: The undesirable and uncontrolled movement of air into a building through cracks and crevices.

Did You Know?

Energy Efficient Ventilation

The tight construction practices of today have substantially reduced the volume of uncontrolled air that enters a building through cracks and crevices. But fresh air is still needed. One answer lies in the installation of dedicated equipment that brings fresh air in, moves stale air out, and allows the two airstreams to exchange heat as they pass through.

As shown here, fresh air is first drawn through a filter or air purifier. In the heat exchanger, heat is transferred from one airstream to the other. In the summer, cool air from the inside precools the outside air before it enters the building. In the winter, warm air from indoors preheats the outside air. Ventilation equipment that can transfer energy in both the heating and cooling modes of operation are called *energy recovery ventilators*. Equipment like this ensures that an adequate supply of fresh, clean air is consistently available indoors, while also reducing the load on the heating and cooling equipment.

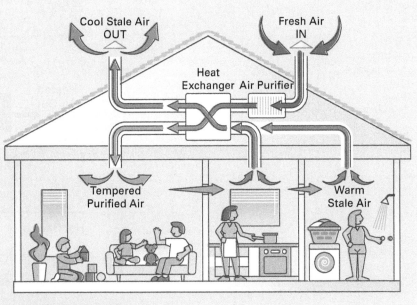

Source: iStock@VectorMine

1.3.0   Air Conditioning

The term *air conditioning* can mean a number of things. Technically, you are conditioning air when you heat or cool it, change the amount of moisture it contains, or simply filter it. But almost universally, the term has come to mean cooling an area to provide comfort.

The most common method of cooling indoor spaces is based on what is known as *mechanical refrigeration*. This method, which came into use in the early twentieth century, is based on a principle known as the *mechanical refrigeration cycle* (*Figure 4*).

Simply stated, the mechanical refrigeration cycle relies on the ability of refrigerants to transfer heat. To do this, mechanical equipment is used to manipulate the refrigerant pressure and temperature to make it colder than the air passing over it. Heat is transferred to the refrigerant as a result. Then the refrigerant is again manipulated so that it is hotter than the air outdoors. Then heat can be transferred out of the refrigerant to the outdoor air. Manipulating refrigerants this way is the foundation of the *vapor compression cycle*.

A mechanical refrigeration system is a sealed system operating under pressure. The four main elements of a mechanical refrigeration system are as follows:

- **Compressor** — The compressor provides the force that compresses refrigerant vapors, creating the pressure differential necessary for the refrigerant to flow and for the vapor-compression cycle to work.
- **Evaporator** — The evaporator is a heat exchanger where the heat in the warm indoor air is transferred to the cold refrigerant. Refrigerant typically changes state from a liquid to a vapor (vaporizes) in the evaporator before returning to the compressor.
- **Condenser** — The condenser is also a heat exchanger. In the condenser, the heat absorbed by the refrigerant is transferred to cooler outdoor air or water. Even though it may be hot outdoors, the temperature of the refrigerant flowing through the condenser is higher, allowing the refrigerant to transfer heat to the air. As the refrigerant transfers the heat away, it *condenses* to a liquid.

Compressor: In a refrigerant circuit, the mechanical device that converts low-pressure, low-temperature refrigerant vapor into a high-temperature, high-pressure refrigerant vapor.

Evaporator: A heat exchanger that transfers heat from the air or liquid flowing over it to the cooler refrigerant flowing through it, typically causing the refrigerant to vaporize.

Condenser: A heat exchanger that transfers heat from a refrigerant to air or water flowing over it, enabling the change of state from a vapor to a liquid.

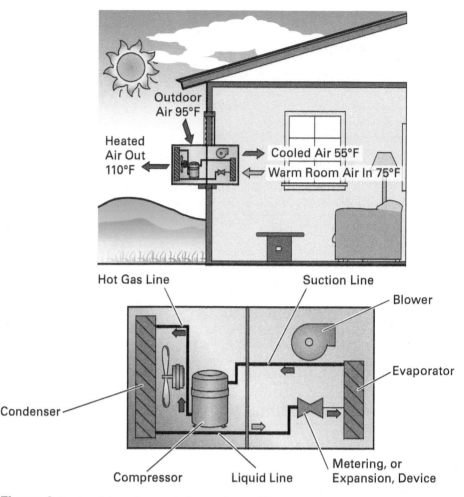

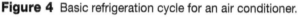

Figure 4 Basic refrigeration cycle for an air conditioner.

Expansion device: A device that provides the pressure drop to reduce high-pressure liquid refrigerant to a lower pressure and significantly reducing the temperature at which it boils. Also known as a *metering device*.

- **Expansion device** — The relatively simple expansion device is essential to the refrigeration cycle. It provides a pressure drop that lowers the pressure and boiling point of the refrigerant as it enters the evaporator. This causes the refrigerant to become a cold liquid/gas mixture and to absorb heat in the evaporator. Expansion devices are also referred to as *metering devices*.

Heat pumps can produce heat by reversing the cooling cycle (*Figure 5*). The basic operating principle of a heat pump is that there is some heat contained in the air outdoors, even though the air may be very cold. In fact, the temperature would have to be –460°F (–273°C) for a total absence of heat to exist. In the heating mode, a special valve, known as a *reversing valve*, switches the flow of refrigerant so that the condenser operates as the evaporator and the evaporator becomes the condenser. Because of this role reversal, the coils in a heat pump are referred to as the *outdoor coil* and *indoor coil* instead of the condenser and evaporator. Heat pumps are also used to extract heat from a water source.

The relationship between temperature and pressure is a critical factor in the vapor compression cycle. As you study the process in NCCER Module 03107, *Introduction to Cooling*, you will learn how the same refrigerant can be very cold at one point and very hot at another. This is possible because of pressure changes caused by compression and the action of an expansion device.

This is a very simple description of the refrigeration cycle with many details yet to be explored. Later in your training, this subject is explored in far greater detail. The relationship between temperature and pressure in the refrigeration cycle will become clear, and you will apply that knowledge regularly. It is essential to understanding and successfully troubleshooting the mechanical refrigeration cycle.

The mechanical refrigeration cycle is largely the same for all air conditioning and refrigeration equipment, from the smallest refrigerator to the system

Hermetic and Semi-Hermetic Compressors

Hermetically sealed compressors are typically used in residential and light commercial air conditioners and heat pumps, as well as in smaller packaged refrigeration units. Semi-hermetic compressors, also known as *serviceable compressors*, are used in larger refrigeration and air conditioning systems. Semi-hermetic compressors can be partially disassembled for repair in the field. Hermetic compressors are sealed and cannot be repaired in the field. Only the largest hermetic compressors are worthy of repair and refurbishment at the manufacturing level. They are otherwise scrapped when they fail.

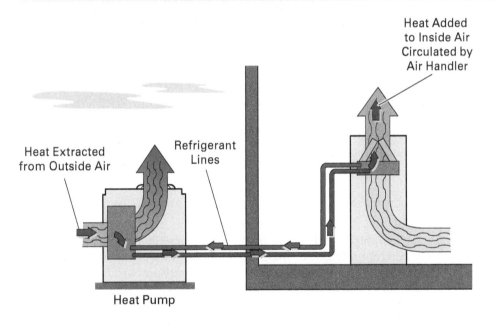

Heat Pump in the Heating Mode

Figure 5 Heat pump operation.

cooling the largest office building. Although there are certainly differences in the construction of these systems, the refrigeration cycle remains fundamentally the same.

1.4.0 Refrigeration

The same mechanical refrigeration cycle is also used in refrigeration equipment such as the coolers and freezers found in supermarkets and convenience stores (*Figure 6*). The most significant difference between comfort cooling and refrigeration is the operating temperatures (*Figure 7*). Applications above 60°F (16°C) are generally considered air conditioning, while those below 60°F are considered refrigeration.

Figure 6 Retail refrigeration equipment.
Source: iStock@danielvfung

Temperature Range

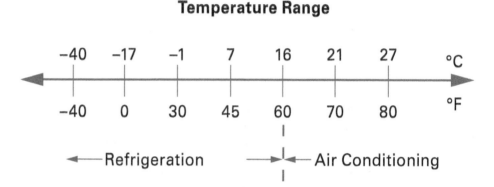

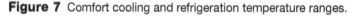

Figure 7 Comfort cooling and refrigeration temperature ranges.

Because of the lower temperatures at which refrigeration equipment operates, refrigerants with different characteristics are often used. Freezers may also use a different refrigerant than refrigerators that operate above freezing. The characteristics of the chosen refrigerant are very important to maximize heat transfer and efficiency in all applications. This is a choice usually made by engineering teams working on behalf of a manufacturer.

Equipment based on the mechanical refrigeration cycle is used for domestic refrigerators as well as for commercial and industrial applications. Warehouses and distribution centers are often equipped with very large coolers and freezers. Food processing plants, such as meatpacking plants, dairies, and seafood processors use refrigeration equipment on a large scale. Refrigeration equipment is also needed for ships, trains, and trucks to bring perishable food to market.

Another important difference between comfort cooling and refrigeration is the need to defrost the evaporator coil in order to eliminate frost buildup. Frost blocks airflow and impedes heat transfer. A refrigeration evaporator coil must often operate below freezing, usually 10°F to 20°F (~6°C to 11°C) colder than the box temperature. To keep a refrigerator around 45°F (~7°C), for example, the evaporator must be near or below freezing. Frost will form on the coil surface when the system runs. At this box temperature though, the frost will melt when the system turns off, since the air in the box is well above freezing. Coils serving freezers, though, are a different matter.

A coil can be defrosted in several ways. Electric heating elements on the coil face are common. Hot refrigerant vapor can also be routed through the coil for a short time. Both approaches are often controlled by timers that periodically initiate a defrost cycle, just like your freezer at home. Sensors can also be used to detect the ice buildup and start a defrost cycle only when it's necessary. Since any heat added to the conditioned space by the defrost cycle must be removed again, defrost cycles should be minimized.

1.0.0 Section Review

1. In a hydronic heating system, the *primary* heat transfer medium is _____.
   a. air
   b. water
   c. natural gas
   d. ammonia

2. Ultraviolet light is a method used to _____.
   a. heat many residences
   b. control bacterial growth
   c. create air movement
   d. produce a cooling effect

3. In the mechanical refrigeration cycle, the purpose of an evaporator is to _____.
   a. convert refrigerant from high pressure to low pressure
   b. transfer heat from the refrigerant to the outdoor air
   c. transfer heat from the indoor air to the refrigerant
   d. provide the pressure difference to circulate refrigerant

4. In a refrigerator that is set for a box temperature of 45°F (~7°C), how would you expect the evaporator coil to be defrosted?
   a. By letting the air in the box melt the frost between operating cycles.
   b. By sending hot refrigerant gas through the coil.
   c. By energizing electric heaters mounted on the coil.
   d. By moving the food out and shutting the system down for a day or two.

2.0.0 Guiding Principles for HVACR Service Technicians

Objective

Describe the principles that guide HVACR installation and service techniques.

a. Identify some common HVACR safety concerns and guidelines.
b. Describe the role of HVACR technicians in energy-efficient system operation and state how LEED principles affect green-building design and construction.
c. Describe common HVACR contractor and technician licensure requirements.
d. Identify important codes and permits.

Performance Tasks

There are no Performance Tasks in this section.

There is more to becoming a successful HVACR service technician than just learning the technical aspects of the craft. The successful technician always follows good safety practices; is conscious of the need to conserve energy and protect the environment; and follows applicable codes and standards. These principles are introduced here, but they are topics that you must learn far more about as your career progresses.

2.1.0 Safety

The subject of on-the-job safety, as defined by Occupational Safety and Health Administration (OSHA) regulations, was covered extensively in NCCER Module 00101, *Basic Safety*. In addition to the general safety practices covered in that module, there are many safety concerns specific to HVACR work. These concerns include the following, but note that this is just a small sample of the safety guidelines you will encounter as your training progresses:

- *Working at height* — HVACR installation and service work often require a technician to work from a stepladder or service equipment on a rooftop. Whenever you are working at height, personal fall protection equipment must be used when working more than 6' (1.8 m). Ladders must be positioned properly and secured.

- *Working around gas furnaces* — Natural gas and propane are both flammable and explosive. Vaporized fuel oil is also highly flammable. Always check gas supply lines for leaks following their installation or service. A gas leak can lead to an explosion with devastating results. The combustion of these gases in a furnace produces several byproducts, including deadly **carbon monoxide (CO)**. A cracked heat exchanger or an improperly installed furnace vent can allow CO to be released into the occupied space. As an HVACR technician, you have a responsibility to consistently consider the safety of your clients.

- *Working around oil furnaces* — Fuel oil is flammable, and when its vapors mix with oxygen, it forms an explosive mixture. Each attempt to restart an oil burner will inject a small volume of oil into the combustion chamber. A homeowner may try to restart a furnace several times before calling for service. To avoid a possible explosion, always check the combustion chamber and allow any accumulated oil to vaporize and leave the chamber before attempting to restart the burner.

There are a number of safety issues associated with refrigerants and oils, including the following:

- In a confined or enclosed space, released refrigerant vapors can displace oxygen and potentially cause suffocation. The flammability of some refrigerants must also be considered. An HVACR technician must be familiar with the properties of the oil and refrigerant present in the system.

- Since many common refrigerants boil at temperatures well below 0°F (–18°C) when exposed to atmospheric pressure, serious injuries can result from contact with liquid refrigerants when they are released from their normally pressurized containers or equipment. As they rapidly boil and change to the vapor state, massive amounts of heat are absorbed, leading to the potential for frostbite (*Figure 8*). Learn to work with gloves early in your career. Today's gloves offer great protection and fit much better than the work gloves of yesteryear.

- **Anhydrous** *ammonia* is used as a refrigerant in large food processing and storage applications. Exposure to substantial amounts of ammonia vapors can cause eye, skin, and mucous membrane irritation. Before working with ammonia, study the safety guidelines carefully and be prepared with the right equipment. You won't encounter it unless you're working in food processing or similar industries.

- When some oils mix with oxygen, an electric arc or open flame can cause an explosion. Oxygen is highly flammable and must never be used to purge a refrigerant circuit or any similar task. Nitrogen is a much safer, non-flammable gas to work with and you will often use it when servicing or assembling refrigerant piping circuits.

Carbon monoxide (CO): A colorless, tasteless, odorless, and toxic gas that is lighter than air and a common byproduct of combustion processes. CO reacts with blood hemoglobin to form a substance that significantly reduces the flow of oxygen to all parts of the body.

Anhydrous: Containing no water.

Symptoms of Frostbite

1st Degree: Skin Whitening, Reduced Sensitivity

2nd Degree: Emergence of Blisters

3rd Degree: Death of Skin and Subcutaneous Tissues

4th Degree: Death of Skin, Soft Tissues and Bones

Figure 8 Symptoms of frostbite.
Source: iStock@Irina Kuznetsova

- Lubricants used in HVACR systems aren't inherently hazardous. But they are generally considered to be hazardous waste and must be disposed of appropriately. Some oils can severely damage roofing membranes used on commercial low-slope roofs. Handle oils with care and avoid spilling them on the roof. Clean up any spills as quickly and as thoroughly as possible.

Again, these precautions represent just a few of the many safety concerns related to HVACR work. Making safety a priority in your career will allow you to analyze each situation carefully and use techniques that minimize any hazards. Completing each workday without injury is an essential goal that every craftworker must share and support.

Refrigerant History: Sulfur Dioxide

Although all refrigerants can be hazardous in the wrong hands, some early refrigerants were quite deadly. Sulfur dioxide (SO_2), a highly toxic gas, was used as a refrigerant in domestic refrigerators in the early 1900s. It has a pungent, suffocating smell and the vapors are heavier than air, making it particularly dangerous. Significant leaks of SO_2 from refrigeration equipment led to some deaths from respiratory arrest. Survivors of a significant encounter can develop bronchitis, pneumonia, and other serious illnesses due to lung damage. SO_2 was replaced as a refrigerant by safer products around 1929. However, it is important to note that SO_2 is still used today in various food processing and fumigation applications.

2.2.0	**Energy Conservation and LEED Green Building Principles**

HVACR systems are significant energy consumers. Per the US Department of Energy (DOE), HVAC systems consume about half of the energy consumed by an average home. With roughly three million HVAC systems replaced in the United States each year, replacing them with more efficient equipment is an important step in reducing energy consumption nationwide.

A number of manufacturers have partnered with the DOE and the US Environmental Protection Agency (EPA) in an ENERGY STAR® program to market units that exceed the minimum efficiency standards. Equipment qualifying for the program are listed at **www.energystar.gov**, and the equipment packaging is clearly marked with the ENERGY STAR logo to draw the attention of consumers.

Building designs may also qualify for ENERGY STAR programs by including HVACR systems and equipment that minimizes energy consumption. Although equipment such as furnaces and **condensing units** are the primary consumers of energy in HVACR systems, many other devices such as programmable thermostats and even light bulbs can be ENERGY STAR certified.

Condensing units: Cooling equipment that typically contains a compressor, condenser coil, condenser fan, and other devices required for compressor control and operation.

Figure 9 Dirty air filter.
Source: iStock@Grandbrothers

Mastic: A flexible sealant that is painted or sprayed to seal duct joints and insulation seams.

A high-efficiency HVACR system only remains effective if it is properly maintained. Clogged coils, dirty air filters (*Figure 9*), improper airflow settings, and incorrect refrigerant charges can substantially reduce energy efficiency as well as performance. Performing periodic maintenance is important, which includes testing and cleaning for peak performance as well as replacing defective or underperforming parts. Periodic and preventive maintenance programs are an important part of all HVACR service businesses. During the months of the year when emergency service requirements are low, maintenance programs often make up a significant portion of a business's cashflow.

In an effort to improve energy efficiency, a number of system add-on devices have been designed and marketed over the years. Many of these devices are designed to recapture energy that would otherwise be wasted. The energy recovery ventilator (ERV) shown in *Figure 10* is a good example. This unit uses cooled or heated air exiting the building to pre-condition fresh air being forced into the occupied space. This saves energy that would otherwise be wasted. Some ERVs work directly through a building ventilation system without being directly connected to the HVAC system.

Another example is an ice storage system (*Figure 11*). Many electric utilities charge a higher rate for electricity used during hours when demand is the highest. To reduce the energy costs of air conditioning a building, large reservoirs of water can be frozen during off-peak hours. The ice serves as a storage medium for cooling capacity that is used to supplement or even replace the mechanical air conditioning system during peak operating hours. This shifts the energy consumption to an off-peak time, saving money and helping balance the utility's energy demand.

There are many similar energy-saving systems and devices in the HVACR industry today. Understanding what they do and how they work will help ensure your success and create additional opportunities for you in the craft.

A building's air distribution system also plays an important role in energy-efficient HVAC system operation. If the supply ductwork leaks, for example, conditioned air can escape into unconditioned spaces, resulting in wasted energy. Leaks in ductwork on the return side also result in energy losses, as well as allowing dust and pollutants to be drawn into the ducts.

A special type of **mastic** is used to seal duct seams (*Figure 12*). Mastic is a flexible sealant that can be applied by painting or spray. Air distribution systems often need to be inspected along with the equipment to ensure the joints remain intact and leak-free. As your training continues, you'll find that air distribution systems are made of other materials besides steel. Duct materials include polyvinylchloride (PVC), fiberglass, and even porous fabrics that disperse the air in the space along their full length (*Figure 13*).

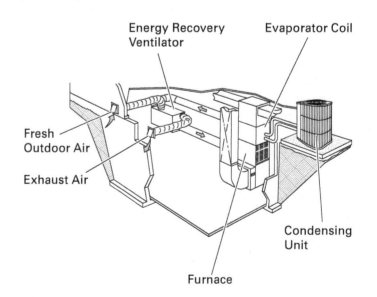

Figure 10 Energy recovery ventilator (ERV).

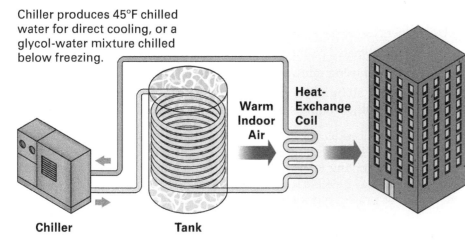

Chiller produces 45°F chilled water for direct cooling, or a glycol-water mixture chilled below freezing.

Chiller   **Tank**

Water tank freezes when glycol mixture is cold enough, or chills warm water passing through when the tank is frozen and the chiller is off.

Figure 11 Ice storage for comfort cooling.

Figure 12 Duct joints sealed with mastic.

Figure 13 Fabric duct system.
Source: iStock@Dmytro Kuznietsov

2.2.1 LEED and Green Building Design

The most energy-efficient and environmentally sound buildings require considerable preparation. In 1993, the Leadership in Energy and Environmental Design (LEED) certification was introduced. The purpose of LEED is to encourage the adoption of **sustainable construction** and building management standards as established by the US Green Building Council (USGBC). Today, it is the most widely used green-building rating system in the world.

All building types can benefit from embracing the program's design principles. The basic concept of LEED is to provide a template for building design and construction that is environmentally and socially responsible. It isn't just about energy efficiency. Building owners can work toward certification in many ways, including the following:

- Using recyclable material during construction
- Minimizing construction waste
- Creating an energy-efficient building envelope that minimizes heating and cooling losses while maximizing occupant health and comfort

Sustainable construction: Construction that has a minimum effect on the land, natural resources, and raw materials throughout a building's life cycle, and generally results in reduced energy consumption as well.

- Ensuring excellent indoor air quality
- Using energy-efficient equipment and appliances
- Installing automated plumbing controls for fixtures that reduce water usage
- Capturing and using rainwater for horticultural use
- Ensuring that all building construction characteristics and operations are sustainable without significant environmental impact

On the HVACR side, meeting the LEED certification standards means selecting energy-efficient equipment and sizing it correctly for the structure and the heating/cooling loads. Preserving the indoor air quality in the most energy-efficient way possible is essential.

A variety of project types are applicable for LEED certification, including new buildings and interiors; existing residential and commercial buildings; and neighborhoods, communities, and cities. LEED credits can be earned in the following categories:

- Location and Transportation (LT)
- Water Efficiency (WE)
- Energy and Atmosphere (EA)
- Materials and Resources (MR)
- Indoor Environmental Quality (EQ)
- Innovative Building Features (IN)
- Regional Priority (RP)

Of these, HVACR is most directly related to the EA and EQ categories. Most LEED-certified commercial buildings also have some form of building management system (BMS) to monitor and control the operation of the HVAC system as well as other vital building systems, such as security and fire protection. A properly programmed and implemented BMS can significantly reduce HVAC-system energy consumption. The environmental and economic impact of an energy-efficient HVAC system is enjoyed as time passes, though the upfront cost may be steep.

One important thing to note is that LEED certification goes beyond the initial design and construction of the building. Once completed, the building and its equipment must also be maintained in accordance with the level of LEED certification awarded. Periodic recertification programs are often implemented.

2.3.0 Licensure and Certification

HVACR installers and service technicians work on equipment that directly affects the health and safety of building occupants. For example, gas furnaces are supplied by natural gas or propane, both of which can be hazardous. Combustion byproducts also represent a significant hazard. The byproducts include CO, so proper venting and periodic maintenance must be performed by knowledgeable technicians. To ensure that companies installing and servicing HVACR equipment are qualified to do so, states and municipalities generally require them to be licensed.

A *mechanical contractor* is a company that specializes in the installation, construction, maintenance, and repair of HVACR equipment. Plumbing is also a craft that falls under the definition of mechanical contracting. Projects on which mechanical contractors work may be large-scale commercial or industrial installations. However, many mechanical contractors focus solely on the residential market.

Most states require mechanical contractors to pass licensure exams and pay an annual fee, enabling them to do business in a given state. The testing and cost

Did You Know?

NCCER Headquarters

The NCCER office building, located in Alachua, FL, and constructed in 2011, was designed and built to the LEED Silver level of certification.

for mechanical licensure in a given state often vary based on the size of projects that a contractor wishes to pursue. Some states have reciprocal agreements that allow licensed contractors in those states to cross over and work in the others. For example, Georgia, Louisiana, Ohio, Tennessee, Texas, Utah, and South Carolina recognize each other's mechanical contractor licenses. Some states require applicants to show evidence of their workers' compensation and liability insurance packages. A contractor may also have to post a **surety bond**.

Surety bond: A funded guarantee that a contractor will perform as agreed.

2.3.1 Technician Licensure and Certification

Of course, every mechanical contractor must rely on skilled craftworkers to accomplish the work. The licensure requirements for HVACR technicians vary widely from state to state. While some states do not require technicians to be licensed or certified, many others do.

Many states that require HVACR technicians to be licensed adhere to a traditional model, with two levels of licensure. A journeyman license allows HVACR technicians to work in their state. However, most states using the model require journeyman to work under the supervision of someone with a master HVACR license. Masters take longer, more challenging tests and are required to have years of experience in the craft before they are eligible to test. This means that a mechanical contracting firm must have at least one person holding a master license. That individual is largely responsible for the work done by the journey-level workers the company employs.

Confusion on the topic of licensure is common, with each state setting their own standards and using their own terminology. It is important to understand that licensure as a journeyman or even a master does not grant the right to conduct business as a contractor. In states that require journeyman and master licensure for HVACR technicians, one must usually hold a master HVACR certification before they can even apply for a mechanical contracting license.

NOTE

Due to the nature of technician and contractor licensure across the United States as well as across the world, it is essential that you become familiar with the requirements of the state or country you reside in before performing any work.

Chlorofluorocarbon (CFC): A compound used as the basis for a class of refrigerants that contain chlorine, fluorine, and carbon.

Hydrochlorofluorocarbon (HCFC): A compound used as the basis for a class of refrigerants that contain hydrogen, chlorine, fluorine, and carbon.

Greenhouse effect: Describes the process through which the sun's warmth and the heat generated from Earth becomes trapped in the lower atmosphere, primarily due to gases in the atmosphere that prevent the heat from being radiated to space.

EPA Certification

Scientific studies proved long ago that the chlorine in chlorofluorocarbon (CFC) and hydrochlorofluorocarbon (HCFC) refrigerants can damage the ozone layer that protects us from the sun's ultraviolet rays. One of the many effects is an increased rate of skin cancer and other skin-related problems. For many years, these refrigerants were routinely released to the atmosphere as necessary when system service was required. In the late 1980s, many countries of the world agreed to work toward eliminating these releases, and to develop and use refrigerants that do not contain chlorine.

Over the years, we have also learned more about the greenhouse effect and the potential for global warming (*Figure 14*). There is substantial evidence to show that certain refrigerants, many of which do not contain chlorine, contribute to global warming by trapping heat around Earth that would normally be radiated into space. Regardless of the science and your own feelings on the topics of ozone layer depletion and global warming, there are two certainties—venting refrigerants to the atmosphere wastes a valuable resource, and refrigerants and similar substances can't harm the environment if they are never released to the atmosphere.

In 1990, the US Congress passed the Clean Air Act (CAA), which called for the phase out of ozone-depleting refrigerants as well as the strict control and labeling of refrigerants. The EPA is responsible for implementing and enforcing the requirements of the act. The CAA had a significant and lasting impact on the HVACR trade. For example, Section 608 of the Act imposed the following restrictions on refrigerants, regardless of their chlorine content:

- Anyone handling refrigerants must be EPA-certified through a written examination. Indeed, even the purchase of refrigerants is restricted to those who are certified. There are four levels of EPA certification, based on the nature of the equipment. The levels range from Type I certification, which allows work on small systems containing less than five pounds of a high-pressure refrigerant, to Type III certification that allows work on systems that use low-pressure refrigerants. The fourth level of certification is the Universal certification. This level is awarded to a technician that successfully passes the tests for the other three levels (*Figure 15*).

- Records must be kept on all transactions involving refrigerants. These records must include contractor purchases, where the refrigerant was used, and repair records for leaking equipment.

- By federal law, refrigerants cannot be released into the atmosphere. Anyone knowingly venting refrigerants is subject to penalties including a stiff fine and loss of certification. The penalties for ignoring some portions of Section 608 of the CAA can even include imprisonment.

These are just a few of the many requirements. As you continue in your training, you will learn everything you need to know to pass the certification exams. There is much to learn! The exams will test your knowledge of the law as well as your understanding of the refrigerant circuit and common service tasks. Note, however, that the exam is not intended to certify you as a technician. Successfully completing the EPA exams will only certify that you have the knowledge needed to uphold the requirements of Section 608 of the CAA. The certification authorizes you to buy and handle refrigerants but does not reflect any level of skill as a technician.

For the HVACR technician, the most challenging requirement is the management of refrigerants currently charged into equipment that requires service. If the refrigerant circuit must be opened for service, the refrigerant inside cannot be vented to the atmosphere. Therefore, it must be captured. This leads to the three Rs of refrigerant handling—recover, recycle, and reclaim:

Recover: To remove refrigerant from a system and temporarily store it in containers approved for that purpose.

Recycle: To circulate recovered refrigerant through filtering devices in the shop or field that remove moisture, acid, and other contaminants.

Reclaim: To process a used or contaminated refrigerant until it is returned to the standards of purity required of new refrigerant.

- *Recovery* — Recovery identifies the process of removing refrigerant from equipment and temporarily storing it in approved containers. Recovery

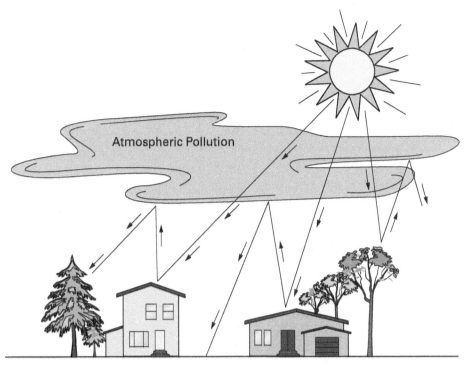

Figure 14 The greenhouse effect.

Figure 15 EPA Section 608 certification card example.

does not provide for any cleaning or filtration of the refrigerant, although it is often filtered as it is removed. But the term only refers to the removal of refrigerant from the equipment. An example of refrigerant recovery equipment is shown in *Figure 16*.

- *Recycling* — Recycling refers to the circulation of recovered refrigerant through filtration devices that can remove moisture, acid, and solid contaminants. Recycling may occur in the field or in a contractor's shop. However, there are a number of restrictions on what recycled refrigerant can be used for, since there is no simple means of determining that the refrigerant is actually pure. For example, it cannot be used in equipment

Figure 16 Refrigerant recovery equipment.

owned by someone other than the owner of the equipment from which it was taken. It can be returned to the same system when service is complete or used in other equipment under the same ownership.

- *Reclamation* — To reclaim a refrigerant means returning it to the standards of purity and cleanliness required of new refrigerant. These standards are outlined in Air Conditioning, Heating, and Refrigeration Institute (AHRI) Standard 700, *Specifications for Refrigerants*. Reclamation is by no means a field or shop procedure. It is a complex process done only by chemical plants and specialized reprocessing facilities. Only reclaimed refrigerant may be resold on the market or to customers.

Going forward, you will learn far more about these three important tasks. That knowledge will be tested as part of this training program and tested again through the EPA Section 608 certification exams. It is critical that everyone in the HVACR industry understand and follow the EPA Section 608 refrigerant regulations. Failing to do so can be very costly to both the craftworker and the employer.

2.4.0 Building Codes and Standards

Building construction and the systems installed by craftworkers are governed by building codes and guided by industry standards. Codes are mandatory directives, while a standard is an industry-accepted practice that is voluntary. However, industry standards are often made a part of building codes and/or project specifications. OSHA also adopts some standards into its regulations. The terminology often used by organizations to indicate this is to say that a standard is "incorporated by reference." When that happens, the standard becomes mandatory rather than voluntary. An example is American Society of Heating, Refrigeration, and Air-Conditioning Engineers (ASHRAE) Standard 62.1, *Ventilation for Acceptable Indoor Air Quality*. This standard is often incorporated into building codes.

Building codes, also called *ordinances*, are designed to protect the health and welfare of the public, ensuring their safety. Code inspectors apply the codes to ensure work is done in an acceptable manner. Building codes have the force of law, which means that contractors must follow the codes or face legal, and perhaps civil, penalties. A contracting firm may always choose to exceed code requirements, but it is obligated to meet them at a minimum. A code describes the minimum acceptable characteristics of building construction details.

No single building code governs all aspects of construction across the United States. Weather conditions can affect how codes are written in different parts of the country. For example, in Florida where hurricanes are likely, local codes include standards that will help a building withstand hurricane-force winds (*Figure 17*). Areas that experience heavy snowfall must have codes that ensure roofs are built to withstand the stress.

Time influences building codes as well. It can take many years for a problem to show up in a material that was considered safe and appropriate for installation at the time. For example, lead-based paint and asbestos, once widely used materials, were found to be unsafe years after their widespread use. As a result, codes were revised to prohibit them. Many older buildings still contain these and similar materials that are no longer considered acceptable. They must be dealt with when major changes are planned.

Contractors and craftworkers must always follow the local and current edition of the code. Local areas typically adopt widely accepted codes and then attach changes to it that are appropriate for the local area. The most widely adopted building code in the United States is the International Building Code (IBC), created and maintained by the International Code Council (ICC). The organization was the result of a merger of various code organizations in the United States in an effort to standardize building codes. The first edition of the IBC was released in 1997, and a new edition is released every three years.

Figure 17 Home built to endure high winds.
Source: iStock@krblokhin

However, many buildings that were constructed in accordance with codes that preceded the IBC, referred to as *legacy codes*, are still in use today. Legacy codes include the following:

- BOCA National Building Code (BOCA/NBC) by the Building Officials Code Administrators International (BOCA)
- Uniform Building Code (UBC) by the International Conference of Building Officials (ICBO)
- Standard Building Code (SBC) by the Southern Building Code Congress International (SBCCI)

Apps for Building Codes

UpCodes specializes in bringing readily accessible, searchable building codes to everyone involved in construction, from architects to subcontractors. The searchable database alone represents a leap forward from having to manually search through thousands of pages of code. Code references are available for specific states and codes through a subscription model, and the database is acceptable through PCs as well as mobile devices.

Source: UpCodes. (n.d.) UpCodes: Features. **https://up.codes/features**. Reprinted with permission.

All HVACR craftworkers need to be familiar with the code governing the area in which they are working. What is acceptable in one city may not be acceptable in another, even when they are in the same state. As the old saying goes: ignorance of the law is no excuse. Even if you are a service technician and are not engaged in the installation process, you can still make a repair or change something about a system that is not acceptable.

There are some national codes that are regularly incorporated into state and municipal building codes by adoption. These codes include the following:

- *National Electrical Code®* (*NEC®*) — This code, specifically identified as NFPA 70®, is published by the National Fire Protection Association (NFPA). The *NEC®* establishes the minimum standards for electrical wiring. This code is rigidly enforced. For that reason, electrical wiring installations for HVACR systems are nearly always performed by licensed electricians that are properly trained and licensed for electrical work.

- *National Fuel Gas Code* — This code, identified as NFPA 54, was also developed by the NFPA. The code covers the installation of gas piping and the venting of appliances such as gas furnaces and water heaters. This is one of two model codes for fuel gases. The other is the International Fuel Gas Code offered by the ICC.

2.4.1 Building Permits

Before construction or installation work begins, a contractor must apply to the local government for a *permit* to do so. To obtain the permit, the contractor must submit a description of the project, including drawings, if applicable. This process ensures that local codes-compliance officers are aware of the job and provides them an opportunity to examine the plans and identify compliance issues before the job begins. Financial penalties result when a contractor fails to get a permit. Repeat offenders face stiffer fines and the loss of their contracting license. A permit typically has to be physically and publicly posted at the jobsite.

As the work progresses, inspectors will visit at various stages of construction. One or more rough-in inspections, for example, are often required so that the inspector can see work and installation techniques that will eventually be concealed. A final inspection must be passed in order for a certificate of occupancy (COO) to be issued for the building. If a system in an existing building is simply being replaced, only a single inspection is usually required.

Here are a few examples of the HVACR-related inspection items that might be on the list of a code-compliance inspector:

- Furnace and boiler venting (*Figure 18*)
- Gas piping
- Duct construction and installation methods
- Pipe supports for refrigerant lines and other piping
- Drain line routing and termination
- Equipment placement
- Location and size of furnace combustion-air intake and exhaust (*Figure 19*)

When an HVACR system is engineered, members of the engineering team also visit regularly to ensure that the work is being done according to their specifications.

2.4.2 Standards

A standard identifies methods and materials that have been agreed upon by industry professionals. Contractors voluntarily refer to and follow standards for the following reasons:

- Standards represent the collective knowledge and experience of architects, designers, engineers, contractors, and other professionals in the industry.

Figure 18 Vented high-efficiency boiler.
Source: iStock@Yliya Zhuravleva

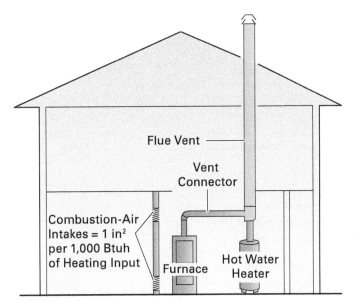

Figure 19 Example of combustion-air intake sizing.

- Standards typically focus on the best and safest methods and materials, and often represent what most codes find acceptable. However, that can never be assumed. Codes may be more stringent than standards, and standards can also exceed applicable code requirements.

- Work that is based on standards will be consistent from project to project, which increases efficiency and builds a good reputation.

- Standards are frequently reviewed and updated to reflect current industry practices.

A number of organizations publish standards that apply to the mechanical contracting industry. One such organization is the Sheet Metal and Air Conditioning Contractors National Association (SMACNA). SMACNA standards are the guiding light of the sheet metal industry for the construction of air distribution systems. Other organizations that publish standards related to the HVACR industry including the following:

- Air Conditioning and Refrigeration Institute (ARI)
- Air Conditioning Contractors of America (ACCA)
- Air Diffusion Council (ADC)
- Air Distribution Institute (ADI)
- Air Movement and Control Association (AMCA)
- American Conference of Governmental Industrial Hygienists (ACGIH)
- American National Standards Institute (ANSI)
- American Society of Heating, Refrigeration and Air-Conditioning Engineers (ASHRAE)
- American Society of Mechanical Engineers (ASME)
- ASTM International (formerly the American Society for Testing and Materials)
- American Welding Society (AWS)
- Fabricators and Manufacturers Association, International (FMA)
- Factory Mutual Engineering Research Corporation (FM)
- Mechanical Contractors Association of America (MCAA)
- National Environmental Balancing Bureau (NEBB)
- Thermal Insulation Manufacturers Association (TIMA)
- Underwriters Laboratories (UL)

2.0.0 Section Review

1. When working at height, personal fall protection equipment *must* be used when working more than _____.
   a. 2' (0.6 m) above ground
   b. 6' (1.8 m) above ground
   c. 12' (3.7 m) above ground
   d. 20' (6.1 m) above ground

2. *Mastic* is a material used to _____.
   a. test for gas leaks
   b. seal ductwork
   c. insulate refrigerant lines
   d. lubricate bearings

3. An EPA Type I certification allows a technician to work on _____.
   a. fossil fuel heating systems
   b. systems with no refrigerant in them
   c. large systems containing more than five pounds of a high-pressure refrigerant
   d. small systems containing less than five pounds of a high-pressure refrigerant

4. Building permits are usually issued by the _____.
   a. local government
   b. state government
   c. federal government
   d. building architect

3.0.0 HVACR Training and Opportunities

Performance Tasks

There are no Performance Tasks in this section.

Objective

Identify HVACR career paths available and describe common training processes.

a. Identify the personal characteristics needed to be a successful craft professional.
b. Describe career opportunities in the various business segments of the HVACR industry.
c. Describe how HVACR training is delivered and the general requirements and structure of an apprenticeship program.

Since ancient times, people have sought ways to make the buildings in which they live, work, and play more comfortable. The HVACR craft requires skilled craftworkers to install, maintain, and repair the equipment that makes this possible. Working in the HVACR industry is challenging and rewarding because the technology is constantly changing to better meet our needs, and the results of your work bring comfort and convenience to your clients.

The HVACR industry has many branches and offers many opportunities for advancement. This section will describe some of those opportunities. But it is also important for you to understand the characteristics needed to be successful in the craft and in the workplace.

3.1.0 Characteristics of Successful Craft Professionals

To be successful, a craft professional must be able to use the required materials, tools, and equipment to finish a task properly and efficiently. An HVACR technician must often be able to adjust methods to meet each situation, without compromising quality. You must keep up with technical advancements and hone your skills in the craft. The sections that follow describe some characteristics that distinguish a professional craftworker.

3.1.1 Professionalism

The word *professionalism* is a broad term that describes the desired overall behavior and attitude expected in the workplace. Professionalism is too often absent from the construction site and the various trades. Many people would argue that professionalism must start at the top for craftworkers to take it seriously. It is true that management should display professionalism and set an example in the workplace, but it is more important that individuals recognize their own responsibility to act professionally. As a trainee, you must find your own motivation to display craftsmanship and professionalism.

Professionalism encompasses all of the characteristics discussed in this section—the total package. It is vital that you do not accept the unprofessional behavior of those around you too casually when it is observed. This is not to say that you should shun an unprofessional worker. Instead, demonstrate professional behavior yourself in spite of it, and serve as an example to effect change. Learn from everything you see, including behavior that is less than professional.

Professionalism is a benefit to both the employer and the employee. The construction industry image reflects the behavior of all its members. Choose to behave professionally, and the industry image will follow.

3.1.2 Honesty

Honesty and personal integrity are important traits. Craft professionals take pride in performing a job well, and in being punctual and dependable. Each job is completed in a professional way, never by cutting corners or reducing materials. A valued professional protects property such as tools and materials belonging to employers, customers, and other trades from damage or theft at the shop or jobsite.

Honesty and success go hand-in-hand. It is not simply a choice between good and bad, but a choice between success and failure. Dishonesty will always catch up with you. Whether it is stealing materials, tools, or equipment from the jobsite or simply lying about work accomplished, someone will find out. You can likely find another employer after a dishonest act has been revealed, but that option will eventually disappear. A craftworker's reputation tends to have a life of its own, and you must protect it. It is as important as any tool in your box.

Honesty means more than just avoiding criminal acts. It means giving a fair day's work for a fair day's pay. It also means accepting responsibility for any errors made. It means carrying out your side of an agreement. Employers place a high value on an employee who prioritizes honesty.

3.1.3 Loyalty

Craftworkers expect their employers to be loyal and look out for their best interests, to provide them with steady employment, and to promote deserving workers to better positions when the opportunity presents itself. Employers have a right to expect their employees to be loyal to them too—to keep the company's best interests in mind, to speak well of them to others, and to keep all matters that pertain to the business confidential. Both employers and employees should keep in mind that loyalty is not something to be demanded. It is something to be earned. Demonstrate loyalty before you expect to receive it.

3.1.4 Willingness to Learn

Each employer and jobsite have their own way of doing things. Employers expect their workers to be willing to learn their ways. As a trainee in the HVACR craft, you will be expected to learn a great deal in a short period of time, and to ask questions when you don't understand (*Figure 20*). As you move into the workplace to apply what you learn, adapting to change and being willing to learn new methods and procedures as quickly as possible will be essential to your success.

Figure 20 Learning on the job.
Source: iStock@monkeybusinessimages

Employability

Generally, before you can get hired as an HVACR service technician, or even as an apprentice, you will have to cross some hurdles. First, you will likely need to pass a drug test.

Most companies not only require drug testing when hiring, but random testing is often conducted as well. There are two practical reasons for this. One reason is that it affects your safety as well as those around you. There are potential hazards everywhere on the job, and a compromised technician is a dangerous one. If you are injured on the job and testing indicates you were impaired in any way, a workers' compensation claim is unlikely to succeed.

Another reason is related to the clients of the company. Many large general contractors and building developers require subcontractors to have a drug testing program as a condition for bidding. A good safety record is also required for contractors to bid on the best projects.

You will also need to present your driving record. Service technicians and installers usually travel to their jobsites in an equipped, company-owned vehicle. The company will need to insure you as a driver. If your driving history indicates you are a risk and the insurance company rejects you, the employer has little choice but to reject you as well. Even when the insurance company accepts, it may be at an unacceptable cost to the employer.

Finally, a background check is usually in order. A history of trouble with law enforcement is often a cause for concern. However, it is important to point out that those who have been incarcerated and have embraced rehabilitation are often given opportunities in this craft, as well as in other construction crafts.

Changes in safety regulations or the purchase of new equipment, for example, makes it necessary for even experienced employees to learn new methods and skills. Humans sometimes resent having to accept change, especially when it comes as a surprise. However, employers will rightfully expect employees to put forth the necessary effort. Construction methods must progress for a business to compete, to stand out as a reliable and competent option, and to earn a profit. It is profit that allows an owner to expand and provide better jobs and benefits.

3.1.5 Willingness to Take Responsibility

Most employers expect workers to see what needs to be done, then do it. It is exhausting to have to ask again and again for a task to be done. It is obvious that, having been asked once, an employee should assume responsibility for it. A craft professional's primary responsibilities should be clearly stated by the employer, providing guidance and clarity to the worker. The craft professional must then take those responsibilities seriously.

Craft professionals need to be engaged enough in their role to know when it is time to assume responsibility for something without direction. Employers are pleased to see employees taking the initiative and embracing responsibility. In addition, a craft professional should always report issues or problems that affect safety or the product in any meaningful way, as well as take responsibility for any personal errors.

3.1.6 Willingness to Cooperate

Cooperating means working together, often as a team. Cooperation is often key to getting things done efficiently. Learn to work as a member of a team with your employer, supervisor, and fellow craft professionals towards the common goal of getting the work done safely, efficiently, and on time (*Figure 21*). Look for ways to contribute that support to the team's goals.

People can work well together most effectively if there is an understanding about what is to be done, when and how it will be done, and what role each individual will play in the effort.

3.1.7 Present and On Time

Tardiness means being late for work, while *absenteeism* means being away from the job for one reason or another. Consistent tardiness and frequent absences are often an indication of poor personal habits, a weak work ethic, and/or a lack of commitment to your craft.

Your working life is significantly regulated by the clock. Each worker is required to be at work at a specified time. Failure to get to work on time results in confusion, lost time, and resentment from team members who do arrive on time. In addition, it may lead to penalties, including dismissal. It is your obligation to be at work at the time indicated. In many cases, a missing team member cripples the team's effort and stalls the project. The company experiences a significant loss when an entire crew is idled.

Try to look at the matter from the point of view of a supervisor. Supervisors cannot keep track of workers who arrive at random times. It is not fair to others for the tardiness of one to be ignored. Acting in the best interests of the team

Figure 21 Learn to work as a team toward a common goal.
Source: Shutterstock.com/CoolKengzz

and the company, a supervisor must address a worker's tardiness. Employers resort to docking pay, demotion, and even dismissal to control tardiness and absenteeism. No employer likes to impose discipline or restrictions on employees. However, to be fair, an employer is sometimes forced to discipline those who fail to meet the standard.

It is sometimes necessary to take time off from work. No one can be expected to work when sick or when there is a serious personal matter at hand. However, it is also possible to get into the habit of allowing unimportant and unnecessary matters to get in the way of the job. This results in lost production and hardship placed on those who try to carry on the work with an incomplete team. Do not allow trivial matters to interfere with your work.

Get a sufficient amount of rest before going to work each day. If you are ill, use the time at home to recover as quickly as possible. You would expect no less from someone you hire to do an important job. If it is necessary to stay home, notify your supervisor as soon as possible so planning for your absence can begin.

A practiced morning routine helps prevent being in a rush and breathless, late arrivals. In fact, arriving a little early shows your interest and enthusiasm for the work, which is appreciated by employers. Tardiness is one of those habits that can easily stand in the way of promotion. A worker that cannot manage his/her own time is not often chosen for a position of leadership.

The most frequent causes of absenteeism are illness, death in the family, accidents, personal business, and dissatisfaction with the job. Some of these causes are legitimate and unavoidable, while others can be controlled, at least to some degree. One can tend to many personal affairs after working hours and plan in advance for those that must be addressed during the workday. A respected, reliable worker will not generally have a problem getting time off to address something important.

If dissatisfaction with the job is the problem, make the necessary plans for change. However, do so in a professional manner, giving your employer as much effort on the last day on the job as was given on the first day.

3.2.0   HVACR Career Opportunities

Career opportunities in the HVACR trade are many and varied. There is a large existing base of equipment that needs service, repair, and replacement. In addition, every time a new residential, commercial, or industrial building is constructed, it typically contains one or more HVACR system elements.

To get an idea of how vast the HVACR trade is, picture the town in which you live. Then think about the fact that almost every building in town contains some form of equipment to provide heating and cooling, as well as air circulation and filtration. They also likely contain some type of refrigeration equipment, even if it's just a domestic refrigerator. Expand that thought to include the entire country, and you'll realize that there are millions of heating, air conditioning, and refrigeration systems at work. New ones are being added every day; old ones are wearing out and being repaired or replaced. From that perspective, you can see the opportunities in the craft are virtually limitless.

Figure 22 provides an overview of career opportunities in the HVACR trade. For the purposes of this discussion, it is convenient to view the HVACR industry as having the following three segments:

- Residential / Light Commercial
- Commercial/Industrial
- Manufacturing

3.2.1 Residential / Light Commercial Segment

Residential / light commercial companies sell, install, and service residential and light commercial equipment and systems such as furnaces and packaged air conditioners. Light commercial facilities such as convenience stores also have refrigeration equipment that must be installed and serviced.

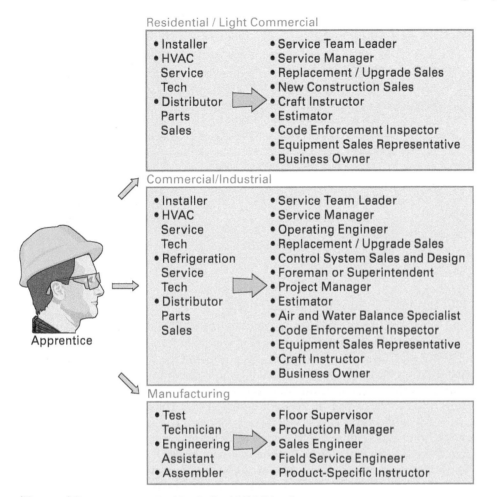

Figure 22 Career opportunities in the HVACR trade.

In this industry segment, you can find anything from one-person installation and service businesses to firms with many employees, including heating and air conditioning specialists, installers, sheet metal workers, and salespeople. In such businesses, HVACR service technicians may work alone or with a partner. They typically respond to service calls for homes and small businesses. They also install furnaces and replacement air conditioning equipment sold by their firm's sales team. In other firms, one group may do replacement installations while another group pursues troubleshooting and maintenance. HVACR technicians to support new home construction are also in demand.

Trainees are often assigned to work with more experienced technicians to complete maintenance work. At this level, a technician is expected to work with a wide variety of products from many different manufacturers. Systems can consist of anything from a window air conditioner to complete, centralized systems with accessories such as humidifiers and specialized filtration. Local and regional distributors provide branded equipment, parts, special tools, and other services for the firms that sell, install, and service HVACR equipment. Distributors of residential and light commercial equipment are often affiliated with a single equipment manufacturer. Contractors that install residential and light commercial systems often provide maintenance and service following the installation through periodic maintenance agreements.

Refrigeration opportunities exist in the light commercial segment as well. Convenience stores and restaurants require walk-in and reach-in refrigeration units. Restaurants also need cooking and kitchen ventilation systems serviced regularly to ensure they perform well. Restaurant kitchens exhaust a lot of air, and they also need to replace the exhausted air in an effective and controlled manner. Most of the air exhausted from a kitchen hood in a restaurant is returned through the same hood assembly, using a special make-up air fan on the roof.

NOTE

Many community-based firms are subsidiaries of nationwide firms known as *consolidators*. A consolidator is an umbrella organization that provides centralized management, purchasing, training, and other functions that give small, local companies the power of a large, national company.

Modern Air Conditioning

Dr. Willis Carrier, founder of Carrier Corporation, is credited with the invention of modern air conditioning. In 1902, he developed a system that could control both humidity and temperature using a non-toxic, non-flammable refrigerant.

Air conditioning started out as a means of solving a problem in a printing facility where heat and humidity were causing problems with the paper. Later systems served similar purposes in textile plants. The concept wasn't applied for comfort until about 20 years later, when Carrier centrifugal chillers were installed in department stores and movie theaters.

Source: The file "Willis Carrier 1915.jpg" is marked with CC PDM 1.0

3.2.2 Commercial/Industrial Segment

Commercial/industrial mechanical contractors install and maintain systems for large office buildings, factories, apartment complexes, hospitals, grocery stores, and similar facilities. The food processing industry requires large refrigeration systems that require constant maintenance and attention.

Large commercial and industrial systems have many components and may require thousands of feet of ductwork and piping. These systems are usually designed by engineers. Many HVACR craftworkers are required for these projects and an individual is more likely to specialize in one or two aspects of the work. Some of the equipment, such as the **chiller** shown in *Figure 23*, are quite large and often require several different crafts to work together. After these systems are installed, air and water balance specialists are contracted to ensure that the design air and water volumes are being delivered to each space or piece of equipment.

For example, where one wall thermostat often controls a residential system, a large commercial system often has centralized digital control systems that are installed and serviced by an HVACR control specialist. These systems, commonly known as building management systems (BMS), require specialized skills to program and maintain. Due to the complexity and abundance of advanced control systems today, many companies focus their attention entirely on the control aspect of the craft.

Chiller: In the HVACR industry, refers to a machine that uses the mechanical refrigeration cycle to cool water for industrial processes or comfort cooling.

Figure 23 Centrifugal chiller.
Source: iStock@BrianBrownImages

Companies that install such equipment are larger mechanical contractors that may work in a multistate area, or even across the world. They are more likely to complete the installation and let the building owner contract with another firm for maintenance. Many companies with large facilities that depend heavily on their HVACR equipment employ their own HVACR maintenance people. Large chillers, boilers, and similar equipment require significant training and experience to service and maintain.

Food processing and mass food storage represent significant opportunities in refrigeration work. Many food processing plants must be refrigerated throughout the work area to ensure the products remain cold as they are processed. Large-scale refrigeration systems and plants can have thousands of feet of piping and many valves that need consistent attention along with the equipment.

3.2.3 Manufacturing Segment

There are hundreds of HVACR equipment and accessory manufacturers. Some of the larger ones cover the entire HVACR spectrum, while others focus on a specific product or market. For example, one may make window air conditioners, another gas furnaces, and yet another might make only chillers designed to support manufacturing processes. Some companies work in niche markets, manufacturing specialized equipment for unique applications.

A major advantage of working for a manufacturer is that the jobs generally pay good salaries, offer benefits, and offer opportunities for mobility and advancement. Many manufacturers have multiple locations, so there are also opportunities for relocation. Many manufacturing companies, especially the larger ones, offer in-house training as well as tuition reimbursement for college courses (*Figure 24*).

Product testing technicians often work in quality control labs. They test manufactured equipment to make sure it performs to specifications. They may also test equipment returned under warranty to determine its status. An engineering assistant or technician works with the engineering staff to help construct and test prototypes of new designs.

Large manufacturers and those that manufacture specialized equipment offer training to their distributors and end users. Equipment-specific training specialists develop and deliver the training to all interested parties. A training specialist must understand every detail of their equipment since they will be questioned at length by their trainees.

Figure 24 HVAC unit assembly.
Source: Shutterstock.com/JU.STOCKER

Manufacturers that design and build specialized equipment or engineered systems generally have a cadre of field service technicians who are called upon to troubleshoot and repair systems that fail under warranty or those that display unique problems. The technicians who perform this work are highly trained, able to work under pressure, and available to travel on short notice. They must also be excellent representatives of their company and able to make wise decisions on their own.

3.3.0 HVACR Training

HVACR craftworkers enter the craft several different ways. While traditional apprenticeships are designed to provide both the skills and knowledge a craftworker needs to reach the journey-level of competency, it is not the only way into the craft.

The construction workforce is trained through a variety of methods. Training in construction crafts is offered by secondary schools (high schools) and postsecondary schools (technical education schools, community colleges, and universities). While instruction in every craft may not be readily available to you this way, the demand for workers and opportunity in the HVACR environment make it a very popular offering. Community colleges and career and technical education (CTE) schools are common, even in rural areas, and the majority offer HVACR programs. Some will offer instruction based on this NCCER program. There are also a variety of online educational programs available for HVACR training.

The education available through local schools, as well as online resources, can often be used to support a local company's apprenticeship program. NCCER craft training programs are uniquely designed to support apprenticeship programs, but they are also used outside of an apprenticeship for education and task training. Registered apprenticeship programs are not available in all areas, but the need for training remains. Employers may still be willing to employ you and pay your education expenses while you learn skills on the job but choose not to assume the responsibilities of operating a registered apprenticeship program.

There are also many craftworkers who enter the craft and learn exclusively on the job from a more experienced worker, without the advantage of classroom training. This generally produces a craftworker with significant limitations who struggles to excel in the craft. However, a motivated individual can take responsibility for their own education outside of the workplace and still find success.

If you are reading this module, you are likely learning through an NCCER-accredited local community college, CTE school, or contractor, or through another source that has adopted this NCCER program for training.

3.3.1 Apprenticeship Programs

Many craftworkers gain their knowledge and skills through a registered apprenticeship program that combines on-the-job learning (OJL) with related classwork. Although most apprenticeship programs, including HVACR apprenticeships, require four years to complete, the result is a well-rounded technician that is capable of performing at the journeyman level.

The Office of Apprenticeship

The Office of Apprenticeship, a division of the US Department of Labor Employment and Training Administration (DOLETA), sets the minimum standards for apprenticeship training in the United States. Apprenticeships are based on mandatory classroom instruction and on-the-job learning (OJL). They typically require a minimum of 144 hours of classroom (or similar) instruction and 2,000 hours of OJL per year. In a four-year apprenticeship program, trainees spend a minimum of 576 hours in the classroom and 8,000 hours in OJL before receiving certificates issued by registered apprenticeship programs. A few construction crafts are designed for two or three years, but most require four years.

The Office of Apprenticeship administers apprenticeship training at the federal level and works closely with state apprenticeship councils (SACs) that administer the programs at the state level. The apprenticeship system has grown up with the United States and, like the country, it is still growing and changing. Programs must generally follow these guidelines:

- Apprentices must usually be at least 18 years old. However, 16- and 17-year-old students may qualify under the Fair Labor Standards Act (FLSA) if certain conditions are met.
- There must be a documented schedule of work processes through which an apprentice is to receive training and experience on the job.
- The program must include organized instruction designed to provide apprentices with knowledge in technical subjects related to their trade.
- There must be a documented, progressively increasing schedule of wages that reward progress.
- There must be proper supervision of OJL with adequate facilities and opportunities for apprentices to observe and practice hands-on skills.
- Apprentices' progress, both in job performance and related instruction, must be evaluated periodically and the appropriate records must be maintained.
- There must be no discrimination in any phase of selection, employment, or training.

Apprenticeship Requirements

For those entering a registered apprenticeship, a high school education is desirable. Courses in shop processes and tools, basic electricity, mechanical drawing, geometry, and algebra are very helpful. Manual dexterity, good physical condition, and quick reflexes are also valued. The ability to solve problems quickly and accurately and to work closely with others is essential. You must also adopt safety as the highest priority.

An apprentice must generally agree to do the following, depending on the state in which you are training (the sponsoring organization may have additional requirements):

- Wear proper safety equipment on the job (*Figure 25*).
- Purchase and maintain tools of the craft as required.

Electrical Theory and Troubleshooting

You will soon learn how important electrical theory and troubleshooting are to an HVACR technician. HVACR service technicians must be able to understand complex electrical diagrams and troubleshoot control circuits that are beyond the capabilities of most journey-level electricians. Electrical troubleshooting is a daily task for the average HVACR service technician, and those that excel at it quickly become valued and respected craftworkers.

Figure 25 PPE for high-voltage work.
Source: Reproduced with Permission, Fluke Corporation

- Submit a monthly OJL report to the apprenticeship committee.
- Report to the committee directly if a change in employment status occurs.
- Attend classroom instruction and adhere to all classroom regulations, including attendance requirements.

Youth apprenticeship programs (YAPs) are also available in some areas, allowing students to begin their apprentice training while still in high school. A student entering the program in 11th grade, for example, may complete as much as two years of an NCCER-based or other program before graduating from high school. In some cases, when time and opportunity exist, students can work in the craft and earn money while still in high school. Upon graduation, the student can enter the industry at a higher level and with more pay than someone just getting started.

NCCER Programs

To address the training needs in the construction crafts, NCCER has developed curricula to support a number of construction-related apprenticeship programs. NCCER uses the standards of the Office of Apprenticeship as the foundation for comprehensive materials to support in-depth classroom instruction and the OJL experience. NCCER curricula provides trainees with industry-driven training and education using a competency-based teaching approach. This means that trainees must show a qualified instructor what they've learned from each training module through exams and performance tasks designed to test their skills.

NCCER programs are built around the knowledge of experienced craft professionals referred to as subject matter experts (SMEs). SMEs take part in the design of the programs and their content. Like every NCCER program, this program was developed by the construction industry, for the construction industry. A continuous review process ensures that the correct training is consistently provided, and the materials remain current and relevant to the craft as time goes by. We hope that every trainee using NCCER training materials enjoys success and advances their skills until they become qualified to serve as an SME and share their knowledge with tomorrow's trainees.

When an NCCER-certified instructor is satisfied that a trainee has demonstrated the required knowledge and skills for a given module, that information can be sent to NCCER and recorded in the Registry. The NCCER Registry provides confirmation of a craftworker's training progress as they move from state to state or company to company. Trainees in a program offered by an NCCER-accredited organization receive a certificate for each level of training completed. *Appendix 03101A* provides examples of NCCER's credential documents. The credentials offered by NCCER through the successful completion of its nationally accepted, standardized training and assessment programs are unique in the construction industry and valid from coast to coast.

3.0.0 Section Review

1. Working in your employer's best interests as they provide steady employment and promote deserving workers is an example of _____.
   a. honesty
   b. loyalty
   c. cooperation
   d. taking responsibility

2. Product testing technicians are *most* likely to be employed by _____.
   a. residential service companies
   b. mechanical contractors
   c. equipment manufacturers
   d. industrial refrigeration contractors

3. The *minimum* number of classroom hours required annually for an apprentice program is _____.
   a. 50 hours
   b. 75 hours
   c. 112 hours
   d. 144 hours

Module 03101 Review Questions

1. In a common residential forced-air furnace, heat is transferred from the _____.
   a. heat exchangers to the air
   b. natural gas or oil to the conditioned space
   c. air to the heat exchangers
   d. refrigerant to the outdoor air

2. Water or steam is often used in heating systems as a(n) _____.
   a. fuel
   b. combustion byproduct
   c. heat-exchange medium
   d. evaporator

3. The equipment that is able to extract heat from the outside air by reversing the refrigeration cycle is called a(n) _____.
   a. furnace
   b. evaporator
   c. heat pump
   d. expansion device

4. One way ventilation can be defined is as a _____.
   a. technique used to prevent outdoor air from entering a building
   b. way of maintaining a constant air temperature
   c. way of removing moisture from indoor air
   d. means of introducing outside air into a building to improve air quality

5. The growth of bacteria and other harmful organisms can be reduced or eliminated by _____.
   a. humidification
   b. standard air filters
   c. ultraviolet (UV) light
   d. cooling the air

6. The term *mechanical refrigeration cycle* refers to _____.
   a. the process by which circulating refrigerant absorbs heat in one location and moves it to another location
   b. the process by which refrigerant moves through a piping system
   c. mobile refrigeration equipment
   d. the process by which refrigerant is reclaimed

7. The expansion device in the mechanical refrigeration cycle _____.
   a. raises the refrigerant pressure entering the evaporator
   b. lowers the refrigerant pressure entering the evaporator
   c. raises the refrigerant pressure at the condenser inlet
   d. lowers the refrigerant pressure at the condenser outlet

8. In a typical mechanical refrigeration system, heat is transferred from the indoor air to the refrigerant at the _____.
   a. compressor
   b. furnace
   c. evaporator
   d. condenser

9. An ERV uses cooled or heated air exiting the building to _____.
   a. pre-condition fresh air being forced into the occupied space
   b. operate pneumatic tools
   c. pressurize other parts of a building
   d. heat domestic water

10. Which of these companies is *most* likely to install HVACR equipment during the construction of a large office building?
   a. Manufacturer
   b. Distributor
   c. Mechanical contractor
   d. General contractor

11. To be resold to other customers, a refrigerant *must* have been _____.
   a. reclaimed
   b. recycled
   c. recovered
   d. refurbished

12. Releasing refrigerants to the atmosphere is _____.
   a. okay, provided that the refrigerant has been recycled
   b. okay, provided that the refrigerant has been reclaimed
   c. prohibited by federal law
   d. prohibited in some states

13. Distributors of residential and light commercial equipment _____.
   a. compete directly with mechanical contractors
   b. often own numerous mechanical contracting firms
   c. manufacture most everything they sell
   d. are usually affiliated with a single equipment brand

14. An air and water balance specialist is *most* likely to be found working _____.
   a. in the residential environment
   b. for a product distributor
   c. in the commercial/industrial environment
   d. in the manufacturing environment

15. The Office of Apprenticeship administers apprenticeship training at the federal level and works closely with _____.
   a. local high schools
   b. apprenticeship graduates
   c. state apprenticeship councils
   d. apprenticeship representatives from manufacturers

Answers to odd-numbered Module Review Questions are found in *Appendix A*.

Cornerstone of Craftsmanship

Jim Koehn

Organization: Central Oregon Community College

How did you choose a career in the industry?

It all started by accident really. I was looking for work just after my 18th birthday and came across a want ad looking for an HVAC helper/warehouseman. I applied and got the job. The rest is history. I worked my way up, starting on the shop floor, then moved into sales. The next chapter included working as an installer, then back to sales again. I eventually moved into the service world and then into management and workforce development. Today, I share what I learned through teaching.

Who inspired you to enter the industry?

I definitely wandered into it! I applied for a job, gave it my all, and developed a successful and enjoyable career in the craft.

What types of training have you been through?

I enjoy learning, so I take advantage of every training opportunity I can get. I've attended product-specific, factory-level programs; engineering and design programs; various troubleshooting classes; and virtually everything in between. I feel there is a responsibility that goes along with these training opportunities. Someone has taken their time to train you, at some cost, and you have a responsibility to learn what is offered. But you also have the responsibility of using that knowledge to the best of your ability. Then, as your career progresses, you have the responsibility of sharing that knowledge with the next generation of craftworkers.

How important is education and training in construction?

Technologies change faster than appropriate training can be developed to support it. Training and education provided by a recognized and approved organization is invaluable. The experts that put the training together, the SMEs that review the content, the project managers, the editors, and the publishers all play an essential role in passing along valuable knowledge and skills to those that will apply it.

How important are NCCER credentials to your career?

NCCER credentials are recognized across the United States as having the highest quality due to the standards and ethics applied to program development. NCCER demonstrates a commitment to quality training and to the individuals and teams that deliver their curricula.

How has training/construction impacted your life?

It has impacted my life in more ways than I can count! I have made friends and become acquainted with many contractors and other professionals through training opportunities, both as a learner and as an educator. Those relationships have proven valuable to me.

What kinds of work have you done in your career?

I've done technical support, sales, service, installation, consulting, engineering and management through my career.

Tell us about your current job.

As an adjunct instructor, I am honored to have the opportunity to share my knowledge and experience with trainees in the craft pipeline. Being an instructor positions me to influence young minds and help them develop skills they will use for the rest of their lives.

What do you enjoy most about your job?

I enjoy watching the students learn a skill, practice it, and master it over time. I get to watch some land their first HVACR job and use the skills they have been taught to further their career.

What factors have contributed most to your success?

Persistence is probably the most important factor. Keep at it! Success doesn't necessarily come fast or quick, but persistence builds character, and you will eventually succeed. Honesty and integrity, of course, have played a role, as they do in any job. I'm willing to learn and accept mistakes along the way, and to learn again from those mistakes. Each time you make a mistake, take a moment and figure out what you can take from the experience. Ask yourself, "What did I learn from that mistake?"

Would you suggest construction as a career to others? Why?

Construction crafts will always be needed. Anytime something needs to be constructed, powered, or heated and cooled, there is a need for professional craftworkers—both men and women. The skills you develop in training are skills that belong to you for the rest of your life.

What advice would you give to those new to the field?

Be patient and learn everything you can!

Interesting career-related fact or accomplishment?

My love of learning and teaching has enabled me to become an influential member of the craft-training community.

How do you define craftsmanship?

Anyone can learn a skill. But taking pride in that skill and polishing it, then passing it on to an apprentice or junior team member demonstrates what craftsmanship is all about.

SCAN ME

Answers to Section Review Questions

Answer	Section Reference	Objective
Section 1.0.0		
1. b	1.1.0	1a
2. b	1.2.0	1b
3. c	1.3.0	1c
4. a	1.4.0	1d
Section 2.0.0		
1. b	2.1.0	2a
2. b	2.2.0	2b
3. d	2.3.1	2c
4. a	2.4.1	2d
Section 3.0.0		
1. b	3.1.3	3a
2. c	3.2.3	3b
3. d	3.3.1	3c

User Update

Did you find an error? Submit a correction by visiting **https://www.nccer.org/olf** or by scanning the QR code using your mobile device.

SCAN ME

Trade Mathematics

Source: iStock@r_mackay

Objectives

Successful completion of this module prepares you to do the following:

1. Convert measurement units from the US standard system to the metric system, and vice versa.
   a. Identify measurement units in the US standard and metric systems.
   b. Convert length, area, and volume values.
   c. Convert weight values.
   d. Convert pressure and temperature values.

2. Solve basic algebraic equations.
   a. Define algebraic terms.
   b. Demonstrate an understanding of the sequence of operations.
   c. Solve basic algebraic equations.

3. Identify and describe geometric figures.
   a. Describe the characteristics of a circle.
   b. Identify and describe types of angles.
   c. Identify and describe types of polygons.
   d. Calculate various values associated with triangles.

Performance Tasks

This is a knowledge-based module. There are no Performance Tasks.

Overview

Math is an essential skill required to advance in the HVACR profession. Math is used when cutting and fitting pipe, when sizing and installing ductwork, and when calculating electrical values such as current flow.

Industry Recognized Credentials

If you are training through an NCCER-accredited sponsor, you may be eligible for credentials from NCCER's Registry. The ID number for this module is 03102. Note that this module may have been used in other NCCER curricula and may apply to other level completions. Contact NCCER's Registry at 1.888.622.3720 or go to **www.nccer.org** for more information.

You can also show off your industry-recognized credentials online with NCCER's digital badges. Transform your knowledge, skills, and achievements into badges that you can share across social media platforms, send to your network, and add to your resume. For more information, visit **www.nccer.org**.

Digital Resources for HVACR

Scan this code using the camera on your phone or mobile device to view the digital resources related to this craft.

1.0.0 Using the Metric System

Performance Tasks	Objective
There are no Performance Tasks in this section.	Convert measurement units from the US standard system to the metric system, and vice versa. a. Identify measurement units in the US standard and metric systems. b. Convert length, area, and volume values. c. Convert weight values. d. Convert pressure and temperature values.

More than 95 percent of the world uses the metric system of measurement. Many HVACR products used in the United States are now manufactured in other countries. The dimensions, weights, temperatures, and pressures in the installation, operating, and maintenance instructions provided with some of these products are stated using metric system values. Even if the literature has been converted to show US standard values, the mounting hardware is likely to remain in metric terms.

The use of the metric system in the United States is becoming more common. Many manufacturers publish their technical manuals and instrument data sheets displaying all values in US standard system units as in the past, along with their metric equivalents in parentheses behind the US standard units. In some cases, only the metric units may be listed, so it is a good time to become familiar with both and understand how to convert from one to the other. *Figure 1* shows some metric system quantities that have become widely recognized.

1.1.0 Units of Measure

Most work in science and engineering is based on the exact measurement of physical quantities. A measurement is simply a comparison of a quantity to some definite standard measure of dimension called a *unit*. Whenever a physical quantity is described, the units of the standard to which the quantity was compared, such as a foot, a liter, or a pound, must be specified. A number alone is not enough to describe a physical quantity.

The importance of specifying the units of measurement for a number used to describe a physical quantity should be apparent because the same physical quantity may be measured using a variety of different units. For example,

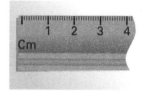

A Two-Liter Bottle of Soda
Instead of a Half-Gallon
Liters = Volume

A Meterstick
Instead of a Yardstick
Meters = Length

A Gram of Gold
Instead of an Ounce
Grams = Weight

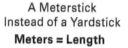
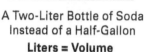

Ice Cubes Freeze at 0° Celsius
Versus 32° Fahrenheit
Celsius = Temperature

Figure 1 Common US standard and metric system values.

length may be measured in inches, feet, yards, miles, millimeters, centimeters, meters, and kilometers.

Today, both the US standard and metric systems are widely used in engineering, construction, and technical work. Therefore, it is necessary to have some degree of understanding of both systems. HVACR work is concerned with the following values, which can be expressed in both the US standard and metric systems:

- Dimensions and distances
- Weight
- Volume
- Pressure
- Temperature

Volume: The amount of space contained within a three-dimensional shape.

Table 1 shows how these values are stated in both the US standard and metric systems.

TABLE 1 Units of Measure in US Standard and Metric Systems

Value	US Standard System	Metric System
Dimensions	Inches and feet	Centimeters and meters
Weight	Ounces and pounds	Grams and kilograms
Volume (dry)	Cubic inches and cubic feet	Cubic centimeters and cubic meters
Volume (liquid)	Quarts and gallons	Liters
Pressure	Pounds per square inch	Pascal and bar
Temperature	Degrees Fahrenheit (°F)	Degrees Celsius (°C) (Sometimes spoken as degrees Centigrade)

Once it is understood, the metric system is actually much simpler to use than the US standard system because it is a decimal system in which prefixes are used to denote powers of 10. The older US standard system, on the other hand, requires the use of conversion factors that must be memorized. For example, 1 mile is 5,280 feet, and 1 inch is $\frac{1}{12}$ of a foot. In contrast, a centimeter is $\frac{1}{100}$ of a meter and a kilometer is 1,000 meters. An additional advantage of the metric system is that it is not necessary to add and subtract fractions. *Table 2* lists some of the more common units in the US standard system.

The metric system prefixes are listed in *Table 3*. From this table, it can be seen that the use of the metric system is logically arranged, and that the name of the

TABLE 2 Common Units in the US Standard System

Unit	Equivalent
12 inches (in)	1 foot (ft)
1 yard (yd)	3 ft
1 mile (mi)	5,280 ft
16 ounces (oz)	1 pound (lb)
1 ton	2,000 lb
1 minute	60 seconds (sec)
1 hour (hr)	3,600 sec
1 US gallon (gal)	0.1337 cubic feet (cu ft)

TABLE 3 Metric System Prefixes

Prefix	Unit			
	Fraction or Whole Number	Decimal	Exponent	Spoken As...
Micro- (µ)	$\frac{1}{1,000,000}$	0.000001	10^{-6}	One-millionth
Milli- (m)	$\frac{1}{1,000}$	0.001	10^{-3}	One-thousandth
Centi- (c)	$\frac{1}{100}$	0.01	10^{-2}	One-hundredth
Deci- (d)	$\frac{1}{10}$	0.1	10^{-1}	One-tenth
Deka- (da)	10	10.0	10^{1}	Tens
Hecto- (h)	100	100.0	10^{2}	Hundreds
Kilo- (k)	1,000	1,000.0	10^{3}	Thousands
Mega- (M)	1,000,000	1,000,000.0	10^{6}	Millions
Giga- (G)	1,000,000,000	1,000,000,000.0	10^{9}	Billions

unit also represents an order of magnitude (via the prefix) that foot and pound cannot.

The most common metric system prefixes are mega- (M), kilo- (k), centi- (c), milli- (m), and micro- (μ). Even though these prefixes may seem difficult to understand at first, most people probably use them regularly. For example, you have probably seen the terms *mega*watts, *kilo*meters, *centi*meters, *milli*volts, and *micro*amps.

<table>
<tr><td>1.2.0</td><td>**Length, Area, and Volume**</td></tr>
</table>

In the field, numbers usually represent physical quantities. In order to give meaning to these quantities, measurement units are assigned to the numbers to represent quantities such as length, **area**, and volume.

Area: The amount of space contained in a two-dimensional object such as a rectangle, circle, or square.

1.2.1 Length

Length typically refers to the long side of an object or surface. With liquids, the measurement of the level of fluid in a tank is basically a measurement of length from the surface of the fluid to the bottom of the tank. Length can be expressed in either US standard or metric units, that is, inches or centimeters.

TABLE 4 Length Conversion Multipliers

Unit	Centimeter	Inch	Foot	Meter	Kilometer
1 millimeter	0.1	0.03937	0.003281	0.001	0.000001
1 centimeter	1	0.3937	0.03281	0.01	0.00001
1 inch	2.54	1	0.08333	0.0254	0.0000254
1 foot	30.48	12	1	0.3048	0.0003048
1 meter	100	39.37	3.281	1	0.001
1 kilometer	100,000	39,370	3,281	1,000	1

Inches

Centimeters

Figure 2 Comparison of inches to centimeters.

Table 4 shows the relationships of the most common units of length. *Figure 2* compares inches to centimeters.

The multipliers in *Table 4* are used to make the conversions from one system to the other.

Example:

An installation plan for an HVACR system manufactured in Europe requires a thermostat to be mounted 122 centimeters (cm) above the floor, but your measuring tape is calibrated in inches. How many inches above the floor should the thermostat be placed?

$$1 \text{ cm} = 0.3937"$$
$$122 \text{ cm} = 122 \times 0.3937"$$
$$122 \text{ cm} = 48.0314" = 48" \text{ (rounded)}$$

or (changing to feet)

$$122 \text{ cm} = 122 \times 0.03281'$$
$$122 \text{ cm} = 4.00282' = 4' \text{ (rounded)}$$

NOTE

Appendix 03102A contains listings of common metric conversion factors.

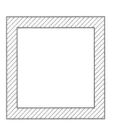

Square Duct

The Meter

Originally, the meter was defined as $^1/_{10,000,000}$ of Earth's meridional quadrant (the distance from the North Pole to the equator). This was how the meter got its name. This distance was etched onto a metal bar that is kept in France. In 1866, the United States legalized the use of the metric system and placed an exact copy of the metal bar into the US Bureau of Standards. In October 1983, scientists redefined the meter to avoid the minor potential of error associated with the metal bar in France. Now a meter is defined as being equal to the distance that light travels in $^1/_{299,792,458}$ of a second, which is equivalent to 39.37".

Rectangular Duct

1.2.2 Area

Area is the measurement of the surface of a two-dimensional object. In HVACR work, it is necessary to know the area of a duct in order to determine the amount of airflow it can support. Duct can be rectangular, square, round (*Figure 3*). Oval duct can also be used to pass through narrow areas, but it is rarely used.

The area of any rectangle is equal to the length multiplied by the width. In the US standard system, the unit of area is the square foot (ft^2) or square inch (in^2). *Figure 4* shows the application of this concept to the square. When converting areas from one measurement system to the other, every dimension must be converted. So, to convert the dimensions of the square shown in *Figure 4*, both the length and the width must be converted, as shown. The area can then be calculated using the metric values.

The same principle applies to a rectangle, as shown in *Figure 5*.

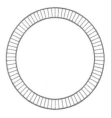

Round Duct

Think About It

Length

If the manufacturer's product data sheet states that a fan coil unit is 1.22 m high and 0.46 m wide, what size opening, in feet, is needed to get this unit inside a building?

Figure 3 Common duct shapes.

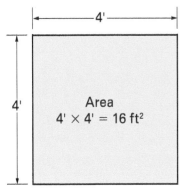

Area = Length × Width
= 4 ft × 4 ft
= 16 ft²

Convert to Square Meters:
1 m = 3.281 ft

Therefore:
Area = (4 ft ÷ 3.281) × (4 ft ÷ 3.281)
= 1.219 m × 1.219 m
= 1.486 m²
16 ft² = 1.486 m²

Figure 4 Converting the area of a square from US standard to metric units.

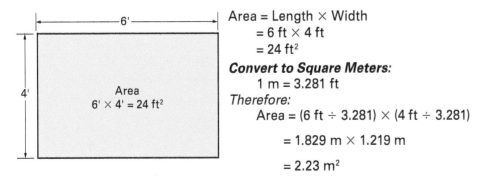

Area = Length × Width
= 6 ft × 4 ft
= 24 ft²

Convert to Square Meters:
1 m = 3.281 ft
Therefore:
Area = (6 ft ÷ 3.281) × (4 ft ÷ 3.281)
= 1.829 m × 1.219 m
= 2.23 m²

Figure 5 Converting the area of a rectangle from US standard to metric units.

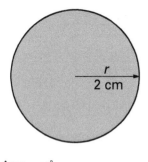

Area = πr^2
    = (3.14) (2 cm $\times$ 2 cm)
    = 3.14 $\times$ 4 cm^2
    = 12.56 cm^2

Convert to Square Inches:

1 in = 2.54 cm

Therefore:

Radius in Inches = 2 $\div$ 2.54
        = 0.7874 in
Area = (3.14) (0.7874 in $\times$ 0.7874 in)
    = 3.14 $\times$ 0.62 in^2
    = 1.95 in^2
12.56 cm^2 = 1.95 in^2

Figure 6 Converting the area of a circle from metric to US standard units.

The area of a circular duct is found using the following formula:

$$Area = \pi r^2$$

Where:

A = the area of a circle
π = a constant of 3.14159 (usually abbreviated to 3.14)
r = the radius (distance from the center to the edge of the circle)

When converting the dimensions of a circle, only the radius has a measured dimension that must be converted. In the example shown in *Figure 6*, square centimeters must be converted into square inches.

1.2.3 Volume

Volume is the amount of space occupied by a three-dimensional object. The volume of a cube or rectangular prism, such as a room in a building, is the product of three dimensions: length, width, and height. In the familiar US standard system, the most common units of volume are cubic feet (ft^3) or cubic inches (in^3). There are a number of situations in which an HVACR technician would need to calculate volume.

Here are some examples:

- In order to calculate the cooling or heating load a given space represents, it is necessary to determine the volume of the space. When calculating the load for a house, the estimator would need to know the volume of each room. Room volume is also a consideration in determining the number of air changes that occur in a given time.

- Assume that the oil needs to be drained from a semi-hermetic compressor. You know the compressor holds 2.5 quarts of oil, but you don't know if the capacity of your container is more than 2.5 quarts. The same principle applies if the oil has been drained into an uncalibrated container and you need to determine how much oil was removed.

- In a hydronic cooling or heating system, it is sometimes necessary to mix glycol (antifreeze) with the water in a specific proportion. To do so, it is necessary to first calculate the amount of water the pipes and equipment components can hold.

Figure 7 shows the three-dimensional measurement of a rectangular prism (space) such as a room. Its volume is calculated by multiplying the length, width,

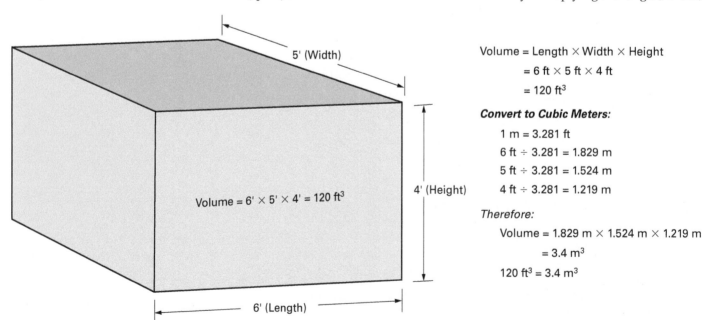

Volume = Length $\times$ Width $\times$ Height
    = 6 ft $\times$ 5 ft $\times$ 4 ft
    = 120 ft^3

Convert to Cubic Meters:

1 m = 3.281 ft
6 ft $\div$ 3.281 = 1.829 m
5 ft $\div$ 3.281 = 1.524 m
4 ft $\div$ 3.281 = 1.219 m

Therefore:

Volume = 1.829 m $\times$ 1.524 m $\times$ 1.219 m
    = 3.4 m^3
120 ft^3 = 3.4 m^3

Volume = 6' $\times$ 5' $\times$ 4' = 120 ft^3

5' (Width)
4' (Height)
6' (Length)

Figure 7 Volume calculation and conversion for a rectangular prism.

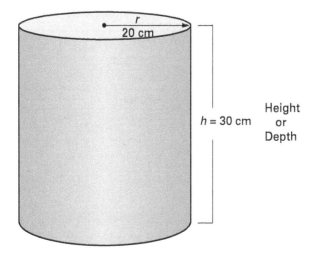

Volume = $(\pi r^2) \times h$

    or

    = Area of the Circle × Height

         Area   Height

    = (3.14) (20 cm) (20 cm) × (30 cm)

    = (3.14) (400 cm²) × (30 cm)

    = (1,256 cm²) × (30 cm)

    = 37,680 cm³

Figure 8 Calculating the volume of a cylinder (metric).

and height. To determine the volume, all the units must be either US standard or metric. To convert from one system to another, you can convert all the units and rework the problem, or you can simply convert the answer to the other system.

This equation works for finding the volume of box-shaped containers but not for tanks and other cylindrical containers, as shown in *Figure 8*. The volume of a cylindrical tank, such as a refrigerant cylinder, can be simplified by treating it as the area of the circle multiplied by its depth (or height). Although this example uses metric values, the formula remains the same for imperial measurements.

In the example shown in *Figure 9*, both the radius and the height have been converted to US standard values before calculating the volume.

1.2.4 Wet Volume Measurements

When the capacity of a space is calculated in cubic feet or cubic meters, it is considered dry volume. When liquid volume is involved, such as determining the amount of a fluid that would fill a container, it is classified as wet volume because liquid measures are used. Common wet measures in the US standard system include pint, quart, and gallon.

The metric system also uses wet measuring units. The liter is the most common measure. A liter is about five percent greater than a quart. Knowing the wet measures for a substance allows easy handling and measuring of fluids, since the fluid conforms to the shape of the container.

Table 5 shows the volume relationships between the liter and the dry volume of the cubic meter in the metric system, and the pint and gallon in the US standard system.

Figure 10 shows that the same metric prefixes that apply to the meter can be used with the liter.

NOTE

The radius of a cylinder or circle equals ½ the diameter. In order to determine the radius of a round container, it is first necessary to determine the diameter, then divide it by 2.

TABLE 5 Volume Relationships

Unit	Cubic Meter	Gallon	Liter	Pint
Cubic meter (m³)	1	264.172	1,000	2,113.38
US gallon (gal)	0.0037	1	3.785	8
Liter (L)	0.001	0.264	1	2.113
US pint (pt)	0.0005	0.125	0.473	1

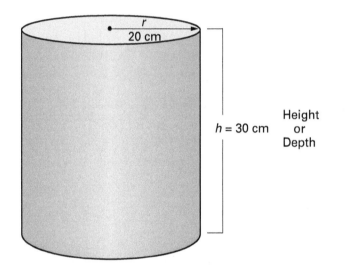

Volume $= (\pi r^2) \times h$

$= (3.14 \times 20^2) \times h$

$= (3.14 \times 400 \text{ cm}^2) \times 30 \text{ cm}$

$= 1{,}256 \text{ cm}^2 \times 30 \text{ cm}$

$= 37{,}680 \text{ cm}^3$

Convert to Cubic Inches:

1 cm = 0.3937 in

20 cm = 7.874 in

30 cm = 11.811 in

Therefore:

Volume $= [(3.14)\,(7.874^2)] \times 11.811 \text{ in}$

$= (3.14 \times 62 \text{ in}^2) \times 11.811 \text{ in}$

$= 194.68 \text{ in}^2 \times 11.811 \text{ in}$

$= 2{,}299.36 \text{ in}^3$

$37{,}680 \text{ cm}^3 = 2{,}299.36 \text{ in}^3$

Figure 9 Converting cylinder volume from metric to US standard units.

Think About It

Volume

Air conditioning equipment located outdoors is mounted on a concrete pad in order to provide stability and protection. While there are prefabricated pads for smaller units, large units often require a poured-in-place pad.

The concrete pad for the unit shown must be 13' × 6' × 6". To order or mix the concrete, the installers must determine how much concrete to order. What is the volume of concrete needed to construct this pad, expressed in cubic feet and cubic yards?

1 kiloliter = 1000 liters

1 deciliter = 0.1 liter

1 centiliter = 0.01 liter

1 milliliter = 0.001 liter

Figure 10 Common metric prefixes used with volumes.

1.3.0   Weight Conversions

Weight is actually the **force** an object exerts on the surface of Earth due to its **mass** and the pull of Earth's gravity. In the US standard system, the most common units for weight are the pound (lb) and the ounce (oz). *Table 6* shows the weight units used in both the US standard and metric systems.

In HVACR work, metric conversions can involve the weight of refrigerant for charging purposes or the weight of equipment for lifting purposes. Imagine that you are installing a new rooftop packaged unit and the equipment specs show the weight of the unit to be 1,200 kg. How do you select the equipment needed to raise the unit onto the roof when the rigging equipment is rated in pounds?

Table 6 shows that 1 kg = 2.205 lb, so multiplying 1,200 by 2.205 yields 2,646 lb. Therefore, any slings or lifting equipment used to raise the unit must have a rated capacity high enough to safely raise this weight.

Force: A push or pull on a surface.

Mass: The quantity of matter present.

1.4.0   Pressure and Temperature

Pressure and temperature are covered together because they are interdependent. If either one changes in a closed system, such as an HVACR system or an automobile tire, the other is affected in proportion. The temperature-pressure relationship is the key to the refrigeration cycle on which most air conditioning

TABLE 6 Weight Equivalents

Unit	Kilogram	Pound	Ounce	Gram
1 kilogram (kg)	1	2.205	35.27	1,000
1 pound (lb)	0.4536	1	16	453.6
1 ounce (oz)	0.0284	0.063	1	28.35
1 gram (g)	0.001	0.002	0.036	1

and refrigeration systems are based. When testing or troubleshooting a system, technicians measure temperature and pressure values at key points in a system to determine if the system is operating correctly.

For clarification, if there is 32 psi of air in the tires of your car, it really means that the air in the tires exerts 32 more pounds of force per square inch on the inside of the tires than the air pressure of the atmosphere exerts on the outside of the tires.

Both liquids and gases are capable of exerting pressure. Gases are compressed under pressure and expand when the pressure is lowered. Liquids are generally considered to be incompressible. The refrigeration cycle refrigerant exists as both a gas and a liquid in different parts of the system.

Tanks of refrigerant provide a good example of liquids and gases exerting pressure. The pressure inside a refrigerant tank depends on the type of refrigerant in the drum and the ambient temperature surrounding the drum. For every refrigerant, there is a specific temperature/pressure relationship. For example, a drum of R-410A refrigerant sitting in the back of an open truck with the sun shining on it may have a drum temperature of 105°F. A temperature/pressure chart for R-410A would show that this corresponds to a pressure of about 339.9 psig (gauge pressure). This means that the drum has a pressure of about 340 pounds pushing outward against its walls for each square inch of drum surface area. At 32°F, the pressure is only 101.6 psig.

Pounds per square inch (psi), the most common unit of pressure in the US standard system, is equivalent to newtons per square meter (N/m^2) in the metric system. One N/m^2 is also referred to as one *pascal*, after Blaise Pascal, a seventeenth-century French philosopher who contributed greatly to mathematics and engineering. The pascal, or Pa, is equal to a force of one **newton (N)** exerted on an area of one square meter. It would take 200,000 pascals (200 kilopascals) to inflate an ordinary automobile tire to about 28 psi. Automobile tire pressure is typically stated in kilopascals (kPa) and psi (*Figure 11*).

In HVACR work, where much higher pressures are involved, the bar (b) is a more convenient measure of pressure when metric values are used. One bar is equal to 14.5 psi and 100 kPa. The bar is the metric value often used on HVACR-related pressure gauges (*Figure 12*) along with psig units. Bars are also used by weather forecasters to describe changes in **atmospheric pressure**. You are probably familiar with **barometric pressure** readings from watching the weather information on the evening news. You will also see the term *millibar*, which is 0.001 b.

The next section relates these metric units to the more familiar pressure units of the US standard system and introduces the concept of pressure measurement at the same time.

NOTE

Even though the atmosphere is not visible and has no surface or texture, it still has mass because it consists of moisture and gases (such as oxygen). Scientists have determined that the atmosphere exerts a pressure of 14.7 pounds per square inch (psi) at sea level. At altitudes above sea level, the pressure decreases because there is less atmosphere. In Denver, Colorado, which is roughly 5,000 feet above sea level, the pressure is 12.1 psi.

Newton (N): The amount of force required to accelerate one kilogram at a rate of one meter per second.

Atmospheric pressure: The standard pressure exerted on Earth's surface. Atmospheric pressure is normally expressed as 14.7 pounds per square inch (psi) or 29.92 inches of mercury (inHg).

Barometric pressure: The actual atmospheric pressure at a given place and time.

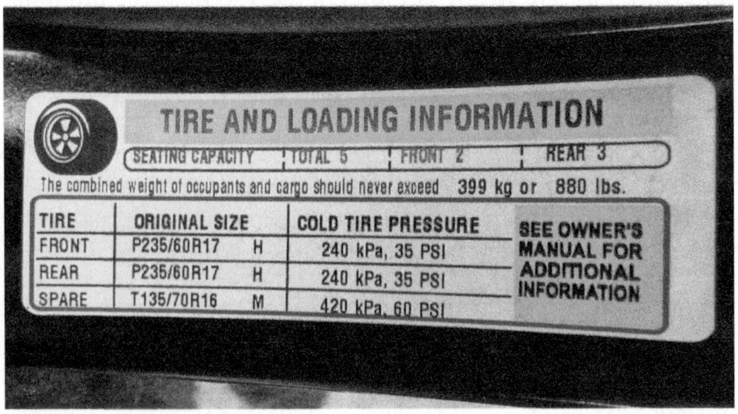

Figure 11 Tire pressure label.

Tire Pressure

If you're an auto racing fan, you probably know that slight variations in tire pressure can affect the handling performance of a race car. A crew chief might change tire pressure by just half of one psi during a pit stop, for example. Compressed air contains moisture, which will heat up as the tire gets hotter and cause the tire pressure to increase. For that reason, race cars use nitrogen instead of compressed air in their tires. Nitrogen is a drier gas and does not react to temperature as significantly.

Source: iStock@maccj

Gauge Manifold Set

The refrigerant gauge manifold set is one of the most common items of service equipment. It is used to monitor system pressures in air conditioning and refrigeration systems. The gauge on the left side of the manifold is called a *compound gauge* because it reads below atmospheric pressure (inHg) and above atmospheric pressure (psig). The right-hand gauge reads high pressures up to 800 psig. Remember that, to convert readings shown on the gauges (psig) to absolute pressure values (psia), 14.7 must be added to the reading.

Source: Ritchie Engineering Company, Inc. / YELLOW JACKET

Source: Ritchie Engineering Company, Inc. / YELLOW JACKET

1.4.1 Absolute Pressure

The standard atmospheric pressure exerted on the surface of Earth is 14.696 psi taken at sea level with the air at 70°F. For most practical applications, this number is usually rounded to 14.7 psi. Atmospheric pressure is also expressed as 29.9213 inches of mercury (inHg), which is equal to 14.696 psi. The rounded-off value for atmospheric pressure is 29.92 inHg.

Weather conditions usually cause a slight variation in the atmospheric pressure. The actual atmospheric pressure is known as the *barometric pressure*. It is usually ignored in measuring hydraulic machinery pressure, but it cannot be ignored when dealing with the low pressures generated by fans and blowers.

Most gauges measure the difference between the actual pressure in the system being measured and the atmospheric pressure. Pressure measured by gauges is called *gauge pressure* (psig). The total pressure that exists in the system is called the **absolute pressure** (psia). Absolute pressure is equal to gauge pressure plus the atmospheric pressure. In other words:

$$psia = psig + 14.7 \text{ psi}$$

Example:

A steam boiler pressure gauge reads 295 psig. Find the absolute pressure.

$$psia = psig + 14.7 \text{ psi}$$
$$psia = 295 + 14.7 \text{ psi}$$
$$psia = 309.7$$

Figure 12 Close-up of an HVACR-related pressure gauge.
Source: Ritchie Engineering Company, Inc. / YELLOW JACKET

Absolute pressure: The pressure measured relative to a perfect vacuum, where there is no measurable pressure. Absolute pressure is expressed in pounds per square inch absolute (psia). Absolute pressure = gauge pressure + atmospheric pressure.

TABLE 7 Pressure Conversions

Multiply	By	To Obtain
Atmospheres	14.7	Pounds per square inch
Pounds per square inch	0.0680	Atmospheres
Pounds per square inch	0.069	Bar
Bar	14.5	Pounds per square inch
Pounds per square inch	27.68	Inches of water (H_2O)
Pounds per square inch	2.31	Feet of water (H_2O)
Pounds per square inch	2.04	Inches of mercury (Hg)
Inches of water	0.0361	Pounds per square inch
Feet of water	0.433	Pounds per square inch
Inches of mercury	0.491	Pounds per square inch

Boiler pressure is sometimes specified in terms of *atmospheres*, where one atmosphere is equal to 14.696 psi (rounded to 14.7 for most purposes). Refer to *Table 7* for some common pressure conversions.

1.4.2 Static Head Pressure

Municipal water systems usually define pressure in terms of the static head or height in feet from the point of use to elevated water reservoirs. Gauge pressure can be converted to static head pressure, and vice versa, using the formula:

$$P = \frac{hd}{144}$$

Where:

P = pressure (psig)

h = height in feet (head)

d = density in lb per cu feet (62.43 for water)

144 = used to convert square feet into square inches

For example, if the water level in a reservoir is maintained at 150 feet above the *point of draw*. The point of draw represents the level from which water is drawn for use. What is the water pressure at the point of draw resulting from this head?

$$P = \frac{hd}{144} = \frac{150 \times 62.43}{144} = 65 \text{ psig}$$

Other terms are necessary to measure extremely low pressures, such as those developed by blowers and fans. These pressures are often measured in inches of water (inH_2O). Sometimes it is necessary to convert from one measure to another. Consult *Table 7* to solve the following problems.

Examples:

1. A steam generator operates at a pressure of 320 atmospheres. What is this pressure in psig?

   psig = 320 × 14.7 = 4,704 psig

2. Blower and fan pressures are usually measured in inH_2O because small differences in pressure can be readily detected. Due to the low pressures involved, the exact atmospheric pressure is usually measured to determine the actual discharge pressure. Calculate the discharge pressure, in psia, if the measured discharge pressure of a blower is 56.55 inH_2O and the barometric pressure is 28.49 inHg.

Step 1   Find the blower discharge pressure in psig.

$$psig = inH_2O \times 0.0361$$
$$psig = 56.55 \times 0.0361$$
$$psig = 2.041455$$

Step 2   Find the actual atmospheric pressure in psi.

$$psi = inHg \times 0.491$$
$$psi = 28.49 \times 0.491$$
$$psi = 13.98859$$

Step 3   Find the absolute pressure of the blower discharge.

$$Absolute\ pressure = gauge\ pressure + actual\ atmospheric\ pressure$$
$$Absolute\ pressure = 2.041455 + 13.98859$$
$$Absolute\ pressure = 16.030045\ or\ approximately\ 16.03\ psia$$

1.4.3 Vacuum

In HVACR work, a **vacuum** is any pressure that is less than the local atmospheric pressure. It is usually measured in terms of inches of mercury (inHg). Actual barometric pressure must always be determined in measuring a vacuum. Absolute pressure in a vacuum is calculated using the following formula:

Absolute vacuum pressure = barometric pressure – vacuum gauge reading

Example:

A vacuum gauge attached to a line reads 17.2 inHg. The barometric pressure reads 29.85 inHg. What is the absolute pressure in the line?

Absolute vacuum pressure = barometric pressure – vacuum gauge reading
Absolute vacuum pressure = 29.85 – 17.2
Absolute vacuum pressure = 12.65 inHg

Vacuum: Any pressure that is less than the prevailing atmospheric pressure; the absence of positive pressure.

1.4.4 Temperature

Temperature is the intensity level of heat and is usually measured on a temperature scale in degrees Fahrenheit (°F) or degrees Celsius (°C). In order to establish the scale, a substance is needed that can be placed in reproducible conditions. The substance used is water. The point at which water freezes at atmospheric pressure is one reproducible condition, and the point at which water boils at atmospheric pressure is another. The four temperature scales commonly used today are the Fahrenheit scale, Celsius scale, Rankine scale, and Kelvin scale (see *Figure 13*). On the Fahrenheit scale, the freezing temperature of water is 32°F and the boiling temperature is 212°F. On the Celsius scale, the freezing temperature of water is 0°C and the boiling temperature is 100°C. Because there are 100 degrees on a standard portion of the Celsius scale, Celsius temperature measurements are sometimes spoken as *degrees Centigrade*. The temperatures at which these fixed points occur were established by the inventors of the scales.

The Rankine scale and the Kelvin scale are based on the fact that where no heat exists, no molecular activity occurs. The temperature at which this occurs is called *absolute zero*, the lowest temperature possible. Both the Rankine and Kelvin scales have their zero-degree points at absolute zero. On the Rankine scale, the freezing point of water is 491.7°R and the boiling point is 671.7°R. The increments on the Rankine scale correspond in size to the increments on the Fahrenheit scale; for this reason, the Rankine scale is sometimes called the *absolute Fahrenheit scale*. Both the Rankine and Fahrenheit scales are part of the English system of measurement.

On the Kelvin scale, the freezing point of water is 273 K and the boiling point is 373 K. The increments on the Kelvin scale correspond to the increments on the Celsius scale; for this reason, the Kelvin scale is sometimes called the *absolute Celsius scale*. Both the Kelvin and Celsius scales are part of the metric system of measurement.

Think About It

Temperature Conversions

Depending on the application, refrigeration systems operate to cool a refrigerated area to temperatures between –40°F and 60°F. Comfort cooling systems operate to maintain temperatures in the conditioned space from 60°F to 80°F. What are the corresponding temperature ranges for refrigeration and air conditioning systems when expressed as degrees Celsius?

The scales of primary importance in HVACR work are the Fahrenheit scale and the Celsius scale. The Rankine scale and the Kelvin scale are used primarily in scientific applications.

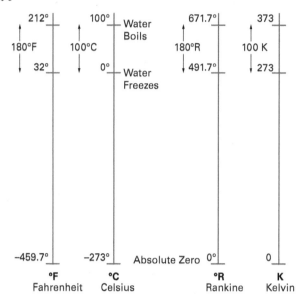

Figure 13 Comparison of temperature scales.

Creating and Measuring a Vacuum

Any air or moisture (non-condensables) trapped in an air conditioning or refrigeration system must be removed before the system can be charged with refrigerant. This requires that a vacuum be drawn on the system using a *vacuum pump*, such as the one shown in the top image.

The vacuum pump creates a pressure differential between the system and the pump. This causes air and moisture vapor trapped in the system at a higher pressure to move into a lower-pressure (vacuum) area created in the vacuum pump. When the vacuum pump lowers the pressure in the system enough, determined by the ambient temperature of the system, moisture trapped in the system boils to a vapor. The water vapor is drawn out of the system with the air and exhausted to the atmosphere. The level of vacuum present in the system can be measured using a *vacuum gauge*, such as the one shown in the bottom image.

Source: Ritchie Engineering Company, Inc. / YELLOW JACKET

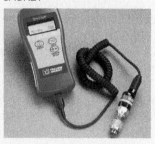

Source: Ritchie Engineering Company, Inc. / YELLOW JACKET

1.4.5 Temperature Conversions

Because both the Fahrenheit and Celsius scales are used in the HVACR industry, it is sometimes necessary to convert between the two. An HVACR technician must be familiar with these conversions.

On the Fahrenheit scale, there are 180 degrees between the freezing temperature and boiling temperature of water. On the Celsius scale, there are 100 degrees between the freezing and boiling temperatures of water. The relationship between the two scales can be expressed as follows:

Digital Thermometers

Thermometers with digital readouts can provide temperature readings in both Celsius and Fahrenheit. Electronic stick thermometers are very common in the toolboxes of HVACR technicians. An infrared thermometer can take accurate temperature readings from a distance, using a beam as an aiming device. Technicians need a variety of thermometer types to address the many scenarios they encounter.

$$\frac{\text{Fahrenheit range (freezing to boiling)}}{\text{Celsius range (freezing to boiling)}} = \frac{180°}{100°} = \frac{9}{5}$$

Therefore, one degree Fahrenheit is $\frac{5}{9}$ of one degree Celsius and conversely, one degree Celsius is $\frac{9}{5}$ of one degree Fahrenheit. Thus, to convert Fahrenheit to Celsius, it is necessary to subtract 32 (since 32°F corresponds to 0°C), then multiply by $\frac{5}{9}$. To convert Celsius to Fahrenheit, it is necessary to multiply by $\frac{9}{5}$, and then add 32.

$$°C = \tfrac{5}{9}(°F - 32)$$
$$°F = (\tfrac{9}{5} \times °C) + 32$$

Practice these calculations to become more comfortable with using them. *Figure 14* shows two examples.

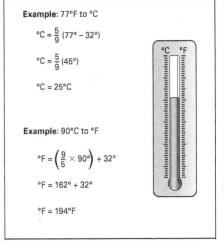

Example: 77°F to °C

$$°C = \tfrac{5}{9}(77° - 32°)$$

$$°C = \tfrac{5}{9}(45°)$$

$$°C = 25°C$$

Example: 90°C to °F

$$°F = \left(\tfrac{9}{5} \times 90°\right) + 32°$$

$$°F = 162° + 32°$$

$$°F = 194°F$$

Figure 14 Sample temperature unit conversions.

1.0.0 Section Review

1. In the metric system, liquid volume is stated in terms of _____.
   a. cubic feet
   b. in^3
   c. liters
   d. pascals

2. What is the area *in square feet* of a circle with a diameter of 3 m?
   a. 7.06 ft^2
   b. 9.42 ft^2
   c. 28.26 ft^2
   d. 76.1 ft^2

3. A condensing unit that weighs 240 kg would equal _____.
   a. 240 lb
   b. 109 lb
   c. 768 lb
   d. 529 lb

4. If the refrigerant pressure at 10°C equals 9.88 bar, what are the corresponding temperature and pressure values in the US standard system?
    a. 50°F and 143.3 psi
    b. 40°F and 143.3 psi
    c. 23°F and 24.4 psi
    d. 50°F and 24.4 psi

2.0.0 Solving Problems Using Algebra

Performance Tasks

There are no Performance Tasks in this section.

Objective

Solve basic algebraic equations.
  a. Define algebraic terms.
  b. Demonstrate an understanding of the sequence of operations.
  c. Solve basic algebraic equations.

Constants: Elements in an equation with fixed values.

Variables: Elements of an equation that may change in value.

Algebra is a branch of mathematics that defines and manipulates equations containing symbols instead of numbers. The symbols may be either **constants** or **variables**. They are connected to each other with mathematical operators such as +, −, ×, and ÷. Knowing how to do calculations in algebra is an important skill for HVACR technicians. Basic algebra is necessary to get through the training program and is essential for doing airflow, electrical current, and voltage calculations. Algebraic calculations are used in many types of troubleshooting.

As in all fields of mathematics, an understanding of algebra requires the knowledge of some basic rules and definitions.

2.1.0 Definition of Terms

Coefficients: Numerals that multiply a variable, e.g., the numeral 2 in the math expression 2*b*.

This section defines basic algebraic terms, including operators, equations, variables, constants, and **coefficients**.

2.1.1 Mathematical Operators

Mathematical operators define the required action using a symbol. Common operators include the following:

- + (Addition)
- − (Subtraction)
- × or • (Multiplication)
- ÷ or / (Division)

2.1.2 Equations

An equation is a collection of numbers, symbols, and mathematical operators connected by an equal sign (=). Some examples of equations are:

$$2 + 3 = 5$$
$$P = EI$$
$$\text{Volume} = l \times w \times h$$

2.1.3 Variables

A variable is an element of an equation that may change in value. For example, here is the simple equation for the area of a rectangle:

$$\text{Area} = l \times w$$

If the area is 12 and the length (*l*) is 6, the equation would read as follows:

$$12 = 6 \times w$$

In this case, it is easy enough to see that the width (w) is equal to 2. The equation then reads as $12 = 6 \times 2$. What is the width if the length is equal to 3?

$$12 = 3 \times w$$

In this case, the width is equal to 4. The equation then reads as $12 = 3 \times 4$. Therefore, in these two equations ($12 = 6 \times w$ and $12 = 3 \times w$), w must be considered a variable because it may change, depending on the value of l.

2.1.4 Constants

A constant is an element of an equation that does not change in value. For example, consider the following equation:

$$2 + 5 = 7$$

In this equation, 2, 5, and 7 are constants. The number 2 will always be 2, 5 will always be 5, and 7 will always be 7, no matter the equation. Constants also refer to accepted values that represent one element of an equation and do not change from situation to situation. One of the most common constants is pi (π). It has an approximate value of 3.14 and represents the ratio of the circumference to the diameter in a circle. It is used in most calculations involving circles.

2.1.5 Coefficients

A coefficient is a multiplier. Consider the following equation:

$$\text{Area} = l \times w$$

In this equation, l is the coefficient of w. It can also be written as lw, without the multiplication sign. No multiplication symbol is required when the intended relationship between symbols and letters is clear. For example:

- $2L$ means 2 times L (2 is the coefficient of L)
- IR means I times R (I is the coefficient of R)

2.1.6 Powers and Roots

Many mathematical formulas used in electronics and construction require that the power or root of a number be found. A power (or **exponent**) is a number written above and to the right of another number, which is called the *base*. For example, the expression y^x means to take the value of y and multiply it by itself x times. The xth root of a number y is another number that when multiplied by itself x times returns a value of y. Expressed mathematically:

$$(\sqrt[x]{y})^x = y$$

Squares and Square Roots — The need to find squares and square roots is common in HVACR mathematics. A square is the product of a number or quantity multiplied by itself. For example, the square of 6 means 6×6. To denote a number as squared, simply place the exponent 2 above and to the right of the base number. For example:

$$6^2 = 6 \times 6 = 36$$

The square root of a number is the divisor which, when multiplied by itself (squared), gives the number as a product. Extracting the square root refers to a process of finding the equal factors which, when multiplied together, return the original number. The process is identified by the radical symbol [$\sqrt{}$]. This symbol is a shorthand way of stating that the equal factors of the number under the radical sign are to be determined. Finding the square roots is necessary in many calculations, including calculating loads and determining the airflow in a duct.

For example, $\sqrt{16}$ is read as the square root of 16. The number consists of the two equal factors 4 and 4. Thus, when 4 is raised to the second power or squared, it is equal to 16. Squaring a number simply means multiplying the number by itself.

Powers and Roots

Powers and roots can be easily calculated using a scientific calculator, such as the one shown here, or a calculator app on a smartphone.

Source: iStock@Amankris

Exponent: A small figure or symbol placed above and to the right of another figure or symbol to show how many times the latter is to be multiplied by itself (e.g., $b^3 = b \times b \times b$).

The number 16 is a perfect square. Numbers that are perfect squares have whole numbers as the square roots. For example, the square roots of perfect squares 4, 25, 36, 121, and 324 are the whole numbers 2, 5, 6, 11, and 18, respectively.

Squares and square roots can be calculated by hand, but the process is very time-consuming and subject to error. Most people find squares and square roots of numbers using a calculator. To find the square of a number, the calculator's square key $[x^2]$ is used. When pressed, it takes the number shown in the display and multiplies it by itself. For example, to square the number 4.235, enter 4.235, press the $[x^2]$ key, and then read 17.935225 on the display.

Similarly, to find the square root of a number, the calculator's square root key $[\sqrt{\ }]$ or $[\sqrt{x}]$ is used. When pressed, it calculates the square root of the number shown in the display. For example, to find the square root of the number 17.935225, enter 17.935225, press the $[\sqrt{\ }]$ or $[\sqrt{x}]$ key, then read 4.235 on the display. Note that on some calculators, the $[\sqrt{\ }]$ or $[\sqrt{x}]$ key must be pressed before entering the number.

Other Powers and Roots — It is sometimes necessary to find powers and roots other than squares and square roots. This can easily be done with a calculator. The power key $[y^x]$ raises the displayed value to the xth power. The order of entry must be y, $[y^x]$, then x. For example, to find the power 2.86^3, enter 2.86, press the $[y^x]$ key, enter 3, press the $[=]$ key, then read 23.393656 on the display.

The root key $[\sqrt[x]{y}]$ is used to find the xth root of the displayed value y. The order of entry is y [INV] $[\sqrt[x]{y}]$. The [INV] or [2nd] key, when pressed before any other key that has a dual function, causes the second function of the key to be operated. For example, to find the cube root of 1,500 or $(\sqrt[3]{1,500})$, enter 1500, press the [INV] $[\sqrt[x]{y}]$ keys, enter 3, press the $[=]$ key, then read 11.44714243 on the display. Note that on some calculators, the $[\sqrt[x]{y}]$ key must be pressed before entering the number.

The answer can easily be checked by using the $[y^x]$ function to raise the answer on the display to the power of 3. For example, $11.44714243^3 = 1,500$.

2.2.0   Sequence of Operations

Complicated equations must be solved by performing the indicated operations in a prescribed sequence. This sequence is: Multiply, Divide, Add, and Subtract (MDAS). For example, the following equation can result in a number of answers if the MDAS sequence is not followed:

$$3 + 3 \times 2 - 6 \div 3 = x$$

To come up with the correct result, this equation must be solved in the following order:

Step 1   Multiply: $3 + \mathbf{3 \times 2} - 6 \div 3$

Step 2   Divide: $3 + 6 - \mathbf{6 \div 3}$

Step 3   Add: $\mathbf{3 + 6} - 2$

Step 4   Subtract: $\mathbf{9 - 2}$

The result is $7 = x$.

2.3.0   Solving Algebraic Equations

Some equations may include several variables. Solving these equations means simplifying them as much as possible and, if necessary, separating the desired variable so that it is on one side by itself, with everything else on the other side. Problems such as these are known as *algebraic expressions*. When an algebraic expression appears in an equation, the MDAS sequence also applies. For example:

$$P = R - [5(3A + 4B) + 40L]$$

The parentheses represent multiplication, so they are worked on first. When working with multiple sets of parentheses or brackets, always begin by eliminating the innermost symbols first, then working your way to the outermost symbols. Thus, in the above equation, the expressions within parentheses $(3A + 4B)$ and the coefficient 5 are multiplied first:

$$P = R - [15A + 20B + 40L]$$

The brackets also represent multiplication, so they are worked on next. The minus sign is the same as a coefficient –1, so each term within the brackets is multiplied by –1:

$$P = R - 15A - 20B - 40L$$

At this point, the equation has been simplified as much as possible, but could it apply to a real-life situation? Suppose you have just installed ductwork in five identical apartments, and you want to determine the profit on the job. If this equation were written out in longhand or spoken, it would look like this:

Profit (P) is equal to the payment received (R) minus five apartments times three pieces of one type of ductwork in each apartment at a certain cost per piece (A) plus four pieces of a second type of ductwork in each apartment at a certain cost per piece (B) plus forty hours of labor times an hourly rate (L).

It makes more sense now, written as an equation:

$$P = R - [5(3A + 4B) + 40L]$$

or

$$P = R - 15A - 20B - 40L$$

Now, plug in numbers for the known values. These are reasonable values to work with:

- R = $7,500, the payment received for the job
- A = $100, for the cost of duct piece A
- B = $150, for the cost of duct piece B
- L = $55, for the hourly cost of labor plus overhead

Here is the result, using those figures:

$$P = \$7,500 - (15 \times \$100) - (20 \times \$150) - (40 \times \$55)$$

Multiplying yields:

$$P = \$7,500 - \$1,500 - \$3,000 - \$2,200$$

Result:

$$P = \$800$$

2.3.1 Rules of Algebra

There are a few simple rules that, once memorized, help simplify and solve almost any equation you encounter as an HVACR technician.

Rule 1:

If the same value is added to or subtracted from both sides of an equation, the resulting equation is valid. For example, consider the following equation:

$$5 = 5$$

If 3 is added to each side of the equation, the resulting equation remains valid (both sides are still equal to each other).

$$5 + 3 = 5 + 3$$

$$8 = 8$$

In the same way, if the same number is subtracted from both sides of an equation, the resulting equation is valid. For example, consider the following equation:

$$5 = 5$$

If 4 is subtracted from both sides of the equation, the resulting equation remains valid.

$$5 - 4 = 5 - 4$$
$$1 = 1$$

Moving variables from one side of an equation to another is done in the same way as moving constants. Recall that when an equation is solved for one particular variable, it means that the variable should be on one side of the equation by itself. For example, consider the following pressure equation used frequently in the trade when working with system pressures in air conditioning and steam boiler systems:

$$\text{Absolute pressure} = \text{gauge pressure} + 14.7$$

To solve this equation for gauge pressure, 14.7 must be moved to the other side of the equation with absolute pressure. To do so, subtract 14.7 from both sides of the equation:

$$\text{Absolute pressure} - 14.7 = \text{gauge pressure} + 14.7 - 14.7$$

It should be clear that the + 14.7 and − 14.7 on the right cancel each other, leaving:

$$\text{Absolute pressure} - 14.7 = \text{gauge pressure}$$

or

$$\text{Gauge pressure} = \text{absolute pressure} - 14.7$$

The equation has been solved for gauge pressure. To take this new equation and solve it for absolute pressure again, simply add 14.7 to each side:

$$\text{Gauge pressure} + 14.7 = \text{absolute pressure} - 14.7 + 14.7$$

Again, the + 14.7 and − 14.7 on the right cancel each other out, leaving:

$$\text{Gauge pressure} + 14.7 = \text{absolute pressure}$$

or

$$\text{Absolute pressure} = \text{gauge pressure} + 14.7$$

Rule 2:

If both sides of an equation are multiplied or divided by the same value, the resulting equation is valid. This example uses Ohm's law, which is used by HVACR technicians to calculate voltage, current, and resistance values for electrical circuits. This equation is as follows:

$$E = IR$$

Where:

$$E = \text{voltage}$$
$$I = \text{current}$$
$$R = \text{resistance}$$

If you know the voltage (E) and the current (I), but need to find the resistance (R), how do you rearrange the equation? To solve this equation for R, I must be moved to the other side of the equation with E. To do so, divide both sides by I:

$$E \div I = IR \div I$$

The two I's on the right cancel each other out, leaving:

$$E \div I = R$$

or

$$R = E \div I$$

The equation has been solved for resistance. To take this new equation and solve it for *E* again, each side is multiplied by *I*:

$$I \times R = E \div I \times I$$

The two *I*'s on the right cancel each other out, leaving:

$$IR = E$$

or

$$E = IR$$

Rule 3:

Like terms may be added and subtracted in a manner similar to constant numbers. For example, consider the following equation:

$$2A + 3A = 15$$

The like terms (2*A* and 3*A*) may be added directly:

$$5A = 15$$

Divide both sides of the equation by 5, leaving:

$$5A \div 5 = 15 \div 5$$

$$A = 3$$

These rules may be used repeatedly until an equation is in the desired form. For example, take a look at the following pressure equation:

$$p = hd \div 144$$

Where:

$$p = \text{pressure}$$
$$h = \text{height}$$
$$d = \text{density}$$

Solve the equation for density (*d*).

Step 1 Remove the fraction by multiplying both sides by 144.

$$p \times 144 = hd \div 144 \times 144$$

The two 144's on the right side cancel each other, leaving:

$$144p = hd$$

Step 2 Divide both sides by *h*.

$$144p \div h = d \times h \div h$$

or

$$d = 144p \div h$$

The equation has been solved for density (*d*).

Fan Airflow Versus Speed

The performance of all fans and blowers is governed by three rules called the *fan laws*. One of these rules states that the amount of air delivered by a fan in cubic feet per minute (cfm) varies directly with the speed of the fan as measured in revolutions per minute (rpm). This is expressed mathematically as:

$$\text{New cfm} = \frac{\text{New rpm} \times \text{Existing cfm}}{\text{Existing rpm}}$$

Given the equation above for calculating a new cfm, how would you solve the equation to determine the new rpm, assuming you know the new cfm, the existing cfm, and the existing rpm?

2.0.0 Section Review

1. In an equation, a value that is subject to change is a(n) _____.
   a. coefficient
   b. variable
   c. exponent
   d. constant

2. When solving an equation, the *first* step is to _____.
   a. multiply
   b. divide
   c. add
   d. subtract

3. Given the equation $E = IR$, rewrite it to solve for I.
   a. $I = E \times R$
   b. $I = E - R$
   c. $I = E \div R$
   d. $I = E + R$

3.0.0 Working with Geometric Figures

Performance Tasks

There are no Performance Tasks in this section.

Objective

Identify and describe geometric figures.
   a. Describe the characteristics of a circle.
   b. Identify and describe types of angles.
   c. Identify and describe types of polygons.
   d. Calculate various values associated with triangles.

Polygons: Shapes formed when three or more straight lines are joined in a regular pattern.

Geometry is the study of various figures. It consists of two main fields: plane geometry and solid geometry. Plane geometry is the study of two-dimensional figures such as squares, rectangles, triangles, circles, and **polygons**. Solid geometry is the study of figures that occupy space, such as cubes, spheres, and other three-dimensional objects. The focus of this section is on the elements of plane geometry.

Geometric objects such as polygons and triangles are formed by lines. A line that forms a right angle (90 degrees) with one or more lines is said to be perpendicular to those lines (*Figure 15*). The distance from a point to a line is the measure of the perpendicular line drawn from that point to the line. Two or more straight lines that are the same distance apart at all points are said to be parallel. Parallel lines never intersect.

3.1.0   Characteristics of a Circle

As shown in *Figure 16*, a circle is a curved line that connects with itself and has these other properties:

- All points on a circle are the same distance (equidistant) from the point at the center.
- The distance from the center to any point on the curved line, called the *radius* (r), is always the same.
- The shortest distance from any point on the curve through the center to a point directly opposite is called the *diameter* (d). The diameter is equal to twice the radius ($d = 2r$).
- The distance around the outside of the circle is called the *circumference*. It can be determined by using the equation: circumference = πd, where π is a constant equal to approximately 3.14 and d is the diameter ($2r$).

• A circle is divided into 360 parts with each part called a *degree*; therefore, one degree = $\frac{1}{360}$ of a circle. The degree is also the unit of measurement commonly used to measure the size of an angle.

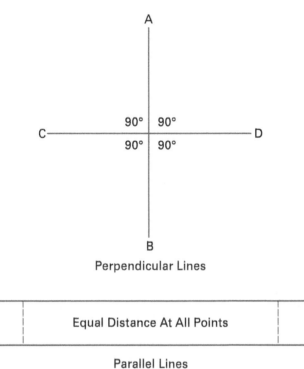

Perpendicular Lines

Equal Distance At All Points

Parallel Lines

Figure 15 Perpendicular and parallel lines.

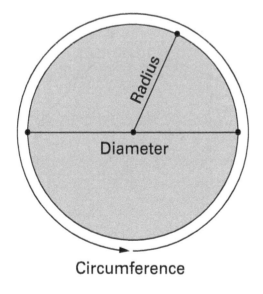

Radius

Diameter

Circumference

Figure 16 Parts of a circle.

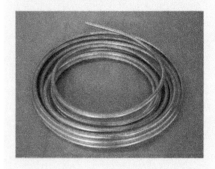
3.2.0   Angles

Two straight lines meeting at a point known as the *vertex* form an angle (*Figure 17*). The two lines are the sides, or rays, of the angle. The angle is the amount of opening that exists between the rays and is measured in degrees. There are two ways commonly used to identify angles. One is to assign a letter to the angle, such as angle *D* shown in *Figure 17*. This is written as ∠*D*. The other way is to name the two end points of the rays and put the vertex letter between them (e.g., ∠*ABC*). When the angle measure is shown in degrees, it should be written inside the angle, if possible. If the angle is too small to show the measurement, it may be placed outside of the angle, with an arrow to the inside.

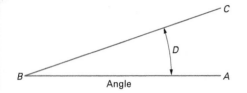

Figure 17 An angle.

The common types of angles are shown in *Figure 18* and *Figure 19* and are defined as follows:

- *Right angle* — A right angle has rays that are perpendicular to each other. The measure of this angle is always 90 degrees.
- *Straight angle* — A straight angle does not look like an angle at all. The rays of a straight angle lie in a straight line, and the angle measures 180 degrees.
- *Acute angle* — An acute angle is less than 90 degrees.
- *Obtuse angle* — An obtuse angle is greater than 90 degrees but less than 180 degrees.
- *Adjacent angles* — When three or more rays meet at the same vertex, the angles formed are said to be adjacent (next to one another). In *Figure 19*, the angles ∠ABC and ∠CBD are adjacent angles. The ray BC is said to be common to both angles.
- *Complementary angles* — Two adjacent angles that have a combined total measure of 90 degrees. In *Figure 19*, ∠DEF is complementary to ∠FEG.
- *Supplementary angles* — Two adjacent angles that have a combined total measure of 180 degrees. In *Figure 19*, ∠HIJ is supplementary to ∠JIK.

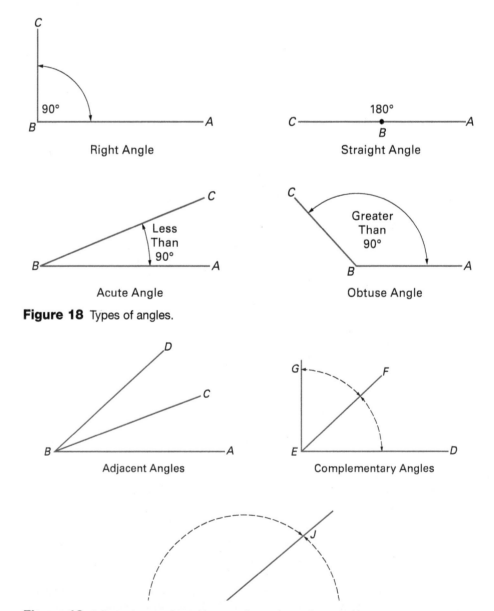

Figure 19 Adjacent, complementary, and supplementary angles.

3.3.0 Polygons

A polygon is formed when three or more straight lines are joined in a regular pattern. Some of the most familiar polygons are shown in *Figure 20*. As shown, they have common names that generally refer to their number of sides. When all sides of a polygon have equal length and all internal angles are equal, it is called a *regular polygon*.

Each of the boundary lines forming the polygon is called a *side* of the polygon. The point at which any two sides of a polygon meet is called a *vertex* of the polygon. The perimeter of any polygon is equal to the sum of the lengths of all sides. The sum of the interior angles of any polygon is equal to $(n - 2) \times 180$ degrees, where n is the number of sides.

For example, the sum of the interior angles for a square is 360 degrees [$(4 - 2) \times 180$ degrees = 360 degrees] and for a triangle is 180 degrees [$(3 - 2) \times 180$ degrees = 180 degrees].

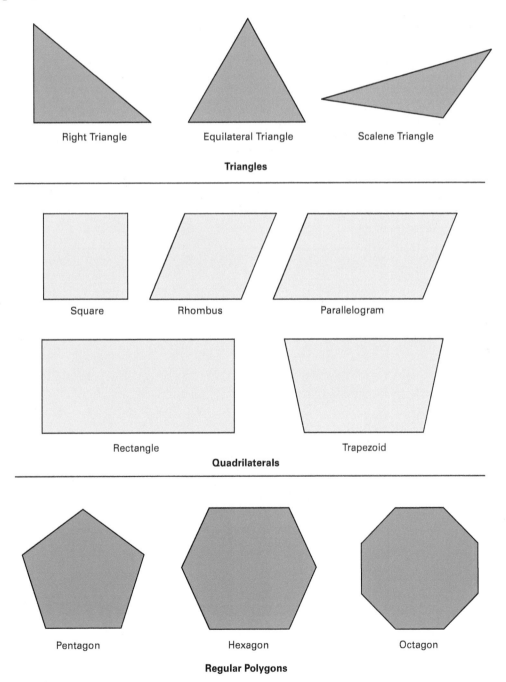

Right Triangle   Equilateral Triangle   Scalene Triangle

Triangles

Square   Rhombus   Parallelogram

Rectangle   Trapezoid

Quadrilaterals

Pentagon   Hexagon   Octagon

Regular Polygons

Figure 20 Common polygons.

3.4.0 Working with Triangles

As mentioned previously, triangles are three-sided polygons. *Figure 21* shows six different types of triangles. A regular polygon with three equal sides is called an *equilateral triangle*. Two types of irregular triangles are the isosceles (having two sides of equal length) and the scalene (having all sides of unequal length). An important fact to remember about triangles is that the sum of the three interior angles of any triangle equals 180 degrees. As shown, all three sides of an equilateral triangle are equal. In this kind of triangle, the three angles are also equal. The isosceles triangle has two equal sides with the angles opposite the equal sides also being equal.

Triangles are also classified according to their interior angles. If one of the three interior angles is 90 degrees, the triangle is called a *right triangle*. If one of the three interior angles is greater than 90 degrees, the triangle is called an *obtuse triangle*. If each of the interior angles is less than 90 degrees, the triangle is called an *acute triangle*. The sum of the three interior angles of any triangle is always equal to 180 degrees. This is helpful to remember whenever you know two angles of a triangle and need to calculate the third.

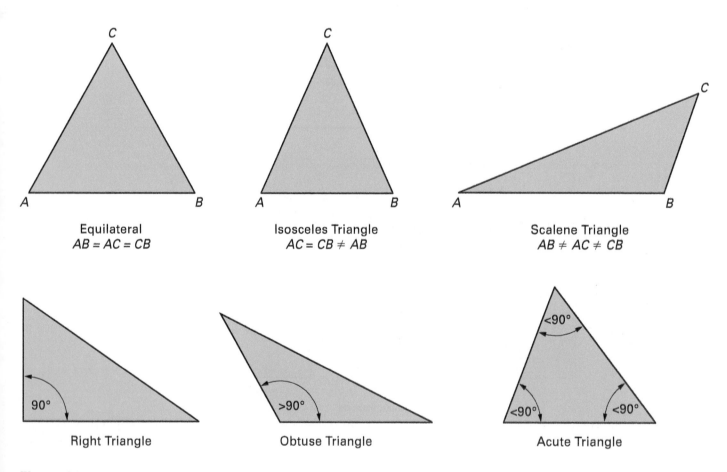

Equilateral
AB = AC = CB

Isosceles Triangle
AC = CB ≠ AB

Scalene Triangle
AB ≠ AC ≠ CB

Right Triangle

Obtuse Triangle

Acute Triangle

Figure 21 Triangles.

3.4.1 Right Triangle Functions

One of the most common triangles is the right triangle. Since it has one 90-degree angle (right angle), the other two angles are acute angles. They are also complementary angles because their sum equals 90 degrees. The right triangle has two sides perpendicular to each other, thus forming the right angle. To aid in writing equations, the sides and angles of a right triangle are labeled (*Figure 22*). Normally, capital (uppercase) letters are used to label the angles, and lowercase letters are used to label the sides. The third side, which is always opposite the right angle (C), is called the *hypotenuse*. It is always longer than either of the other two sides. The other sides can be remembered as *a* for altitude and *b* for base. Note that the letters used to label corresponding sides and angles are opposite each other. For example, side *a* is opposite angle *A*, and so forth.

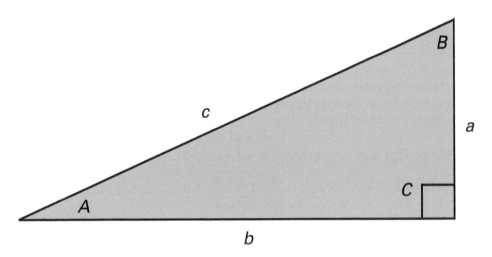

Figure 22 Common labeling of angles and sides in a right triangle.

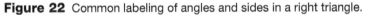

Trigonometry

Fifty years before Pythagoras, Thales (a Greek mathematician) figured out how to measure the height of the pyramids using a technique that later became known as *trigonometry*. Trigonometry recognizes that there is a relationship between the size of an angle and the lengths of the sides in a right triangle. Pythagoras eventually developed the Pythagorean theorem around the work of Thales.

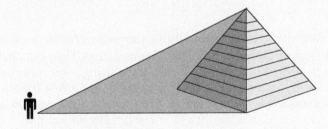

3-4-5 Rule

The 3-4-5 rule is based on the Pythagorean theorem and has been used in building construction for centuries. It is a simple method for laying out or checking right angles and requires only the use of a tape measure.

The numbers 3-4-5 represent dimensions in feet that describe the sides of a right triangle. Right triangles that are multiples of the 3-4-5 triangle are used (e.g., 9-12-15, 12-16-20, 15-20-25, etc.). The specific multiple used is determined by the relative distances involved in the job being laid out or checked.

An example of the 3-4-5 rule using the multiples of 15-20-25 is shown here. In order to square or check a corner as shown in the example, first measure and mark 15' down the line in one direction, then measure and mark 20' down the line in the other direction. The distance measured between the 15' and 20' points must be exactly 25' to ensure that the angle is a perfect right angle.

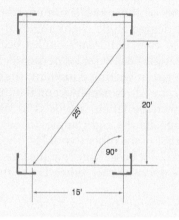

3.4.2 Right Triangle Calculations Using the Pythagorean Theorem

If the lengths of any two sides of a right triangle are known, the length of the third side can be determined using a rule called the *Pythagorean theorem*. It states that the square of the hypotenuse (*c*) is equal to the sum of the squares of the remaining two sides (*a* and *b*). Expressed mathematically:

$$c^2 = a^2 + b^2$$

The formula may be rearranged to solve for the unknown side as follows:

$$a = \sqrt{c^2 - b^2}$$
$$b = \sqrt{c^2 - a^2}$$
$$c = \sqrt{a^2 + b^2}$$

For example, a right triangle has an altitude (side *a*) equal to 8' and a base (side *b*) equal to 12'. To find the length of the hypotenuse (side *c*), proceed as follows:

$$c = \sqrt{a^2 + b^2}$$
$$c = \sqrt{8^2 + 12^2}$$
$$c = \sqrt{64 + 144}$$
$$c = \sqrt{208}$$
$$c = 14.422$$

To determine the actual length of the hypotenuse using the formula above, it is necessary to square both sides and add them together. Then calculate the square root of that sum. Fortunately, this is easy to do using a scientific calculator. On many calculators, you simply key in the number and press the square root [√] key. On some calculators, the square root does not have a separate key. Instead, the square root function is the inverse of the [x^2] key, so you have to press [INV] or [2nd], depending on the calculator, followed by [x^2], to obtain the square root.

3.4.3 Converting to and from Decimal Feet

When using trigonometric functions to calculate numerical values for lengths, distances, and angles, the answers are normally expressed as a decimal. Construction drawings and specifications often express such measurements in feet and inches. For this reason, it is often necessary to convert between these two measurement systems. Conversion tables are available in many reference books. If conversion tables are not readily available, you should become familiar with the methods for making such conversions mathematically.

To convert values given in decimal feet into equivalent feet and inches, use the following procedure. For this example, 45.3646' will be converted to feet and inches.

Think About It

Converting Dimensions

When using a calculator to determine the length of the hypotenuse for a right angle, the answer shown on the calculator is 6.354'. What is the length of the hypotenuse when expressed in feet and inches?

Step 1 Subtract 45' from 45.3646' = 0.3646'.

Step 2 Convert 0.3646' to inches by multiplying 0.3646' by 12 = 4.3752".

Step 3 Subtract 4" from 4.3752" = 0.3752".

Step 4 Convert 0.3752" into eighths of an inch by multiplying 0.3752" by 8 = 3.0016 and produces the fraction $\frac{3.0016}{8}$ or, when rounded off, $\frac{3}{8}$. Therefore, 45.3646' = 45'-4$\frac{3}{8}$".

To convert values given in feet and inches (and inch-fractions) into equivalent decimal feet values, use the following procedure. For example, convert 45'-4$\frac{3}{8}$" to decimal feet.

Step 1 Convert the inch-fraction to a decimal. This is done by dividing the numerator (top number of the fraction) by the denominator (bottom number of the fraction). For example, $\frac{3}{8}$" = 0.375".

Step 2 Add the 0.375" to 4" to obtain 4.375".

Step 3 Divide 4.375" by 12 to obtain 0.3646'.

Step 4 Add 0.3646' to 45' to obtain 45.3646'. Therefore, 45'-4$\frac{3}{8}$" = 45.3646'.

3.0.0 Section Review

1. The formula for calculating the circumference of a circle is _____.
   a. πr^2
   b. 2π
   c. πd
   d. $I \times R$

2. An angle that is greater than 90 degrees but less than 180 degrees is a(n) _____.
   a. right angle
   b. acute angle
   c. supplementary angle
   d. obtuse angle

3. Which of the following is *not* considered a polygon?
   a. Triangle
   b. Circle
   c. Rectangle
   d. Trapezoid

4. If a right triangle has a hypotenuse of 15" and a height of 12", what is the length of the base?
   a. 9"
   b. 19.2"
   c. 81"
   d. 9'

1. One meter is equal to _____.
    a. 0.3937'
    b. 39.37'
    c. 3.281'
    d. 3,281'

2. A room has a length of 3.568 m and a width of 2.438 m. The area of this room *in square feet* is _____.
    a. 20 ft^2
    b. 71.36 ft^2
    c. 93.64 ft^2
    d. 384 ft^2

3. A duct has inside dimensions of 16" × 12". The area *in square feet* is equal to _____.
    a. 1.33 ft^2
    b. 1.92 ft^2
    c. 13.3 ft^2
    d. 28 ft^2

4. The value of π, to two decimal places, is _____.
    a. 1.00
    b. 3.14
    c. 3.33
    d. 3.45

5. How many liters of liquid can a 55-gallon drum hold?
    a. 55 L
    b. 110 L
    c. 175 L
    d. 208 L

6. A crate containing a boiler is marked as weighing 175 lb. What is its weight in *kilograms*?
    a. 68.5 kg
    b. 79.38 kg
    c. 96.33 kg
    d. 125 kg

7. Find the absolute pressure if the gauge pressure reads 164 psig.
    a. 14.7 psia
    b. 149.3 psia
    c. 164 psia
    d. 178.7 psia

8. A temperature of 0°C is equal to _____.
   a. −17.8°F
   b. 32°F
   c. 100°F
   d. 144°F

9. A temperature of 68°F is equal to _____.
   a. 20°C
   b. 36°C
   c. 41°C
   d. 100°C

10. The square root of 12,500, expressed mathematically as $\sqrt{12{,}500}$, is equal to _____.
    a. 1.11803
    b. 11.1803
    c. 111.803
    d. 1118.03

11. 6^2 is the same as _____.
    a. 4
    b. 8
    c. 12
    d. 36

12. Given the equation $P = E^2/R$, show the equation required to solve for R.
    a. $R = E^2/P$
    b. $R = E^2 \times P$
    c. $R = P/E^2$
    d. $R = \sqrt{PE}$

13. The distance from the center of a circle to any point on the curved line is called the _____.
    a. circumference
    b. radius
    c. diameter
    d. cord

14. A triangle in which all sides are of unequal length is called a(n) _____.
    a. equilateral triangle
    b. isosceles triangle
    c. right triangle
    d. scalene triangle

15. Which of the following is *not* a form of the Pythagorean theorem?
    a. $c^2 = a^2 + b^2$
    b. $a = \sqrt{(c^2 - b^2)}$
    c. $b = \sqrt{(c^2 - a^2)}$
    d. $c = \sqrt{(a^2 - b^2)}$

Answers to odd-numbered Module Review Questions are found in *Appendix A.*

Cornerstone of Craftsmanship

Allen Heath
Commercial Equipment Sales
Epting Distributors

How did you choose a career in the industry?

My dad worked in the sheet metal industry, so I was around that type of work growing up. He was also a welder, but he didn't seem to like either craft a lot. My high school offered those vocational programs too, but HVACR stood out to me because there seemed to be many paths available within the craft.

Who inspired you to enter the industry?

My dad and my vocational instructor, Bill Moon. I attended his classes at Donaldson Career Center. Bill had a small HVAC company at the time, so we were able to work at night and attend classes during the day.

What types of training have you been through?

I began in vocational school while in high school. I also took a variety of courses at Greenville Technical College. I worked for Trane for a number of years, and they offer a lot of training programs for both their employees and contractors. They offered programs in a variety of locations, so I attended week-long programs in places like Lacrosse, WI; Pueblo, CO; and Clarksville, TN. Each of those locations provided an opportunity to have some fun as well!

How important is education and training in construction?

Training and education are absolutely critical to any industry. The HVACR industry requires a strong foundation in chemistry, math, and physics. You can't grow in the industry without an understanding of these subjects, and more. Training has made me better at my craft, kept me at a high level of pay, provided access to leadership positions, and supported my family for 43 years.

What kinds of work have you done in your career?

As an adult, I've worked in mechanical contracting as an HVACR service technician and managed parts and supplies for major manufacturers. I've also worked as an instructor and led service teams. The HVACR industry is vital in many environments and has opened doors to some interesting places. Nuclear power plants and precious metal refineries were some of the most interesting. As a service technician, I encountered something different every day. As an instructor, I watched the knowledge and skill of students grow.

Tell us about your current job.

I had retired from Trane, but that lasted about three weeks! A recent opportunity came up and I decided to pursue it. Today, I'm a commercial equipment sales professional for a major HVACR wholesaler.

What do you enjoy most about your job?

In sales, I get to visit and support friends and colleagues in the area I have known for years through my other industry jobs.

What factors have contributed most to your success?

I think the opportunities I've encountered were a direct result of my willingness to lead, take a few risks with my career, try new things, and work hard each day.

Would you suggest construction as a career to others? Why?

I would definitely recommend it. The construction industry allows our world to continue growing. If you have a construction-related skill, you will always have a job.

What advice would you give to those new to the field?

Never stop learning and growing. It is also important to remember that safety is a personal responsibility. You must take it seriously!

Interesting career-related fact or accomplishment?

I'm proud to say that I was named the top technician out of a group of 32 talented peers. For an HVACR service technician, that is an admirable achievement. But I prefer to say that my greatest accomplishment is a list of names. Each name is a trainee I have worked with who is still contributing to the industry today. They are successful in their own careers, some now in positions of leadership or owning companies of their own.

How do you define craftsmanship?

Craftmanship is a combination of both artwork and industry. To me, it means being professional and accurately solving whatever problem you are confronted with. Getting a client's equipment running again in a reliable state requires craftsmanship.

Answers to Section Review Questions

Answer	Section Reference	Objective
Section 1.0.0		
1. c	1.1.0	1a
2. d	1.2.2; *Figure 6*	1b
3. d	1.3.0	1c
4. a	1.4.0; 1.4.5	1d
Section 2.0.0		
1. b	2.1.3	2a
2. a	2.2.0	2b
3. c	2.3.1	2c
Section 3.0.0		
1. c	3.1.0	3a
2. d	3.2.0	3b
3. b	3.3.0	3c
4. a	3.4.2	3d

User Update

SCAN ME

Did you find an error? Submit a correction by visiting **https://www.nccer.org/olf** or by scanning the QR code using your mobile device.

Section Review Calculations

Section 1.0.0

Question 1

$d = 3$ m

$r = \frac{1}{2}d$

$r = \frac{1}{2} \times 3 = 1.5$ m

1 m = 3.281 ft

$r = 1.5$ m $\times 3.281$ ft/m $= 4.922$ ft

Area $= \pi r^2$

Area $= 3.14 \times 4.922^2$

Area $= 3.14 \times 24.226$

Area $= \textbf{76.1 ft}^2$

Question 3

1 kg = 2.205 lb

$240 \times 2.205 = \textbf{529.2 lb}$

Question 4

$°F = (\frac{9}{5} \times °C) + 32$

$°F = 1.8 \times 10 + 32$

$°F = \textbf{50°F}$

1 bar = 14.5 psi

9.88 bar = **143.3 psi**

Section 2.0.0

Question 3

Divide both sides by R, leaving:

$E/R = I \times R \div R$

The two R's on the right of the equation cancel each other, leaving:

$E/R = I$

$I = E \div R$

Section 3.0.0

Question 4

$b = \sqrt{(c^2 - a^2)}$

$b = \sqrt{(225 - 144)}$

$b = \textbf{9"}$

Basic Electricity

Source: Courtesy of Greenlee Tools, Inc.

Objectives

Successful completion of this module prepares you to do the following:

1. State the fundamentals of power generation and identify common electrical safety practices.
    a. State how electrical power is generated and distributed and identify the two kinds of current produced.
    b. Identify general electrical safety practices and state electrical lockout/tagout procedures.
    c. State OSHA electrical safety requirements and identify common electrical PPE.
2. Explain basic electrical theory.
    a. Define common electrical units and apply Ohm's law and the power formula.
    b. Describe the differences between series and parallel circuits and calculate circuit values for each type.
3. Identify the electrical measuring instruments used in HVACR work and describe how to use them.
    a. Describe how voltage is measured.
    b. Describe how current is measured.
    c. Describe how resistance is measured and how continuity is determined.
4. Identify electrical components used in HVACR systems and describe their functions.
    a. Identify and describe various load devices and explain how they are represented on circuit diagrams.
    b. Identify and describe various control devices and explain how they are represented on circuit diagrams.
    c. Identify and describe common electrical diagrams used in HVACR work.

Performance Tasks

Under supervision, you should be able to do the following:

1. Draw a connection diagram for a circuit that includes the following components:
    - A power control switch
    - 120VAC/24VAC control transformer
    - A control relay with a 24VAC coil
    - (2) 120VAC lights, controlled by the relay and wired in parallel
2. Assemble the circuit based on the connection diagram developed in *Performance Task 1*, powered by a GFCI-protected power source.
3. With the circuit de-energized, check circuit components and relay contacts for continuity.
4. With the circuit de-energized, measure and record the resistance of the transformer windings, relay coil, and lights.
5. Energize the circuit, turning on the lights, and measure and record the total circuit current.

Digital Resources for HVACR

SCAN ME

Scan this code using the camera on your phone or mobile device to view the digital resources related to this craft.

6. Measure the voltage provided by the power source to the transformer primary.

7. De-energize and disable the circuit power source. Verify that power is disabled with a voltmeter.

Overview

An HVACR system needs an electrical power source to operate. Most of the problems an HVACR technician encounters when faced with a service call involve the electrical system, so an understanding of electrical theory, power and control components, and circuits is essential. Technicians must have a firm foundation in these topics to get started in the craft, allowing them to understand the circuits of complex systems and troubleshoot them when problems arise.

Industry Recognized Credentials

If you are training through an NCCER-accredited sponsor, you may be eligible for credentials from NCCER's Registry. The ID number for this module is 03106. Note that this module may have been used in other NCCER curricula and may apply to other level completions. Contact NCCER's Registry at 1.888.622.3720 or go to www.nccer.org for more information.

You can also show off your industry-recognized credentials online with NCCER's digital badges. Transform your knowledge, skills, and achievements into badges that you can share across social media platforms, send to your network, and add to your resume. For more information, visit www.nccer.org.

1.0.0 Fundamentals of Electricity

Performance Tasks

1. Draw a connection diagram for a circuit that includes the following components:
   - A power control switch
   - 120VAC/24VAC control transformer
   - A control relay with a 24VAC coil
   - (2) 120VAC lights, controlled by the relay and wired in parallel

2. Assemble the circuit based on the connection diagram developed in *Performance Task 1*, powered by a GFCI-protected power source.

Objective

State the fundamentals of power generation and identify common electrical safety practices.

a. State how electrical power is generated and distributed and identify the two kinds of current produced.

b. Identify general electrical safety practices and state electrical lockout/tagout procedures.

c. State OSHA electrical safety requirements and identify common electrical PPE.

Power: The rate of doing work, or the rate at which energy is dissipated. Electrical power is measured in *watts*.

Loads: Devices that convert electrical energy into another form of energy (heat, mechanical motion, light, etc.). Motors are one of the most common electrical loads in HVACR systems.

Air conditioning, heating, and refrigeration systems use electricity as a source of **power** for compressors, fan motors, and other **loads**. Even a simple gas- or oil-fired furnace needs electricity to operate. Electrical switching devices, sensors, and indicators are used in the control circuits of all systems, in addition to major loads such as fan motors. Control systems can also be quite complex. Many systems use microprocessor controls, and many systems used in commercial buildings are managed from a centralized computer system with the appropriate software.

Many of the problems an HVACR technician encounters on a service call involve the electrical system at some level. Therefore, it is important to understand electrical circuits and be able to interpret complex circuit diagrams. Without these skills, it is impossible to diagnose electrical problems accurately and reliably.

1.1.0 Power Sources and Current Types

There are many sources of electrical power. A great deal of the electricity is generated at coal-fired power plants like the one shown in *Figure 1*. Nuclear power plants (*Figure 2*) are also a major source of electricity. Steam produced in these power plants is used to drive huge turbine generators, which produce the electricity. Other sources of electricity include turbine generators powered by natural gas, as well as hydroelectric power plants (*Figure 3*) where water flowing over dams is used to drive the turbines. Alternative and sustainable power sources, like the sun and wind, continue to be developed to reduce our dependence on fossil fuels and nuclear power.

Figure 1 Coal-fired power plant.

Figure 2 Nuclear power plant.
Source: NRC

The electrical power produced at the source travels through long-distance transmission lines (*Figure 4*). The **voltage** may be as high as 765,000 **volts (V)**. High voltages are used because they result in lower line losses over the long distances that the electricity must travel. A device known as a **transformer** is used to lower the voltage to more usable levels as it reaches electrical substations and eventually homes, offices, and factories.

Voltage: The driving force that makes current flow in a circuit. Voltage, often represented by the letter *E* in formulas, is measured in *volts*. Also known as *electromotive force (emf)*, voltage represents the difference of electrical potential.

Volts (V): The unit of measurement for voltage, represented by the letter *V*. One volt is equivalent to the force required to produce a current of one ampere through a resistance of one ohm.

Transformer: Two or more coils of wire wrapped around a common core. Used to raise and lower voltages.

Figure 3 Hydroelectric power generation.
Source: U.S. Army Corps of Engineers

Going Green

Alternative Energy Sources

To reduce dependence on fossil fuels such as oil and coal, initiatives are being taken all over the world to develop alternatives. From these initiatives, many secondary or alternative methods of producing electrical power have emerged. Among the most common alternative sources are wind and solar energy. Power-generating facilities that burn organic material known as *biomass* are also becoming more common. Biomass describes any material that will decompose, such as wood, plants, leaves, or garbage, and can be burned. Methane gas produced by decomposition in landfills is also being captured and used as an energy source.

Source: Courtesy of DOE/NREL

The voltage in a typical residential home is usually about 240V. At the wall outlet where small household appliances are plugged in (such as televisions and toasters), the voltage is about 120V (*Figure 5*). Larger electrical loads such as electric stoves, clothes dryers, and air conditioning systems usually require the full 240V. Commercial buildings and factories may receive anywhere from 208V to 575V. This depends on the amount of power their machines consume and their voltage requirements.

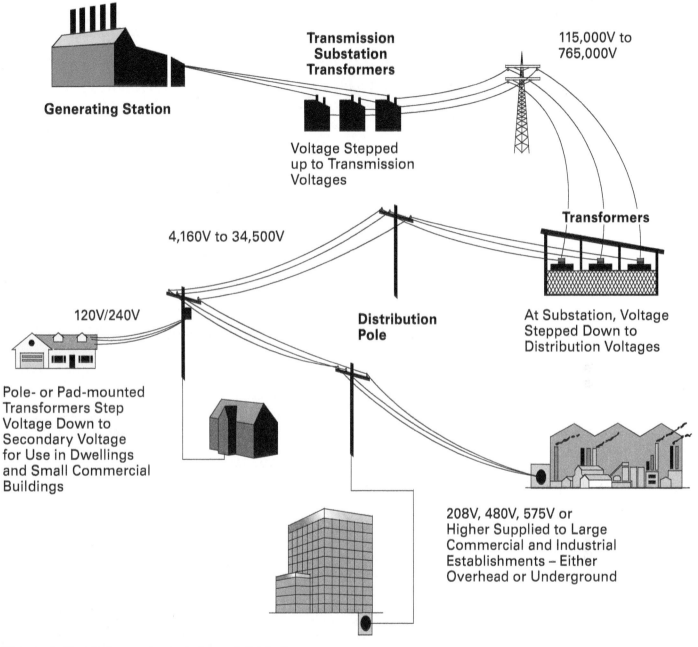

Figure 4 Electrical power transmission and distribution.

1.1.1 AC and DC Voltages

The kind of electric **current** produced by a battery is known as **direct current (DC)**. Automobiles, smartphones, laptops, calculators, and flashlights are good examples of common devices that use DC voltage. All of these examples are battery-powered and must be able to move from place to place. The electricity supplied by the local utility is **alternating current (AC)**.

Almost all HVACR equipment requires AC power. However, many HVACR equipment components, such as electronic circuit boards and new motor designs, do require DC voltage. In these cases, units use a special electronic device to convert AC power to DC power, instead of using batteries that need to be replaced or recharged. Known as a **rectifier**, this device is often built into each component that requires DC power. The charger that allows a cell phone or other portable device to be plugged into a wall socket for charging is a rectifier, coupled with a transformer to also reduce the voltage.

Current: The rate at which electrons flow through an electrical circuit.

Direct current (DC): An electric current that flows in only one direction, such as the current provided by a common battery.

Alternating current (AC): An electrical current that changes direction on a cyclical basis.

Rectifier: A device that converts AC voltage to DC voltage.

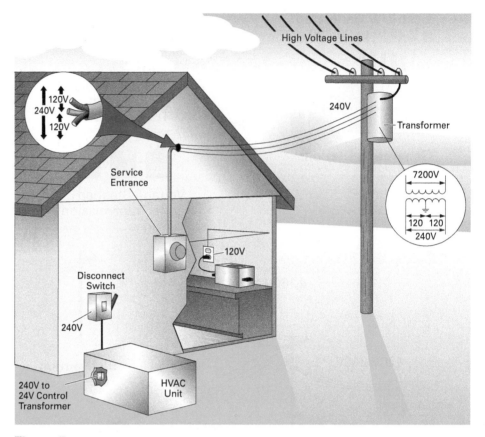

Figure 5 Residential power distribution.

One of the basic principles of AC components is that the flow of an electric current through a **conductor** produces a magnetic field around the conductor. If the conductor is coiled around an iron bar, the strength of the magnetic field is multiplied. The result is an **electromagnet** (*Figure 6*) that attracts and repels other magnetic objects, just like an iron magnet. This is the basis on which electric motors, transformers, and similar components operate. The opposite is also

Conductor: Relevant to electricity, a material through which an electrical current easily passes.

Electromagnet: A coil of wire wrapped around a soft iron core that, when energized, creates a powerful magnetic field.

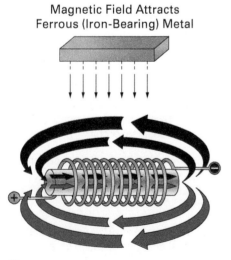

Figure 6 Magnetic field generation.

Electromagnets

Electromagnets can be extremely powerful. For example, they are powerful enough to lift a vehicle at the scrap yard. One advantage of an electromagnet is that it can be turned on and off. Electromagnets are often used at scrap yards and processing facilities that recycle steel and other metals. When handling scrap, aluminum and other non-ferrous materials are not attracted to the magnet, providing a basic means of sorting metals. The same principle of creating a magnetic field around an electrically energized winding applies to many components that are used in HVACR electrical circuits.

Source: Shutterstock.com/AngelPet

true; that is, if a conductor is rotated in a magnetic field, a voltage is induced into the conductor. This is the principle on which a generator works.

The electricity produced by a power generating station and distributed across the country to consumers is AC. In a typical power plant, coal or nuclear materials produce steam, which in turn causes the shaft of a giant *turbine* (*Figure 7*) to rotate. The turbine shaft is connected to the rotor of a generator, which also rotates. As the rotor turns, a current is induced into the *armature*. As the armature changes position in relation to the magnetic field, the magnitude of voltage increases and decreases (alternates) in a pattern shown in *Figure 8*. So, while DC voltage is plotted as a straight line, AC is constantly changing. The rate at which it changes is so rapid, though, that it appears to the user to be consistent at 70.7 percent of the maximum (peak) voltage induced.

Power Transformers

The voltage produced by the motor-generator set in a power station may be in the 13,000V range. Giant transformers like the one shown here are used to step up the voltage to the level carried by the high-voltage transmission lines. The voltage is carried over the high-voltage transmission lines to substations where other transformers step down the voltage to the level required for local distribution. Pole-mounted transformers like the one shown here, as well as pad-mounted models for utility and industrial purposes, step the voltage down further to the level needed for homes and businesses.

Figure 7 A power-generating turbine under construction.
Source: iStock@industryview

(A) Large Transformer   (B) Pole Transformer

1.2.0   Electrical Safety

Working with electrical devices always involves some degree of danger. Technicians must be constantly alert to the potential for electric shock and take the necessary precautions. The amount of electrical current that passes through the human body determines the outcome of an electrical shock. The higher the voltage, the greater the current, and the greater the chance of receiving a fatal shock. Electrical safety is thoroughly covered in NCCER Module 00101, *Basic Safety*. Some key points are presented here to reinforce the importance of safety when working on electrical equipment.

WARNING!

Voltages of 600V or more are almost ten times as likely to kill as lower voltages. On the job, you spend most of your time working on or near much lower voltages. However, lower voltages can also kill or cause significant injuries. Electrical shocks can also cause secondary incidents, such as a fall. Many electrical accidents can be prevented by following proper safety practices, including the use of **ground fault circuit interrupters (GFCIs)** and proper grounding.

Ground fault circuit interrupters (GFCIs): Devices that interrupt and de-energize an electrical circuit to protect a person from electrocution in the event of a ground fault.

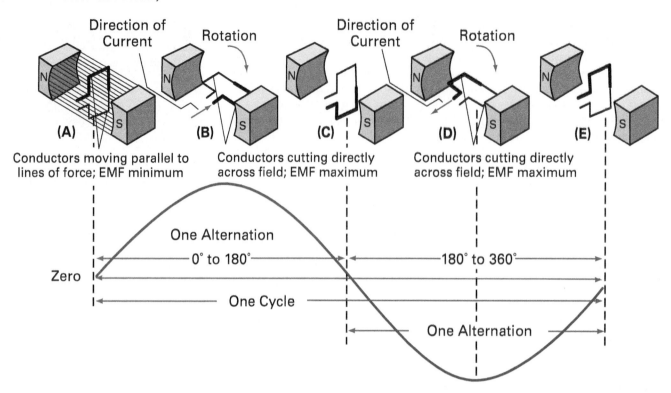

(A) AC Voltage Sine Wave

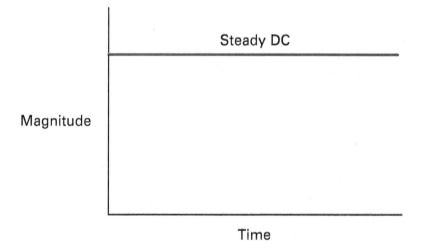

(B) DC Voltage

Figure 8 AC voltage cycle versus DC voltage.

Did You Know?

Electrical Safety

In the United States, electrocution is one of the four leading causes of death in the construction workplace, making it one of OSHA's Fatal Four hazards. According to the Electrical Safety Foundation International (ESFI), of the six primary occupational categories, the construction industry had the highest rate of electrical fatalities. Between the years 2011 and 2019, the fatality rate was 0.7 per 100,000 workers. It doesn't have to be that way, and you need to be part of the change.

Electrical current flows along the path of least **resistance** to return to its source. If a person comes in contact with a live conductor and a grounded object, that person becomes a load in the electrical circuit. The potential for shock increases dramatically if the skin is damp because the moisture is a very effective conductor. Broken skin (a cut, for example) will also reduce the resistance human skin offers. Even currents of less than one-third of an **ampere (A)** can severely injure or kill a person. *Table 1* describes the effects of different levels of current flowing through the human body.

TABLE 1 Effects of Current on the Human Body

Current Value	Typical Effects
1 mA	Perception level. Slight tingling sensation.
5 mA	Slight shock. Involuntary reactions can result in other injuries.
6 mA to 30 mA	Painful shock, loss of muscular control.
50 mA to 150 mA	Extreme pain, respiratory arrest, severe muscular contractions. Death possible.
1,000 mA to 4,300 mA	Ventricular fibrillation, severe muscular contractions, nerve damage. Typically results in death.

This information has been included in this training module to help you gain respect for the environment where you work and to stress how important safe working habits really are. Since electricity is not something you can see, technicians must give constant attention to the existence of energized electrical circuits and recognize the potential for serious injury.

1.2.1 General Safety Practices

HVACR technicians work with potentially deadly levels of electricity every day. They can do so because good safety practices become second nature to them. Here are some general safety practices to follow whenever you are working with electricity:

- Unless it is essential to work with the power on, always de-energize the system at the source. Lock out and tag out the disconnect switch in accordance with company or site procedures and OSHA requirements. Lockout and tagout procedures are presented in the next section.
- Use a voltmeter to verify that the power to the unit is disabled. Remember that even though the power may be switched off, there is still electrical potential at the input side of the disconnect switch or circuit breaker. Also note that some pieces of equipment may have more than one power supply or component that is energized from another source.
- Use personal protective equipment (PPE) such as rubber gloves when necessary.
- Use insulated hand tools and double-insulated power tools.
- Use a GFCI (*Figure 9*) when using power tools.
- Short wiring to ground before touching de-energized wires.
- If possible, do not kneel directly on the ground when making voltage measurements. The more insulation there is between your body and the ground, the safer it is.
- Remove all metal jewelry such as rings, watches, and bracelets.
- When it is necessary to test an energized circuit, keep one hand outside the unit if possible. This reduces the risk of completing a circuit through your upper body that might cause current to pass through your heart.
- When working with electric power tools, inspect the tool and its power cord to make sure they are safe to use.

NOTE

One-third of an ampere is generally written as 0.33A, and the term *ampere* is usually abbreviated and spoken as *amp*. A *milliamp (mA)*, reflected in the values shown in *Table 1*, represents 0.001 amp, or $^1/_{1,000}$ amp.

Resistance: An electrical property that opposes the flow of current through a circuit. Resistance is measured in *ohms* and represented by the letter *R* in formulas.

Ampere (A): The basic unit of measurement for electrical current, represented by the letter *A*; typically spoken and written as *amp*.

NOTE

OSHA regulations state that employees have a duty to follow the safety rules put in place by their employers. The amount an employee can lawfully collect from worker's compensation or disability insurance due to an injury may be restricted if the employee was in violation of safety rules when injured. A company also may terminate employment or take other disciplinary action if safety rules are violated.

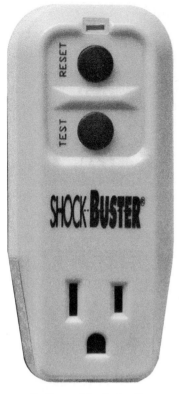

Figure 9 Ground fault circuit interrupter (GFCI).

The *National Electrical Code*® (*NEC*®), together with the electrical code for your local area, defines the minimum requirements for the installation of electrical systems. The *NEC*® specifies the minimum provisions necessary for protecting people and property from electrical hazards. An electrician generally installs the electrical circuit from the building's electrical service to the disconnecting device for the HVACR equipment. When installing an HVACR system, the HVACR technician may be required to connect the power from the disconnecting device to the HVACR unit. When doing so, be sure to follow all applicable code requirements.

1.2.2 OSHA Lockout/Tagout Rule

In addition to many other safety rules, OSHA enforces electrical safety through *29 CFR 1926.417, Lockout and Tagging of Circuits*. This rule covers the specific procedure to be followed for the "servicing and maintenance of machines and equipment in which the unexpected energization or startup of the machines or equipment, or releases of stored energy, could cause injury to employees. This standard establishes minimum performance requirements for the control of such hazardous energy."

The purpose of the OSHA procedure is to make sure that machinery is isolated from all potentially hazardous energy sources, and the power source is locked out and tagged before workers perform any service or maintenance activities where the unexpected energization, startup, or release of stored energy could cause injury.

> **WARNING!**
>
> The OSHA procedure provides only the minimum requirements for the lockout/tagout procedure. Also note that there are other potential energy sources that may need to be locked out, such as pneumatic and hydraulic energy sources. Craftworkers should also consult the required lockout/tagout procedures established by the employer. Remember, lives depend on this procedure. It is critical that the correct procedures be used.

To prepare for a lockout/tagout, make a survey to locate and identify all isolating devices (*Figure 10*). Be certain which switch(es), valve(s), or other energy-isolating devices apply to the equipment to be locked and tagged.

The following procedure outlines the general steps for lockout/tagout. Note that this procedure is provided only as an example. Each employer will designate who is qualified to conduct the procedure. Not every employee has received the proper training to do so.

> **WARNING!**
>
> It is essential to remember there is often more than one type of energy applied to a system or its equipment. Even a single type of energy may have more than one source. An example of this is a packaged HVAC unit with one electrical power source for the unit and a separate power source for an installed electric-heating package. Failure to recognize and render all energy sources safe can result in serious injury or death.

Step 1 Notify all personnel affected by the lockout/tagout and why it is necessary.

Step 2 Identify all the potential sources of energy connected to the equipment. A residential condensing unit, for example, will be powered by an electrical disconnect switch. But there will also be a 24V control power source connected, which typically originates from the furnace or air handling unit.

Step 3 If the machine or equipment is running, shut it down using the normal procedure. For example, press the stop button, open the toggle switch, or turn the system thermostat off.

Disconnecting means may be mounted on or within the HVAC equipment. Disconnect must be located within sight of equipment.
NEC Section 440.14

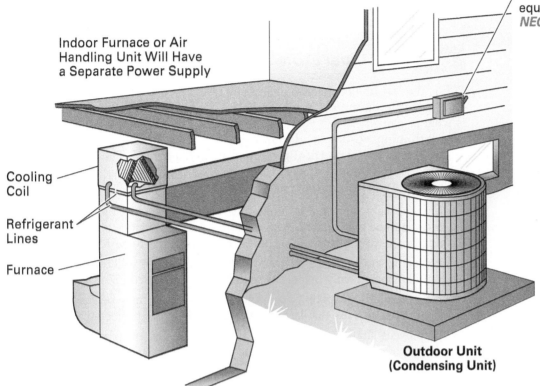

Indoor Furnace or Air Handling Unit Will Have a Separate Power Supply

Cooling Coil

Refrigerant Lines

Furnace

Outdoor Unit
(Condensing Unit)

Figure 10 Residential condensing unit power supply.

Step 4 Operate the switch, valve, or other energy-isolating device(s) so that the equipment is isolated from its energy source(s). Stored energy must be dissipated or restrained by repositioning, blocking, or bleeding down a pressurized source. Examples of stored energy are springs, elevated machine members, rotating flywheels, hydraulic systems, and air, gas, steam, or water pressure.

Step 5 Use a voltmeter that has been tested on a known source to confirm that electrical power has been de-energized. After making sure that no personnel are exposed, operate the start button or other normal operating controls to make certain the equipment will not operate. This will confirm that the energy sources have been disconnected.

Step 6 Lock out and tag the energy-isolating devices with your assigned lock and tag.

All equipment must be locked and tagged to protect against accidental operation when such operation could cause injury to personnel. Never try to operate any switch, valve, or energy-isolating device when it is locked and tagged, and do not attempt to defeat the lockout device.

After service and/or maintenance is complete, power can be restored to the system. Use the following basic procedure to restore power to the equipment:

Step 1 Check the area around the machines or equipment to make sure that all personnel are at a safe distance.

Step 2 Remove all tools from the machine or equipment and reinstall any guards.

Step 3 Confirm that all equipment operating controls are in the neutral or Off position.

CAUTION

Return operating controls to their neutral or Off position after confirming that all energy sources have been disabled.

Step 4   Double-check that all personnel are in the clear, then remove all lock-out/tagout devices.

Step 5   Turn on the disconnect switch or other energy-isolating device to restore energy to the equipment. Use a voltmeter to ensure that power has been restored and the voltage is as required for the equipment.

1.3.0   Electrical Protective Equipment

PPE is the last line of defense against injury. If an electrical hazard cannot be eliminated by design or by physically installing a practical barrier, your PPE is what remains between you and the hazard. This, of course, is in addition to proper training, wisdom, and thinking before you act.

OSHA addresses electrical protection in *CFR 1910.335*. The design requirements for specific types of electrical protective equipment are covered in *CFR 1910.137*. The OSHA requirements include the following:

- Craftworkers working in areas where there are potential electrical hazards shall be provided with, and shall use, appropriate electrical protective equipment to protect specific parts of the body and to perform the work.
- Protective equipment shall be maintained in a safe, reliable condition and shall be periodically inspected or tested, per *CFR 1910.137*.
- If the insulating capability of protective equipment may be subject to damage during use, the insulating material shall also be protected.
- Employees shall wear nonconductive head protection wherever there is a danger of head or face injury due to contact with exposed energized parts.
- Employees shall wear protective equipment for the eyes and face wherever there is a danger of injury to the eyes or face from an electric **arc**, or from flying objects resulting from an electrical energy release.
- When working near exposed energized conductors or circuit parts, each employee shall use insulated tools or handling equipment if the equipment could contact these surfaces.
- Ropes and handlines used near exposed energized parts shall be nonconductive.
- Protective shields, protective barriers, or insulating materials shall be used to protect employees from shock, burns, or other injuries while working near exposed energized parts.
- Fuses must only be removed using an appropriate fuse-pulling device that is insulated for the circuit voltage (*Figure 11*). Be sure to check the specifications and ensure the puller is rated for the voltage with which you are working.

Arc: A visible flash of light and release of heat that occurs when electrical currents cross an air gap.

> **WARNING!**
>
> Fuses should never be installed or removed while the circuit is energized. This could cause an electrical arc or allow direct contact with the circuit, resulting in a serious injury or death.

Figure 11 Fuse puller.
Source: Photo 19254279 © Dave Willman | Dreamstime.com

Most electrical safety gear and PPE are differentiated by the level of protection they provide. Some of the most common equipment used by HVACR technicians will be reviewed in the sections that follow.

1.3.1 Classes of Rubber Protective Equipment

The American National Standards Institute (ANSI) and ASTM International (formerly known as American Society for Testing and Materials) developed a classification system for rubber protective equipment. The maximum AC voltage for each class and the identifying tag colors are shown in *Table 2*.

HVACR technicians, even in the industrial environment, are not typically exposed to voltages exceeding 600VAC.

TABLE 2 Rubber Protective Equipment Classification

Classification (Tag Color)	Max Voltage
Class 00 (beige tag)	500V
Class 0 (red tag)	1,000V
Class 1 (white tag)	7,500V
Class 2 (yellow tag)	17,000V
Class 3 (green tag)	26,500V
Class 4 (orange tag)	36,000V

Rubber protective equipment must be tested periodically. Before use, you must ensure the equipment has a current test date stenciled on it. The class rating tag must also be clearly visible. Gloves must be electrically tested every six months and any time the insulating value is questioned. Insulating gloves must also be inspected by the user each time before use.

Rubber protective equipment is available in two types. Type 1 designates equipment that is manufactured of natural or synthetic rubber. Type 2 designates equipment that is ozone resistant, made from any elastomer or combination of elastomeric compounds. Type 2 rubber protective equipment is more rigid than Type 1 equipment and more difficult to work in as a result.

For most HVACR work, Type 1 products are the appropriate choice, although they are not resistant to *ozone*. Ozone is a form of oxygen that is produced from electricity and is present in the air surrounding a conductor at high voltages. Normally, ozone is found at voltages of 10kV and higher, such as those found in electric utility transmission and distribution systems. Type 1 rubber protective equipment can also be damaged by ultraviolet rays. Therefore, it is best not to expose them to the sun's rays when they are not in use.

1.3.2 Rubber Gloves

Rubber gloves are used to prevent contact with energized circuits. A separate leather cover protects the rubber glove from punctures and other damage. The leather protectors also provide the wearer with a level of protection against **arc flash**. Glove dust can be used to absorb moisture and prevent the gloves from sticking. A pair of protective gloves and leather covers are shown in *Figure 12*, along with an example of glove dust.

Voltage-rated rubber gloves have gauntlets and are available in various sizes. Ensure you have the correct size, as it is awkward to work in gloves that are too big. For the best possible protection and service life, observe the following general guidelines when using protective gloves in electrical work:

Arc flash: The intense light and heat produced by an electrical energy release such as an *arc fault* that occurs when an electrical connection occurs through the air to ground or by a connection between different phases of power.

- Always wear leather protectors over your gloves. Leather protectors shield the gloves against contact with sharp or pointed objects that may damage the rubber. Leather protectors also provide burn protection.

- Always wear rubber gloves right-side out (serial number and size to the outside).

- Always keep the gauntlets up. Rolling them down eliminates valuable protection. Tuck the sleeves of your shirt or protective suit into the gauntlets.

- Always inspect and field check gloves before using them.

- Use small amounts of manufacturer-approved glove dust or cotton liners inside the rubber gloves. This helps to absorb some of the perspiration that can damage the gloves over years of use. It also makes it easy to pull them on and remove them.

- Wash the rubber gloves in lukewarm, clean, fresh water after each use. Dry the gloves inside and out prior to returning to storage. Never use any type of cleaning solution on the gloves.

- Store the gloves in a cool, dry, dark place that is free from ozone, chemicals, oils, solvents, or other materials that could damage the gloves. Do not store gloves near hot pipes or in direct sunlight. Store them in their natural shape in a bag or box inside their leather protectors. They should be undistorted, right-side out, and unfolded. Storing gloves inside out will reduce their service life.

- Gloves can be damaged by many chemicals, especially petroleum-based products such as oils, gasoline, hydraulic fluid inhibitors, and hand creams. Use rubber gloves only for their intended purpose, not for handling chemicals or other work. This also applies to the leather protectors. If contact is made with these or other petroleum-based products, the contaminant should be wiped off immediately. If any signs of physical damage or chemical deterioration are found (e.g., swelling, softening, hardening, or stickiness), do not use the gloves.

- Never wear watches or rings while wearing rubber gloves; this can cause damage on the inside. Never wear anything conductive.

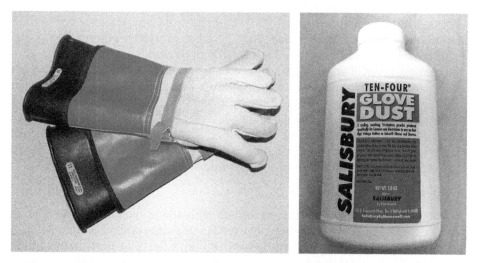

Figure 12 Rubber gloves with leather protectors, and glove dust.

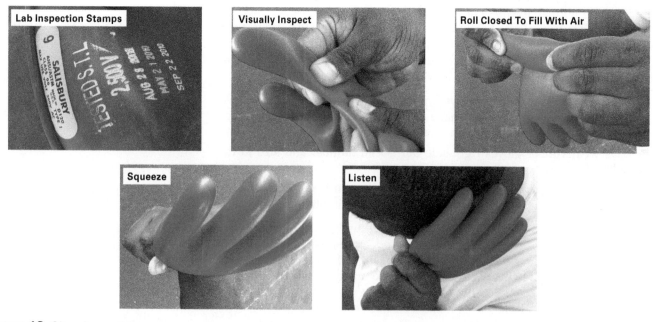

Figure 13 Glove inspection.
Source: Mike Powers

Before rubber gloves are used, a visual inspection and an air test should be made. This is done prior to use and as many times during the day as you feel necessary. Refer to *Figure 13*. To perform a visual inspection, lightly stretch small areas of the glove, checking for defects such as the following:

• Embedded foreign material
• Deep scratches
• Pinholes or punctures
• Snags or cuts

Gloves can be inspected by rolling the outside and inside of the protective equipment between the hands. Squeeze the inside of the gloves together to bend the outside area and create enough stress to expose cracks, cuts, and similar defects. Once the entire surface has been checked, turn the glove inside out and repeat the procedure. Remember not to leave the gloves inside out as that places stress on the preformed rubber.

Remember that even the smallest amount of damage to a rubber glove compromises its insulating ability. If the gloves are suspect, turn them in for professional evaluation. If the gloves are obviously defective or damaged, turn them in for disposal. Never leave a damaged glove lying around where others may mistake it for a usable product.

After a visual inspection, other defects may be found by applying an air test, as shown in *Figure 13*. Use the following general steps to perform an air test in the field:

Step 1 Roll the glove down from the gauntlet to trap air inside.

Step 2 Trap the air by squeezing the gauntlet tightly with one hand. Use the other hand to manipulate and increase the pressure in one area at a time, squeezing the palm, fingers, and thumb to check for weaknesses and defects.

Step 3 Hold the glove up to your ear to try to detect any escaping air.

Gloves may also be inspected using a glove inflator (*Figure 14*). Glove inflators allow for more thorough and faster field testing. The glove is secured to the housing using a nylon strap and inflated by pumping the bellows of the inflator against a firm surface. The inflated glove is then checked for leaks.

1.3.3 Other Protective Apparel

Besides rubber gloves, there are other types of protective gear for electrical work, such as hard hats, safety glasses, and face shields.

OSHA requires that all employees exposed to high-voltage hazards use electrically rated (Class E) hard hats. Class E hard hats have the highest rating and are tested to withstand 20,000VAC for up to three minutes. The other electrically rated hard hat is a Class G hard hat, which is tested to withstand 2,200VAC for one minute. Class G hard hats are typically required for HVACR craftworkers. The class information can be found on the label inside the hat. Labels found on safety equipment should never be removed.

Class E and G climbing-style hard hats are also available. Both classes can be equipped with eye protection, arc-rated face shields, chin straps, lights, and hearing protection. A rule of thumb is to replace a hard hat every five years. However, hard hats exposed to severe conditions (e.g., chemicals, constant sunlight, or high temperatures) should be replaced every two years.

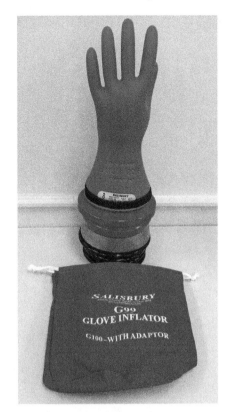

Figure 14 Glove inflator.

CAUTION

To avoid damage, never use compressed gas for the air test or inflate gloves beyond the manufacturer's recommendations.

OSHA requires you to wear safety glasses at all times whenever you are working on or near energized circuits. Face protection (a shield) is worn over your safety glasses and is required whenever there is danger of electrical arcs or flashes, or from flying or falling objects resulting from an electrical energy release. Of course, there are other tasks that require a face shield for protection, such as grinding.

Your clothing must also be carefully selected. Never wear synthetic-fiber (acrylic, polyester, or nylon) clothing for HVACR work. These materials will melt when exposed to high temperatures and, because they burn hotter, they will increase the severity of a burn. Remember that you will also be working with torches in the future, so synthetic fibers in clothing are best avoided altogether.

HVACR technicians should generally wear flame-resistant (FR) pants and shirts, leather boots or shoes, safety glasses, and hard hats in accordance with OSHA regulations. FR clothing is categorized by *arc rating*. All electrical equipment has the capacity to deliver a calculated amount of incident energy. This energy is measured in calories per square centimeter (cal/cm^2) and is determined by a **Hazard Risk Assessment**. Higher arc ratings indicate greater protection from arc flash and energy releases.

When working with energized equipment, arc-rated suits with rated face and head protection are required in some applications. These special suits offer greater protection than any kind of traditional clothing alone. Your organization's safety officer will determine what level of protection is needed for the tasks you perform.

There is a lot to know and understand about electrical safety and the proper use of electrical PPE. OSHA and the construction industry have determined that mandating and monitoring the use of electrical PPE is needed to ensure the safety of all craftworkers that work with or near significant electrical hazards. Employers are responsible for ensuring the needed PPE is available, but it is your responsibility to use it according to the OSHA guidelines.

Hazard Risk Assessment: A process, often conducted by safety personnel, used to identify and evaluate the potential hazards associated with a task and develop ways to control or eliminate them.

1.0.0 Section Review

1. The voltage supplied to a commercial building from a power transformer may be as high as _____.
   a. 765,000V
   b. 2,400V
   c. 575V
   d. 120V

2. Which of the following is a recommended safety practice when performing measurements on live electrical circuits?
   a. Kneel on the ground when using a voltmeter.
   b. Keep both hands inside the unit while working.
   c. Avoid wearing gloves of any kind around energized circuits.
   d. Use insulated hand tools and double-insulated power tools.

3. Periodic inspection and testing of electrical protective equipment are required by _____.
   a. ANSI
   b. OSHA
   c. NFPA
   d. NEMA

2.0.0 Basic Electrical Theory

Objective

Explain basic electrical theory.

a. Define common electrical units and apply Ohm's law and the power formula.

b. Describe the differences between series and parallel circuits and calculate circuit values for each type.

Performance Tasks

1. Draw a connection diagram for a circuit that includes the following components:
   - A power control switch
   - 120VAC/24VAC control transformer
   - A control relay with a 24VAC coil
   - (2) 120VAC lights, controlled by the relay and wired in parallel

2. Assemble the circuit based on the connection diagram developed in *Performance Task 1*, powered by a GFCI-protected power source.

3. With the circuit de-energized, check circuit components and relay contacts for continuity.

4. With the circuit de-energized, measure and record the resistance of the transformer windings, relay coil, and lights.

5. Energize the circuit, turning on the lights, and measure and record the total circuit current.

6. Measure the voltage provided by the power source to the transformer primary.

7. De-energize and disable the circuit power source. Verify that power is disabled with a voltmeter.

An HVACR electrical circuit generally consists of four main elements:

- A voltage source
- A load device that uses the electricity from the source
- One or more control devices
- Conductors to carry the current

When a circuit is complete, an electrical current will flow through the conductors and devices, causing the load device to perform work. The source produces a voltage that causes current to flow while the load offers resistance to current flow and consumes power to do work. Voltage, current, resistance, and power are the building blocks of electricity.

Did You Know?

Lightning

A lightning strike can carry a current of 20,000A and reach temperatures of up to 54,000°F (~30,000°C). It is no wonder that being in the path or vicinity of a lightning strike is often deadly.

Source: NASA

2.1.0 Power Definitions and Common Formulas

An electrical current is established by the movement of electrons from one point to another. Electrons are negatively charged atomic particles that exist in all matter. When a difference in the number of electrons exists between two points, electrons flow away from the point with more electrons to the point with fewer electrons. This difference in electrical potential between the two points is called *voltage*. A voltage must be present for electrons to flow.

Note that, when physically observing electron flow, the current flows from negative to positive. However, electricians and electrical engineers conventionally view current as flowing in the opposite direction, from positive to negative. This concept is an artifact from the time when Benjamin Franklin first began experimenting with electricity. Even so, it works particularly well with electronic and microprocessor theory and is now the generally accepted view for current flow.

Lightning is a good example of natural electron flow. A storm cloud has a negative charge with respect to Earth. Lightning occurs when the difference in potential between the cloud and Earth becomes so great that the air between them acts as a conductor.

In the common 12-volt, direct-current (VDC) car battery, a chemical reaction causes one of the poles to be negative with respect to the other. If a conductor is connected between the negative (–) and positive (+) poles of the battery, electrons flow from the negative pole to the positive pole (*Figure 15*).

To make use of the potential energy in the battery, resistive devices such as light bulbs are connected between the two battery poles. Resistance is the property that causes a material to resist or impede current flow. A resistive device in a circuit represents an electrical load. When the switch closes, the moving

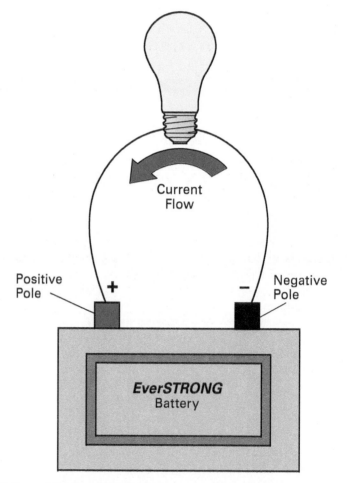

Figure 15 Current flow in a DC circuit.

electrons flow through the bulb, which converts the electrical energy into light. The bulb consumes the electrical energy in the process. The rate of energy consumption is known as *power* and is expressed in **watts (W)**. For example, a 60W light bulb is designed to consume 60 watts of power.

Current is expressed in amps (A), voltage is expressed in volts (V), and resistance is expressed in **ohms (Ω)**. *Amps* is a shortened way to say *amperes*. The AC voltages HVACR technicians work with usually range from 24VAC to 600VAC. The occasional DC voltages HVACR technicians encounter will most often be low voltages in the 5VDC–15VDC range.

Higher DC voltages, typically up to about 600VDC, are found in variable-frequency drive (VFD) controls that can modulate the speed of AC-powered motors. Electronically commutated motors (ECMs) that are now used consistently in residential and light-commercial products are also DC-powered. Occasionally, technicians may encounter a very high-voltage DC circuit, such as the ignition circuit for a gas or oil furnace, which can produce a 10,000VDC spark. This high voltage is needed to allow the voltage to jump across a gap, creating an arc. However, even though the voltage is very high in these devices, the current is generally very low.

Currents in some HVACR circuits may be as low as a few microamps (millionths of an amp). Currents in the milliamp range are more common. Some circuits, especially motor circuits, may consume very high currents in the range of 100A or more. From a safety point of view, the important thing to remember is that even currents in the milliamp range can be hazardous to you.

Resistances may be large values expressed in thousands of ohms (kilohms or kΩ) or even millions of ohms (megohms or MΩ). A resistance value can also be only a few milliohms (mΩ), or thousandths of an ohm. For example, a **thermistor** (temperature-sensitive resistor) used as a temperature-sensing device falls into this category.

> **Watts (W)**: The unit of measurement for power consumed by a load.
>
> **Ohms (Ω)**: The basic unit of measurement for electrical resistance, represented by the symbol Ω (omega).
>
> **Thermistor**: A semiconductor device that changes resistance with a change in temperature, typically used as a temperature sensor.

Voltage

Eighteenth-century physicist Alessandro Volta theorized that when certain objects and chemicals come into contact with each other, they produce an electric current. Believing that electricity came from contact between metals only, Volta coined the term *metallic electricity*. To demonstrate his theory, Volta placed two discs, one of silver and the other of zinc, into a weak acidic solution. When he linked the discs together with wire, electricity flowed through the wire.

Thus, Volta introduced the world to the battery, known at the time as the *Voltaic pile*. Volta needed a unit to measure the strength of the electric push. The volt, named after Volta, is that measure. Voltage has other names as well, including *electromotive force* (*emf*) and *potential difference*. Voltage can be viewed as the potential to do work.

Voltage, current, resistance, and power are all mathematically related. If any two of the values are known, the other two can be calculated. For example, if it is known how much voltage is available and how much power the load consumes, the amount of current the circuit will draw can be calculated. This can be done using the formulas discussed in the following sections.

2.1.1 Ohm's Law

Ohm's law is a set of simple formulas for calculating voltage, current, and resistance. The formulas were derived by Georg Ohm, a German physicist, in 1827. The values in the formula are expressed as E, I, and R, respectively. The letter E is used to represent voltage based on Ohm's original concept of voltage as electromotive force (E). Similarly, the letter I for current is based on the intensity of electron flow.

Figure 16 shows the three variations of Ohm's law. The pyramid style provides a way to remember the formula visually. If you know two of the values, cover the unknown value with your finger to see how to solve the problem—by

	Letter Symbol	Unit of Measurement
Current	I	Amperes (A)
Resistance	R	Ohms (Ω)
Voltage	E	Volts (V)

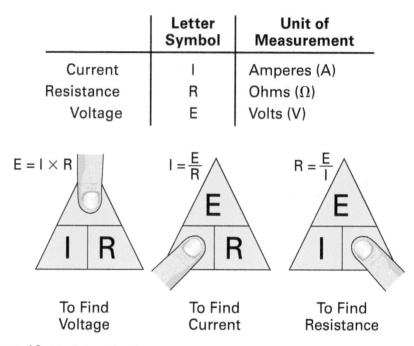

To Find Voltage   To Find Current   To Find Resistance

Figure 16 Ohm's law triangle.

multiplication or division of the two knowns. For example, if the voltage and resistance are known, covering the *I* reveals that *E* is divided by *R* to determine current (*I*).

For example, in a 120V circuit containing a 60Ω load, the current flow is 2A. This can be determined by using Ohm's law, as follows:

$$I = \frac{E}{R} = \frac{120V}{60\Omega} = 2A$$

The pyramid also shows that current (I) is multiplied by resistance (R) to determine voltage (E). You can check your work that way, by multiplying 2A by 60Ω, which equals 120V.

Note that Ohm's law applies only to resistive circuits. Because of the nature of alternating current, circuits containing motors, **relay** coils, and other **inductive loads** do not act the same as purely **resistive loads**. Other formulas are required to work with inductive circuits and loads.

2.1.2 The Power Formula

The power consumed by most load devices determines their rating, rather than the amount of resistance they offer. For example, an electric heating element may have a rating of 5,000W, or 5kW. The equations based on Ohm's law will not help to determine how much current the heater will draw because two variables in the Ohm's law formula—resistance and current—are missing. In this case, the power formula must be used.

Figure 17 shows the variations of the power formula, as well as Ohm's law variants since they are closely related. In each quadrant of the power formula, the three variations of the equations that determine the value in white are shown. The version of the power formula needed looks like this:

$$I = \frac{P}{E}$$

The *P* in the power formula represents power. In a 240V circuit, the 5,000W heater would draw 20.8A (5,000W ÷ 240V). This is a common electric heating-element capacity, and a common voltage at which they operate. For a load like this, the wire must be sized to accommodate this much current and more, for safety's sake. The wire sizing rules of the *NEC*® are the basis for wire sizing and the tables provide guidance for every possible combination of load, voltage, and environmental conditions.

Relay: A magnetically operated control device consisting of a coil and one or more sets of contacts.

Inductive loads: Electrical loads that operate by converting electrical current into magnetism, which is often used to do mechanical work. An electric motor is a good example of an inductive load.

Resistive loads: Electrical loads that consume only active power, with no magnetic field involved. Heating elements found in electric furnaces, stoves, and toasters are examples of resistive loads.

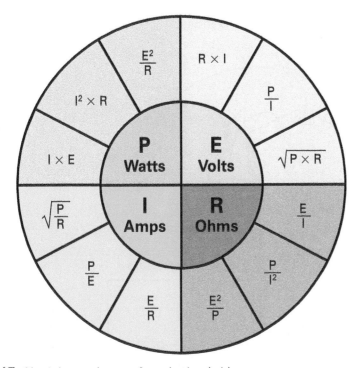

Figure 17 Ohm's law and power formula visual aid.

If the heater were to operate at 120V, the power formula indicates that it would draw 41.7A (5,000W ÷ 120V). That means the wire would have to be sized to accommodate twice as much load, resulting in significantly larger and more costly wiring. Current is the dominant factor in wire sizing. Overloaded electrical conductors can cause a fire. Properly applied overcurrent protection devices, such as circuit breakers and fuses, prevent this from happening.

Electric motors are often rated in equivalent horsepower instead of watts. A swimming pool pump, for example, might be powered by a two-horsepower (hp) electric motor. There is a simple conversion from horsepower to watts: 1 hp = 746W. Therefore, a 2 hp motor would consume 1,492W. In a 120V circuit, it would draw more than 12A as shown here:

$$I = \frac{P}{E} = \frac{1,492W}{120V} = 12.43A$$

The ability to perform this type of calculation is important because adding new loads to or changing the existing loads in an existing circuit is often something a contractor must consider.

2.2.0　Electrical Circuits

A simple electrical circuit is shown in *Figure 18*. An electrical circuit is a closed loop that contains a voltage source, a load, and conductors to carry current. It often contains a switching device to open and close the circuit as desired. Otherwise, there is no way to control the circuit.

An electrical conductor is a material that readily carries an electric current. Most metals are good electrical conductors as well as thermal conductors, meaning that they also transfer heat well. Copper is the most common electrical conductor used for wiring. Gold and silver are better conductors but are too expensive to use, except in micro-miniature electronic circuits. Water is also an excellent conductor, due in part to the minerals it contains. That is why it is dangerous to work with electricity or use electrical appliances in or around a wet environment.

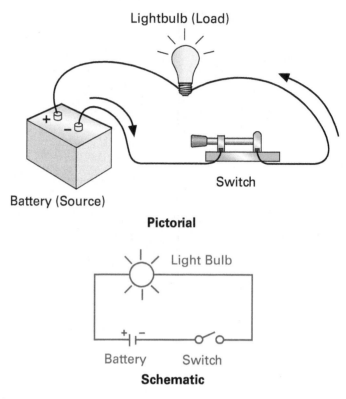

Figure 18 A simple electrical circuit.

Insulator: A substance that inhibits the flow of electrical current; the opposite of conductor.

An **insulator** is the opposite of a conductor. It inhibits the flow of electricity. Rubber, fiberglass, and porcelain are good insulators. Electrical tools are often insulated with rubber to help protect against electrical shock. Special tools used by power line workers, called *hot sticks*, are usually constructed of fiberglass. Although wood is a good insulator, it also absorbs water which makes it a poor choice as an insulator.

Circuits typically need both conductors and insulators—conductors to move the current and insulators to keep electrical paths from crossing. *Figure 19* shows simple examples of two primary kinds of circuits found in HVACR systems—series circuits and parallel circuits. The most common by far is the parallel circuit, in which all the loads are connected across the power source.

2.2.1 Series Circuits

A series circuit provides only one path for current flow. For that reason, the current is the same at any point in the circuit. The total resistance of the circuit is equal to the sum of the individual resistances. The 12V series circuit in *Figure 19* has two 30Ω loads. The total resistance (R_T) is therefore 60Ω. The total amount of current (I_T) flowing in the circuit is 0.2A, which is calculated as follows:

$$I_T = \frac{E_T}{R_T} = \frac{12V}{60\Omega} = 0.2A$$

Horsepower

The term *horsepower* originates from the time when horses were used extensively to do work. It was determined that a good horse could raise the equivalent of 33,000 pounds to a height of one foot in one minute using a pulley system. This can be stated as 33,000 foot-pounds per minute of work.

If there were five 30Ω loads, the total resistance would be 150Ω. The current flow is the same through all the loads. However, the voltage measured across any load (voltage drop) varies depending on the resistance of that load. The sum of the voltage drops equals the total voltage applied to the circuit. In the case of the five 30Ω loads, each would drop the applied voltage by one-fifth. This is because all five of the loads are equal. Note that only the loads are considered in determining the type of circuit.

Circuits containing only loads in series are uncommon in HVACR work. An important trait of a series circuit is that if the circuit is open at any point, no current will flow. For example, if five light bulbs are connected in series and one of them blows, all five lights will go off because the **continuity** of the circuit has been broken.

Continuity: Having a continuous path for current to follow. The absence of continuity indicates an open circuit.

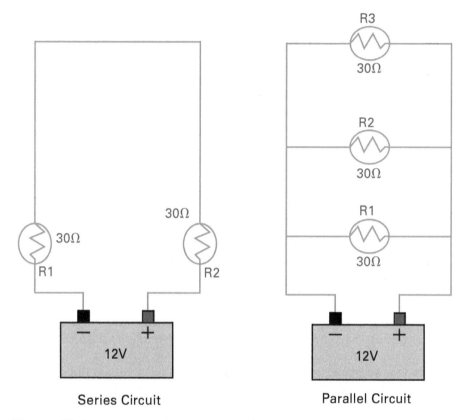

Figure 19 Series and parallel circuits.

Is It a Series Circuit?

When the term *series circuit* is used, it refers to the way the loads are connected. The same is true for parallel and series-parallel circuits (a combination of both circuit types). You will rarely, if ever, find loads in an HVACR system connected in series, or in a series-parallel arrangement. The simple circuit shown here illustrates this point. At first glance, you might think it is a series-parallel circuit. On closer examination, you can see that there are only two loads (the fan relay coil and the contactor coil), and they are connected in parallel. Therefore, it is a parallel circuit. The control devices are wired in series with the loads, but only the loads are considered in determining the type of circuit.

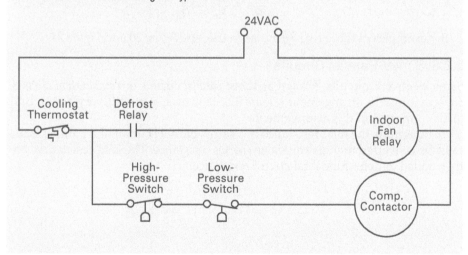

2.2.2 Parallel Circuits

In a parallel circuit, each load is connected directly to the voltage source. Therefore, the voltage drop is the same through all loads because each load sees an equal applied voltage. The source sees the circuit as two or more individual circuits containing one load each.

In the parallel circuit shown in *Figure 19*, the power source sees three circuits. Each contains a 30Ω load. The current flow through any load varies depending on the resistance of that load. Thus, the total current drawn by the circuit is the sum of the individual currents through the three parallel loads.

The total resistance of a parallel circuit is calculated differently from that of a series circuit. In a parallel circuit, the total resistance is less than the smallest of the individual resistances. For example, each of the three 30Ω loads draws 0.4A at 12V; therefore, the total current is 1.2A. This is calculated as follows:

$$I = \frac{E}{R} = \frac{12V}{30\Omega} = 0.4A$$

$$0.4A \text{ per load} \times 3 \text{ loads} = 1.2A$$

Now Ohm's law can be used to calculate the total resistance (R_T):

$$R_T = \frac{E_T}{I_T} = \frac{12V}{1.2A} = 10\Omega$$

This example is simple because all the resistances have the same value. When the resistances are different, the current must be calculated for each load individually. The sum of the individual currents is equal to the total current.

Parallel circuits have an advantage over series circuits because they offer multiple paths. Therefore, if the continuity of any one path is broken, current can still flow in the other paths. In other words, parallel circuits continue working even if one load branch of the circuit opens. Household circuits are wired in parallel. This is why series circuits are very rarely found in HVACR systems and equipment. Most such circuits are found in packaged electronic devices that are not serviced in the field.

The following formulas can be used to convert parallel resistances to a single resistance value:

- For two resistances connected in parallel:

$$R_T = \frac{R_1 \times R_2}{R_1 + R_2}$$

- For three or more resistances connected in parallel:

$$R_T = \frac{1}{\dfrac{1}{R_1} + \dfrac{1}{R_2} + \dfrac{1}{R_3}}$$

For examples of these two formulas in use, see *Figure 20* and *Figure 21*.

2.2.3 Series-Parallel Circuits

Some electronic circuits, known as *series-parallel circuits* or *combination circuits*, contain a hybrid arrangement (*Figure 22*). However, you will very rarely find loads connected in this arrangement.

To determine the total resistance of a series-parallel circuit, the parallel loads must be converted to their equivalent series resistance. The load resistances are then added to determine total circuit resistance.

NOTE

Because series-parallel circuits are uncommon, you will not have to determine their characteristics very often. However, these calculations are presented in more detail in NCCER Module 26104, *Electrical Theory*.

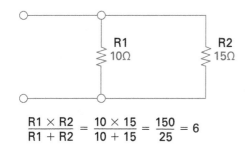

$$\frac{R1 \times R2}{R1 + R2} = \frac{10 \times 15}{10 + 15} = \frac{150}{25} = 6$$

Figure 20 Parallel circuit with two loads.

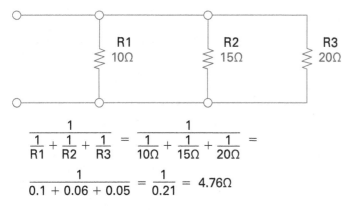

$$\frac{1}{\dfrac{1}{R1} + \dfrac{1}{R2} + \dfrac{1}{R3}} = \frac{1}{\dfrac{1}{10\Omega} + \dfrac{1}{15\Omega} + \dfrac{1}{20\Omega}} =$$

$$\frac{1}{0.1 + 0.06 + 0.05} = \frac{1}{0.21} = 4.76\Omega$$

Figure 21 Parallel circuit with three loads.

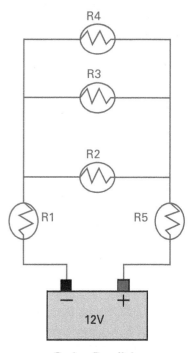

Series-Parallel
Circuit

Figure 22 Series-parallel circuit.

2.0.0 Section Review

1. Which of the following is measured in watts?
   a. Current
   b. Voltage
   c. Resistance
   d. Power

2. The formula $I \times R$ is used to solve for which electrical value?
   a. Voltage
   b. Current
   c. Resistance
   d. Power

3. How much current should a 1.5 hp, 240V motor consume?
   a. 0.0063A
   b. 0.21A
   c. 4.66A
   d. 160A

4. The total resistance of the circuit shown in *Figure SR01* is _____.
   a. 3Ω
   b. 10Ω
   c. 60Ω
   d. 90Ω

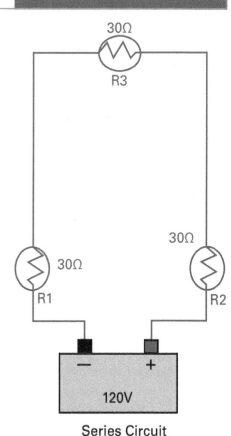

Series Circuit

Figure SR01

3.0.0 Electrical Measuring Instruments

Performance Tasks

1. Draw a connection diagram for a circuit that includes the following components:
   - A power control switch
   - 120VAC/24VAC control transformer
   - A control relay with a 24VAC coil
   - (2) 120VAC lights, controlled by the relay and wired in parallel

2. Assemble the circuit based on the connection diagram developed in *Performance Task 1*, powered by a GFCI-protected power source.

3. With the circuit de-energized, check circuit components and relay contacts for continuity.

4. With the circuit de-energized, measure and record the resistance of the transformer windings, relay coil, and lights.

5. Energize the circuit, turning on the lights, and measure and record the total circuit current.

6. Measure the voltage provided by the power source to the transformer primary.

7. De-energize and disable the circuit power source. Verify that power is disabled with a voltmeter.

Objective

Identify the electrical measuring instruments used in HVACR work and describe how to use them.

a. Describe how voltage is measured.
b. Describe how current is measured.
c. Describe how resistance is measured and how continuity is determined.

Test meters are used to measure voltage, current, and resistance. The most common test meter is the volt-ohm-milliammeter (VOM), commonly known as a **multimeter** (*Figure 23*). The multimeter combines a voltmeter (for voltage measurement), an ohmmeter (for resistance measurement), and an ammeter (for current measurement) into a single instrument. Some can also measure temperature and a few other things, given the necessary accessories or probes. An **analog meter** is so-called because a pointer moves in proportion to the value being measured, and the value is determined by reading a scale displayed behind the needle. Analog meters are rarely found in field use today. A **digital meter**, or DMM, displays the result numerically on the screen. Digital meters are the common choice of today's technicians.

The multimeter shown in *Figure 23* can make current measurements as is, but only in a low range. They are often limited to current measurement in the milli-amp range. HVACR technicians typically measure current beyond the capacity of the multimeter without an accessory. A multimeter can be combined with a **clamp-on ammeter** in a single product, as shown in *Figure 24*. This instrument can handle much larger current measurements. Other technicians prefer to use a standard multimeter with a compatible accessory clamp-on ammeter (*Figure 25*). Both options offer equal performance, and the choice usually comes down to personal preference.

Multimeter: A test instrument capable of reading voltage, current, and resistance. Also known as a *volt-ohm-milliammeter (VOM)*.

Analog meter: A meter that uses a needle to indicate a value on a scale.

Digital meter: An instrument that provides a direct numerical reading of a measured value.

Clamp-on ammeter: An ammeter with operable jaws that are placed around a conductor to sense the magnitude of the current flow.

Figure 23 Digital multimeter.
Source: Reproduced with Permission, Fluke Corporation

Figure 24 Clamp-on multimeter.
Source: Reproduced with Permission, Fluke Corporation

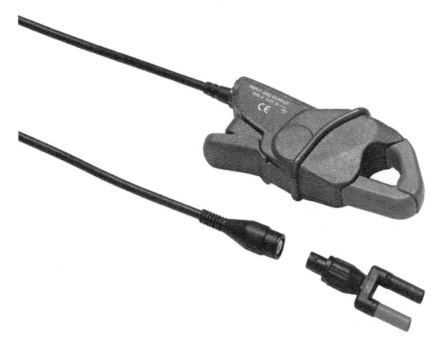

Figure 25 Clamp-on ammeter accessory for a multimeter.
Source: Reproduced with Permission, Fluke Corporation

A third option for current measurement is a stand-alone clamp-on ammeter. They look much like the multimeter in *Figure 24* but do not have the ability to measure voltage or resistance. It is not unusual for an HVACR technician to have a standard multimeter and a separate clamp-on ammeter. Many technicians have several different meters, and there are times when having several clamp-on ammeters in addition to a multimeter can be helpful in testing and troubleshooting.

The primary functions of multimeters are discussed in the sections that follow, with basic guidance in taking electrical measurements.

3.1.0 Measuring Voltage

Voltage measurements are very common for HVACR technicians. All multimeters have a rotary switch or similar control that is used to select the test function (voltage, current, resistance, or other function), as well as the range of values within a given function.

A typical multimeter rotary switch is shown in *Figure 26*. Most digital meters today are *auto-ranging*; that is, they automatically adjust the range based on the voltage sensed. In this case, the rotary switch is only used to pick the correct function.

Some meters, however, require the user to select the voltage range. When this is the case, select a range that includes the voltage you expect to measure, or a higher range if you are unsure. If the range of the value to be measured is unknown, the highest range should be used to avoid damaging the meter. Once an initial reading is taken, the range can be adjusted accordingly for greater precision. When using a voltmeter to check for power to a load, the leads are connected in parallel with (across) the load to be tested, as shown in *Figure 27*.

The basic steps to set up a multimeter and take your first voltage measurement follow. The steps are based on the multimeter shown in *Figure 23*. It is important to review the operating instructions for the meter you will be using during training and to complete the Performance Tasks for this module, but all multimeters are used in a similar way.

Before using a voltmeter to take an important measurement, it should be tested on a device you know is energized. You should also have an idea of the voltage supplied to that device. Once you know the meter works and is reading the correct voltage, you can use it for the important measurement. A good, safe place to make the test measurement is a wall receptacle.

Figure 26 Multimeter rotary switch.
Source: Reproduced with Permission, Fluke Corporation

To ensure it is energized, plug in a lamp or other device and be sure it works. In the United States, the voltage should be about 110VAC to 120VAC. When you test it and it shows the expected voltage, you know the meter is operational. You can then trust it when it says there is no voltage in an important, safety-related test immediately thereafter.

After donning the appropriate PPE, including safety glasses and a glove set rated for the voltage present, follow these basic steps to test meter operation:

Step 1 Insert the black lead into the COM jack. The COM jack is usually color-coded black as well as marked.

Step 2 Insert the red lead into the VΩ jack. This jack is color-coded red on most meters. There is no difference in the two leads other than their color. But for some measurements you'll make in the future, keeping the leads straight is important.

Step 3 Turn the rotary switch to ∠. The V is for voltage, and the symbol above it represents the sine wave of AC power. You'll be making an AC power measurement, so this is the proper setting. The display should now be active, letting you know the meter is on.

> **WARNING!**
>
> If the meter is set to measure resistance or other values other than voltage, and the probes are touched to energized components, severe meter damage and possible personal injury can result. Although meters are equipped with fuses for protection, meters can and do occasionally experience catastrophic failure.

Step 4 Insert the test leads into the receptacle openings—black lead first, red second. Although it will make no difference here, get in the habit of placing the black lead first, and the red lead second.

> **WARNING!**
>
> With the test leads inserted into the receptacle, the metal tips become electrically energized. Do not touch the metal portion to avoid being shocked. Do not allow the metal portions of the probes to touch each other while energized as this will cause a **short circuit** and trip the circuit breaker serving the circuit. A momentary arc will also occur, which will damage the test leads at a minimum and possibly cause an injury.

Step 5 Read the voltage measurement on the display. Ensure that the voltage is as expected. When finished, remove the red lead from the receptacle first, then remove the black lead.

Once you have verified that the meter is functional and trustworthy, check the voltage at a power disconnect switch for an air conditioning unit. The power wiring usually enters the disconnect from the bottom or back and is routed to the left or upper terminal block. The right or lower terminal block is where the wires powering the unit are connected. The following are what you can expect when testing the voltage in a disconnect:

- When the power source feeding the disconnect, such as a circuit breaker at another panel, is switched to off, you should not find any energized terminals or surfaces inside.

- When the power source is turned on but the disconnect switch is turned off, you should have voltage across the incoming terminals, but no voltage across the outgoing terminals.

- When both the power source and the disconnect switch are turned on, you should have voltage across the upper terminals as well as the lower terminals.

After donning the appropriate PPE, including safety glasses and a glove set rated for the voltage present, follow these steps to take a voltage measurement at a disconnect switch:

Step 1 Locate the source of power for the disconnect switch (such as a circuit breaker in a central panel) and switch it to the Off position. However, there is no guarantee that it works! As you continue, work as if the disconnect is still energized until proven otherwise.

Step 2 Ensure that the disconnect handle is in the Off position (down), or that the pull-out insert has been removed. The insert acts as a jumper to connect and disconnect power in some disconnects.

Step 3 Open the disconnect switch door and lock it in the open position. This may require you to lift it up to a 90-degree angle and push it back slightly toward the wall.

Step 4 If there is a metal or non-conductive shield blocking access to the terminals, carefully remove it. Do not make contact with anything behind the shield in the process. Some non-conductive shields (generally made from heavy paper) have openings that allow the voltage to be tested without removing them.

Step 5 Examine the wiring before proceeding. Visually trace the path of the wires coming in and identify which ones are connected to the incoming power terminals. These are from the power source. Confirm that you are looking at the incoming power terminals by checking the wiring coming from the equipment as well. They should be connected to the other set of terminals. Also look for any signs of burning, melted insulation, or significant discoloration of the terminals that would suggest a problem.

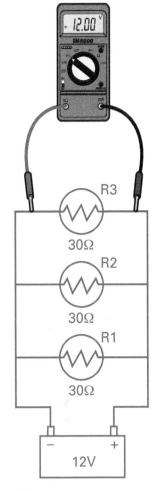

Figure 27 Checking voltage across a load device.

WARNING!

The terminal layout of power disconnects can vary. Ensure that you are checking the incoming power terminals. With the disconnect off or the insert removed, there will be no power at the outgoing power terminals, regardless of the source circuit-breaker position. Checking the wrong set of terminals may lead you to incorrectly conclude the disconnect switch is de-energized.

Step 6 Ensure the meter is still on and set to the correct position ($\angle$) for an AC voltage reading.

Step 7 Place the probe of the black lead firmly against one incoming wiring terminal and the probe of the red lead against the other incoming terminal.

Step 8 Check the meter display; there should be no voltage present.

Step 9 Repeat the procedure by checking the outgoing terminals. There should be no voltage detected.

Step 10 Step away from the disconnect and ask a co-worker or fellow trainee to turn the power source back on. Ensure that no one approaches the disconnect switch at this time.

Step 11 Carefully repeat *Step 7*. This time, the meter should display the line voltage at the incoming terminals.

Step 12 Step away from the disconnect and, once again, ask someone to turn the power source off. Then check the incoming power terminals to ensure the disconnect is de-energized.

Step 13 Carefully reassemble any shield that may have been removed. Install the jumper (if equipped) in the correct position—some have both an On and Off position. Then close the disconnect door. Alternatively, close the door and bring the handle to the On position.

NOTE

Disconnects with a switch handle on the exterior have an interlock that prevents the switch being placed in the On position while the door remains open, as an extra measure of safety.

Step 14 Restore power to the disconnect. If practical, restart the system to ensure power has been restored to the equipment and it starts as expected.

From a safety point of view, never forget that you are often working with or near electrically energized wiring and components. Electricity is an invisible hazard that you must be consistently aware of throughout your career.

3.1.1 Voltage Testers

A voltage tester (*Figure 28*) is a very simple tool used only to check for the presence of voltage. It can be used as a quick troubleshooting tool or as a safety device to make sure the voltage is turned off before touching terminals or conductors.

When the probes are touched to the circuit, the light on the instrument will turn on if any voltage is present. Instruments like these are available in several voltage ranges, some of which are limited to only 90V. It is important to know about the circuit you are checking and the voltage you expect from it. It's best to choose a tester that measures up to 1,000V.

Note that this type of compact, simple instrument is not generally used for work requiring accuracy. Technicians use these to detect the simple presence or absence of voltage. For troubleshooting tasks, a multimeter is far more effective and informative. But a voltage tester is a great confirmation tool.

Wireless Meters

Today's most advanced multimeters use modules placed at the point of measurement that transmit the information, such as voltage or current, to a compatible multimeter. Multiple readings can be shown on the meter display, and the information can also be shared with other devices. Wireless modules offer greater safety by allowing you to maintain a safe distance from energized surfaces, although they are also capable of making traditional measurements using probes.

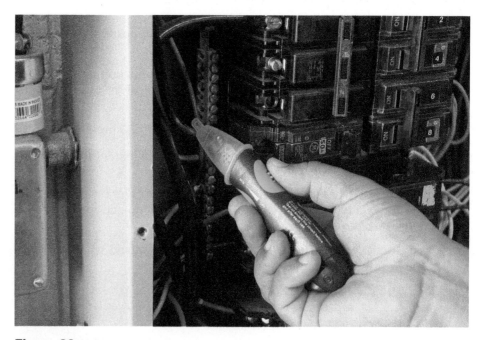

Figure 28 Voltage tester.
Source: Courtesy of Greenlee Tools, Inc.

3.2.0 Measuring Current

Clamp-on ammeters like the one shown in *Figure 24* are used by HVACR technicians to measure AC current. The jaws of the ammeter are placed around a conductor. Current flowing through the wire creates a magnetic field, which induces a proportional current in the ammeter jaws. Multimeters with an accessory clamp-on ammeter work the same way.

Not all ammeters or multimeters have the same capacity to measure current. When the measurement of high current levels is expected, make sure the range of the ammeter is suitable for the task. Most clamp-on ammeters, though, can handle the currents you are likely to measure in HVACR work.

A good place to take your first current measurement, under supervision, is at a functioning condensing unit. Before beginning, don the proper PPE—safety glasses at a minimum. To use a clamp-on ammeter, follow these basic steps:

WARNING!

Electrically energized power circuits often carry large currents and are hazardous. Keep your hands and all other parts of your body away from energized surfaces and wiring. Avoid reaching into a panel with both hands at the same time. Ensure that you are in a stable position before reaching toward the conductor with the ammeter.

Step 1 Ensure that the unit thermostat is switched to the Off position and turn off the power to the unit before removing any panels.

Step 2 Remove the outer access panel, then the inner panel protecting the controls.

NOTE

The low-voltage control circuit, energized at 24VAC, will still be active. Do not touch or short-circuit any control wiring to avoid damage to the control transformer.

Step 3 Remove any test leads connected to the meter and turn the rotary switch to the proper function. This should be labeled with the letter *A*. The display should indicate the meter is on and awaiting an input.

Step 4 Examine the tips of the jaws and ensure they meet properly and firmly. If they do not, it will affect the accuracy of the reading.

Step 5 Press the jaw release lever to open the jaws and then close them around one of the two power wires connected to the **contactor** (*Figure 29*). Make sure no other wires are passing through the jaws. More than one conductor passing through the jaws will cause an erratic reading.

Contactor: Control device consisting of a magnetic coil and one or more sets of contacts used as switching devices in line-voltage, high-current circuits.

Step 6 Have a second person go to the thermostat, set it to Cool, and turn the setting down below the room temperature. At the condensing unit, you should hear the contactor snap closed, but nothing else should happen since line voltage has not been restored.

Step 7 While watching the ammeter and ensuring you are clear of all energized surfaces, restore power to the condensing unit at the disconnect switch. Be prepared, as the compressor and fan will likely start immediately.

Inrush current: A significant rise in electrical current associated with energizing inductive loads such as motors.

Step 8 The ammeter reading will jump to a high reading at first, then settle back to a stable value as the system stabilizes. The initial surge of current is called **inrush current** and is normal.

Step 9 Document the reading on the display once the system is stable.

Step 10 Carefully open the jaws of the ammeter and remove it from the conductor.

Step 11 To ensure your safety, turn the power off to the unit. Then secure the access panels.

CAUTION

Before restoring power, wait at least five minutes. This allows time for the system pressures to equalize. Compressors are not designed to start when there is a large pressure differential.

Step 12 After a five-minute waiting period, restore power to the condensing unit.

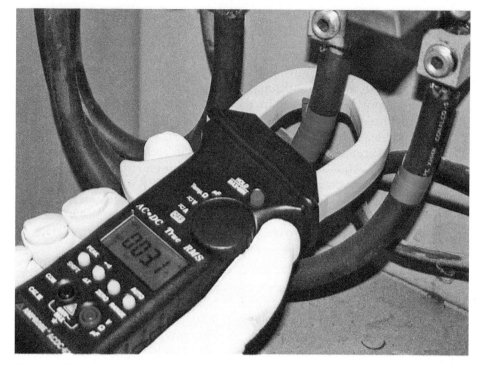

Figure 29 Clamp-on ammeter in use.
Source: Photo courtesy of Tim Dean

Some clamp-on ammeters don't read very small currents accurately. When very low currents are being measured with a clamp-on ammeter, it may be necessary to wrap a single conductor multiple times around the meter jaws, as illustrated in *Figure 30*. This has the effect of multiplying the current reading. The reading is then divided by the number of wraps to determine the current flow. For example, if the wire passes through the meter jaws five times, as shown in *Figure 26*, and the reading is 5A, the actual current is 1A (5A ÷ 5 wraps of wire). This is done when the reading is expected to be near or below the accurate range of the meter to make the reading more accurate. Using ten wraps is also a common practice, for even greater accuracy.

In addition to good safety practices, remember the following guidelines when measuring current:

- If the ammeter jaws are dirty or misaligned, the meter will not read correctly. The jaws must meet and interlock completely.

- If using an analog ammeter, always start at a higher range than necessary to avoid damaging the meter. Then adjust to a range that provides the needed level of detail or accuracy.

- Do not clamp the ammeter jaws around two different conductors at the same time; an inaccurate reading will result.

3.2.1 Inline Current Measurement

Inline current measurements, using probes at each end of the measured circuit, are rarely needed. However, it is required occasionally, especially when troubleshooting some burner ignition systems. The practice is used to measure very small current values. Several milliamps can be measured accurately using wraps around the ammeter jaws, as described in the previous section. But measurements in the microamp (μA) range require an inline measurement.

Most multimeters are capable of measuring tiny currents, but it is important to ensure that the measurement will fall within the maximum current range for the meter. Many are capable of inline readings up to 10A, but not all. It should also be noted that inline current readings can be inherently hazardous. While working with tiny currents in the microamp range is not a problem, currents above 1A can be deadly.

WARNING!

Inline current measurements require leads to be connected directly to the meter in series with an energized circuit. This results in temporary, exposed wiring connections that may carry a current sufficient to shock or electrocute you. Use extreme caution and ensure that all connections are secure, and the leads are well controlled.

An inline current measurement means that the ammeter must be placed in series with the circuit. To do so, the circuit must be de-energized. The arrangement is shown in *Figure 31*. If the ammeter was placed across a closed circuit, some of the current would flow through the existing circuit instead of the meter, and the current reading would be incorrect.

3.3.0 Measuring Resistance

An ohmmeter contains an internal battery that acts as a voltage source. Therefore, resistance measurements must always made on de-energized circuits and components.

While ohmmeters are sometimes used to measure the precise resistance of a load, such as motor windings, they are more often used to check continuity in a circuit. A wire or closed switch usually offers no significant resistance to current flow.

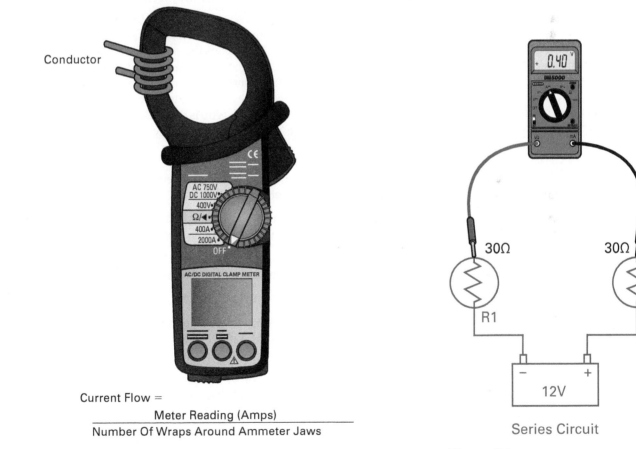

Current Flow =

Meter Reading (Amps)
_____
Number Of Wraps Around Ammeter Jaws

Figure 30 Measuring low currents.

Series Circuit

Figure 31 Inline current measurement.

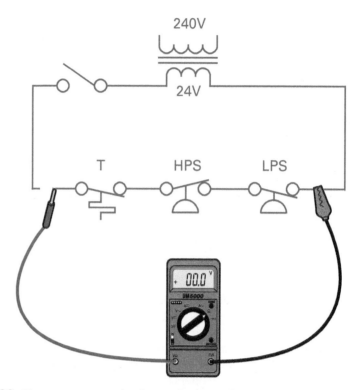

Figure 32 Ohmmeter connection for continuity testing.

With the ohmmeter connected as shown in *Figure 32*, and the three switches closed, the current produced by the ohmmeter battery will flow unopposed and the meter should show zero resistance. This means the circuit has continuity (it is continuous). If a switch is open, however, there is no path for current and the meter will see infinite resistance (it does not have continuity). In other words, the reading on the meter will not change when the leads are connected to the circuit. That is a sign that one or more of those switches are open or a wire is disconnected.

A fuse offers practice at checking continuity. Before installing a fuse, it is a good idea to check it for continuity first, ensuring that it is intact and not open. Checking fuses is a common troubleshooting task as well. Follow these basic steps to check a fuse for continuity, again remembering that you must follow the instructions for the meter you are using:

Step 1   Obtain a common cartridge fuse, like the one shown in *Figure 33*. The type and class don't matter. Even a small glass fuse or automotive fuse will work.

Step 2   Insert the black lead into the COM jack.

Step 3   Insert the red lead into the VΩ jack.

Step 4   Turn the rotary switch to Ω. As soon as the setting is made, a digital meter should respond with OL shown on the display. It begins looking for a completed circuit as soon as it is energized.

Step 5   You can test the meter by simply touching the two test lead tips together. The meter should display 0 ohms, and many will also provide an audible signal for continuity.

Step 6   Place one test lead on one end of the fuse, and the other on the opposite end. The meter should respond the same way as it did when tested—0 ohms displayed and possibly an audible signal. This indicates that the fuse is intact.

Figure 33 Cartridge fuse.

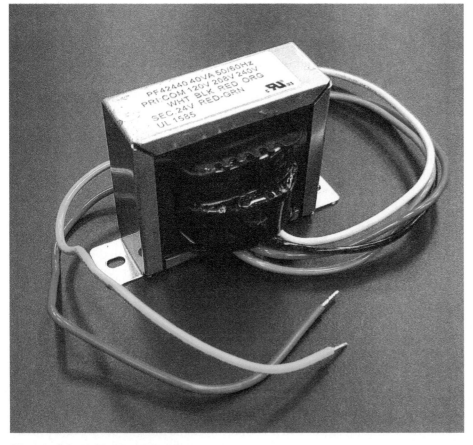

Figure 34 Control transformer.

Now you can test a load device of some type to measure resistance. A small 24VAC control transformer for a furnace or air conditioner is a good place to start (*Figure 34*). Begin where you left off in the previous exercise.

Step 7 Multiple wires often extend from the transformer although some may have terminals instead. Since you have not yet studied transformers in detail, your instructor will help you identify the correct wires to test. Transformers are constructed of coils of copper wire called *windings* that do offer some resistance.

Step 8 Touch the leads of the ohmmeter to the bare wire ends identified by your instructor, representing one of the windings. There should be continuity (unless the transformer is damaged) and the audible signal (if equipped) will sound. But the display should read a small resistance now instead of 0 ohms. Record the reading.

Step 9 Now test the other transformer winding, using the leads identified by your instructor. You should get a different resistance reading this time. Record this reading as well. As you learn more about transformers, consider the difference in the two resistance readings and why that is the case.

Step 10 Turn the meter rotary switch to Off and disconnect the leads.

Figure 35 Handheld, battery-operated megohmmeter.

Source: Reproduced with Permission, Fluke Corporation

3.3.1 Megohmmeters

Megohmmeters are special ohmmeters that are designed to read extreme resistance values. The name comes from the fact that the reading is generally in megohms (MΩ), or millions of ohms. A common ohmmeter is not capable of reading such high resistance values. Typically, a common ohmmeter will indicate a circuit with megohms of resistance is simply open (no continuity).

In the HVACR field, megohmmeters (*Figure 35*) are sometimes used to test the condition of motor windings, including the motor in compressors. Megohmmeters develop high DC voltages and apply that voltage to a motor winding. A megohmmeter will report the many millions of ohms of resistance there should be between the motor windings and ground. This reading helps to indicate the potential for a grounded wiring to occur, providing a means of predicting the longevity of the motor. Motor windings that are well insulated will show an extreme amount of resistance to ground—enough resistance to prevent current from following the path.

When oil and refrigerant in a circuit become contaminated with acids that can form over time, the compressor motor's resistance to ground may decrease. This is due to the breakdown of the winding insulation through chemical attack. Lower than normal readings, therefore, often indicate that the motor winding insulation has broken down to some degree.

To test a compressor or motor winding, it is best to operate the compressor until it reaches its normal operating temperature. As soon as the compressor is shut down, power is disconnected from the system and the power leads to the compressor are physically disconnected. With the compressor electrically isolated from everything else, one probe of the megohmmeter is connected to a winding connection, while the other is connected to an unpainted point on the motor frame or compressor body. Both connections must be secure and provide good electrical contact; some paint may need to be removed to ensure a good electrical path. The megohmmeter is then energized, applying a high DC voltage to the circuit momentarily.

Handheld, battery-operated megohmmeters have become more popular in recent years, with many capable of generating as much as 1,000VDC (1kVDC) from the battery alone. Other megohmmeters can generate up to 15kVDC. Older megohmmeters have a hand crank to manually generate the voltage. For general HVACR work, the handheld, battery-operated models provide sufficient test voltages.

WARNING!

Never use a megohmmeter on a motor in a vacuum. Air conditioning and refrigeration compressors are often evacuated before refrigerants are added. Conducting a test while the motor windings are exposed to a vacuum can result in motor damage and possible personal injury. In addition, the test voltages generated for high-resistance testing are hazardous. To safely use a megohmmeter, it is essential to follow the manufacturer's instructions and acknowledge the warnings.

Predictive Maintenance with Megohmmeters

Megohmmeters are valuable tools that can aid in troubleshooting, but they are more often used for what is known as *predictive maintenance*. For valuable and critical motors, readings can be taken when the motor is new and then compared to future readings on a periodic basis. This process will reveal any trend toward an electrical failure.

When the resistance readings fall below what is considered to be a safe level, it indicates that an electrical failure is developing. The motor can then be replaced on a planned basis, rather than be allowed to fail unexpectedly and interfere with plant operations. Motors related to a critical application or life safety are usually tested regularly and proactively replaced when the megohmmeter readings fall. Since predictive maintenance is rarely applied to inexpensive, disposable motors, tested motors are usually sent to a motor repair facility where they are refurbished.

3.0.0 Section Review

1. When testing a circuit with a voltmeter, _____.
   a. the circuit must be broken to avoid an erroneous reading
   b. the power switch for the test unit must be turned off
   c. a clamp-on voltmeter must always be used
   d. the meter is placed in parallel with the component or circuit

2. Which type of measurement requires the meter to be connected in series with the circuit under test?
   a. A voltage measurement
   b. A resistance measurement
   c. A very small current measurement
   d. A very large current measurement

3. The ohmmeter in *Figure SR02* should read _____.
   a. the resistance of the 24V transformer winding
   b. zero resistance (closed circuit)
   c. infinite resistance (open circuit)
   d. the resistance of the three switches

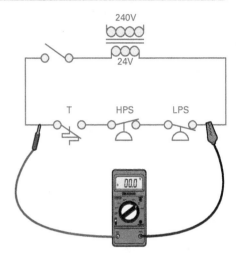

Figure SR02

4.0.0 Common HVACR Electrical Components

Objective

Identify electrical components used in HVACR systems and describe their functions.

a. Identify and describe various load devices and explain how they are represented on circuit diagrams.

b. Identify and describe various control devices and explain how they are represented on circuit diagrams.

c. Identify and describe common electrical diagrams used in HVACR work.

Performance Tasks

1. Draw a connection diagram for a circuit that includes the following components:
   - A power control switch
   - 120VAC/24VAC control transformer
   - A control relay with a 24VAC coil
   - (2) 120VAC lights, controlled by the relay and wired in parallel

2. Assemble the circuit based on the connection diagram developed in *Performance Task 1*, powered by a GFCI-protected power source.

3. With the circuit de-energized, check circuit components and relay contacts for continuity.

4. With the circuit de-energized, measure and record the resistance of the transformer windings, relay coil, and lights.

5. Energize the circuit, turning on the lights, and measure and record the total circuit current.

6. Measure the voltage provided by the power source to the transformer primary.

7. De-energize and disable the circuit power source. Verify that power is disabled with a voltmeter.

Circuit diagrams, also called *wiring diagrams* or *wiring schematics*, use symbols to represent electrical components and connections. In this section, you will learn about electrical components and how they are shown on wiring schematics.

Electrical components generally fall into two categories: *load devices* and *control devices*. Some devices, such as contactors, contain both an energy-consuming element (load) and a control element (switch contacts).

Some common electrical symbols used on wiring diagrams are shown in *Appendix 03106A*. Many are also shown along with the devices discussed in the sections that follow. However, diagram symbols can vary significantly.

4.1.0 Loads

Any device that consumes electrical energy and does work in the process is considered a load. Loads convert electrical energy into some other form of energy, such as light, heat, magnetism, or mechanical energy.

Loads have resistance and consume power. An electric motor is an example of a load, as is the burner on an electric stove. As electrical current flows through a burner element, it is converted into heat. Even the coil of a relay, although it consumes little power, represents a load.

4.1.1 Motors

Electric motors convert electrical energy into mechanical energy. Motors are one of the most common and substantial loads found in HVACR systems. Motors are primarily used to drive compressors, fans, and blowers.

Figure 36 shows examples of three common motors. *Figure 37* shows how the same motors may be depicted on schematic wiring diagrams. Because it is likely that several motors will appear on a typical HVACR circuit diagram, letter codes such as IFM (indoor fan motor), OFM (outdoor fan motor), and COMP (compressor) are used to help identify the role of a motor. Other components that appear on circuit diagrams, such as relays and safety devices, are also identified by letter codes. The letter codes are generally defined in a *legend* found on the wiring diagram.

Although many thousands of these motors remain in service, new HVACR equipment is often equipped with electronically commutated motors, or ECMs (*Figure 38*). These motors are more energy-efficient and versatile. The motor assembly is powered with AC but internally rectifies the voltage to DC before it is fed to the motor.

In most residential and commercial air conditioning systems, the compressor motor is the largest single electrical load. In a residential system, the indoor blower motor and outdoor condenser fan motor may draw only a few amps of current while the compressor may consume five to ten times more.

4.1.2 Electric Heaters

Electric heaters are also called *resistance heaters* because they represent resistive loads. Resistive loads are the simplest type of load, where there is no reliance on magnetism to do the work. Another good example of a resistive load is a common light bulb.

Figure 36 Examples of electric motors.

The common diagram symbol for an electric heater is the same symbol used for a resistor. An electric resistance heater assembly and diagram are shown in *Figure 39*. Electric heaters, though, come in a wide variety of forms. The assembly shown in *Figure 39* would likely be found in a heat-pump air handling unit. *Figure 40* shows a small electric heater that might be found wrapped around the base of a compressor to keep it warm. Toasters and hair dryers also contain similar electric heaters.

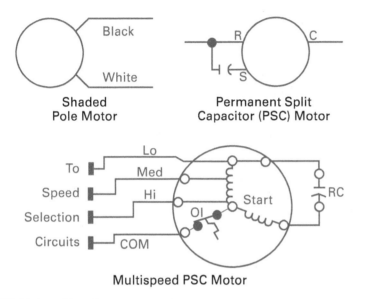

Figure 37 Motor wiring diagram examples.

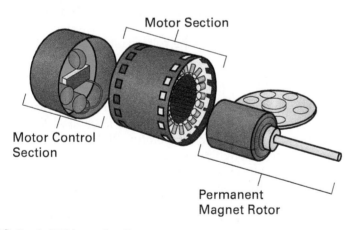

Figure 38 Basic ECM construction.

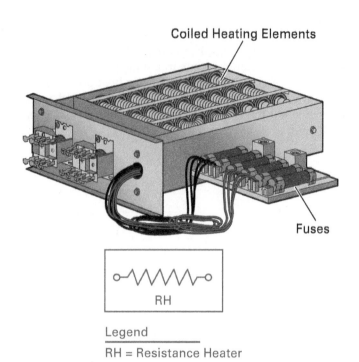

Figure 39 Electric heat assembly.

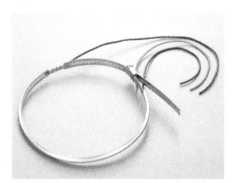

Figure 40 Compressor crankcase heater.

Source: Backer Springfield

There are many types of control devices. Their primary function is to control the operation of loads. Many control devices also control the action of other controls. Anyone working on HVACR systems will encounter these devices on a daily basis.

Although some control devices operate at the same voltage as the loads, the controls found inside most residential and commercial equipment operate at the low voltage of 24VAC. However, some of today's control systems use low-voltage DC power to communicate, based on the voltage returned to, or by, a solid-state control.

4.2.1 Switches

Switches stop and start the flow of power to other control devices or loads. Electrical switches are classified according to the force used to operate them. Examples include manual, magnetic, temperature, light, moisture, and pressure switches. Drawing symbols often reflect the way the switch is operated.

The simplest type of switch is one that makes (closes) or breaks (opens) a single electrical circuit. More complicated switches control several circuits. The switching action is described by the number of poles (electrical circuits to the switch) and the number of throws (circuits fed by the switch). *Figure 41* shows some of the common switching arrangements:

- *Single-pole, single-throw (SPST)* — Power enters at one terminal and can be connected to, or disconnected from, one terminal. It can control a single circuit.

- *Double-pole, single-throw (DPST)* — Power enters at two separated terminals and can be connected to, or disconnected from, two separate terminals. A double-pole switch can control two circuits.

- *Double-pole, double-throw (DPDT)* — Power enters at two separated terminals and can be connected to, or disconnected from, two pairs of terminals. It can also control two circuits.

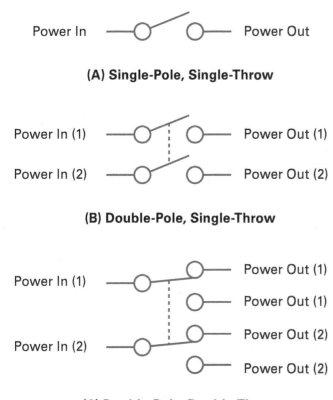

(A) Single-Pole, Single-Throw

(B) Double-Pole, Single-Throw

(C) Double-Pole, Double-Throw

Figure 41 Switch contact arrangements.

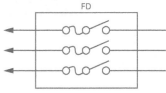

Figure 42 Fused power-disconnect switch.

One good example of a manual switch is a power-disconnect switch (*Figure 42*). They are found both indoors and outdoors, in accessible locations and within sight of the equipment they serve. They provide craftworkers with a means of de-energizing equipment for service. The disconnect in *Figure 42* also houses fuses for circuit protection, but not all disconnect switches do.

Some disconnects, like the one shown in *Figure 42*, have a handle on the outside of the enclosure to turn the power on and off. Others have a plug assembly that is pulled out to turn the power off. You may also spot a disconnect switch that houses a circuit breaker.

Thermostats are primary control devices in heating, cooling, and refrigeration systems. A *thermostatic switch* opens and closes in response to changes in temperature. While a cooling thermostat closes its contacts on a temperature rise, a heating thermostat closes its contacts on a temperature fall (*Figure 43*).

The thermostat shown in *Figure 43* is an electronic model. Mechanical thermostats are very rarely used today although some remain in service. Electronic models sense temperature using special temperature-sensitive resistors called *thermistors*. The electronic circuitry translates the resistance to a temperature and then determines if heating or cooling is needed. Thermistors are covered in more detail later in this module. Other temperature-controlled switches may provide thermal protection for motors, or act as safety switches in furnaces.

Pressure switches are used in a variety of ways. One common use is to shut off a system compressor if the refrigerant pressure is outside of a normal range. Pressure switches can also control fan motors or pumps, as well as many other devices. *Figure 44* shows some examples of pressure switches. They have many possible forms and uses in the HVACR world.

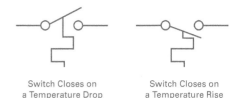

Switch Closes on a Temperature Drop

Switch Closes on a Temperature Rise

Figure 43 Thermostat and thermostatic switch action.
Source: iStock@koinseb

HPS   Opens On Pressure Increase

LPS   Opens On Pressure Drop

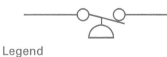

Legend

HPS = High-pressure Switch
LPS = Low-pressure Switch

Figure 44 Pressure switches.

Time-delay fuses: Fuses with a built-in time delay to accommodate the inrush current of inductive loads.

Case History

Faulty Disconnect Switch

A service technician was in a crawl space performing pre-winter maintenance on an oil furnace. He had de-energized the furnace using the power-disconnect switch, but he did not verify the absence of power at the unit using a meter. Unfortunately, the disconnect was wired improperly, allowing current to flow to the furnace even when the switch was in the Off position. As a result, the technician was electrocuted.

The Bottom Line: Don't assume that every disconnect or power switch works even if it has worked properly 100 times before. Always verify the absence of power by using a reliable voltmeter that is consistently tested on a known voltage source before it is used to identify a de-energized circuit.

4.2.2 Fuses and Circuit Breakers

Fuses and circuit breakers protect components and wiring against damage from current surges or short circuits. A short circuit occurs when current flow bypasses a load. An example would be when conductors touch due to damaged insulation, or if you cross two sources of power with a metal tool. Because there is no load, current flow is uninhibited and can quickly create enough heat to arc and start a fire.

As shown in *Figure 45*, a fuse contains a metal link that melts when the current exceeds the rated capacity of the fuse (e.g., 20A), which is its design capacity to withstand current. Fuses come in a variety of shapes and sizes, but their rating is based on current. Because the link is in series with the circuit, the circuit opens when the strip melts. The fuse will "blow" more quickly in the presence of a short circuit than it will in the presence of a minor current overload. Once a fuse has blown, it must be replaced.

Cartridge fuses are very common, but plug fuses are becoming increasingly rare. You will also encounter very small glass cartridge fuses and blade fuses like those used in automobiles. These small fuses are usually related to control-circuit and printed-circuit board protection.

Motors draw far more current during startup than they do when they are running normally because more energy is required to make the rotor and anything connected to it rotate from a dead stop. It is a characteristic associated with most inductive loads. Even relay coils exhibit this elevated current draw when they are first energized since they are inductive loads. Once a motor has reached its normal operating speed, its current drops sharply. The inrush current is often four to six times that of its normal running current. It is important to note that inrush current occurs and recedes very quickly, often in the blink of an eye.

To accommodate the inrush current when a motor starts and protect the motor, different fuses are used. Otherwise, the fuse would blow each time a motor tries to start since the inrush current usually exceeds the fuse's current limit. Special delayed-opening fuses, called **time-delay fuses** or *slow-blow fuses*, are used for HVACR equipment and similar inductive loads. The electrical industry refers to them as *dual-element fuses*, the construction of which is shown at the bottom of *Figure 45*. These fuses will not blow unless the current surge is sustained. Note that resistive loads, such as electric heaters, do not experience an inrush of current when they are first energized and should not be served by this type of fuse.

Circuit breakers (*Figure 46*) serve the same basic purpose as fuses. The big advantage of circuit breakers is that they operate like switches and can be reset when they trip. Fuses must be replaced once they open. If a circuit breaker trips more than once, however, it is a sign of a recurring problem. Special HACR (heating, air conditioning, and refrigeration) circuit breakers are used for air conditioning equipment for the same reason that time-delay fuses are used. They provide the necessary time delay for inductive loads to start and for current flow to stabilize.

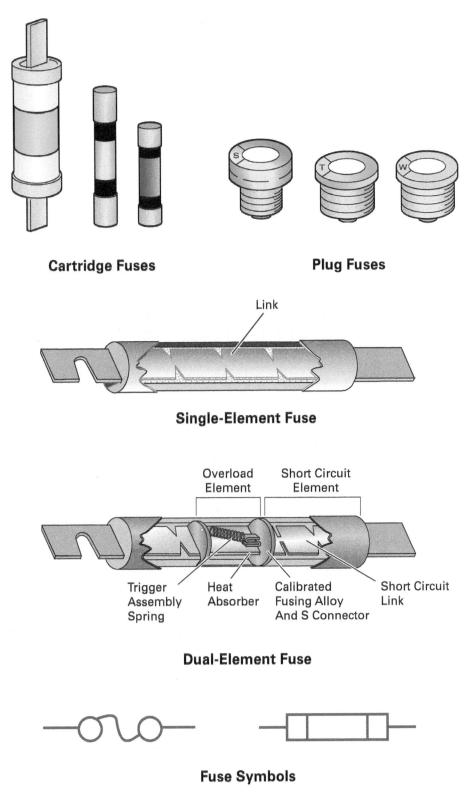

Cartridge Fuses   **Plug Fuses**

Link

Single-Element Fuse

Overload Element   Short Circuit Element

Trigger Assembly Spring   Heat Absorber   Calibrated Fusing Alloy And S Connector   Short Circuit Link

Dual-Element Fuse

Fuse Symbols

Figure 45 Fuses.

A *thermal-trip circuit breaker* contains a spring-loaded metal element that opens when high current flow develops excessive heat, causing it to trip. *Magnetic-trip circuit breakers* contain a small coil of wire that develops a strong enough magnetic field when the current is excessive to open the contacts magnetically.

A problem known as a **ground fault** occurs when a live conductor touches another conducting substance, such as the metal frame of a unit, that has a path to ground. GFCI devices are installed to detect and quickly isolate such problems for safety's sake.

Ground fault: An unintentional, electrically conducting connection between an energized conductor and another conductor connected to earth, such as an equipment frame, or directly to earth. The fault current flows to the earth ground, as well as through a person or other conductive object in the path.

Thermal Trip

Magnetic Trip

Figure 46 Circuit breaker panel.

4.2.3 Solenoid Coils

Solenoid coil: An electromagnetic coil of wire used to control a mechanical device such as a valve or a relay.

A **solenoid coil** (*Figure 47*) works like an electromagnet. When current flows through the coil, magnetism is produced and used to attract or repel a magnetic shaft that passes through its core. Solenoid coils are used to open and close valves and position switching devices like relays. A solenoid-operated valve is often used to start and stop the flow of refrigerant.

4.2.4 Relays, Contactors, and Motor Starters

Motor starters: Magnetic switching devices used to control heavy-duty motors.

Relays, contactors, and **motor starters** are magnetically controlled switches consisting of a solenoid-style coil and one or more sets of movable contacts that open and close in response to current flow in the coil. In other words, they are electrically operated switches. What generally separates them is the magnitude of the load they carry. While contactors can typically carry the same electrical load as a motor starter, a motor starter incorporates overload protection. You can think of a motor starter as a contactor with built-in overload protection. Relays can't carry high levels of current.

Figure 48 shows one example of a control relay, along with several relay schematics. When current flows through the relay coil, a magnetic field is created, just like the solenoid coil mounted on a valve. The magnetic field acting on an armature draws the contacts together. Some contacts close when the relay is energized, providing a path for current. They are referred to as *normally open* (*NO*) contacts because they default to open when the relay is de-energized. Other contacts open when the relay is energized. These are called *normally closed* (*NC*) contacts since they default to closed when the relay is de-energized. NC contacts are often shown on a drawing with a slash through them.

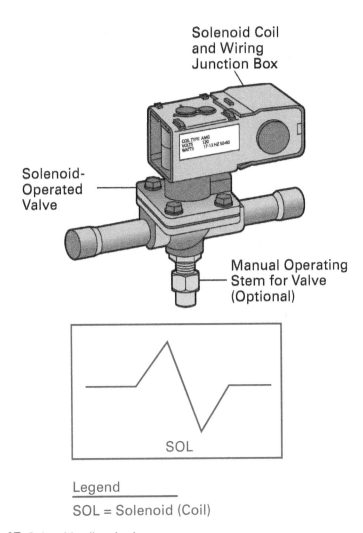

Solenoid Coil and Wiring Junction Box

Solenoid-Operated Valve

Manual Operating Stem for Valve (Optional)

SOL

Legend

SOL = Solenoid (Coil)

Figure 47 Solenoid coil and valve.

The coil and contacts of a relay are often shown at different locations on a wiring diagram. This is because the relay coil is usually a part of the low-voltage (24VAC) control circuit, powered by the transformer, while the contacts are part of a line-voltage circuit. On the drawing, the coil and contacts have the same name or identifier. The identifier *CR* is used in *Figure 48*, which stands for *control relay*. When there are a number of control relays, they are numbered CR1, CR2, and so on. The acronym *DFR* is often used to identify a defrost relay in a heat pump, while *FR* usually identifies a fan relay. The drawing legend provides this information. The same pole and throw descriptions used for switches also apply to relays and contactors.

A *time-delay relay* uses a thermal element or digital control to delay its response to an event. Delays can range from a few seconds to hours. A typical use in HVACR is to keep a blower running for one to three minutes after the heating burners or air conditioning unit has shut down. The blower extracts residual heat or cooling effect, slightly improving efficiency. Similarly, a time-delay relay might delay the start of a blower until the furnace heat exchangers have warmed up. This keeps cool air from being blown into the conditioned space.

A contactor is basically a heavy-duty relay designed to carry larger currents through its contacts. They are commonly used to control the flow of power to compressors and similar electrical loads.

Motor starters are used to start motors, as the name implies. Many motors require external protection, depending on their capacity in horsepower. The *NEC*® determines what motors and motor applications require a motor starter. A motor starter combines a contactor with overload protection devices. When excessive current is encountered, a built-in overload relay opens the circuit to

Plug-In Relays

Relays are the most common switching devices found in air conditioning and heating systems. The figure shows several types of relays designed to plug into matching bases. The wiring is connected to the base assembly, and the relay simply plugs into it. No wiring connections or tools are needed to replace the relay. Since many relays of this type have a clear housing, allowing the technician to see the contacts inside, they are sometimes referred to as *ice cube relays*.

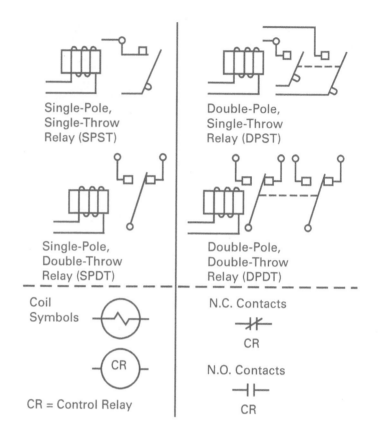

Figure 48 Relay and relay schematics.
Source: HVACGuru

the contactor coil and stops the motor. This characteristic alone makes a motor starter different from a standard contactor.

Figure 49 shows an example of how a relay and contactor might be used in a control circuit. The coils of both the contactor and the relay are shown on the 24VAC side of the diagram (lower portion), while the contacts of both devices are shown on the line-voltage side of the diagram (upper portion).

- The coil of the relay, labeled as *IFR* (indoor fan relay), can be energized two ways. When the *Fan* switch is in the *Auto* position as shown, the relay is energized when the thermostat is switched to *On* and the cooling switch closes on a rise in temperature.

- Changing the *Fan* switch from *Auto* to *On* will energize the relay coil immediately.

- The contactor coil is also energized when the thermostat is switched to *On* and the thermostat closes. The NO contacts of the relay and contactor close when they are energized.

- Once the IFR contacts close, the indoor fan starts. When the contactor contacts close, both the compressor and the outdoor fan motor (OFM) start.

Remember that diagrams like this one show the position of the switches and contacts when de-energized unless otherwise stated.

4.2.5 Transformers

A transformer is used to raise (step up) or lower (step down) a voltage. Step-down transformers are very common in HVACR work. Step-up transformers are typically used in the power industry, boosting the voltage of the power-generating facility for transmission to distant consumers. Only very small step-up transformers are used in the HVACR field, typically for the purpose of generating an ignition spark for oil and gas burners. Beyond this, however, step-up transformers are rarely needed in HVACR applications.

The vast majority of transformers encountered in HVACR work are step-down transformers, like the control transformer shown in *Figure 50*. A control transformer steps the line voltage down to 24VAC. Low-voltage circuits are far less hazardous to work with and are easily insulated. Low-voltage components are also less expensive to manufacture.

A transformer consists of two or more coils of wire wound around a common iron core. When current passes through one coil, called the *primary winding*, the magnetic field develops cuts through the other coil, called the *secondary winding*. The primary winding represents the input, and the secondary winding represents the output.

The primary-winding magnetic field creates, or induces, current in the secondary winding through induction. Depending on the number of turns of wire in the secondary winding, the voltage induced in the secondary is stepped down or stepped up from that in the primary.

The number of turns in the primary and secondary windings determine exactly what happens to the secondary voltage. Refer to *Figure 51*. Note the number of turns drawn on each side of the two vertical bars, which represent the iron core. There are 15 turns on the line-voltage side, and 3 on the low-voltage side. This is a ratio of 15:3, or 5:1. Therefore, the secondary voltage will be one-fifth of the voltage applied to the primary winding. When 120V is applied, a 24V output is generated in the secondary winding.

Control transformers are sized by their *volt-amp (VA) rating*. The VA rating is the product of the output voltage of the transformer multiplied by the current it is capable of sustaining (i.e., V × A). For example, the 40VA transformer in *Figure 50* can sustain a current load of roughly 1.7A (40VA ÷ 24V = 1.66A). To find the current any transformer can sustain, divide the VA rating by the design secondary voltage. In HVACR equipment, that will be 24VAC in nearly every case.

To protect the transformer, some control circuits contain a fuse or circuit breaker in the secondary circuit wiring. Many small transformers have a fuse built into them. In many cases, that fuse is buried beneath the outer insulation and can only be accessed by destroying the material. They are generally soldered into the winding and are challenging to replace. The fuse is primarily added for fire protection and not designed for convenient field repair, although it can be done. Fuses and miniature circuit breakers in the control circuit are far more practical and easier to service.

> **NOTE**
>
> Equipment wiring diagrams rarely depict the winding ratio of transformers accurately.

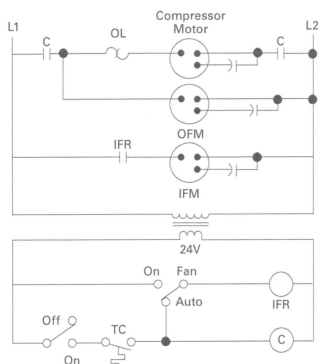

Figure 49 Contactor (C) and a relay (IFR) in a typical HVACR circuit.

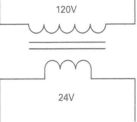

Figure 50 Control transformer.

4.2.6 Overload Protection Devices

Overload devices stop the flow of current when safe current or temperature limits are exceeded. *Figure 52* shows examples of external overload devices that are often used on small compressors. Devices that are actuated (activated) by excessive temperature are referred to as *thermal overloads*. A thermal overload device reacts to excessive temperatures adjacent to it.

Current overloads react to a predetermined current value. However, temperature does play a role. By design, as current passes through the device, it heats up. Too much current results in it becoming too hot, and an internal thermal device opens the contacts. However, there are also current overloads that are solid-state and monitor current in another way.

Overload devices may be embedded in the windings of motors, including compressor motors. They are also found in motor starters.

Protective devices are placed into two categories. **Line duty** devices are directly in line with the voltage source powering the load, experiencing the same current flow. **Pilot duty** devices control other devices in a circuit that can handle more current; they are not designed to handle large currents. Instead, they open when an overload condition occurs, forcing a line duty device to open. Line-duty device contacts, like a contactor, can carry far more current than pilot-duty device contacts.

Line duty: Describes a control that is capable of carrying the same current that a load consumes, installed in the load's power circuit.

Pilot duty: Describes a control device that cannot carry the current of a significant load, and is therefore installed in a control circuit, typically controlling the action of a line duty device.

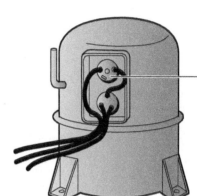

External Line Break
Bimetal Overload
(Klixon)

Current and Temperature (Thermal Overload Device)

Contacts

Bimetal Snap Disc

Line Duty

Motor

OL

Current

Snap Disc and Heater

Pilot Contacts

Pilot Duty

C

C   OL   Motor   C

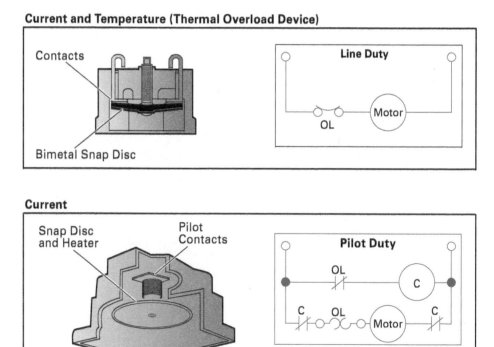

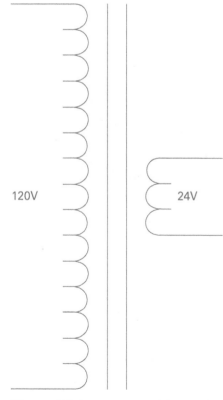

120V

24V

Figure 51 Transformer winding ratio.

Figure 52 Overload protection devices.

Some overload devices reset automatically when the cause of the overload has been removed. Others must be manually reset, usually by pressing a button on the device. If the overload device is embedded in the motor windings, it will reset automatically.

4.2.7 Thermistors

A thermistor is a special type of resistor. Its resistance changes in response to the surrounding temperature. A thermistor serves as an input to an electronic control device such as a thermostat. When the resistance of the thermistor changes beyond a setpoint, the control circuit reacts by opening or closing a set of contacts. In a more sophisticated control arrangement, an electronic device may send out a small current or voltage proportional to the thermistor's resistance.

Figure 53 shows a thermistor used to measure temperature on the inlet side of an air damper. The thermistor is an input to a controller that opens and closes the damper depending on the temperature.

4.2.8 Electronic Controls

Modern comfort-system thermostats are a perfect example of electronic controls that now dominate many aspects of HVACR system control. The thermostat shown in *Figure 54* is fully programmable. This means that operating schedules for time and temperature can be made for each day of the week. In addition, an installer menu allows it to be configured for a variety of applications and operating preferences. It is also Wi-Fi® enabled, allowing the user to access basic thermostat settings and functions from any internet-connected device, such as a smartphone, tablet, or computer.

In many of today's HVACR products, electromechanical control devices like switches and relays have been replaced or combined with electronic controls. Electronic circuits use solid-state timing, switching, and sensing devices to control loads and protective circuits. Electronic circuits operate on DC voltage rectified from an AC power input to the board. Electronic circuits are reliable because they have no moving parts.

Most electronic circuits consist of micro-miniature components mounted on sealed modules or printed circuit (PC) boards. One example, an integrated furnace control (IFC), is shown in *Figure 55*. The copper ribbon circuits on the back of the board serve the same purpose as wiring, providing the conductive path between components.

However, note that traditional electromechanical relays are often found on the boards. These relays have DC solenoid coils, but they are otherwise like an AC-operated relay used to complete a control circuit. Note that part of the

Figure 53 Thermistor in service.

Figure 54 Programmable, Wi-Fi® enabled electronic thermostat.

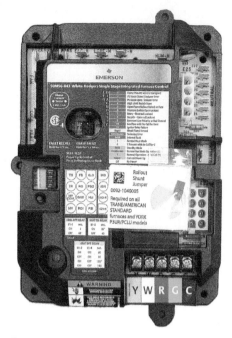

Figure 55 Integrated furnace control (IFC).

board is sealed for its protection, and all external connections are clearly identified. Many such controls are configured at installation for compatibility, using microswitches or tiny jumpers across terminals.

Programmable controllers (*Figure 56*) provide a wider range of programmability than the average electronic control. They are equipped with microprocessors like those of a standard computer. A programmable controller can receive a large amount of data and then make decisions based on its programming. They offer versatility and precision in handling multiple control actions and processes.

Microprocessor-controlled systems can be self-diagnosing. When there is a problem, the microprocessor evaluates information from sensors distributed around the system to determine the source. It may also test itself in various ways. A digital readout, flashing light codes, or another method of communication is then used to tell the technician what function is inoperative and/or what component is not responding as expected.

In the field, PC boards are usually treated as black boxes. In other words, if the technician finds that a function of the board or controller has failed, the entire device is replaced rather than repaired. Unlike conventional circuits, it is not necessary to analyze the circuit to figure out which component has failed.

Many of these devices are quite valuable. Some PC boards and controllers can be turned in for partial credit against a new part, and the manufacturer will then refurbish the failed device. Electronic controls are often condemned without proper testing due to improper troubleshooting techniques. Their operation is often misunderstood as well. It is essential to have the right information on hand when testing or troubleshooting any electronic device. Don't be too quick to condemn them, as they are very reliable.

4.3.0 Electrical Diagrams

HVACR technicians encounter a variety of electrical diagrams. The diagrams provided in this section will allow you to make a connection between various components and their appearances on electrical diagrams. One thing that makes them challenging is poor standardization in their appearance among manufacturers.

An electrical diagram is like a road map. If it can be read and interpreted, a technician can determine how the components are supposed to act when the system is running properly. It also enables a technician to quickly identify where to begin troubleshooting. Diagrams supplied by manufacturers come in a variety of formats. *Figure 57* provides two examples, both of which might be provided inside a piece of equipment. It shows a connection diagram and a ladder diagram.

A connection diagram shows how the wiring is physically connected. It also may show the color of each wire, which is not generally shown on ladder diagrams. The wiring diagram is helpful for troubleshooting or tracing wiring.

In a ladder diagram, the wire colors and physical connection information are removed, and the pictorial views of some components are replaced with standard electrical diagram symbols. This diagram makes it easier to envision the flow of power and trace circuits. The power source is presented as the upright legs of the ladder, and each load line together with its related control devices is represented as a rung of the ladder. Electrical diagrams of all types generally have legends that identify each of the components by their abbreviations.

The diagrams provided by the manufacturer may include a variety of features. *Figure 58* is an example of wiring information from a manufacturer. They often include ladder diagrams, connection diagrams, and/or component-location diagrams. The legend and notes are provided at the bottom. Similar diagrams are often pasted on the inside of a unit access panel.

Ladder diagram: A simplified schematic diagram in which the load lines are arranged like the rungs of a ladder between vertical lines that represent the voltage source.

Figure 56 Microprocessor-based programmable controller.

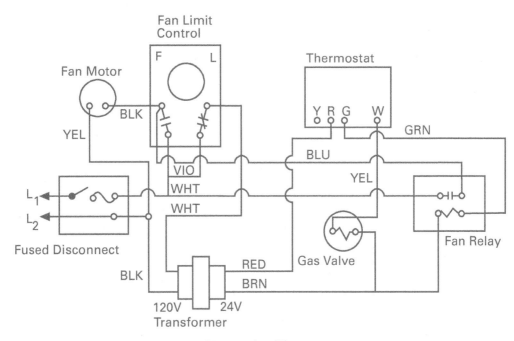

Connection Diagram

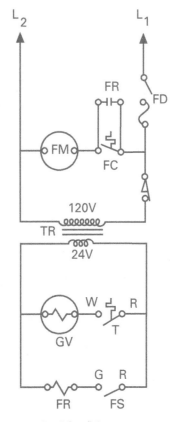

Ladder Diagram

Legend					
FC	Fan Limit Control	FR	Fan Relay	GV	Gas Valve
FD	Fused Disconnect	FS	Thermostat Fan Switch	TR	Transformer
FM	Fan Motor				

Figure 57 Examples of electrical diagrams.

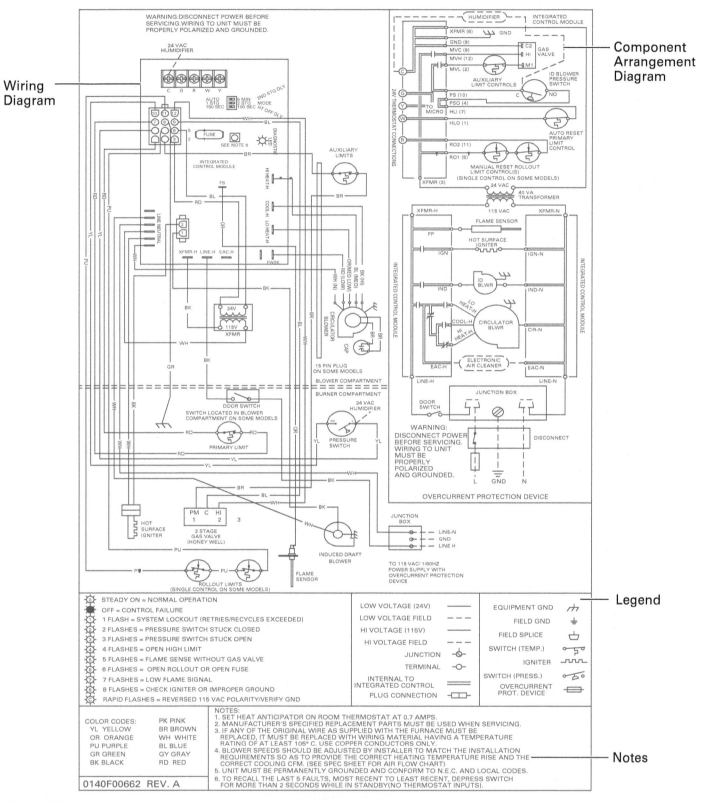

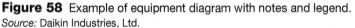

Figure 58 Example of equipment diagram with notes and legend.

Source: Daikin Industries, Ltd.

Hopscotch Troubleshooting

Each switch or set of relay contacts in a circuit represents a condition that must be met in order to energize the compressor contactor (CC). This must occur before the compressor can operate. If the thermostat is calling for cooling and the compressor is not running, it is likely that something is keeping the control circuit open. When you are troubleshooting the circuit with a voltmeter, start by placing your meter probes across the entire circuit, as shown in the illustration. This verifies that voltage is applied to the circuit. Then, move the hot (red) probe to the next component in the series chain (position 2) and read the voltage. Keep doing this until the meter registers no voltage. The last component you jumped (or its related wiring) will be the defective component. This technique is referred to as *hopscotch troubleshooting*.

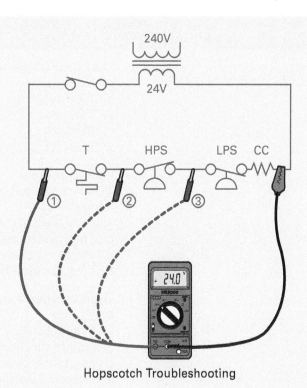

Hopscotch Troubleshooting

4.0.0 Section Review

1. Which of the following devices is a load that converts electrical energy into mechanical energy?
   a. Transformer
   b. Electric heater
   c. Motor
   d. Light bulb

2. Which of the following is a switching device commonly used to control the flow of power to a compressor?
   a. Contactor
   b. Relay
   c. Transformer
   d. Solenoid coil

3. The type of circuit diagram that has the power source shown as upright legs with connections between them is a _____.
   a. component-location diagram
   b. wiring diagram
   c. ladder diagram
   d. simplified schematic

1. Central air conditioning systems in homes generally require a power source of _____.
   a. 1,500 volts
   b. 750 volts
   c. 240 volts
   d. 120 volts

2. What characteristic do devices like flashlights and smartphones that use DC power instead of AC power, what characteristic do they often have in common?
   a. They often operate at a much higher voltage than AC devices.
   b. They usually require more current when they operate.
   c. They are battery-powered and mobile.
   d. They are usually used to produce supplemental heat.

3. The *worst* effect a current of 150 mA is likely to have on the human body is _____.
   a. a slight tingling sensation
   b. a slight shock and involuntary action
   c. painful shock and loss of muscular control
   d. extreme pain and possible death

4. One milliamp is equal to _____.
   a. $1/10$ of an amp
   b. $1/100$ of an amp
   c. $1/1{,}000$ of an amp
   d. $1/1{,}000{,}000$ of an amp

5. Class 0 (red tag) rubber gloves are used when working with voltages less than _____.
   a. 500 volts
   b. 1,000 volts
   c. 5,000 volts
   d. 7,500 volts

6. How much current is drawn by a 120VAC circuit containing a 1,500W load?
   a. 125 mA
   b. 0.08A
   c. 12.5A
   d. 180,000A

7. How much power is consumed by a 120V parallel circuit that contains a 100Ω load and an 80Ω load and draws 2.7A?
   a. 44W
   b. 80W
   c. 120W
   d. 324W

8. If a load is supplied by 120V and draws 5A of current, how much power does it consume?
   a. 24W
   b. 115W
   c. 125W
   d. 600W

9. The circuit shown in *Figure RQ01* is a(n) _____.
    a. series circuit
    b. parallel circuit
    c. open circuit
    d. short circuit

10. If a circuit containing five 50Ω resistors has a total resistance of 10Ω, the circuit is a _____.
    a. series circuit
    b. parallel circuit
    c. series-parallel circuit
    d. combination circuit

11. Which of the following statements about voltage testing is *true*?
    a. When accuracy is required, use a voltage tester.
    b. Voltage testers are essentially the same as multimeters.
    c. A multimeter is better than a voltage tester for troubleshooting.
    d. Voltage testers generally offer recording functions.

12. Placing the jaws of a clamp-on ammeter around two separate conductors will result in _____.
    a. an inaccurate reading
    b. a more accurate reading
    c. doubling the reading
    d. damaging the meter

13. Measuring current in the microamp (μA) range requires a(n) _____.
    a. clamp-on ammeter
    b. inline current measurement
    c. ohmmeter
    d. megohmmeter

14. The purpose of a legend on a circuit diagram is to _____.
    a. show the history of the drawing
    b. identify the abbreviated diagram components
    c. show the physical locations of components
    d. identify the wire colors used on the diagram

15. Which of these components is considered a resistive load?
    a. A switch
    b. A thermistor
    c. A thermostat
    d. An electric heater

16. A DPDT switch can control _____.
    a. 1 circuit
    b. 2 circuits
    c. 3 circuits
    d. 4 circuits

17. Fuses are rated based on their _____.
    a. length
    b. current capacity
    c. shape
    d. diameter

18. The symbol shown in *Figure RQ02* represents a _____.
    a. transformer
    b. solenoid coil
    c. thermal-trip circuit breaker
    d. magnetic-trip circuit breaker

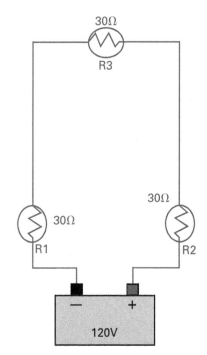

Figure RQ01

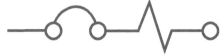

Figure RQ02

19. A device that acts as an electromagnet to position a valve or relay is called a _____.
    a. pressure switch
    b. thermostat
    c. solenoid coil
    d. thermistor

20. A transformer is used to _____.
    a. change electrical energy into mechanical energy
    b. raise or lower a voltage
    c. absorb excess current in a circuit
    d. increase the resistance of a circuit

Answers to odd-numbered Module Review Questions and Calculations are found in *Appendix A*.

Cornerstone of Craftsmanship

Dave Jensen
Operations Manager
Epic Controls LLC

How did you choose a career in the industry?

I was always interested in refrigeration and HVAC as a career. After serving four years in the USAF, I attended a technical college to learn the craft. Soon thereafter, I started working in the HVACR field.

Who inspired you to enter the industry?

I was actually self-inspired. It always seemed like an interesting field to me. I enjoyed working with my hands as well, so it was a good fit for me.

What types of training have you been through?

I've attended many factory schools, including programs offered by Johnson Controls, Carrier, York, and Goodman. I've also completed some training that was offered online.

How important is education and training in construction?

I think it's very important. Once you learn a craft and are competent in the required skills, there will always be a job available for you. The demand for talent in the HVACR craft will always be there.

How important are NCCER credentials to your career?

Although I don't possess training credentials from NCCER, I am an NCCER-certified instructor. NCCER's HVACR program has taught me some new things about our industry, and the NCCER Instruction Certification Training helped me become a better instructor.

How has training/construction impacted your life?

My entire career has been built around the HVACR industry, and I've taken advantage of the many positions that were offered to me.

What kinds of work have you done in your career?

I've worked as a service technician and as a service manager in the craft. I've also served as sales manager and branch manager for a well-known HVACR wholesaler, R.E. Michel.

Tell us about your current job.

I am the operations manager for Epic Controls LLC, an organization that specializes in digital control systems for HVACR systems. I also currently teach the HVACR two-year degree program at Central New Mexico Community College and an HVACR program on behalf of Associated Builders and Contractors (ABC) of New Mexico.

What do you enjoy most about your job?

I really enjoy teaching service and installation technicians as well as our clients.

What factors have contributed most to your success?

To be successful, you must learn to work with other people. I'm sure that my willingness and ability to work with others has been a factor. Attending the factory-sponsored training programs has also been a factor.

Would you suggest construction as a career to others? Why?

Yes, I would certainly recommend it. There are so many different directions you can take. In the HVACR craft alone, you can pursue service, installation, controls, sales, and management. But there are many construction crafts available, and each of those offers different paths. There is something for everyone.

What advice would you give to those new to the field?

Learn your craft and perform your work to the best of your ability. It matters. It's also important to set goals for what you want to learn and accomplish in your career so that you're consistently reminded of where you wish to go and make progress in that direction.

Interesting career-related fact or accomplishment?

Being a certified instructor is an important accomplishment for me. But I'm also proud to be a problem solver—people call me with technical questions all the time. To be seen as a reliable resource is inspiring to me.

How do you define craftsmanship?

The result of learning craft skills and practicing them to the best of your ability.

Answers to Section Review Questions

Answer	Section Reference	Objective
Section 1.0.0		
1. c	1.1.0	1a
2. d	1.2.1	1b
3. b	1.3.0	1c
Section 2.0.0		
1. d	2.1.0	2a
2. a	2.1.1; Figure 16	2a
3. c	2.1.2	2a
4. d	2.2.1	2b
Section 3.0.0		
1. d	3.1.0	3a
2. c	3.2.1	3b
3. c	3.3.0	3c
Section 4.0.0		
1. c	4.1.1	4a
2. a	4.2.4	4b
3. c	4.3.0	4c

Section Review Calculations

Section 2.0.0

Question 3

First, convert the motor's horsepower to watts:

$$\text{Power in watts} = 1.5 \text{ hp} \times 746 \text{ W/hp}$$
$$\text{Power in watts} = 1{,}119\text{W}$$

Use the power in watts and the applied (total) voltage with the power formula to determine the current:

$$I = \frac{P}{E} = \frac{1{,}119\text{W}}{240\text{V}} = \textbf{4.66A}$$

Question 4

Since this is a series circuit, add all of the resistances to determine the total resistance (R_T):

$$R_T = R_1 + R_2 + R_3$$
$$R_T = 30\Omega + 30\Omega + 30\Omega$$
$$R_T = \textbf{90}\boldsymbol{\Omega}$$

Introduction to Heating

Source: iStock@Evgeniy1

Objectives

Successful completion of this module prepares you to do the following:

1. Explain the fundamental concepts of heating and combustion.
   a. Describe the heat transfer process and how temperature, humidity, and heat are measured.
   b. Describe the combustion process and identify common fuels and their characteristics.
2. Identify and describe gas furnaces, their components, and their basic installation and maintenance requirements.
   a. Describe various types of gas furnaces.
   b. Identify and describe common gas furnace components and controls.
   c. Describe the basic installation and maintenance requirements of gas furnaces.
3. Describe hydronic and electric heating systems.
   a. Describe hydronic heating systems and identify common components.
   b. Describe electric heating equipment and identify common components.

Performance Tasks

Under supervision, you should be able to do the following:

1. Identify and describe the function of the primary components in an induced-draft furnace or condensing furnace.
2. Check and record the temperature rise, manifold gas pressure, and flame quality on an operating gas furnace.
3. Using the furnace manufacturer's installation instructions, determine if a furnace installation has the required clearances.

Overview

Most homes and businesses need some type of heating system. Installing and servicing furnaces is a big responsibility. Because flames and combustible fuels are often involved, there is always the potential for fire or explosion. A well-trained technician minimizes those hazards, and fossil fuels are consistently used to heat many thousands of buildings safely and reliably.

There are practical reasons for choosing one heating system over another. When most heating systems are properly installed according to the manufacturer's instructions and serviced by qualified technicians, they will operate satisfactorily for many years.

Digital Resources for HVACR

Scan this code using the camera on your phone or mobile device to view the digital resources related to this craft.

1.0.0 Heating Fundamentals

Performance Tasks

There are no Performance Tasks in this section.

Objective

Explain the fundamental concepts of heating and combustion.

a. Describe the heat transfer process and how temperature, humidity, and heat are measured.

b. Describe the combustion process and identify common fuels and their characteristics.

People rely on heating systems in homes, schools, and places of business to maintain comfortable conditions. There are many types of heating systems. Many of them use natural gas or oil as a fuel, while many others use electricity as the energy source.

Before introducing the various heating systems, it is important to first understand basic heating terms and processes. Two basic processes you'll need to be familiar with are *heat transfer* and **combustion**.

Combustion: The process by which a fuel is burned in the presence of oxygen.

1.1.0   Heat Transfer

For heat transfer to occur, there must be a difference in temperature between two objects. A larger difference in temperatures results in more heat being transferred, and at a more rapid pace. Heat always moves from a warmer region to a cooler region. Any object, including the human body, transfers heat if a surface or the air around it is cooler than the object.

The reason humans feel cold in the winter is because the body is losing heat to the cooler air around it. To keep the heat from leaving us, there are two choices: wear layers of insulation (clothing) or heat the surrounding air so there is a smaller difference in temperature. In hot weather, your body feels hot because it cannot readily transfer heat to the warmer air. At that point, the body cools itself through the evaporation of moisture from the skin.

Heat can be transferred through **conduction**, **convection**, and **radiation**. Heating systems use one or more of these methods to warm the air in the conditioned space, as shown in *Figure 1*.

In addition to these three methods of heat transfer, this section will also review the concepts of temperature, humidity, and heat measurement.

Conduction: The process by which heat is directly transferred between materials in contact when there is a difference in temperature.

Convection: The air movement caused by the tendency of hotter air to rise and colder, denser air to fall, resulting in heat being transferred.

Radiation: In the context of heat transfer, the transmission of heat by electromagnetic waves passing through air; also referred to as *thermal radiation*.

Thermal conductivity: The measure of a material's capacity to conduct heat, expressed as the volume of heat movement over some unit of time. Metals typically offer high levels of thermal conductivity, while materials like spun fiberglass and foam insulation do not conduct heat well.

1.1.1 Conduction

Conduction is the flow of heat from one material to another through direct contact. The rate at which a material transmits heat is known as its **thermal conductivity**. The amount of heat transferred by conduction is determined by the size of the contact area, the conductivity of the materials, and the temperature difference between the materials. A simple example of conduction is the transfer of heat from hot coffee to a cold cup. The cup becomes warmer while the coffee becomes colder.

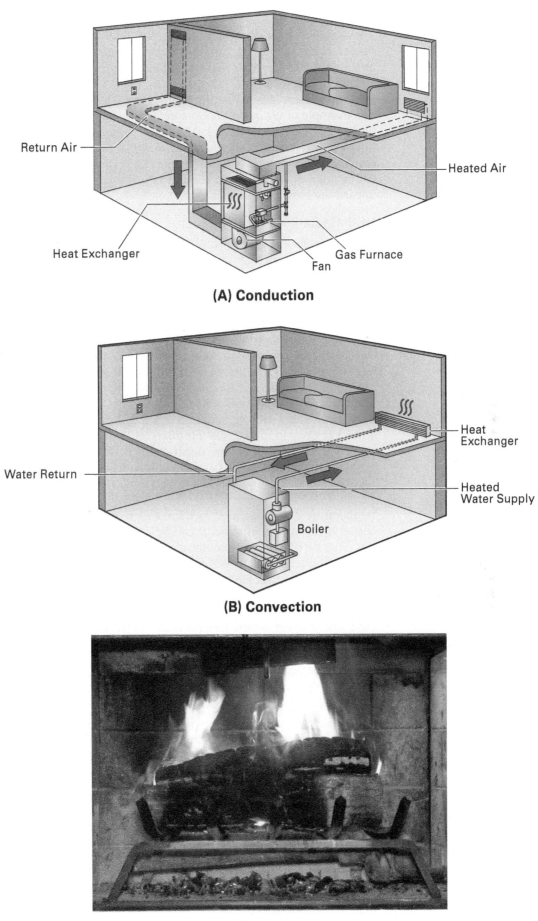

(A) Conduction

(B) Convection

(C) Radiation

Figure 1 Methods of heat transfer.

1.1.2 Convection

Convection occurs through natural air motion due to warmer air rising and denser, cooler air falling. For convection to occur, there must be a difference in temperature between air masses. A larger temperature difference means more air movement.

Hot-water and steam heating terminals, such as the one shown in the convection example of *Figure 1*, rely on convection. The **heat exchanger**—a baseboard unit in this case—warms the air around it, causing it to rise and establish airflow. As it rises, cool air moves in to replace it, and consistent convection is established. The air leaving the terminal gives up its heat to the cooler room air, furnishings, and walls. As it cools, it falls back toward the floor and heads back toward the source of heat.

Heat exchanger: A component used to transfer heat between fluids while keeping the fluids physically separated.

1.1.3 Radiation

Radiation refers to the transfer of heat through space by electromagnetic waves. Through radiation, heat can pass from one object to another without warming the air between them. The amount of heat transferred by radiation depends upon the area of the radiating body, the temperature difference between the two objects, and the distance between the heat source and the object being heated.

The heat from the sun or a fireplace is a good example of radiant heat. Radiant heat tends to be more intense nearest the source. It heats only solid objects directly in its path. When you are facing a campfire, for example, the front of your body can be hot while your back is cold.

One of the most common and effective applications of radiant heat is the tubular gas-fired radiant heater (*Figure 2*). This equipment is highly effective in large, open areas such as automotive garages and warehouses. Heating the air is not a priority in these areas. Humans and other solid objects, however, can be kept comfortably warm. This significantly limits the heat lost when large doors are opened. Smaller models are used on patios and porches.

Figure 2 Gas-fired, tubular radiant heater.
Source: Copyright Detroit Radiant Products Co

1.1.4 Temperature Measurement

Temperature can be defined as the measure of heat present in a substance, reported on a numerical scale. It can be accurately measured using a thermometer. There are two thermometer scales in frequent use. One is the US standard system where temperature is measured in *Fahrenheit*. The other scale is called the *Celsius*, or centigrade, scale. It is important to understand both scales since both are used in various parts of the world. Refer to *Figure 3*.

The Celsius thermometer is the easier of the two to comprehend. The freezing point of water on the Celsius scale is 0°C and the boiling point of water is 100°C. These temperatures are valid at sea level. At lower or higher altitudes, these values change slightly. The freezing point of water on the Fahrenheit scale, again at sea level, is 32°F and the boiling point of water is 212°F. To reduce confusion about which scale is in use, a capital C or a capital F is placed after the degree symbol (°).

There are times when you will need to change from one temperature scale to the other. There are many apps available for this purpose that make it very simple. They are all based on the following formulas:

$$°C = \tfrac{5}{9}(°F - 32)$$
$$°F = (\tfrac{9}{5} \times °C) + 32$$

To make calculator use easier, you can substitute 0.555 for $\tfrac{5}{9}$ in the first equation. In the second equation, you can substitute 1.8 for $\tfrac{9}{5}$. These changes will make little or no difference in the result for your purposes as a technician:

$$°C = 0.555(°F - 32)$$
$$°F = (1.8 \times °C) + 32$$

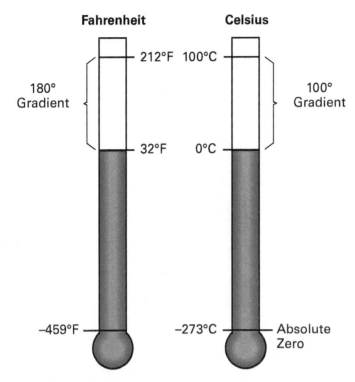

Figure 3 Fahrenheit and Celsius temperature scales.

1.1.5 Humidity Measurement

Convective heat transfer is affected by the moisture content of the air. You often hear the term **relative humidity** in weather reports. This term describes the volume of moisture in the air relative to its water-holding capacity. A relative humidity of 50 percent means that the air contains half the moisture it is capable of holding at its current temperature.

Air reaches its **dew point** when its relative humidity reaches 100 percent, meaning that it is saturated and can't hold any more moisture. If the air cools below its current temperature, the moisture literally begins falling out of it (condenses), forming what we refer to as dew.

Warm air can hold more water than cool air. As a result, if the temperature of a volume of air increases, without any change in its actual moisture content, the relative humidity falls. For example, 80°F (27°C) air with a relative humidity of 50 percent contains 76 grains of moisture per pound of dry air. Raising the temperature of the air to 85°F (29°C), without adding or removing any moisture, results in a relative humidity of 42 percent. This is because the 85°F air can hold more moisture than air at 80°F, so it contains a lower percentage of its total capacity.

The body feels cooler when the humidity is low because the evaporation rate of perspiration from the skin increases, causing a greater cooling effect. Since heat is required to change perspiration to a vapor, that heat comes from your body. Therefore, homes are often equipped with a humidifier to add moisture during the winter months when the air is dry, increasing the feeling of warmth.

HVACR technicians often measure the humidity in a space using a *sling psychrometer* or an *electronic psychrometer*. The sling psychrometer (*Figure 4*) is a simple mechanical device equipped with two identical thermometers. One is called a *dry-bulb thermometer*, while the other is called a *wet-bulb thermometer*. The only difference is that the bulb of the wet-bulb thermometer is pushed inside a wetted wick. A small amount of water is added to the reservoir on the end, and the wick keeps the bulb of the thermometer moist. The design of the tool allows you to extend the thermometer assembly and rotate it rapidly in the air.

Relative humidity: The ratio of moisture present in a given sample of air compared to the volume of moisture it can hold when saturated (at dew point), expressed as a percentage.

Dew point: The temperature at which air becomes saturated with water vapor (100 percent relative humidity).

(A)

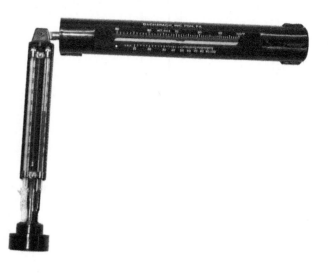

(B)

Wet-Bulb
Temperature
(°F, WB)

Wick

Dry-Bulb Temperature
(°F, DB)

Water Reservoir
(Screw Cap for Filling)

Figure 4 Sling psychrometer.
Source: Courtesy of Bacharach, Inc. (4B)

Psychrometric chart: A comprehensive
chart that presents the important
properties of air and water vapor as they
relate to each other.

Figure 5 Electronic psychrometer.
Source: Reproduced with Permission, Fluke
Corporation

The dry-bulb thermometer reads the actual air temperature. The wet-bulb thermometer is affected by the evaporation of water as it rotates in the air and will cool as a result. By comparing the wet-bulb temperature to the dry-bulb temperature, the relative humidity can be determined. The psychrometer has scales built in to determine the relative humidity. You can also use a **psychrometric chart** to determine the humidity. *Psych charts*, as they are often called, will be covered later in your training. As primitive as this device may sound, it is quite accurate and based directly on the science behind air and its relationship with water.

An electronic psychrometer (*Figure 5*) is easier to use, but it may require periodic calibration to ensure accuracy. Sling psychrometers never require calibration. However, electronic models can provide more functions. Most can also measure air temperature and report the dew point of the air without the use of charts.

1.1.6 Heat Measurement

The unit of measurement for heat in the US standard system is the British thermal unit (Btu). One Btu is the amount of heat required to raise the temperature of one pound of water by 1°F (at sea level). This is an important definition to remember.

It is also important to be aware of the *rate* of heat production; that is, how fast heat is produced. The unit for this is *Btus per hour* (*Btuh*). For example, the capacity of a furnace may be stated as 110,000 Btuh.

Furnaces are largely selected based on their Btuh capacity. Actually, a fossil-fuel furnace has two ratings. The *input rating* is the actual amount of heat contained in the volume of fuel admitted to the burner. The *output rating* reflects the amount of usable heat that is produced. There are losses related to the combustion of the fuel, such as the amount of heat that exits through the furnace vent. Therefore, the input rating will always be higher than the output rating.

The difference between the input and output reflects the efficiency rating of the furnace. For example, if a furnace has an input rating of 150,000 Btuh and an efficiency rating of 90 percent, its output rating is 135,000 Btuh (90 percent of 150,000 Btuh). The remaining 10 percent is lost as the byproducts of combustion are vented outdoors, and a significant amount of heat goes with them.

1.2.0 Combustion

Combustion refers to the burning of a fuel. In the HVAC industry, we burn fuel to extract the heat. You may recall from fire safety training that there are three components necessary for combustion to take place:

- *Fuel* — The fuel can be a gas, such as natural gas; a liquid, such as fuel oil; or a solid, such as coal. Two elements that all fuels have in common are carbon and hydrogen.
- *Heat* (a source of ignition) — A pilot burner, an electric spark, or a **hot surface igniter (HSI)** can be used to ignite a gas burner, while a continuous electric arc is typically used to ignite fuel oil.
- *Oxygen* — Oxygen must be present for the combustion of fuel to take place.

Combustion is a chemical reaction between fuel, heat, and oxygen; therefore, all three components must be present. Natural gas will burn as long as a source of ignition and oxygen are both present. Liquid fuel oil is **atomized** (converted to a fine spray) before it is burned. This ensures an effective ratio of oxygen and fuel.

Hot surface igniter (HSI): A ceramic device that becomes extremely hot when current is passed through it, used to ignite a fuel/air mixture.

Atomized: Broken into tiny pieces or fragments, such as a liquid being broken into tiny droplets to create a fog-like spray.

1.2.1 Incomplete Combustion and Carbon Monoxide

Complete combustion takes place when carbon combines with oxygen to form carbon dioxide (CO_2). CO_2 is nontoxic and is released to the atmosphere. Hydrogen combines with oxygen to form water vapor (H_2O), which is also released to the atmosphere from a fossil-fuel appliance.

Incomplete combustion results from a lack of oxygen and causes undesirable products to form. These include carbon monoxide (CO) and pure carbon (soot), as well as other compounds. CO is a toxic gas. Soot coats the surfaces of the combustion chamber and heat exchanger, reducing heat transfer.

Since CO is odorless, tasteless, colorless, and non-irritating, it can be very difficult to detect without special instruments. As a result, it causes more fatalities than you might expect. While CO is odorless, *aldehydes* formed during incomplete combustion have a distinct, sharp odor that irritates the membranes of the nose and throat. If aldehydes are present, it is virtually certain that CO is also present. On the other hand, the absence of noticeable aldehydes does not mean that CO is also absent. The odor of aldehydes can alert both consumers and service technicians to a problem, but it can't be depended upon as an alarm.

CO enters the bloodstream through the lungs. The symptoms of CO inhalation include confusion, lightheadedness, vertigo, and headaches. Flu-like symptoms are often reported, leading people to believe their health issue may be viral or bacterial. Severe cases of inhalation lead to death. CO can even have severe effects on an unborn child, and it affects animals too. Long-term exposure to CO can lead to memory loss and depression.

HVACR technicians must take CO very seriously and maintain furnaces in peak operating condition. Consumers rely on their technician for protection against this invisible hazard. CO poisoning can often be prevented with the use of CO detectors (*Figure 6*), and technicians can help customers by encouraging their installation. The National Fire Protection Agency (NFPA®) recommends installing a CO detector on each floor of a home with fossil-fuel equipment and outside each sleeping area. CO detectors may be required by code in some areas.

Fossil-fuel equipment must be adjusted so that complete combustion takes place. Sufficient air must be provided to the burner for proper combustion. Furnaces are normally adjusted to provide from 5 percent to 50 percent excess combustion air to guard against incomplete combustion.

1.2.2 Combustion Efficiency

When fuel is burned in a furnace, a certain amount of heat is lost through the vent, or *flue*, that carries combustion byproducts out of the building. Venting is, of course, necessary, but the losses should be kept to a minimum. This allows the

Figure 6 Carbon monoxide (CO) detector.
Source: iStock@dehooks

furnace to provide its maximum efficiency. Air entering a combustion chamber at room temperature or below is quickly warmed to flue gas temperatures that range from 100°F to 600°F (38°C to 316°C), depending upon the furnace design.

Technicians can determine the efficiency of a furnace based on the temperature and CO_2 content of the flue gases. Knowing the efficiency of the furnace also makes it possible to calculate the actual furnace output. Combustion analysis equipment (*Figure 7*) can be used by service technicians to analyze the content of the flue gases from both gas- and oil-burning systems to determine the actual combustion efficiency.

The industry standard for defining furnace efficiency is **Annual Fuel Utilization Efficiency (AFUE)**. The AFUE considers operating efficiency as well as combustion efficiency on an annualized basis. In the 1990s, most **natural-draft furnaces** on the market were replaced by **induced-draft furnaces**. These gas furnaces use a small blower to help move combustion byproducts through the heat exchanger and flue vent. This allowed engineers to design more complex and effective heat exchangers, increasing the AFUE.

The increase in efficiency primarily results from the heat exchanger having more surface area in contact with the hot air. With a blower to help force air through, the air path through the heat exchanger can be longer and more complex, resulting in improved heat transfer. Induced-draft furnaces generally have AFUEs that range from 78 percent to 85 percent.

The **condensing furnace** has the highest AFUE for gas furnaces—90 percent or greater. These furnaces use a special coil as a secondary heat exchanger to extract additional heat from the flue gases before they are vented outdoors. The maximum AFUE reached is around 98 percent, which means that all but 1.5 percent of the heat available is delivered to the conditioned space. In addition to secondary heat exchangers, these furnaces use smart gas valves and advanced electronic controls to achieve an exceptional level of efficiency. Since most of the heat is removed from the flue gases, the vent piping can be constructed of plastic. This is certainly never possible with natural- or induced-draft furnaces, which typically have flue gas temperatures of 400°F (~200°C) or more.

Over the years, the US Department of Energy (DOE) has made a number of changes to the national efficiency standards. To accommodate the needs of different areas, the US states have been broken into three regions—the North, the South, and the Southwest regions. In the North, most new gas furnaces are currently required to have an AFUE of 90 percent. This is because they are expected to operate far more than they do in the South and Southwest regions.

Annual Fuel Utilization Efficiency (AFUE): Industry standard for defining furnace efficiency, expressed as a percentage of the total heat available from the relevant fuel. The AFUE considers the thermal efficiency of a unit as well as the losses of efficiency that occur during startup, warmup, and shutdown.

Natural-draft furnaces: Gas furnaces that rely on the natural tendency of hot air to rise to support combustion and assist the passage of combustion byproducts through the heat exchanger.

Induced-draft furnaces: Gas furnaces in which a motor-driven fan creates (induces) a draft through the heat exchanger, assisting the passage of combustion byproducts and enabling more effective heat exchanger designs.

Condensing furnace: A furnace equipped with a secondary heat exchanger that extracts additional heat from flue gases that would otherwise be vented to the atmosphere.

Figure 7 Combustion analysis equipment.
Source: Bacharach products make the heating, ventilation, air-conditioning, refrigeration (HVAC-R), and process

Since induced-draft furnaces cannot reach a 90 percent AFUE, condensing furnaces are required. In the South, new furnaces must have an AFUE of only 80 percent. If the furnace is part of a packaged heating and cooling unit that is installed outdoors (identified as *weatherized equipment*), an AFUE of 81 percent is required, regardless of state.

1.2.3 Flame Characteristics

The type of flame and the intensity with which it burns have a direct relationship with the efficiency of a heating unit. Oil burners operate with a yellow flame. Natural-draft gas burners have a blue flame with slight yellow/orange tips, while the flame from induced-draft models is blue with very little or no colored tips at all. The difference is mainly due to the way air is mixed with the fuel.

A yellow flame in a gas burner is a sign of incomplete combustion. The burner should be checked for a sufficient supply of air. A proper gas flame is produced when approximately 50 percent of the combustion air (**primary air**) is mixed with the gas before ignition. The balance of the air, called **secondary air**, is supplied during combustion to the area above the flame.

Incomplete combustion can be caused by a lack of primary or secondary air, or by contact between the flame and a cool surface. It is also possible to provide too much air to a burner, resulting in a flame that is less than ideal for operating efficiency. However, this does not result in incomplete combustion.

1.2.4 Fuels

Fuels are available in three forms: gases, liquids, and solids. *Table 1* shows the heating values of common fuels. Gases include natural gas, manufactured gas, and liquefied petroleum (LP). Liquids include fuel oil, although kerosene is occasionally used. There are six grades of fuel oil, with No. 2 being the most common for home heating systems. Only large furnaces and boilers use heavier grades of oil, which have higher heating values. Solids include coal and wood. This module focuses mainly on natural gas, which is the heating fuel most likely to be encountered in the United States.

The three common types of gaseous fuels are described further here:

- *Natural gas* — Natural gas is the most common of the gaseous fuels. It comes from the earth and often accumulates in the upper part of oil wells. Natural gas can also be found in formations of sedimentary rock like sandstone and shale. It is colorless and nearly odorless. An odorant, such as a sulfur compound, is added so leaks can be easily detected. The heating value of natural gas varies from 900–1,200 Btus/ft^3, depending on the local supplier. For consistency, most calculations regarding the capacity of gas piping are based on a Btu content of 1,000 Btus/ft^3. The principal component of natural gas is methane.

- *Manufactured gas* — Manufactured gases are combustible gases that are normally produced from solid or liquid fuel and are used mostly in industrial processes. They are produced mainly from coal, oil, and other hydrocarbons, and are comparatively low in heat content. Manufactured gases are not economical space-heating fuels.

Primary air: Air that is introduced into a burner and mixes with the gas before combustion occurs.

Secondary air: Air that is introduced into a burner above or around the combustion process to further support combustion and venting of the combustion byproducts.

TABLE 1 Heating Values of Common Fuels

Fuel	Heat Released
Solids	
Wood	6,200 Btus/lb (avg)
Bituminous Coal	12,000–15,000 Btus/lb
Anthracite (Hard Coal)	13,000–14,000 Btus/lb
Oil*	
Grade 1	137,000 Btus/gallon
Grade 2	140,000 Btus/gallon
Grade 4**	141,000 Btus/gallon
Grade 5	148,000 Btus/gallon
Grade 6 (Bunker C)	152,000 Btus/gallon
Gas	
Natural***	900–1,200 Btus/ft^3
Manufactured	500–600 Btus/ft^3
Liquefied Petroleum (LP)	1,500–3,200 Btus/ft^3

*Grades are determined by ASTM International.
**This grade is not commonly used.
***Check with the local gas company for specific values.

- *Liquefied petroleum (LP)* — LP is a byproduct of the oil-refining process. It is stored in its liquid form under pressure, but it vaporizes when released from storage at a much lower pressure for combustion. There are two types of LP gas: propane and butane. Propane is more useful as a space-heating fuel because it boils at −40°F (−40°C) and can be readily vaporized for heating in a northern climate. Butane vaporizes at about 32°F (0°C). Propane has a heating value of about 2,500 Btus/ft^3, while butane has a heating value of approximately 3,200 Btus/ft^3. Although it is not commonly used in comfort systems, butane does have the highest heat content per cubic foot. LP gas is usually propane with a small amount of butane added.

WARNING!

Propane and butane vapors are considered more dangerous than those of natural gas because they have a higher specific gravity. This means the vapors are heavier than air and tend to accumulate near the floor and in low spots, increasing the danger of an explosion if an ignition source is present.

LP gas is an alternative in areas where natural gas is not available. Since natural gas is delivered in its vapor form by pipeline, many rural areas do not have access to it. Manufacturers of gas appliances make LP conversion kits that adapt natural gas furnaces for LP use. The kits contain replacement burner **orifices** and different springs that are installed in the gas valve. The springs change the output pressure of the gas valve.

Orifices: Small, precisely drilled holes.

NOTE

Oil furnaces are fairly common in some parts of the United States, most notably in rural areas. Oil heating and burners are covered later in the HVACR program.

Did You Know?

Home Energy Consumption

The US Energy Information Agency tracks information related to energy use in residences across the United States. The graph shown here reflects residential energy consumption, by energy source, from 1950 through 2020. The use of both electricity and natural gas have grown substantially over the years. The use of fuel oil has contracted, but its use, as well as the use of renewable sources, has remained relatively stable over the years.

The Energy Information Center offers a great deal of helpful energy-related information that goes far beyond the residential environment. For more information, visit their website at **https://www.eia.gov/**.

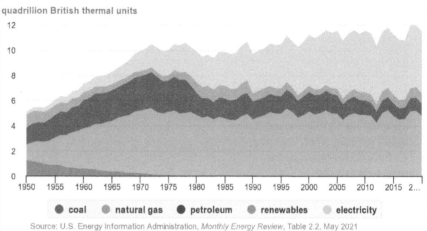

U.S. residential sector energy consumption by energy source, 1950 to 2020

quadrillion British thermal units

● coal   ● natural gas   ● petroleum   ● renewables   ● electricity

Source: U.S. Energy Information Administration, *Monthly Energy Review*, Table 2.2, May 2021
Note: Electricity excludes losses in electricity generation and delivery. Petroleum includes heating oil, liquefied petroleum gas (propane), and kerosene. Renewables includes wood, geothermal energy, and solar energy.

Source: U.S. Energy Information Administration (October 2020)

1.0.0 Section Review

1. The type of heat you experience when sitting around a campfire is _____.
   a. heating by convection
   b. heating by conduction
   c. heating by radiation
   d. heating by light energy

2. The *most* common gaseous fuel for furnaces in the United States is _____.
   a. manufactured gas
   b. propane
   c. butane
   d. natural gas

2.0.0 Gas Furnaces

Performance Tasks

1. Identify and describe the function of the primary components in an induced-draft furnace or condensing furnace.

2. Check and record the temperature rise, manifold gas pressure, and flame quality on an operating gas furnace.

3. Using the furnace manufacturer's installation instructions, determine if a furnace installation has the required clearances.

Objective

Identify and describe gas furnaces, their components, and their basic installation and maintenance requirements.

a. Describe various types of gas furnaces.
b. Identify and describe common gas furnace components and controls.

c. Describe the basic installation and maintenance requirements of gas furnaces.

Forced-air gas furnaces are the most common method of heating in the United States. Gas furnaces can be installed in attics, basements, equipment rooms, or crawl spaces.

In a forced-air gas furnace, return air from the space being heated is passed over a heat exchanger where the heat from combustion is transferred to the air. The air is then sent to the space through the supply ductwork.

A forced-air furnace uses a blower to move the air over the heat exchangers and circulate it through the distribution system (return and supply ductwork). If cooling is installed with a forced-air furnace, the cooling coil is normally placed in the airstream at the outlet of the furnace. Since the air leaving a cooling coil is generally at or near its dew point, positioning the coil at the furnace outlet avoids sending moist air through the blower and across the furnace heat exchanger.

2.1.0 Types of Gas Furnaces

There are four basic configurations of forced-air gas furnaces. An upflow furnace (*Figure 8*) takes room air in through the bottom and discharges it from the top. The blower is located below the heat exchanger.

A horizontal furnace (*Figure 9*) is used in attics or crawl spaces where the height of the furnace must be kept to a minimum. Air enters at one end of the unit and moves horizontally over the heat exchanger, exiting at the opposite end.

The lowboy furnace (*Figure 10*) occupies more floor space but is shorter than the upflow design. It is best suited for basement installations with limited headroom. Both the return duct and supply duct are connected to the top of the unit. Lowboy furnaces are almost exclusively oil-fired units.

The counterflow, or *downflow*, furnace (*Figure 11*) is commonly used in structures with an under-floor air distribution system. The blower is located above the heat exchanger and the return air duct is connected at the top. The supply duct is connected at the bottom. The return air duct is generally located in the attic, with a return air grille in the ceiling. The furnace itself works well for garage or first-floor closet installations.

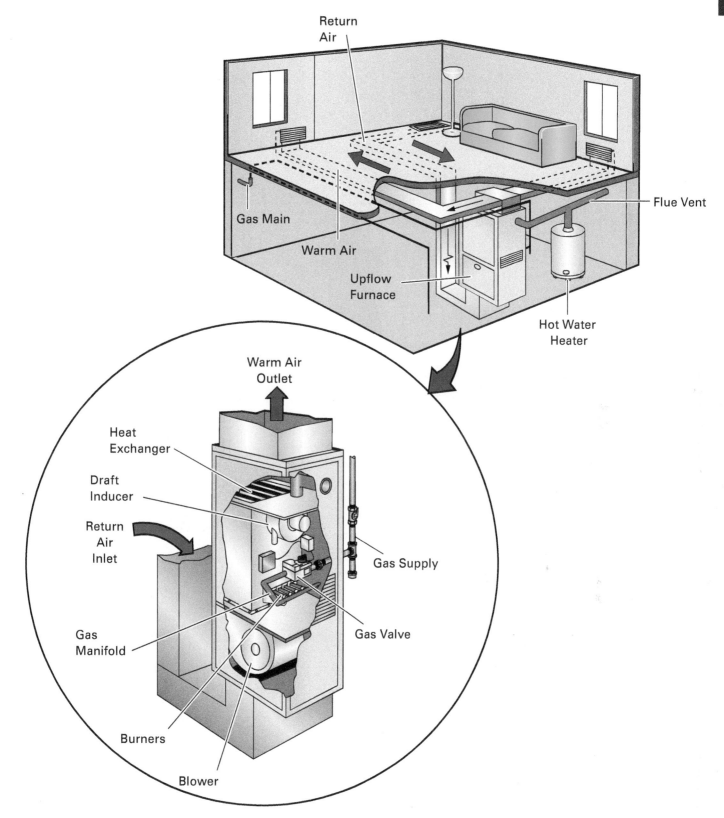

Figure 8 Upflow furnace construction.

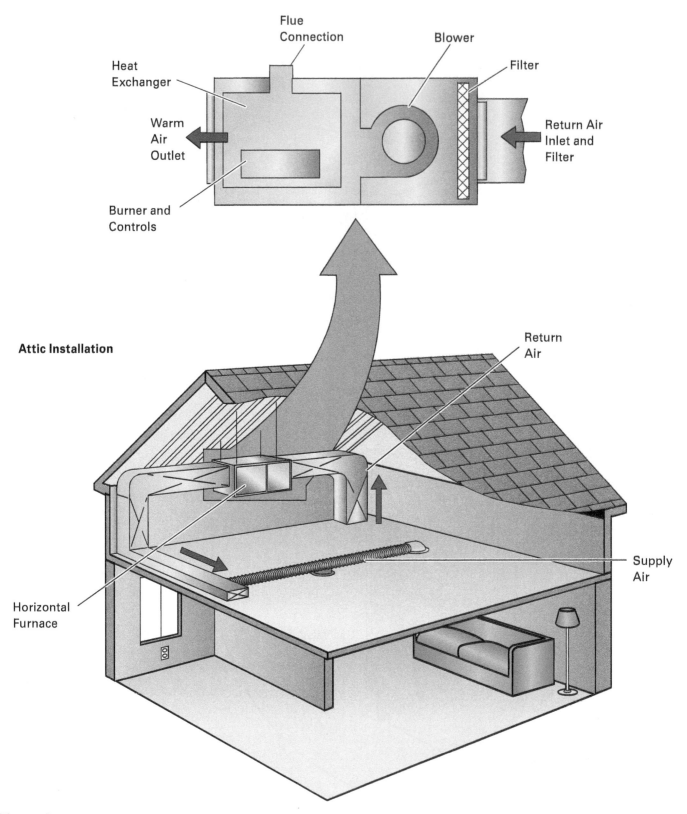

Attic Installation

Figure 9 Horizontal furnace construction.

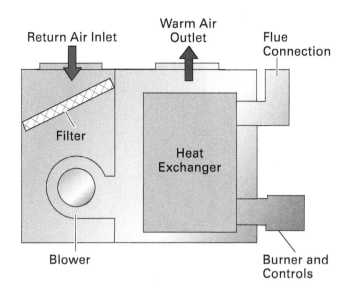
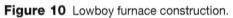

Figure 10 Lowboy furnace construction.

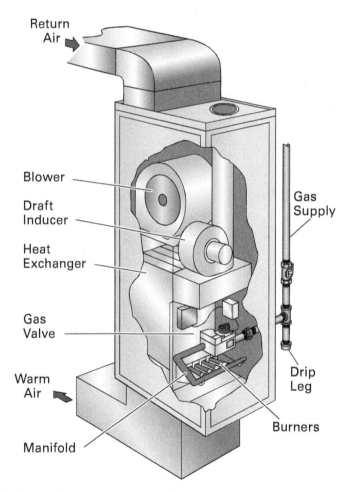

Figure 11 Counterflow furnace construction.

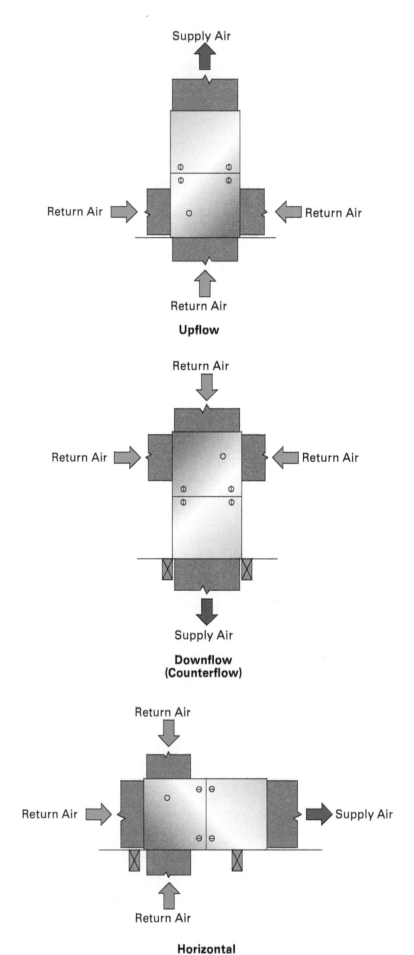

Figure 12 Multipoise furnace installation positions.

The **multipoise furnace** was developed by manufacturers to increase the flexibility of a single furnace model. A multipoise furnace can be configured in the field by the installer for upflow, counterflow, or horizontal use. *Figure 12* shows an example of a multipoise furnace installed in three different positions.

Most multipoise furnaces are shipped from the factory ready for upflow installation. The furnace must be field converted for counterflow or horizontal airflow applications. For most furnaces, the conversion is fairly easy.

Converting a condensing furnace is more complex. Combustion air and vent piping have different attachment schemes depending on the position in which the furnace is installed. Because condensing furnaces also collect **condensate** from the flue gases, the condensate drain and drain trap must be properly arranged for each position. The furnace manufacturer's conversion procedure must be followed carefully.

> **WARNING!**
>
> Failure to precisely follow the manufacturer's installation instructions for multipoise furnaces can be hazardous since venting performance and condensate drainage can be affected. Improper installation can lead to hazardous conditions or furnace failure.

Multipoise furnace: A furnace that can be configured for upflow, counterflow, or horizontal installation, relative to airflow.

Condensate: Water produced by the condensation of water vapor.

2.2.0 Gas Furnace Components and Controls

A modern furnace contains many components and control devices. These include safety controls that stop the flow of gas if the furnace overheats. Components and controls in a gas-fired furnace can be conveniently grouped into those related to airflow; those related to the flow of fuel gas and combustion air; and those related to the combustion process. *Figure 13* shows the interior of a condensing furnace and identifies its major components.

2.2.1 Heat Exchangers

The heat exchanger is where heat transfer takes place, as the name implies. Heat exchangers are heated by the burning gas, with combustion confined to the inside of the heat exchanger. Room air flowing on the outside of the heat exchanger picks up the heat as it passes over it. Of course, the room air and the combustion byproducts must never be allowed to mix. This can happen when a heat exchanger develops a leak due to cracks or corrosion.

Heat exchangers are usually made of a stamped or rolled aluminized-steel product. There are several types of heat exchangers. The types shown in A, B, and C of *Figure 14* are commonly found in gas-fired furnaces. The burners fire directly into the heat exchanger and the hot gases flow through the heat exchanger and out the other end to a collector box that directs the gases to the vent system. Types A and B are found in induced-draft furnaces, where the induced-draft blower helps push combustion air through the more complex passages.

The heat exchangers shown in D and E of *Figure 14* are typical of those found in oil-fired furnaces. The inside of the cylindrical primary heat exchanger (D) is in direct contact with the flame. The secondary heat exchanger (E) extracts additional heat from the hot flue gases. Oil does not burn as cleanly as gas. Therefore, the long, thin heat exchangers used for gas heating would likely become clogged if used in oil-fired furnaces.

The number of heat exchanger sections depends on the furnace capacity. In a gas furnace, each heat exchanger section typically has its own burner in or beneath it. A low-capacity furnace might have two sections fed by two burners, while a high-capacity furnace might have five or more sections. The heat exchanger sections terminate in a single collector box that directs the flue gases into the vent piping.

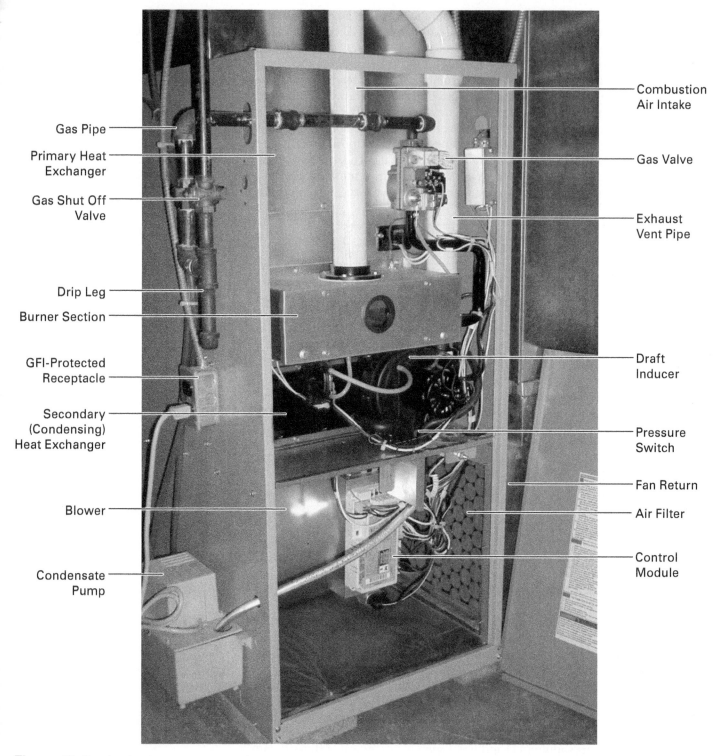

Gas Pipe

Primary Heat Exchanger

Gas Shut Off Valve

Drip Leg

Burner Section

GFI-Protected Receptacle

Secondary (Condensing) Heat Exchanger

Blower

Condensate Pump

Combustion Air Intake

Gas Valve

Exhaust Vent Pipe

Draft Inducer

Pressure Switch

Fan Return

Air Filter

Control Module

Figure 13 Condensing furnace components.

Condensing Furnaces

Condensing furnaces are equipped with a secondary heat exchanger made from stainless steel, or from plain steel with a corrosion-resistant coating. After the hot air from combustion has passed through the primary heat exchanger, where a great deal of heat has already been extracted, it moves on to the secondary heat exchanger. The secondary heat exchanger appears much like a cooling coil. As the flue gases move through the coil, enough heat is extracted from them that water vapor in the gases condenses. Since water vapor must give up heat to condense from a vapor to a liquid, additional heat is extracted from the process.

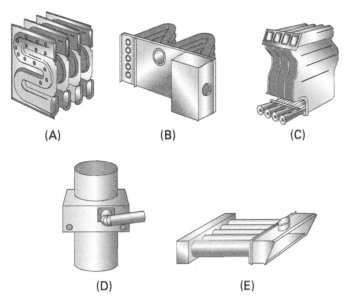

(A)   (B)   (C)

(D)   (E)

Figure 14 Heat exchanger designs.

The condensate that forms must be collected and drained away. It is often acidic and must be disposed of in accordance with local codes. When the condensate cannot be drained away by gravity, a condensate pump is installed, as shown in the bottom left corner of *Figure 13*.

2.2.2 Fans and Motors

Figure 15 shows the flow of combustion air and conditioned air through a basic furnace. The blower wheel and motor, commonly referred to as the *blower assembly*, are responsible for circulating the air through the ductwork. Most furnaces today contain two blowers. The main blower circulates conditioned air through the duct system, and the induced-draft blower pushes or draws combustion air through the heat exchanger and flue vent. The induced-draft blower is not related to the movement of conditioned air through the structure.

Most induced-draft blowers have single-speed motors. However, a variable-speed **electronically commutated motor (ECM)** is often used to power the draft inducer in furnaces that have staged or modulating gas valves. The burner input on these furnaces can vary. The ECM enables the induced-draft blower to operate at different speeds to match the fuel input more closely.

Much of the successful performance of a heating system depends on the proper operation of the blower assembly. It draws return air from the conditioned spaces, then forces it over the heat exchanger and through the distribution system to the space to be heated. Outside air is sometimes mixed with the return air to improve indoor air quality. This was rarely necessary in older homes. However, as construction standards improve and infiltration is reduced, forcing or drawing fresh outdoor air into the structure is becoming more common. For commercial installations, it is required.

Although the goal of a furnace is to heat the air passing through to warm a space, the furnace requires this airflow to take the heat away so that it doesn't overheat. While an excess amount of air flowing through is of little consequence to its operation, insufficient airflow will result in an overheated furnace. This is an important concept to understand, and it applies to all fossil-fuel and electric heating equipment.

Centrifugal blower wheels with forward-curved blades are used in virtually all furnaces. Air enters through both sides of the wheel and is compressed at the outlet by centrifugal force. The volume of airflow is measured in cubic feet per minute (cfm). An exception to the use of a centrifugal blower wheel is found in gas-fired unit heaters. Propeller fans are still often used in these units. They provide reasonable performance when there is no ductwork to resist the flow of air.

NOTE

When condensing furnaces are installed in attics, it may be necessary to provide freeze protection to prevent freezing of the condensate and condensate pump (if equipped).

Electronically commutated motor (ECM): A DC-powered motor that operates at varying speeds based on the programming of its AC-powered electronic control module that also converts AC to three-phase DC. ECMs operate more efficiently than standard motors.

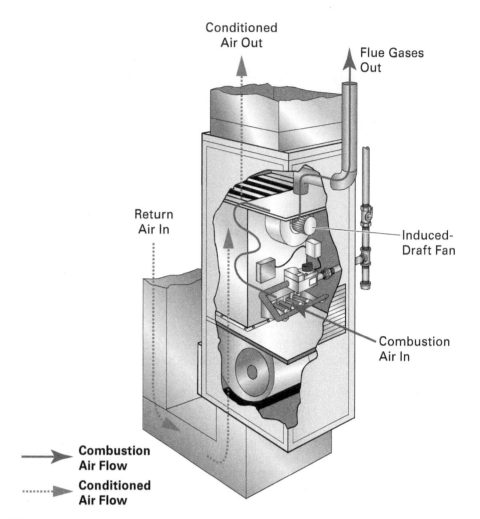

Conditioned
Air Out

Flue Gases
Out

Return
Air In

Induced-
Draft Fan

Combustion
Air In

→ **Combustion
Air Flow**

·····▶ **Conditioned
Air Flow**

Figure 15 Furnace air flow.

Blower wheels and motors must be matched to each other and to the system to deliver the required amount of air. It is often necessary to adjust airflow settings during an installation because the resistance of air distribution systems varies.

Most furnace blowers today are *direct-driven*, meaning that the blower wheel is attached directly to the shaft of the motor (*Figure 16*). Blower speed is often changed by simply changing the speed of a multispeed electric motor. This type of motor has a series of electrical taps (connection points) for the different windings, each of which provides a different rotational speed. The speeds chosen for heating and cooling are often different, with the higher speeds used for cooling.

ECMs that modulate their speed electronically have become common for blowers in furnaces, especially in "deluxe" models. They can provide a more precise air volume under varying conditions that results in quieter and more efficient operation, as well as enhanced comfort.

2.2.3 Air Filters

Air filters can remove dust, pollen, mold, and other particles from the air circulating through the building. Air filters are commonly located at the return duct connection to the furnace, or at the return air inlet(s) located in a central area of the building. This approach provides more filter area and keeps the return air ducts clean too.

Filters are available in different levels of efficiency (*Figure 17*). In this case, *efficiency* refers to their ability to capture small particles. Particle size is measured in *microns*. A micron is $^1/_{1,000}$ of a millimeter, or 0.001 millimeters.

Common filters can remove particles of dust, dirt, pollen, and molds as small as 10 microns. That may sound pretty good, but it also means that many particles, in that size range and smaller, are passing through. Inexpensive

Figure 16 Direct-drive furnace blower assembly.

(A) Electrostatic
Permanent Filter

(B) Pleated Filter

(C) Bag Filter

(D) HEPA Filters

Figure 17 Air filters.
Source: Parker HVAC Filtration (17B–D)

woven-fiberglass filters available in retail stores offer poor filtration and are rarely available today. Many permanent washable filters, which are made of metal and/or polyurethane foam, aren't much better.

High-efficiency particulate arrestance (HEPA) filters can capture particles down to about 0.3 microns. That covers mold, pollen, and most bacteria. However, that level of filtration isn't necessary in common areas. They are used in laboratories, surgical rooms, and similar applications.

Today's most common air filter has a sturdy cardstock frame with a pleated media, as shown in *Figure 17* (B). Bag filters can offer more efficiency as well as a massive amount of surface area for collection. An *electronic air cleaner* can capture very fine particles, including those that make up smoke and vapors. Electronic air cleaners can be obtained as standalone units, or they can be added to a furnace as an accessory.

A dirty filter reduces airflow, sometimes causing comfort problems. The furnace may also run hotter than it would normally. Filters need to be replaced or cleaned periodically. The operating conditions dictate the timing for replacement or cleaning.

Turn off the furnace before removing an air filter. The blower must be disabled while the filter is being replaced, for several reasons. If the filter is located in or adjacent to the blower wheel, you can be injured if it starts unexpectedly. Regardless of its location, some debris is likely to be shaken from the filter as you remove it. If the blower is running, it will quickly pull the debris into the airstream. Remember that the filter is not only designed to protect the occupants and help keep the premises cleaner. They are also there to keep the ducts and the heat-transfer surfaces clean.

Some thermostats are equipped with a "Check Filter" alert. A timer in the thermostat simply tracks the elapsed time and activates the warning at a

NOTE

Air filtration and the related minimum efficiency reporting value (MERV) system is presented in NCCER Module 03204, *Air Quality Equipment*.

specified interval, such as 90 days. The period between reminders is an adjustable option. Technically, the thermostat has no means of determining the condition of the air filter.

Filters must be well-fitted to the cabinet or rack in which they are installed. If the filter fits loosely, dirty air can easily bypass it. If a filter is replaced with one that is significantly more efficient, it will also be more resistant to airflow. Adjustments to the airflow may be required to accommodate it.

2.2.4 Gas Valves

The function of a gas burner is to safely produce complete combustion. To do this, the furnace must regulate the gas flow, create the proper mixture of gas and air, and safely ignite the gas. A common gas burner assembly has four major components: a gas valve; a manifold and orifices; burners; and an ignition device. Each of these components are discussed separately in the sections that follow.

Older systems incorporated a gas control valve and a gas pressure regulator as separate components. The series of valves and controls in a gas supply line is called a *gas train*. Today, these functions are combined into a single gas valve assembly. Modern gas valves are also **redundant gas valves**; that is, they contain two independent valves in series. If one fails, the other will shut off the gas flow. The redundant gas valve assembly shown in *Figure 18* is just one of the many valves found in today's furnaces. Furnace gas valves are carefully engineered and are extremely reliable.

There are several types of automatic gas valves used in equipment. Redundant gas valves are used in virtually all residential and commercial products. Large burners use two or more single-function *solenoid gas valves* (*Figure 19*) to control gas flow, along with independent gas pressure regulators and gas-pressure safety switches.

A *diaphragm-operated valve* (*Figure 20*) uses gas pressure above and below the diaphragm to assist in its movement. When the coil is energized (a solenoid coil is still required to operate the valve), the gas supply to the area above the

Redundant gas valves: Gas-flow control valves that contain two gas valves in series.

NOTE

Commercial and industrial gas trains often have one or more gas pressure regulators as well as multiple gas valves that provide On-Off control of the gas flow to the burner. The structure of the gas train for large burners is often determined by insurance regulations.

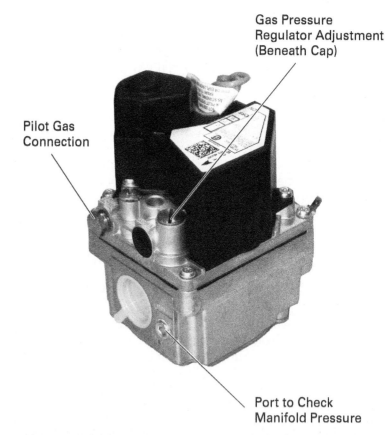

Gas Pressure
Regulator Adjustment
(Beneath Cap)

Pilot Gas
Connection

Port to Check
Manifold Pressure

Figure 18 Redundant gas valve.

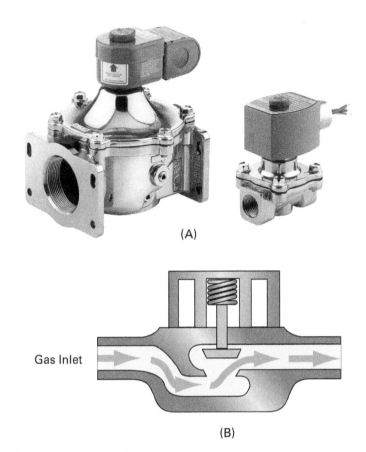

(A)

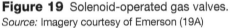

(B)

Figure 19 Solenoid-operated gas valves.
Source: Imagery courtesy of Emerson (19A)

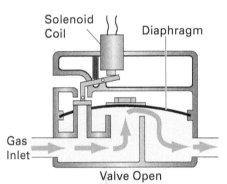

Figure 20 Diaphragm gas valve operation.

diaphragm is closed off and that area is vented to the atmosphere. The pressure of the gas in the lower section then pushes the diaphragm up, allowing gas to flow through the valve. When the coil is de-energized, the upper section is again exposed to the gas pressure. With equal pressure above and below the diaphragm, the natural bow in the diaphragm forces it down to shut off the main gas flow.

Gas valves on some furnaces are designed to control two levels, or *stages*, of heat. The gas pressure at the burner is reduced when only the first, or low, stage of the gas valve is energized. Two-stage furnaces fire only the first stage when the heating demand is small. However, on very cold days, if the temperature does not rise as expected when operating on the low stage alone, the valve switches to high-stage operation. Turning the thermostat up several degrees above room temperature will also initiate high-stage operation. The higher firing rate also helps a system recover quickly from an energy-saving setback in temperature. Separate gas-pressure adjustments can be made for each stage.

Modulating gas valves generally vary the burner input rate from around 40 percent of full capacity up to 100 percent, often in 1 percent increments. This allows the burner input to closely match the actual heating demand. This significantly reduces temperature variations, as the furnace attempts to synchronize itself with the demand.

A modulating gas valve uses a **stepper motor** to control its position. A DC signal is used to tell the stepper motor to change position, and what position to assume. This signal comes from an electronic control that responds to the heating demand from the room thermostat.

2.2.5 Manifold and Orifices

In a gas furnace, gas fuel is supplied at a low pressure to a *manifold*. The manifold pressure is controlled by a pressure regulator inside the gas valve assembly. The typical manifold pressure for a natural-gas, induced-draft furnace is $3\frac{1}{2}$" of water column (w.c.). This means that the gas pressure is sufficient to raise a

Stepper motor: A small DC-powered motor that moves in discrete steps and can be commanded to any position, with each step representing a small fraction of a complete 360-degree rotation.

column of water just $3^1/_2$". This equates to only 0.126 pounds per square inch (psi) of pressure. A furnace operating on propane generally has a gas manifold pressure around 11" w.c.

As the gas leaves the manifold through a series of small orifices and into the burners, it mixes with the primary combustion air. The rate at which gas is supplied to each burner is controlled by the size of the orifice and the applied gas pressure.

The burner manifold (*Figure 21*) delivers gas equally to all the burners in a furnace. It is made of $^1/_2$" to 1" black iron pipe or steel tubing, with threaded openings called **spuds** positioned along its length. Each orifice is screwed into a spud, as shown in *Figure 21*. The orifice, a precision hole drilled into a piece of brass or similar metal, determines how much gas is delivered to the burner. The size of the orifice depends on the type of fuel gas, the pressure in the manifold, and the gas input desired for each burner. They are selected by the equipment manufacturer. A number may be stamped on each spud to show the correct orifice size, usually in thousandths of an inch.

Spuds: Threaded metal fittings that screw into the gas manifold, into which a gas orifice is installed.

Gas Valve Inlet Pressure

Most furnace gas valves are designed for a maximum inlet pressure of only $^1/_2$ psig (~14" w.c.) from the source. Gas mains run at much higher pressures. The pressure must be reduced to a usable pressure at the gas meter for a building. In many cases, the pressure is reduced to 2 psig or 5 psig. It is then reduced again, to just below $^1/_2$ psig, before it enters each appliance (furnace, water heater, etc.). In other cases, the pressure is reduced to slightly less than $^1/_2$ psig at the meter, then fed directly to the appliance gas valve.

Higher gas pressures allow for a greater volume of gas to flow through smaller pipe, and smaller pipe sizes result in lower installation costs. However, the gas pressure must be limited inside occupied structures for safety reasons.

WARNING!

Never change the orifice sizes in a gas burner unless directed by the manufacturer. Changing the orifice size can result in incomplete combustion and unstable operation.

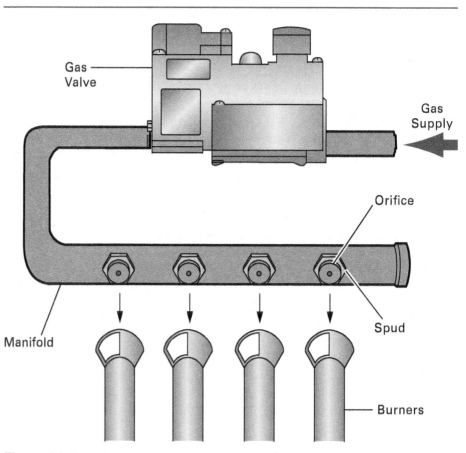

Figure 21 Typical gas manifold.

Gas burners require air to be mixed with the gas before combustion. This air is called *primary air*. Primary air (*Figure 22*) amounts to approximately one-half of the total air required for proper combustion. Too much primary air causes the flame to lift off the burner surface. Too little primary air causes a lazy yellow flame.

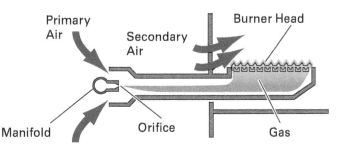

Figure 22 Combustion airflow in a natural-draft burner.

Combustion air supplied to the burner over and around the burning fuel is called *secondary air*. If too little secondary air is present, incomplete combustion can occur. As discussed earlier, most gas burner units are designed for approximately 50 percent of the combustion air to be secondary. For correct flame generation, the proper ratio between primary and secondary air must be maintained. The burner system is specifically designed to do that, with little or no adjustment required.

2.2.6 Gas Burners

Gas burners can be either single-port (in-shot) or multiport. Single-port burners (*Figure 23*), found in induced-draft furnaces, each direct a single flame into the heat exchanger, like a torch. The flame is generated just outside the heat exchanger opening, in the furnace **vestibule**. *Figure 24* provides an example of single-port burners in operation. Note the clean, blue flame.

An induced-draft blower is located near the opposite end of the heat exchanger from the burner. It creates a slight negative pressure in the combustion chamber, drawing in (inducing) combustion air. This negative pressure also draws the flame from the single-port burner into the heat exchanger. One advantage of an in-shot burner design is that it can be positioned in a horizontal or vertical position since it does not depend on a natural draft to operate.

Multiport burners distribute the flame through slots or holes along the length of the burner. Unlike single-port burners, multiport burners are located inside the heat exchanger, as shown in *Figure 22*. This style is found in natural-draft furnaces.

2.2.7 Ignition Devices

There are several approaches to igniting the air-gas mixture in the burner. Refer to *Figure 25*.

Older furnaces used a **standing pilot**, which is a small gas flame that remains lit all the time. When gas begins to flow to the main burner, the pilot flame immediately ignites it. The pilot assembly contains a **thermocouple** that converts the heat from the pilot flame into a small electrical voltage—about 30 millivolts (mV). The voltage generated holds a safety valve open inside the main gas valve. If the pilot flame stops burning, the safety valve closes. This prevents gas from entering the combustion chamber without a verified source of ignition. This style of ignition is still very common in water heaters, but you will rarely encounter it in modern furnaces.

A *thermopile* is used when more voltage is needed. A thermopile consists of thermocouples wired in series. Thermopiles generate about 750 mV. Gas furnaces that have no blower (such as wall and floor furnaces) don't need a power supply and operate using only a thermopile to power the gas valve and its control circuit.

An intermittent ignition approach saves energy because there is no pilot flame burning constantly, even in the summer months. However, a pilot flame may still be used to light the main burner. When the thermostat calls for heat, a special step-up transformer produces a high-voltage arc (10,000V or more) that

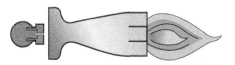

Figure 23 Single-port, or in-shot, gas burner.

Vestibule: The compartment of a gas furnace that houses the gas valve, manifold, and pilot assembly.

Figure 24 Single-port burners in operation.
Source: iStock@pricelessphoto

Standing pilot: A gas pilot flame that remains continuously lit.

Thermocouple: A device comprised of wire made from two different metals joined at the end that generates a scalable voltage based on the temperature of the junction.

Flame rectification: A method of proving the existence of a pilot flame by applying an AC current to a flame rod, which is then rectified to a DC current as it flows back to a ground source. Monitoring for a DC current flow provides the means of proving a pilot flame has been established.

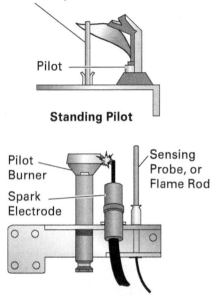

Standing Pilot

Electric Spark Ignition

Hot Surface Igniter

Figure 25 Gas furnace ignition devices.

Orifice Size

It may be necessary to increase or decrease the size of a burner orifice to obtain the correct burner input. The correct way to do this is to replace the orifice with one recommended by the manufacturer for the conditions. Changing the size of a gas orifice is not something to be done without guidance from the equipment manufacturer.

Never try to drill out an orifice to increase its size. Field drilling can roughen the inside of the orifice, disrupting the smooth flow of gas. The drilling must be performed with great precision, which is best done in a machine shop or manufacturing environment.

ignites a pilot flame, or ignites the main burner directly. To prove a flame exists, intermittent pilots use a process called **flame rectification**. The flame itself produces a tiny current, in the microampere range, that passes through a probe called a *flame rod*. If the current doesn't exist, the gas valve is closed.

Most modern furnaces now use a device known as a *hot surface igniter*, or *HSI* (*Figure 26*). It is placed next to a burner in place of a pilot flame or spark igniter. When the burner starts, current flows through the igniter, causing it to become extremely hot. The intense heat ignites the gas. They come in various shapes and mounting styles.

With an HSI, a flame sensor must be placed near the burners to prove ignition occurs and act as a safety shutoff device. If the main burner does not ignite within a preset period of time, the gas valve closes. The system may try to ignite several times before locking out the gas valve entirely. When that happens, the control circuit must be reset manually. The flame sensor may be a separate part or be integrated into the HSI itself.

The first HSIs were very fragile. If they accidentally touched a hard surface while being removed or installed, they were likely to shatter. Manufacturers eventually developed an HSI with a silicon-nitride element. This material is more durable than the earlier types and is less likely to shatter. Most technicians do not handle the ceramic portion of the element—convention taught that oils from hands would damage the element when heated. While manufacturers typically disagree, it is always best to minimize touching the element and handle and igniter by its base.

2.2.8 Gas Furnace Safety Controls

Gas furnaces are equipped with several devices designed to shut off the furnace if a hazardous condition exists. One of these is the high-temperature limit switch. *Figure 27* shows one example of a high-temperature limit switch. The switch is thermally actuated and is strategically mounted near a heat exchanger surface. Note the information on the label. L140–30F means that the limit opens at 140°F (60°C) and closes when it drops 30°F to 110°F (43°C).

Limit switches must be located where they can sense the heat exchanger temperature accurately. Overheating can occur due to a blocked flue vent or a failed blower motor. Electrically, they are wired in series with the gas valve. If the furnace overheats, the switch opens, de-energizing the gas valve. These switches may reset (close) automatically when the temperature drops below its setpoint, although some may need to be reset manually.

Flame rollout switches (*Figure 28*) are high-limit switches placed in the vestibule. If there is insufficient combustion air or the vent is restricted, the burner

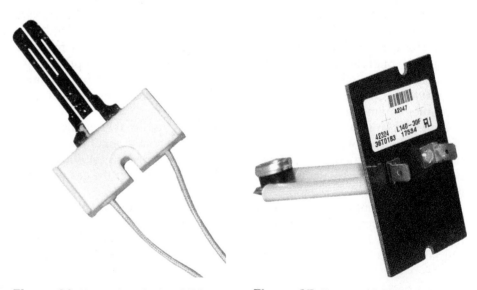

Figure 26 Hot surface igniter (HSI).

Figure 27 Furnace high-temperature limit switch.

flames may roll out of the combustion chamber and into the vestibule. The roll-out switch senses the excessive heat, opens, and disables the gas valve.

A combustion-airflow pressure switch (*Figure 29*) is used to monitor combustion airflow through the furnace. If the pressure drops below the setpoint, indicating a failure of the blower or a blockage, the switch opens and disables the gas valve. The pressure switch must be closed before the gas valve will open. The switch cycles each time the furnace cycles, opening when it shuts down and closing after the induced-draft blower starts and establishes the required pressure.

Rollout
Switch

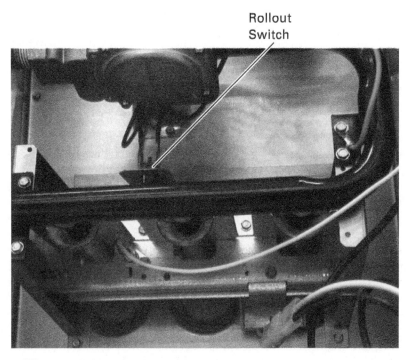

Figure 28 Flame-rollout safety switch.

Figure 29 Combustion-airflow pressure switch.
Source: HVACGuru

2.3.0   Gas Furnace Installation and Maintenance

Proper installation of a gas furnace is important for the safety of building occupants and reliable operation of the furnace. The following factors must be considered during the installation of a furnace:

- Location
- Clearances
- Vent type and route
- Combustion air source

Applicable national, state, and local codes must be followed when installing a furnace. However, the manufacturers' instructions may be more stringent than the local codes. This is done to make sure that codes are not inadvertently violated and to ensure that products are applied in the best way possible. In addition, stringent requirements help protect manufacturers when problems occur with a poor installation.

Proper handling of furnaces is especially important. The cabinets are made of sheet metal, which can be damaged if the furnace is dropped or mishandled. Also, some components, such as ceramic igniters, can be damaged. Many parts can shift or be dislodged when a furnace is mishandled. Although a dent in the cabinet may not affect unit operation, it affects how a client feels about a newly installed product and the quality of your work.

A furnace should be installed in a location that minimizes the length of duct runs when possible. There should be plenty of space around the furnace to permit access for service and repair. The manufacturer specifies minimum clearances to combustible materials and for service access. Flammable materials must not be placed near the heat exchangers, burners, or flue vent. Choosing a location that is safe and accessible is well worth a little extra ductwork. However, choosing a location is rarely a decision made by the installing technician.

Furnaces should never be located near a source of aerosol sprays, bleaches, detergents, air fresheners, or cleaning solvents. Even low concentrations of vapors from these materials can corrode a furnace. The fumes or vapors are drawn into the combustion chamber along with the combustion air. If they cannot be avoided, combustion air must be brought in from the outdoors. Manufacturers may refuse to honor warranties if a furnace is exposed to corrosive vapors.

To achieve additional heating capacity, as well as additional airflow, furnaces can be *twinned*. Two furnaces are placed side-by-side, and their controls are integrated so that they act as a single unit. The heating sections can be operated in stages, although the two blower motors should always operate as one. It should be noted, however, that the merging of the two airstreams typically results in a loss of some total airflow. For example, two twinned furnaces that would each produce 1,200 cfm of airflow alone may produce roughly 2,100–2,200 cfm as a twinned pair.

2.3.1 Furnace Venting

The flue gases leaving the furnace normally contain CO_2 and water vapor. CO may be present if combustion is incomplete. CO_2 results when complete combustion is present. It is not toxic, but it can displace oxygen in an enclosed area if the concentration is large enough. In extreme cases, it can cause asphyxiation. Because of the potential danger, especially from CO, flue gases must be vented in strict accordance with local and national codes.

Natural- and induced-draft furnaces both produce hot flue gases. These gases must be vented through a metal vent or a masonry chimney (*Figure 30*). It is often necessary to use a double-wall metal vent product. This is required when the vent must pass close to combustible materials, such as when it passes through a floor or roof.

Assumptions Can Cause Real Trouble

It is a common mistake to assume that a code compliance inspector will accept an installation with a flaw, no matter how unavoidable the flaw may be. For example, when replacing an existing furnace, there may be less clearance on the sides than the manufacturer or local codes require. There may not be a practical solution to the problem. However, the inspector is not obligated to accept the installation regardless of the implications.

Whenever an unavoidable code-compliance issue is noted during an installation, it is always best to reach out to the inspector and discuss the problem right away. Inspectors do have the authority to accept certain issues, but they are not often comfortable with your assumption that it will be accepted. This is an occasion when it is far better to ask for permission instead of forgiveness!

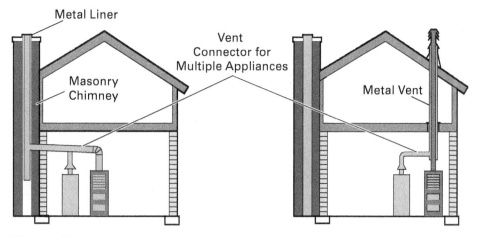

Figure 30 Natural- and induced-draft furnace venting.

The flue gases from a furnace contain water vapor. When the flue gases enter a cold chimney, the water condenses and reacts with other combustion byproducts to form compounds that attack the mortar in the chimney, causing it to deteriorate. For that reason, if a masonry chimney is used, it must be lined and have the correct dimensions. Even if the chimney has a tile liner, it is sometimes necessary to add a metal liner to meet code requirements. This is especially true if the chimney rises on the outside of the building, as a tile liner takes longer to warm up than a metal liner.

When sizing and selecting a vent system, the following factors must be considered:

- The distance from the furnace to the chimney
- The temperature of the flue gases
- Capacity and characteristics of other gas appliances, such as a water heater, that share the same vent or chimney

Condensing Furnaces

Condensing furnaces can be vented through the roof or an outside wall using polyvinyl chloride (PVC pipe). PVC products can typically handle temperatures up to 140°F (60°C). Plastic pipe can be used because these furnaces extract so much heat from the combustion process, the flue gas temperatures remain below this temperature.

For proper condensing-furnace operation, the PVC vent must be sized and installed according to the furnace installation instructions. Requirements related to the diameter of the combustion air and vent pipes, as well as the maximum number of elbows that can be, must be strictly followed. It is also important to note that the combustion air must usually be piped in from the same general location where the vent line is terminated (*Figure 31*). Vent piping must be pitched to allow any condensate that forms to drain away. A special termination fitting may be required for the vent and combustion air intake outdoors. Different furnaces and furnace manufacturers have different installation requirements. Never assume they are all the same.

NOTE

Venting requirements are covered in greater detail in NCCER Module 03202, *Chimneys, Vents, and Flues*.

CAUTION

Condensation is a major concern with induced-draft furnaces. The flue gases are cooler than those of natural-draft furnaces, so condensation is more likely to occur in the vent system. The condensation can drain back to the furnace and corrode the heat exchangers. Furnaces must be properly and carefully vented in accordance with the manufacturer's instructions and locally accepted codes to minimize condensation. A careful inspection of the vent and heat exchangers must be part of any maintenance inspection.

Common-Vented Furnace and Water Heater

Venting a furnace with a gas water heater is a common practice. It eliminates the need to have two separate chimneys. An added benefit is that the water heater provides an additional heat source for the vent, keeping it warm and reducing the possibility of condensation occurring inside.

The Gas Appliance Manufacturers Association (GAMA) provides tables and instructions that help installers determine how to best vent a furnace. Manufacturers typically offer simplified tables and specific instructions for their products in their installation instructions.

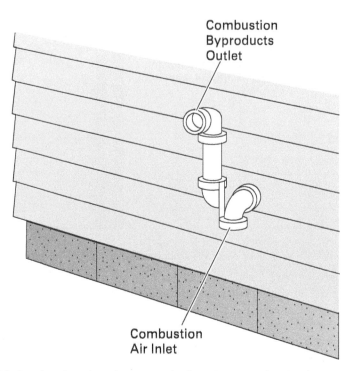

Figure 31 Combustion air and vent terminations for a condensing furnace.

When installing a condensing furnace to replace a gas furnace that was vented with a gas water heater, the venting arrangement for the water heater must be reconsidered. The existing vent will now only be serving the water heater, while it was sized to vent both units. The existing vent may be oversized for the water heater alone, which may prevent it from venting properly.

This situation can be corrected by lining the chimney with a smaller liner, or by installing smaller metal vent piping. Correcting the problem can add a significant amount of labor and material costs to the project.

2.3.2 Combustion Air

Many furnaces obtain combustion air from the area surrounding the furnace. This approach is common with natural-draft and induced-draft gas furnaces because the indoor air used for combustion can usually be made up by normal infiltration. If the furnace is installed in a confined or tightly sealed space, such as a closet, the space must be ventilated to allow air to get inside. However, even if the available space and building construction meet the standards for using indoor air for combustion, there are other factors that must be considered.

Condensing furnaces need a lot of combustion air. For that reason, they must have combustion air piped in from the outdoors. The intake is typically located near and just below the exhaust vent as shown in *Figure 31*. Care must also be taken when locating the combustion air source for a condensing furnace. For example, it should not be located near a swimming pool because there can be heavy concentrations of chlorine in the air.

WARNING!

Installation of combustion-air piping requires special attention to local codes. Improper installation can result in incomplete combustion.

2.3.3 Gas Furnace Maintenance

Gas furnaces require periodic cleaning and inspection of the gas burner, pilot assembly, and other components. At the beginning of each heating season, certain maintenance tasks should be performed. For example, air filters should be cleaned or replaced. The blower motor and blower wheel should also be cleaned. The heat exchangers can sometimes be cleaned gently with a soft, flexible wire brush attached to a drill. However, this must be done very carefully to avoid damaging the heat exchangers. Cleaning the interior of a heat exchanger is rarely, if ever, required as long as complete combustion is taking place.

The most important inspection items and operational characteristics are discussed in the sections that follow.

Combustion Process

The operation of the burners and pilot flame (if any) should be checked. The burners of a natural-draft furnace should produce a steady blue flame with slight yellow/orange tips. A yellow flame indicates a lack of primary air. A lack of primary air to the burners may also produce other symptoms, including *flashback* to the spud and soot deposits on the burners and heat exchangers. Flashback refers to the ignition of gas upstream of the burner. Flashback can also be caused by low gas pressure.

A jumpy blue flame that lifts off the burner of a natural-draft furnace usually indicates too much primary air is being forced in. A flickering or distorted flame could also be a sign of a crack in the heat exchanger. An open crack or hole allows the main blower to affect the flame. When this is the case, the flickering flame may only show itself when the indoor blower is running.

If there is any indication of a cracked heat exchanger, shut down and inspect the furnace. A cracked heat exchanger can release CO and other combustion byproducts into the building, posing a serious risk to occupants. A furnace with a cracked heat exchanger should not be operated. The furnace must be tagged as unsafe and/or inoperable, and the gas supply to it must be disabled.

WARNING!

A furnace that is deemed unsafe to operate due to a cracked heat exchanger or other serious defect must be locked out and tagged, and the fuel supply physically disconnected from the furnace.

Venting

The byproducts of combustion that are vented to the outdoors include CO_2, nitrogen, and hydrogen. If the furnace heat exchanger or vent piping is damaged, flue gases may enter the building. For this reason, proper venting of the furnace, a thorough inspection of the heat exchangers and flue vent, and proper adjustment of the flame are essential parts of the service technician's job.

Manifold Pressure

If the burners are not receiving the correct gas pressure, incomplete combustion may occur. The heat exchangers also may not heat up enough to prevent condensation inside. The manufacturer's installation instructions supplied with the furnace specify the correct manifold pressure.

For most natural-gas furnaces, the correct manifold pressure is $3\frac{1}{2}$" w.c. For LP gas, the manifold pressure is higher, usually around $9\frac{1}{2}$" w.c. Of course, the gas pressure supplied to the gas valve must be higher than the outlet pressure for it to properly regulate the manifold pressure. A gas valve inlet pressure of 7" w.c. to 9" w.c. for natural gas is usually the minimum requirement. It is important to check the furnace documentation for the proper manifold and gas valve inlet pressure. This information is often found on labels inside the unit as well.

One way to check the manifold pressure is to connect a **U-tube manometer** to the manifold pressure port on the gas valve (*Figure 32*). Gas valves are equipped with a port specifically for this purpose. A point of connection may also be located on the manifold itself. A manometer measures pressure by the amount of water that is displaced. In *Figure 32*, note that the manometer reads $1\frac{3}{4}$" w.c., both up and down. The two values are added, resulting in a pressure reading of

NOTE

The manifold pressure on today's more advanced gas furnaces may differ. For example, some furnace products are designed to operate at a slight negative pressure in the manifold. Always check the furnace data plate and/or service manual to determine the correct manifold pressure.

U-tube manometer: A U-shaped instrument that measures small pressures by displacing a column of liquid.

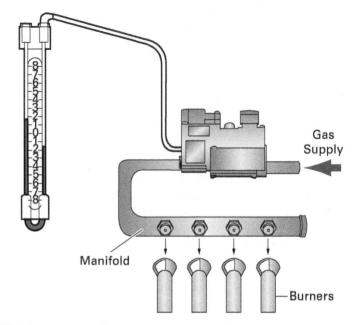

Figure 32 Measuring manifold pressure.

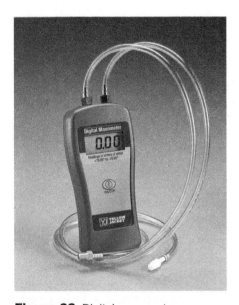

Figure 33 Digital manometer.
Source: Ritchie Engineering Company, Inc. /
YELLOW JACKET

$3\frac{1}{2}$" w.c. The pressure can also be measured using a digital manometer (*Figure 33*).

Note that there must be sufficient pressure at the inlet of the gas valve to account for pressure drop through the valve itself and to allow it to operate properly. If only $3\frac{1}{2}$" w.c. is provided at the inlet of the valve, it will not be able to maintain this same pressure in the manifold. This is due to pressure drop as it flows through the valve. An inlet pressure of 7" w.c. for natural gas and 12" w.c. for propane should be sufficient for the gas valve to provide the desired manifold pressure. Higher inlet pressures are fine as long as the gas valve is rated for higher pressures. This information can be found on the gas valve label.

If the pressure needs to be adjusted, the pressure-regulator adjusting screw on the gas valve is used to make the adjustment. It is usually located on top of the valve, with a cap covering the adjusting screw (*Figure 34*). The screw changes the pressure applied to a spring, which in turn changes the pressure applied to an internal diaphragm.

Most gas valves can be used for both natural and LP gas. The regulator spring in the gas valve can be changed, which changes the available adjustment range. Without first changing the spring, most gas valves cannot be adjusted to support LP use.

Temperature Rise

The temperature rise of a furnace is defined as the temperature difference between the supply and return air streams. In other words, it is the temperature change of the air as it passes over the heat exchanger. Operating outside of the specified temperature range can cause some operating problems, including overheating.

The approved temperature rise of a furnace is usually printed on the unit data plate. For example, the temperature rise stated on a data plate may be 40°F to 70°F (22°C to 30°C). It will always be stated as a range of temperatures.

If the temperature rise is too low, it usually indicates there is too much airflow. A temperature rise that is higher than the stated range usually means there isn't enough airflow. Although it is rare, the problem can also be caused by the burner being under- or over-fired.

To measure the temperature rise, begin by running the furnace for at least 10 minutes after the indoor blower has cycled on. This allows all the furnace and heat exchanger surfaces to reach a normal operating temperature. Then measure and record the return air and supply air temperatures at the appropriate

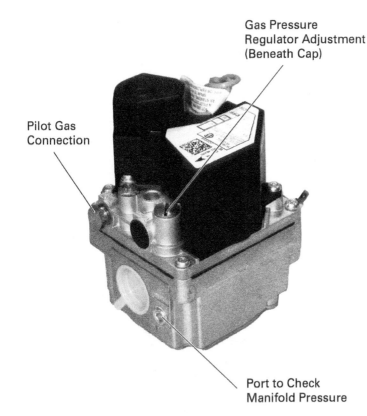

Gas Pressure
Regulator Adjustment
(Beneath Cap)

Pilot Gas
Connection

Port to Check
Manifold Pressure

Figure 34 Gas valve pressure-regulator adjustment access.

locations as shown in *Figure 35*. The difference in the inlet and outlet temperature represents the temperature rise. If the rise is outside of the stated range, troubleshooting will be necessary to determine the cause.

The return air measurement should be taken as close to the furnace inlet as possible. Inserting the probe through a small hole in the furnace return-air **plenum** is usually a good approach.

The supply air temperature measurement requires some additional thought. If the probe is inserted into the supply duct too close to the furnace, where a "line of sight" exists between the probe and the heat exchanger, an inaccurate reading will result. The temperature probe will be affected by radiant heat from the heat exchangers. Here are three options for measuring the supply air temperature:

- If a refrigerant coil is installed on the outlet of the furnace, it will help prevent radiant heat from reaching the probe, and the measurement can be made close to the outlet of the coil.

- If the supply duct is routed straight away from the furnace without a nearby change in direction, measure the temperature about 6' (~2 m) away to minimize the effect of radiant heat.

- If there is an immediate turn in the supply duct, taking the measurement one or two feet away from the turn is usually adequate. The change of direction in the duct will prevent radiant heat from affecting the temperature reading.

Always ensure that the blower motor is operating at the desired heating speed when taking temperature rise measurements. If the blower is switched to On at the thermostat, it typically operates at the speed selected for the cooling mode. This is generally a higher speed than used for heating. Thus, it would cause inaccurate temperature rise measurements. Ensure the thermostat Fan switch remains in Auto, which will result in the fan operating at its normal heating speed.

Also be sure that the entire air distribution system is intact. If the temperature rise is measured with the ductwork detached or incomplete, the airflow will be higher than normal and may change significantly once the duct system is complete.

Plenum: An enclosure placed at the supply and/or return air openings of a unit to which ducts are connected.

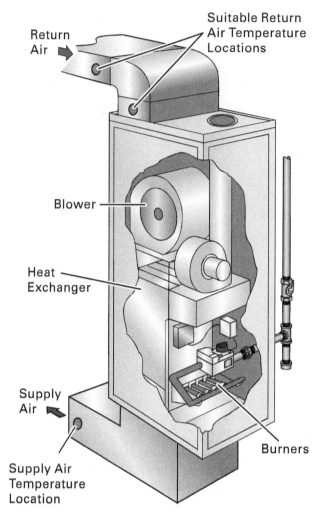

Return Air

Suitable Return Air Temperature Locations

Blower

Heat Exchanger

Supply Air

Supply Air Temperature Location

Burners

Figure 35 Temperature rise measurements.

2.0.0 Section Review

1. The style of furnace that has both duct connections on the top is called a(n) _____.
   a. upflow model
   b. counterflow model
   c. horizontal model
   d. lowboy model

2. If too little primary air is supplied to a gas burner, the flame will _____.
   a. lift off the burner surface
   b. turn yellow
   c. turn blue
   d. turn orange

3. Gas furnaces should *not* be vented directly into masonry chimneys because _____.
   a. they are generally too big
   b. mortar causes flue gases to condense
   c. they are usually too far from the furnace
   d. condensation can cause the mortar to deteriorate

3.0.0 Hydronic and Electric Heating Systems

Objective

Describe hydronic and electric heating systems.

a. Describe hydronic heating systems and identify common components.
b. Describe electric heating equipment and identify common components.

Performance Tasks

There are no Performance Tasks in this section.

Although gas-fired, forced-air furnaces are the most common method of heating in the United States, hydronic heating systems are also very popular. They are typically found in colder regions. Heat pumps, a form of all-electric heat, are commonly used in areas of the country with moderate climates to provide both heating and cooling. Electric heating systems based on resistance heaters alone are also used, but the operating cost is significantly higher than that of a heat pump.

In this section, both hydronic and electric heating systems are explored.

3.1.0 Hydronic Heating Systems

Thus far, this module has focused on forced-air gas-heating systems. They heat air, which is then circulated throughout the structure. A *hydronic heating system* (*Figure 36*) heats water instead, and then circulates it throughout the structure.

The advantages of hydronic heating over forced-air heating generally include:

- More even temperatures and improved comfort
- Piping requires less space than ductwork
- Quieter operation

Some disadvantages of hydronic heating include a higher installation cost than forced-air systems. There is also no practical way to use the same system for cooling.

A boiler like the one shown in *Figure 37* heats the water. The term *boiler* may be a little misleading. While a steam boiler does heat water to create steam, a water boiler does not boil the water. Boilers can be gas- or oil-fired, or even electric.

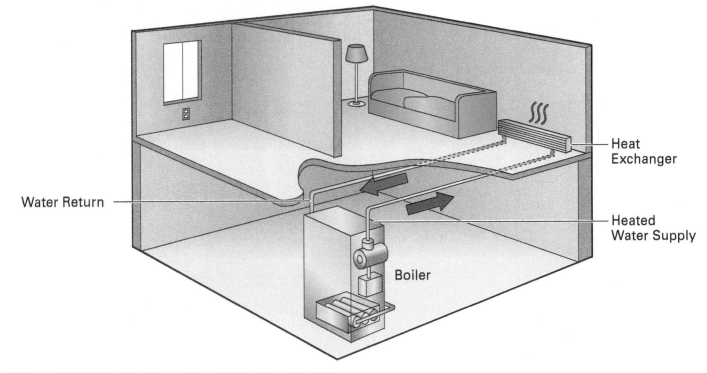

Figure 36 Hydronic heating system with baseboard radiation.

Figure 37 An induced-draft, gas-fired boiler.
Source: Utica Boilers

NOTE

This module focuses on residential hydronic systems. Other modules that expand on hydronic systems include NCCER Module 03203, *Introduction to Hydronic Systems*, and NCCER Module 03306, *Steam Systems*.

Sectional boiler: A boiler consisting of two or more similar sections that contain water, with each section usually having an equal internal volume and surface area. Boiler sections are generally made from cast iron and are shipped in pieces to be assembled on-site.

Gas-fired boilers are also available in high-efficiency condensing versions that extract additional heat from the flue gases. *Figure 38* shows the interior of a condensing boiler. The model shown has an AFUE rating of 95 percent.

Today's residential and light-commercial boiler designs store very little water and are designed to provide the needed heating capacity on demand. Older models could be compared to tea kettles, where gallons of hot water were maintained at a temperature ready for distribution. The change results in a more compact package that can even be wall-mounted (*Figure 39*). For larger commercial installations, multiple boilers can be installed and staged to respond based on demand.

Boilers can produce hot water or steam. Steam boilers are rarely installed in new residential applications today, due to cost and maintenance requirements. However, if an existing home has a steam boiler that needs to be replaced, it is often replaced with another steam boiler. A conversion from one steam to water, or vice-versa, is costly in an existing home because of the differences in system piping and other components.

3.1.1 Major Boiler Components

The major components of a boiler include the burner and burner controls, the boiler sections where water is heated, and a pump to circulate the warmed water throughout the structure. The pump may be mounted on the boiler or be a separate component. The burner and its safety and operating controls are similar in design and function to those used in forced-air furnaces.

Traditional boiler heat exchangers are made of cast iron. Less expensive boilers of a traditional design have one-piece steel heat exchangers. Large boilers are created using a number of cast iron heat exchanger sections. This type of boiler is known as a **sectional boiler**.

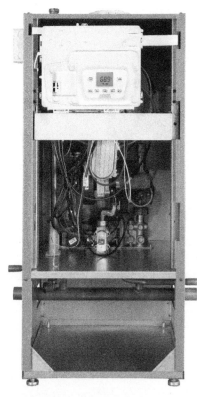

Figure 38 Condensing boiler.
Source: Utica Boilers

Figure 39 Wall-mounted condensing boiler.
Source: Utica Boilers

Newer, more efficient boilers use finned copper tubing to create a heat exchanger. This construction provides faster heat transfer. They hold less water onboard and they can respond faster than cast iron models. Wall-mounted boilers may use heat exchangers made from stainless steel. An aluminum-silicon (AlSi) metal alloy is also being used. The burner flame is applied to the outside of the boiler sections or heat exchanger, warming the water within. Some boilers contain a separate heat exchanger that allows the boiler to also provide domestic hot water.

The circulator pump moves the water through the pipes to the heating terminals throughout the structure. The terminals transfer the heat to the air. The cooler water then returns to the boiler as the process repeats itself.

3.1.2 Boiler Components and Controls

One of the most important controls found on a boiler is the aquastat (*Figure 40*). This temperature-controlled switch controls burner operation on a traditional boiler to maintain the water in the boiler within a specified temperature range. Aquastats come in a variety of styles, depending on the application. Aquastats also serve as limit switches, turning off the burner if the water exceeds the limit setting. Since high-efficiency boilers do not store water, their burner-control approaches are more complex.

Water expands when heated, and the pressure increases when there is no room for expansion in a closed space. If left unchecked, rising pressure in a boiler can cause it to leak or burst. To prevent this, boilers are equipped with a pressure-relief valve (*Figure 41*). The valve opens and relieves the pressure. These valves are mechanical and have no connection to the electrical control circuit.

WARNING!

The discharge of all pressure relief valves must be piped to a safe location, as the discharged water can be dangerously hot.

A low-water level control ensures that water remains above the minimum needed in the boiler. If the water level is too low, the control prevents the burner from operating by breaking the burner control circuit.

A circulating pump (*Figure 42*) is sometimes factory-installed on the boiler itself. However, since the pumping requirements for each system varies, pumps are often provided and installed by the contractor. Smaller pumps for circulating water are generally referred to as *circulators* rather than pumps, as the pressure and flow volume is lower than that of typical base-mounted pumps. Circulators are installed directly in the piping, rather than being supported by a base.

(A) Electromechanical

(B) Digital

Figure 40 Typical aquastats.
Source: Robertshaw Controls Company (40B)

Around the World
Condensing Boilers

Condensing boilers have long been common in Europe, where heating systems are designed to operate at lower water temperatures (less than 150°F, or 65°C) and fuel efficiency has been a higher priority. In North America, the higher water temperatures traditionally used in heating system design made the use of condensing boilers less desirable. However, improvements in condensing boiler technology have greatly helped overcome this problem, allowing the use of condensing boilers to now be common in North America. They are significantly more efficient than previous designs.

Source: iStock@matteogirelli

Figure 41 Pressure relief valve.

(A) Standard Inline Circulator   (B) ECM-Driven Circulator

Figure 42 Hydronic circulators.
Source: Courtesy of Taco, Inc.

3.1.3 Other Components of a Hydronic Heating System

Other specialized components are needed in a hydronic heating system. *Figure 43* provides an overview of common hydronic system components. In residential applications, the heat is often delivered to the conditioned space by finned baseboard terminals. Another popular method is embedding tubing in the floor (*Figure 44*). This system applies heat evenly and maintains comfortable floors in cold climates.

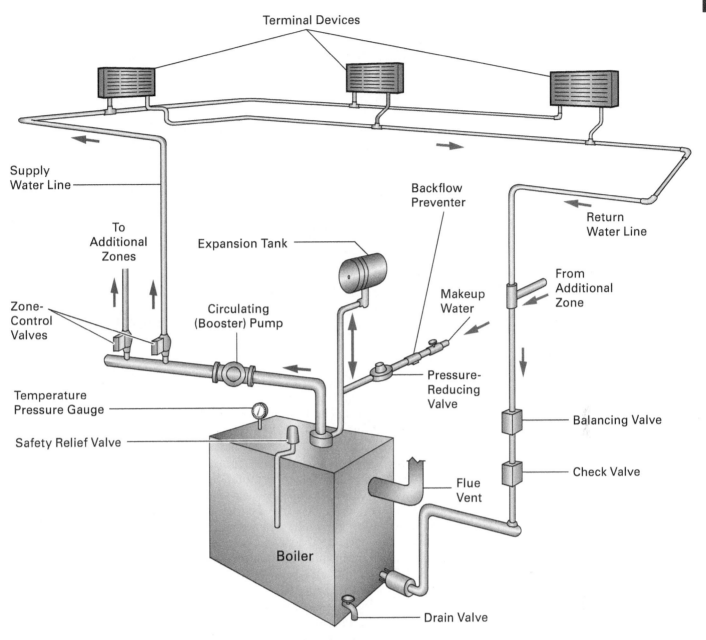

Figure 43 Simplified hydronic heating system with multiple heating zones.

Different rooms, or *zones*, within the same structure can be kept at different temperatures. This is accomplished by installing a zone control valve in the piping that controls the flow of hot water to each zone. Each zone control valve is opened and closed by its own room thermostat. This allows hot water to flow to the heat-transfer device (baseboard unit, floor piping) in the zone only when needed. Here is how a zone valve controls a circulating pump in a typical system:

- A room thermostat for a zone calls for heat, commanding a motor-driven zone valve to open.
- The zone valve is equipped with normally open switch contacts that close when the valve opens.
- The closure of the switch contacts completes the control circuit to energize the circulating pump, and to start the boiler in many cases.

The pump remains running until the thermostat is satisfied, the zone valve closes, and the control switch in the valve opens the circuit. If other zone valves have since opened, the pump will remain running until all zones are satisfied. Alternatively, each zone has its own small circulating pump that operates

Figure 44 Hydronic floor-heating system.
Source: iStock@filmfoto

independently of the other pumps. There are several control and pumping schemes used in hydronic systems.

Air must be eliminated from hydronic systems since it interferes with water circulation in closed systems. When systems are first filled with water, a great deal of air must be purged. However, small amounts of air enter the system any time additional water is added, since air is entrained in water from the municipal water supply. As a result, air elimination must be a continuous process.

When the system is idle (no flow), air tends to rise to the highest point it can access. In large systems, there may be many high points. Air vents are typically positioned at these high points. When a manual vent is opened to purge a system, it should be left open until a steady stream of water is obtained. Automatic air vents (*Figure 45*) are installed to eliminate air without constant attention from a technician. The outlet of an automatic air vent must often be piped to a safe drain.

A makeup water valve connected to the water supply automatically replaces any water lost due to leaks or air venting. It is a pressure-regulating valve and is set to maintain a specified water pressure in the hydronic system. A typical pressure setting from the factory is 12 psig, but most are adjustable. The makeup water valve may also be close-coupled to a pressure-relief valve as shown in *Figure 46*. The relief valve protects the boiler and other system components from over-pressurization. This valve is in addition to the pressure-relief valve mounted on the boiler itself. Since many traditional boilers are rated for a maximum pressure of 30 psig, this is a common setting for both relief valves. *Backflow preventers* are installed upstream of the makeup water valve to ensure that water from the hydronic system cannot accidentally move backwards and enter the potable water supply.

Figure 45 Automatic air vents.
Source: Courtesy of Taco, Inc.

Figure 46 Water-pressure regulator and pressure-relief valve combination.
Source: Courtesy of **www.wattswater.com**

Since water expands when heated, hydronic systems include some type of expansion tank. It is mounted above the boiler in *Figure 43*. The expanding water causes the air inside the tank to be compressed, providing a flexible cushion. This prevents the pressure in the system from reaching unsafe levels due to normal expansion.

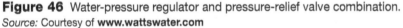

3.2.0 Electric Heating

All-electric heating can be provided in a number of ways. Heat pumps are the most common all-electric heating systems in use. Self-contained electric heating units are also available.

3.2.1 Electric Furnaces

In an electric furnace (*Figure 47*), air passes directly over the heating elements. From the outside, an electric furnace looks like a simple metal enclosure with access doors. Since there is no combustion, there is no need for a vent to carry the byproducts of combustion away. This feature allows for greater installation flexibility.

The amount of heat supplied by an electric furnace depends upon the number and size of the individual heating elements. The elements are often sequenced on in stages.

Although heating elements are available in many different ratings, individual elements in central heating systems are typically rated at 4–5 kilowatts (kW). However, remember that the actual heating capacity of a resistance heating element is dependent upon the voltage provided. As the voltage increases, the heating capacity of the element increases along with the wattage.

The major components of an electric furnace are the heating elements, the blower assembly, controls, and the outer cabinet. Like furnaces, a filter is often provided at the return air inlet.

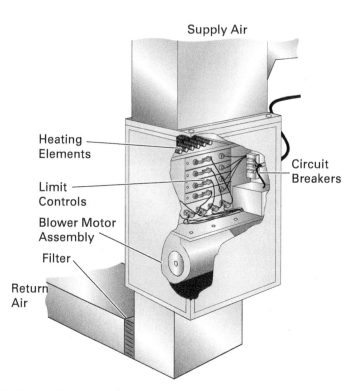

Figure 47 Electric furnace.

The function of the heating elements is to provide the heat required for the space. The heating elements (*Figure 48*) are typically made from a nickel/chromium alloy (Nichrome). The heating element wire is normally spiraled and threaded through a metal holding rack, which has ceramic insulators to prevent the wires from touching the metal. The elements are a larger version of those found in household appliances such as electric clothes dryers and toasters.

The circuit for each heating element may contain a **fusible link** in addition to a high-temperature safety switch. The switch opens in response to an overheating condition, and automatically resets when the temperature falls below its setpoint. A common setting for the limit switch is to open at approximately 160°F (~71°C), and to close when the temperature drops below 125°F (~52°C). The fusible link is used for added protection in case the switch fails to open. The fusible link melts to permanently open the circuit when exposed to excessive heat, although excessive current could also cause it to overheat and melt. A technician must replace the link before the related heating element will function. The furnace is also be protected by standard fuses or circuit breakers.

Fusible link: An electrical safety device that melts to open a circuit when exposed to excessive heat or a current that causes excessive heat.

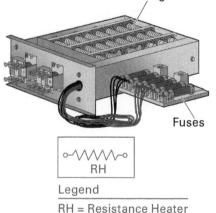

Figure 48 Electric resistance-heating elements.

Did You Know?

Converting Watts to Btuh

It's very simple to determine exactly how much heat is being produced by one or more electric heating elements. 1 kW of electric heat produces 3,412 Btuh. Remember that Watts = Volts × Amps, or per the power formula, $P = I \times E$. If a heating element is operating at 230VAC and a current of 17A, the watts are calculated as follows:

Watts = 230VAC × 17A
Watts = 3,910W, or 3.91 kW

To determine heat output, multiply the number of kilowatts by 3,412—the number of Btuh produced per kilowatt:

Btuh = 3.91 kW × 3,412
Btuh = 13,341 Btuh (rounded)

Note that, if the incoming voltage is 208VAC, the heat output drops to 12,065 Btuh. The incoming voltage is a key factor in electric heat output.

The blower in an electric furnace is driven by a multi- or variable-speed direct-drive blower. Filters, humidifiers, and cooling coils can be added to an electric furnace in much the same manner as gas and oil furnaces. Therefore, electric furnaces can support year-round comfort like other furnaces.

A residential electric-furnace power supply is usually 208/240VAC. Power is supplied by three wires: two hot conductors and one grounding conductor. Unlike a 120V power supply, there is no neutral wire. Fused disconnect switches are often provided near the installation, in addition to the circuit breaker found in the electrical panel. All external wiring must conform to the *National Electrical Code®* (*NEC®*).

3.2.2 Heat Pumps

Heat pumps were briefly introduced in NCCER Module 03101, *Introduction to HVACR*. A heat pump can extract heat from the outdoor air by reversing the mechanical refrigeration cycle. Air-to-air heat pumps are the most common, but water-to-air, water-to-water, and ground-source heat pumps are also popular.

Air-to-air heat pumps begin to lose heating capacity as the outdoor temperature falls, so they are more common in moderate climates. Since the temperature of groundwater supplies are far more consistent, water-to-air units provide more consistent capacity throughout the heating season.

A heat pump requires a special valve known as a *reversing valve*, as shown at the bottom of *Figure 49*. They are also referred to as *four-way valves* since they have four piping connections. The valve changes the refrigerant flow path. In addition to the reversing valve, a heat pump requires a defrost control to remove ice from the outdoor coil during subfreezing weather. Ice is removed by briefly changing over to the cooling mode, sending hot refrigerant vapor to the outdoor coil.

Refrigerant metering devices must be provided for both the indoor and outdoor coils. Each coil must serve as an evaporator coil in one mode or the other. Note in *Figure 49* that check valves are also provided to ensure proper flow

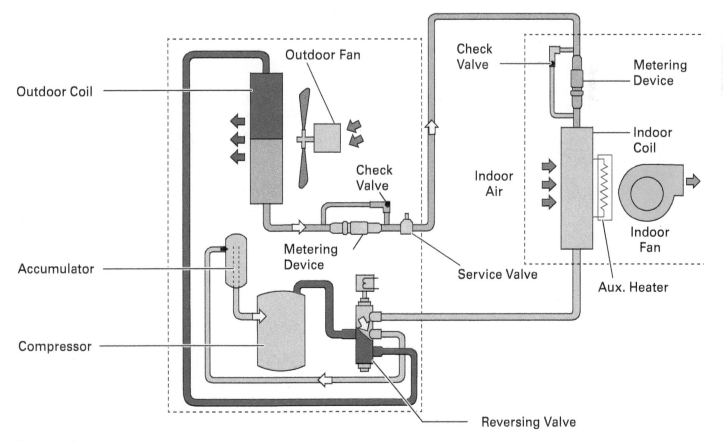

Figure 49 Schematic diagram of a heat-pump refrigerant circuit.

through or around the metering device at each coil. In newer equipment, the metering device and check valve are usually a single device.

Most heat pumps cannot provide enough heat for a structure alone. They are most often supplemented by electric heating elements that operate when heat pumps can't keep up with the heat loss. Picture an air-to-air heat pump matched with an electric furnace, but the furnace also contains a refrigerant coil. These units are referred to as *fan coil units*. The heating elements also provide a backup if the heat pump fails, and they provide some heat when the heat pump is operating in the defrost mode. Supplemental heat for heat pumps may be required by local code, but there are practical reasons for it to be installed as well.

Water-source heat pumps reject or collect heat using water instead of air. They provide more consistent heating and cooling capacity than air-source units. The water can come from several sources.

The source of water could be pumped from a pond or lake, for example. There are, however, inherent problems with using dirty water from these sources. There may also be problems that arise from returning warmer water to ponds and lakes, such as significant algae growth.

Geothermal heat pump: A heat pump that transfers heat to or from the ground, using the earth as a source of heat in the winter and a heat sink in the summer. The system uses water as the heat-transfer medium between the earth and the refrigerant circuit.

A **geothermal heat pump** also uses water as a means of heat transfer, but not from a surface source. Groundwater from a well is a cleaner, more reliable option. The water is generally returned to the ground through a second well some distance away. Coils or long lengths of tubing can also be placed in the ground or along the bottom of a lake or pond (*Figure 50*). Water is circulated through the tubing to absorb or reject heat from the ground or water. This results in a closed-loop system, where the same clean water is recirculated through the loop. This significantly reduces problems with contaminants in the water.

Water-source heat pumps are very efficient, due to the stability of ground and groundwater temperatures. In both the heating and cooling mode, they can provide very efficient operation along with consistent capacity, regardless of outdoor air temperatures.

Cooling tower: A heat-rejection unit that moves air through falling water to cool it.

In commercial installations, water-source heat pumps may be connected to a boiler and a **cooling tower** instead of to a groundwater source or buried piping.

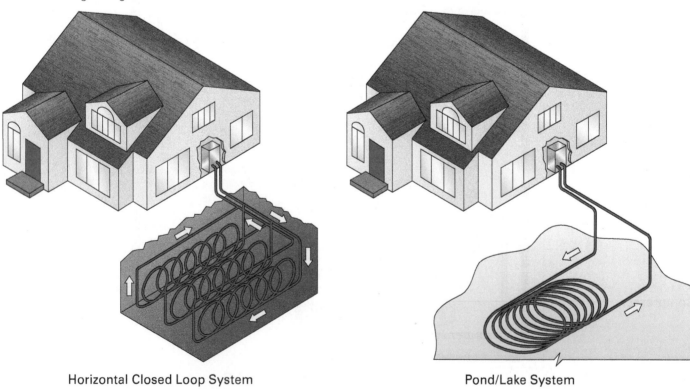

Horizontal Closed Loop System   Pond/Lake System

Figure 50 Water-source heat pump closed-loop examples.

The boiler and cooling tower heat or cool the water in a loop serving many heat pump units, keeping the loop temperature consistent year-round.

Dual-Fuel Systems

Heat pumps are sometimes combined with fossil-fuel furnaces. This type of system is referred to as a **dual-fuel system**. However, the furnace and the heat pump cannot operate at the same time. Electric heaters can operate at the same time as the heat pump because the heating elements are downstream of the indoor coil; the heat they produce does not pass over the coil. When added to furnaces, the indoor coil is typically located downstream of the furnace. If the two operate at the same time, the heat from the furnace would quickly increase the refrigerant pressure of the heat pump beyond a safe level.

An outdoor thermostat is used to control which of the two operates. It enables the furnace and disables the heat pump when the outdoor temperature falls below the point where the heat pump alone cannot keep up with the heating load. This temperature, called the **balance point**, can be determined using a graph of the structure's heat loss. It is then compared to a graph of the heat-pump heating capacity at various temperatures, as shown in *Figure 51*.

At the balance point, the heat pump's heating capacity matches the heat loss of the structure. If the temperature drops below the balance point, the heat pump cannot keep up on its own. In a dual-fuel system, the heat pump is disabled, and the furnace takes over. At or above the balance point, the heat pump can maintain the indoor setpoint alone, although it must run continuously to do so when the outdoor temperature is near the balance point.

Note that this discussion applies to air-to-air heat pumps. Remember that the capacity of water-source heat pumps is not significantly affected by the outdoor temperature. Thus, their heating capacity is more consistent throughout the heating season.

Also note that economics can play a role in selecting a temperature for the system changeover to occur. It may be more economical to switch to the fossil-fuel furnace at a higher temperature if the energy cost is lower.

Dual-fuel system: A heating system typically comprised of a heat pump and a fossil-fuel furnace. Heat pumps supplemented by electric heat are not considered dual-fuel systems.

Balance point: An outdoor temperature value that represents the matching of a structure's heat loss with the capacity of a given heat pump to produce heat. At the balance point, a heat pump must run continuously to maintain the indoor setpoint, without falling below or rising above it.

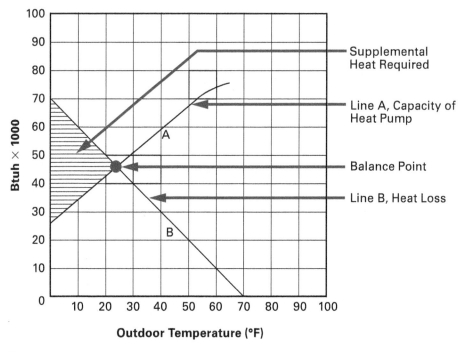

Figure 51 Balance point graph.

3.0.0 Section Review

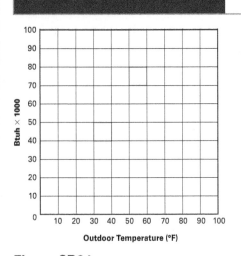

Figure SR01

1. The makeup water valve in a hydronic heating system is used to _____.
   a. replace water lost from the system
   b. release excessive pressure in the boiler
   c. control individual zone temperatures
   d. fill the expansion tank

2. Refer to *Figure SR01*. Assume a heat pump has heating capacities of 40,000 Btuh at 17°F and 70,000 Btuh at 47°F. The home has zero heat loss at 70°F, but it reaches a heat loss of 80,000 Btuh at 0°F. Plot these two lines on the figure. What is the resulting balance point temperature?
   a. 20°F
   b. 26°F
   c. 31°F
   d. 34°F

Module 03108   Review Questions

1. When a coffee cup gets warm when hot coffee is poured in, it represents heat transfer by _____.
   a. radiation
   b. convection
   c. conduction
   d. capillary action

2. When you become warm while sitting some distance from a fire, heat is being transferred by _____.
   a. radiation
   b. convection
   c. conduction
   d. combustion

3. A temperature of 140°F is equal to _____.
   a. 12°C
   b. 60°C
   c. 133°C
   d. 202°C

4. Humidifiers are used to _____.
   a. reduce condensation
   b. add moisture to the air
   c. keep furniture dry
   d. lower the room temperature

5. Carbon monoxide (CO) is produced when _____.
   a. incomplete combustion is present
   b. oil and oxygen are mixed
   c. there is too much water vapor in the flue gas
   d. there is excess air in the combustion chamber

6. A yellow flame in a gas burner indicates that _____.
   a. a proper amount of combustion air is present
   b. there is too much primary air
   c. there is too much secondary air
   d. incomplete combustion is at work

7. Most calculations regarding the capacity of gas piping are based on natural gas having a heat content of _____.
   a. 750 Btus/ft^3
   b. 1,000 Btus/ft^3
   c. 1,500 Btus/ft^3
   d. 1,800 Btus/ft^3

8. Which furnace configuration is represented in *Figure RQ01*?
   a. Lowboy
   b. Horizontal
   c. Counterflow
   d. Upflow

9. In a condensing furnace, the secondary heat exchanger _____.
   a. is used only for room-air dehumidification
   b. is used to reheat the flue gases before they are vented
   c. extracts heat by condensing moisture from the flue gases
   d. extracts heat by cooling the room air that passes over it

10. The type of blower used to circulate room air through a furnace is a(n) _____.
    a. centrifugal blower wheel with backward-curved blades
    b. centrifugal blower wheel with forward-curved blades
    c. propeller fan
    d. axial fan

11. Venting of the flue gases through PVC pipe is an option _____.
    a. only for condensing furnaces
    b. only for natural-draft furnaces
    c. for all furnaces with an induced-draft blower
    d. for any furnace not connected to a masonry chimney

12. A manometer is used to measure _____.
    a. the volume of water is in the flue gas
    b. the gas pressure in the manifold
    c. the flame rectification current
    d. the ratio of combustion primary air to secondary air

13. Sectional heat exchangers in a large boiler are usually constructed from _____.
    a. steel
    b. cast iron
    c. stainless steel
    d. finned copper tubes

14. The typical pressure setting from the factory for a makeup water valve is _____.
    a. 100 psig
    b. 45 psig
    c. 30 psig
    d. 12 psig

15. The actual heating output of an electric heating element largely depends on the _____.
    a. voltage provided
    b. temperature of the return air stream
    c. relative humidity of the return air stream
    d. temperature difference of the return and supply air streams

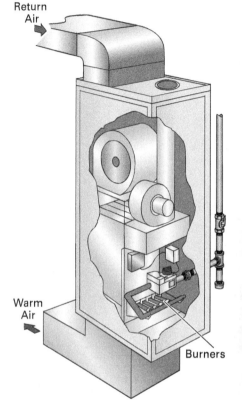

Return Air

Warm Air

Burners

Figure RQ01

Answers to odd-numbered Module Review Questions are found in *Appendix A*.

Answers to Section Review Questions

Answer	Section Reference	Objective
Section 1.0.0		
1. c	1.1.3	1a
2. d	1.2.4	1b
Section 2.0.0		
1. d	2.1.0; *Figure 10*	2a
2. b	2.2.5	2b
3. d	2.3.1	2c
Section 3.0.0		
1. a	3.1.3	3a
2. b	3.2.2	3b

User Update

Did you find an error? Submit a correction by visiting **https://www.nccer.org/olf** or by scanning the QR code using your mobile device.

SCAN ME

Introduction to Cooling

Source: iStock@DenBoma

Objectives

Successful completion of this module prepares you to do the following:

1. Explain the fundamental concepts of the refrigeration cycle.
   a. Describe the relationship between heat transfer and the state of substances.
   b. Describe pressure/temperature relationships.
   c. Describe basic refrigerant flow and the changes of state occurring in the refrigeration cycle.
   d. Identify common instruments used to measure pressure and temperature.
2. Identify common refrigerants and their identifying characteristics.
   a. Identify fluorocarbon refrigerants.
   b. Describe the use of ammonia as a refrigerant.
   c. Identify various refrigerant containers and their safe handling requirements.
3. Identify the major components of cooling systems and their function.
   a. Identify various types of compressors.
   b. Identify different types of condensers.
   c. Identify different types of evaporators.
   d. Describe devices used to meter refrigerant flow.
   e. Recount basic refrigerant piping concepts.
   f. Identify various accessories used in refrigerant circuits.
4. Identify the common controls used in cooling systems and how they function.
   a. Identify common primary controls.
   b. Identify common secondary controls.

Performance Tasks

Under supervision, you should be able to do the following:

1. Measure and record the dry bulb and wet bulb temperatures of the supply and return air streams in an operating cooling system.
2. Connect a refrigerant gauge manifold and properly calculate subcooling and superheat on an operating system.
3. Identify refrigerants using cylinder color codes.
4. Identify the compressor, condenser, evaporator, metering device, and accessories in a cooling system.

Digital Resources for HVACR

SCAN ME

Scan this code using the camera on your phone or mobile device to view the digital resources related to this craft.

Overview

To service cooling and heat pump systems, you must have a clear understanding of the refrigeration cycle and the function of its primary components. Technicians can apply this knowledge to all refrigerant circuits. Despite the differences in the many possible refrigerants and applications, the basic principles presented here apply to all direct-expansion refrigerant circuits.

Industry Recognized Credentials

If you are training through an NCCER-accredited sponsor, you may be eligible for credentials from NCCER's Registry. The ID number for this module is 03107. Note that this module may have been used in other NCCER curricula and may apply to other level completions. Contact NCCER's Registry at 1.888.622.3720 or go to **www.nccer.org** for more information.

You can also show off your industry-recognized credentials online with NCCER's digital badges. Transform your knowledge, skills, and achievements into badges that you can share across social media platforms, send to your network, and add to your resume. For more information, visit **www.nccer.org**.

1.0.0 Refrigeration Cycle Fundamentals

Performance Tasks

1. Measure and record the dry bulb and wet bulb temperatures of the supply and return air streams in an operating cooling system.
2. Connect a refrigerant gauge manifold and properly calculate subcooling and superheat on an operating system.

Objective

Explain the fundamental concepts of the refrigeration cycle.

a. Describe the relationship between heat transfer and the state of substances.
b. Describe pressure/temperature relationships.
c. Describe basic refrigerant flow and the changes of state occurring in the refrigeration cycle.
d. Identify common instruments used to measure pressure and temperature.

In nature, nothing is completely without heat. Saying that something is cold simply means that it has less heat energy than something else. Thermal energy can neither be created nor destroyed in a closed system, but it can be transferred from one object or place to another. That statement virtually defines air conditioning and refrigeration processes.

Refrigeration is a term that describes the transfer of heat from one place, where it is not wanted, to another place where the added heat is insignificant or unimportant. Air conditioning units and refrigerators simply transfer heat out of a space or object and move it elsewhere (*Figure 1*). In most cases, the heat is transferred outdoors. Domestic refrigerators and small commercial reach-in cases reject heat to the space. This heat must be moved outdoors by the building cooling system.

A refrigerant circulates through a refrigerant circuit to enable the process. The refrigerant absorbs heat from the refrigerated space and carries it elsewhere. Various refrigerants are covered later in this module and program. Even water can be considered a refrigerant when it is used to transfer heat.

1.1.0 Heat and the States of Substances

To understand the refrigeration cycle, you must understand the nature of heat. Like light and electricity, heat is a form of energy. Like other forms of energy, heat can do work, and it can be measured and controlled. Its ability to do work depends on two characteristics: the temperature (intensity) and the heat content (quantity).

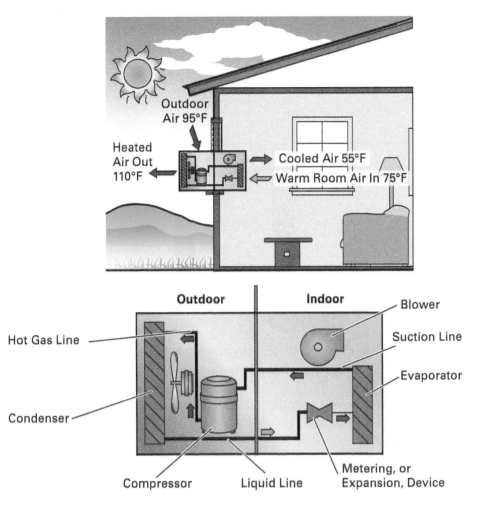

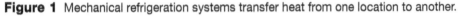

Figure 1 Mechanical refrigeration systems transfer heat from one location to another.

1.1.1 Temperature

Temperature provides a way to compare the intensity of heat contained in objects or substances. The intensity of heat is measured in degrees with some sort of thermometer. Two different temperature scales are commonly used for measuring temperature outside of the laboratory. The temperature scale most often used in the United States is the Fahrenheit scale (*Figure 2*). The other scale, used worldwide and often encountered in the United States, is the Celsius, or Centigrade, scale.

The topic of temperature conversion has been covered in previous modules. However, it is important to remember how the two scales relate to each other. For example, consider the following statement: *There is usually a 3°F difference between the first and second stage of cooling.* This is not a measured temperature of 3°F, but a differential temperature. To convert this value to °C, simply multiply it by $5/9$, or 0.55. A differential value in °C is converted to °F by multiplying it by $9/5$, or 1.8.

1.1.2 Heat Content

Heat content refers to the amount of heat energy contained in a substance. Heat content in the US standard system in the HVACR industry is measured by the **British thermal unit (Btu)**. One Btu is the amount of heat needed to raise the temperature of one pound (lb) of water by 1°F.

British thermal unit (Btu): The amount of heat needed to raise the temperature of one pound of water by one degree Fahrenheit.

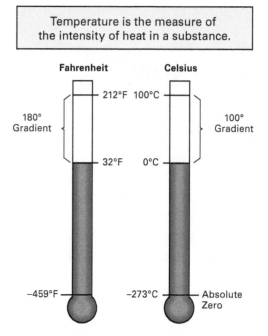

Temperature is the measure of
the intensity of heat in a substance.

Figure 2 The Fahrenheit and Celsius temperature scales.

For example, to heat 1 lb of water from 40°F to 50°F (a difference of 10°F), 10 Btus of heat must be added to the water. If it is 10 lb of water, raising the temperature 10°F would require 100 Btus. When the process is reversed, the same math holds true. If 1 lb of water is cooled by 10°F, 10 Btus of heat were transferred out of the water. In the metric system, heat content is measured in *joules*.

1.1.3 Sensible and Latent Heat

Depending on their heat content and temperature, substances can exist in three states: solid, liquid, and vapor. Using water as an example, the three states are ice (solid), water (liquid), and steam (vapor).

In a closed refrigerant circuit, the refrigerant inside exists in either a liquid or a gaseous state. For this reason, these two states are the ones you will be most concerned with moving forward.

In a closed system, when a substance such as water changes from one state to another, something peculiar happens. Assume there is 1 lb of ice, and the ice has been refrigerated enough to reach a temperature of 0°F (–18°C). If heat is then added back to the ice, a thermometer would show its temperature rising until it reaches 32°F (0°C).

Refer to *Figure 3*. At 32°F, the ice begins the process of changing back to water. While heat is continually added and the ice melts, the thermometer reading will remain fixed at 32°F instead of continuing to rise as expected. In fact, the ice remains at 32°F until the entire pound has melted.

The increase in temperature from 0°F to 32°F (–18°C to 0°C) registered by the thermometer is called **sensible heat**. It is considered sensible because it can be sensed by a thermometer, and you can feel the change. Sensible heat is heat that is added to, or removed from, a substance without any change of state involved.

But what about the heat that was added to the ice *after* the thermometer reached 32°F, when a change in state to a liquid began? The ice continued to absorb heat, without any sensible change in temperature, until all the ice melted. A significant amount of heat was transferred to the ice while the thermometer did not show a change in temperature. This heat is called **latent heat**.

Latent heat is heat energy that is added to, or removed from, a substance during a change of state. While the substance is actively changing states, no measurable change in temperature occurs. All the heat energy transferred during that time is devoted solely to the change of state. The combination of sensible heat and latent heat together represent the total heat, or **enthalpy**, of a substance.

Sensible heat: Heat energy that can be measured by a thermometer or sensed by touch.

Latent heat: The heat energy absorbed or rejected when a substance is in the process of changing state (e.g., solid to liquid, liquid to gas, or the reverse of either change) without a change in temperature.

Enthalpy: The total heat content (sensible and latent) of a refrigerant or other substance.

If heat is continually added to the pound of water after all the ice has melted, the thermometer will show an increase in temperature until the temperature eventually reaches 212°F (100°C). At this point, the water begins to boil (at sea level) and starts to change to its vapor state. We call it *steam*. As additional heat is added, the thermometer reading will remain steady at 212°F until all the water has become steam. During this time, latent heat is at work. Once all the water has been converted to steam, the thermometer will once again register a rise in sensible heat.

Figure 3 provides a graphical analysis of the process described here. The preceding discussion has been confined to one temperature scale to maintain clarity. To review the process using metric values, refer to *Figure 4*.

It takes a lot more heat to cause a change in state than is needed for a sensible change in temperature. For water, it requires 144 Btus of latent heat to change 1 lb of ice at 32°F (0°C) into 1 lb of water at 32°F. Consider that this is 144 times the heat needed to raise the temperature of water just one degree.

The change from water to steam requires an even greater amount of latent heat. It takes 970 Btus to change 1 lb of water at 212°F (100°C) to steam at 212°F. The takeaway is that an enormous amount of heat must be released or absorbed to change the state of a substance. In the HVACR craft, we use this fact to our advantage.

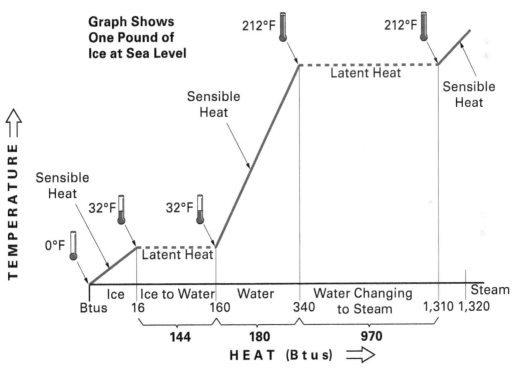

Sensible Heat:

Heat energy that can be measured by a thermometer or sensed by touch.

Latent Heat:

The heat energy absorbed or rejected when a substance is in the process of changing state without a change in temperature.

Figure 3 Changing the state of water.

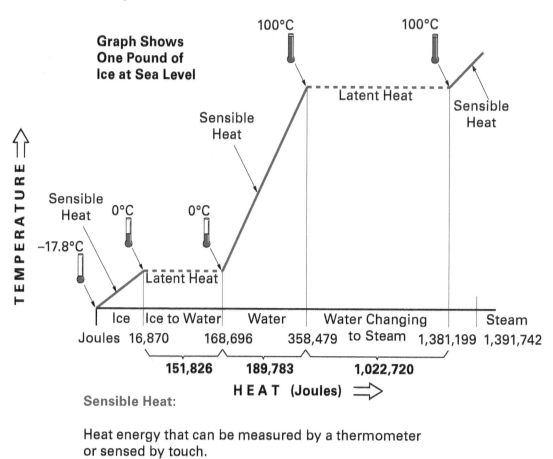

Figure 4 Changing the state of water (metric values).

Latent heat of fusion: The heat released from or absorbed by a liquid when changing to or from a solid, e.g., ice to water or water to ice.

Latent heat of vaporization: The heat absorbed by a liquid when changing to a vapor state, e.g., water to steam.

Saturation temperature: The temperature at which liquid changes to a vapor at a given pressure, which is also its boiling point. In a saturated condition, both liquid and vapor are likely present in the same space, as is the state of refrigerant stored in a cylinder.

Latent heat of condensation: The heat released from a vapor when changing to its liquid state, e.g., steam condensing to water.

Superheat: The increase in temperature of a refrigerant vapor after it has completed the change of state from a liquid to a vapor. Only after the phase change is complete can the vapor begin to increase in temperature as additional heat is absorbed.

Figure 5 graphically illustrates important terminology related to the change of state. Although water is still being used as an example, the terminology also applies to refrigerants and similar substances. Note that there are several names for latent heat, depending on the nature of the change of state.

- **Latent heat of fusion** — The heat absorbed or rejected in changing to, or from, a solid state, e.g., ice to water or water to ice.
- **Latent heat of vaporization** — The heat absorbed during the change from a liquid to a vapor, e.g., water to steam. The temperature at which this process begins is the boiling point. In refrigeration work, the boiling point is also known as the **saturation temperature**.
- **Latent heat of condensation** — The heat rejected from a vapor in changing back to a liquid state, e.g., steam to water.

Two other terms shown in *Figure 5* are very important for HVACR technicians to understand. Service technicians are concerned with these values on nearly every service or maintenance call. **Superheat** is the sensible (measurable) heat added to a vapor after the change of state from a liquid is complete. For example, water boils at 212°F (100°C) and remains at this temperature throughout the change of state. Once the last of the water has become steam and more heat is added, the sensible temperature of the steam begins to rise. Once the temperature increases beyond 212°F, the steam is described as *superheated*. If the steam temperature rises 5°F more to 217°F (102.8°C), then it is said to have 5°F (2.8°C) of superheat.

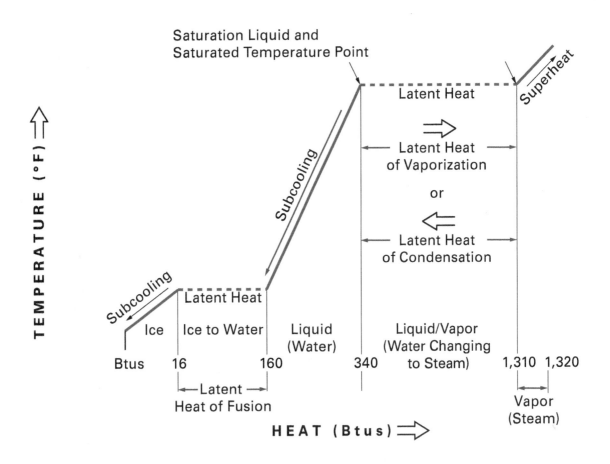

Figure 5 Terminology related to changing states.

Subcooling represents the sensible heat lost from a liquid after it has condensed from a vapor. Steam, if cooled down to 212°F, will begin to condense and change back to water. After that process is complete, the sensible temperature will begin to fall. If the water is cooled until it is 5°F cooler, to 207°F (97.2°C), it has been subcooled 5°F (2.8°C). Ice can be subcooled too. If it is cooled 5°F below the freezing point to 27°F (–2.8°C), the ice has been subcooled 5°F.

The importance of these terms and their meaning cannot be overstressed. Understanding them and the related concepts is essential to understanding how the refrigeration cycle transfers heat. As you continue working through this module and follow a refrigerant through its circuit, their importance will become clear.

Subcooling: The reduction in temperature of a refrigerant liquid after it has completed the change in state from a vapor. Only after the phase change is complete can the liquid begin to decrease in temperature as additional heat is removed.

1.1.4 Specific Heat Capacity

Specific heat is the amount of heat required to raise the temperature of one pound of a substance one degree Fahrenheit. Thus far, water has been used to demonstrate the process of changing states. It was shown that water has a specific heat of one Btu per pound, per each degree Fahrenheit of temperature change. This is written as 1 Btu/lb/°F. When the discussion is clearly related to specific heat, it may be written simply as 1.00.

Different substances have different specific heat values. The specific heat of water provides the baseline for all other substances. *Table 1* shows the specific heat values for some common substances.

Specific heat: The amount of heat required to raise the temperature of one pound of a substance by one degree Fahrenheit. Expressed as Btu/lb/°F. The specific heat of water (H_2O) is the standard at a specific gravity of 1.00.

TABLE 1 Specific Heat Values

Substance	Btu/lb/°F
Water	1.00
Ice	0.50
Air (dry)	0.24
Steam	0.48
Aluminum	0.22
Brass	0.09
Lead	0.03
Iron	0.10
Mercury	0.03
Copper	0.09
Alcohol	0.60
Kerosene	0.50
Olive oil	0.47
Pine	0.67

Specific heat values also vary from one state to another. As shown in *Table 1*, water has a specific heat of 1.00. Ice, however, has a specific heat of 0.50. The result is that it takes twice as many Btus of heat to raise one pound of water by one degree as it does to raise one pound of ice by one degree. If one Btu of heat is added to one pound of ice, the temperature increases 2°F.

This fact was illustrated in *Figure 3* and *Figure 4*. Because the specific heat of ice is 0.50, it took only 16 Btus of heat to raise the temperature of the ice from 0°F to 32°F.

1.1.5 Heat Transfer

Heat transfer is the movement of heat from one place to another, either within a substance or between substances. Heat always flows from a warmer location to a cooler location, like water running downhill. Remember that the three methods of heat transfer are conduction, convection, and radiation.

The rate at which heat is transferred varies for different substances. Some support the transfer of heat, while others restrict it. Materials that transfer heat readily and quickly are called **conductors**. **Insulators** are materials that resist heat transfer. Cork, fiberglass, and polyurethane foam are good examples of insulators.

Most metals are good conductors of heat. Copper and aluminum are used in refrigerant circuits because they are good thermal conductors. Thermal conductors are often good conductors of electricity as well.

The rate of heat transfer describes how fast heat can be added to or removed from an object or between objects. We often measure it in Btus per hour (Btuh). But since many thousands of Btus are often involved, a larger unit of measure is very useful. The larger unit is a **ton of refrigeration**.

One ton of refrigeration is defined as 12,000 Btus per hour (12,000 Btuh). The ton is commonly used in refrigeration work to describe the heating or cooling load for a space, or the nominal capacity of equipment. The ton is based on the amount of heat required to melt one ton of ice in a 24-hour period, as shown in *Figure 6*.

NOTE

Conduction, convection, and radiation are covered in detail in NCCER Module 03108, *Introduction to Heating*.

Conductors: Relevant to heat transfer, materials that readily accept or transfer heat by conduction.

Insulators: Materials that resist heat transfer.

Ton of refrigeration: Unit for measuring the rate of heat transfer. One ton is defined as 12,000 Btus per hour, or 12,000 Btuh.

1.2.0 Pressure/Temperature Relationships

An understanding of pressure/temperature relationships is essential for HVACR technicians. Before presenting the details of this relationship, we'll review the concept of pressure and its measurement.

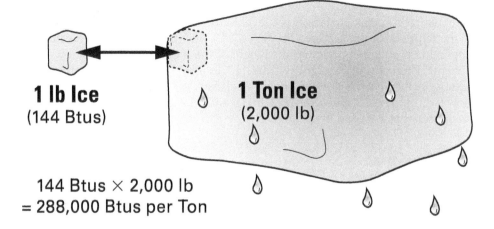

1 lb Ice
(144 Btus)

1 Ton Ice
(2,000 lb)

144 Btus × 2,000 lb
= 288,000 Btus per Ton

$$\frac{288,000 \text{ Btus}}{24 \text{ Hours Melt Time}} = 12,000 \text{ Btuh} = \textbf{1 Ton Refrigeration}$$

Figure 6 One ton of refrigeration illustrated.

1.2.1 Pressure

Pressure is defined as *force per unit area*. The state of a substance determines the direction pressure is exerted (*Figure 7*). Pressure may be exerted in one direction, several directions, or all directions. Using the three states of water as an example, ice (a solid) exerts pressure only in a downward direction. The same is true for all solids. As a liquid, water exerts pressure against the sides and bottom of the container. As a gas, it exerts pressure equally on all the surfaces of a container.

Pressure is expressed in pounds per square inch (psi) in the United States. The *kilopascal* (kPa) is another common unit of pressure measurement. The *pascal* is the official unit of metric pressure measurement. However, since the pascal is such a small unit of measure, the kilopascal is more commonly used in HVACR work. One kilopascal is equal to 1,000 pascals. Even so, a kilopascal is still much smaller than a pound per square inch. 1 psi is equal to about 6.9 kPa.

A variety of pressure units are encountered in the HVACR industry. For example, a pressure below **atmospheric pressure** (vacuum) is usually measured in inches of mercury (inHg). The next section explains how this unit of measure is determined.

Technicians must be able to accurately convert pressure values. There are many apps and websites that can convert all units of pressure measurement quickly and accurately.

Atmospheric pressure: The pressure exerted on Earth's surface by the weight of the atmosphere, equal to 14.7 psi, or 29.92 inHg, for 70°F air at sea level.

1.2.2 Absolute and Gauge Pressure

Earth is surrounded by a blanket of air. Our atmosphere primarily consists of oxygen, nitrogen, and water vapor. It has weight, and it exerts a force called *atmospheric pressure* on all things on Earth's surface. Atmospheric pressure can be measured with a **barometer**. For this reason, atmospheric pressure is often referred to as *barometric pressure*.

Figure 8 shows a simple mercury-tube barometer. The top of the tube is sealed, while the open end at the bottom rests in a dish of mercury. The tube doesn't touch the bottom. It is positioned just off the bottom, allowing mercury to freely come and go. Air pressure pushing down on the mercury in the dish causes the column to rise in the tube. An increase in atmospheric pressure pushes the mercury higher.

Barometer: An instrument used to measure atmospheric pressure, typically in inches of mercury (inHg).

Ice (Solid)

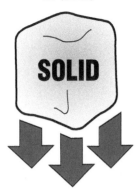

Water (Liquid)

Water Vapor (Gas)

Figure 7 The state of a substance determines the direction pressure is exerted.

Pressure Affects Boiling Point

As altitude increases, atmospheric pressure decreases. Because temperature and pressure are directly related, a liquid will boil at lower temperatures as the altitude increases. For example, in a high-altitude location like Denver, Colorado, where the elevation above sea level is approximately 1 mile (~1,600 meters), the atmospheric pressure is about 83 percent of sea level. In Denver, water boils at about 203°F (95°C).

Visualize a column of air with a cross section of one square inch (1 in²). It extends from Earth's surface at sea level to the farthest edge of the atmosphere. If the temperature at sea level is 70°F (21°C), and the weight of the column is applied to the dish of mercury, the height of the mercury column will be 29.92" (76 cm). The column's weight will be 14.7 lb (6.67 kg). This means that every square inch of Earth's surface (at sea level) has 14.7 lb of air pressure pushing down on it. These values of 29.92 inHg and 14.7 psi at sea level at 70°F are standards that are used frequently.

Did You Know?

Predicting Weather

Barometers measure atmospheric air pressure. Generally, an increased pressure indicates that the weather is pleasant. A reduced pressure indicates that the weather is cloudy and/or rainy.

Changes in barometric pressure are used to predict changes in weather conditions. For example, if it is raining outside, but the barometer is rising, it indicates that better weather is on the way. In the midst of a hurricane, the barometric pressure is an indicator of intensity. Extremely low pressures exist in powerful hurricanes. While traditional barometers work very well, sophisticated weather stations capture far more data and can transmit the information wirelessly.

Source: iStock@sarawutk

Source: iStock@JIRAROJ PRADITCHAROENKUL

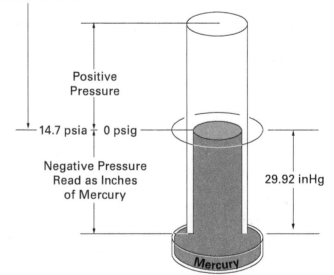

Figure 8 A mercury-tube barometer.

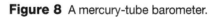

There are two scales based on the unit of pounds per square inch. One pressure scale, called the **absolute pressure** scale, is based on the barometer measurements just described. On this scale, pressures are expressed as pounds per square inch absolute (psia). Zero represents a total absence of pressure (a perfect vacuum). Using the absolute pressure scale, standard atmospheric pressure is correctly expressed as 14.7 psia.

Another scale, called the **gauge pressure** scale, is used for most practical work. Gauge pressure scales use the atmospheric pressure as the zero point. When a gauge reads zero, the 14.7 psi exerted by the atmosphere has already been considered.

Positive gauge pressures are expressed in pounds per square inch gauge, or psig. Negative pressures, those below 0 psig, are expressed in inches of mercury vacuum (inHg vacuum). Note that any confined space that is below atmospheric pressure is said to be in a vacuum. A *deep vacuum* would generally describe a space that is at, or approaching, 0 psia.

Gauge pressures can easily be converted to absolute pressures by adding 14.7 to the gauge pressure value. For example, a gauge pressure of 10 psig equals an absolute pressure of 24.7 psia, as shown here:

$$10 \text{ psig} + 14.7 \text{ psia} = 24.7 \text{ psia}$$

A comparison of the gauge and absolute pressure scales is shown in *Figure 9*.

A Perfect Vacuum

No matter how hard we try, we simply can't create a perfect vacuum. Even the naturally occurring, high-quality vacuum of space is not perfect, with some hydrogen atoms floating around in each cubic meter of space at a minimum. A perfect vacuum is defined as a space that has no matter at all. Also consider that a perfect vacuum can only exist at a temperature of 0 K—absolute zero. This is another unachievable concept with our current knowledge.

Although you won't be concerning yourself with creating a perfect vacuum, using a vacuum pump to remove the air and moisture from a refrigerant circuit before adding the refrigerant is a common task for an HVACR technician.

1.2.3 The Relationship Between Pressure and Temperature

Our discussion of water and its changes of state has been based on an important factor. The only pressure considered has been the atmospheric pressure at sea level. But pressure and temperature have a special relationship when it comes to the states of matter.

The temperature at which a liquid or gas changes state depends on the pressure. For example, as the pressure on water falls below the atmospheric pressure, its boiling temperature decreases as well. If the pressure increases, the boiling point increases. Using our water example, water boils at 212°F (100°C) at standard atmospheric pressure (14.7 psia, 0 psig, or 101 kPa). Because the pressure is only about 11.6 psia at 5,000' above sea level, water boils there at about 203°F (95°C).

Now think about a pressure cooker. If the pressure is increased to about 29.7 psia (15 psig + 14.7 psia atmospheric pressure), water boils at about 250°F (121°C). *Figure 10* shows the pressure/temperature relationship for water and its boiling point.

The refrigeration cycle is directly related to the effect pressure has on liquids and vapors. The temperature at which refrigerants change from one state to another can be manipulated by changing the pressure. Remember that increasing the pressure of a liquid raises the temperature at which it boils. Reducing the pressure lowers the boiling point. Controlling the pressure means controlling the state of the refrigerant.

Every refrigerant has its own unique pressure/temperature relationship. A typical pressure/temperature (P/T) chart that covers two common refrigerants is shown in *Table 2*. The chart shows the saturation pressure for each refrigerant at a given temperature. The saturation temperature is also the boiling point for the refrigerants. All the pressures shown are pounds per square inch gauge, except for those shown in italics. Values in italics represent inches of mercury below atmospheric pressure (vacuum).

Absolute pressure: Positive pressure measurements that start at zero atmospheric pressure, expressed in pounds per square inch absolute (psia). Gauge pressure plus the atmospheric pressure (14.7 psi at sea level at 70°F) equals absolute pressure.

Gauge pressure: Pressure measured on a standard gauge, expressed as pounds per square inch gauge (psig), that is calibrated to read zero with standard atmospheric pressure applied.

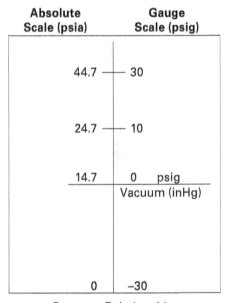

Pressure Relationship
psia vs psig

Figure 9 Relationship between absolute and gauge pressure.

TABLE 2 P/T Saturation Chart for HFC-134a and HCFC-22

Temperature		HFC-134a	HFC-410A
°F	°C	psig (*inHg*)	psig
–40	–40.0	*14.8*	11.6
–35	–37.2	*12.5*	14.9
–30	–34.4	*9.8*	18.5
–25	–31.7	*6.9*	22.5
–20	–28.9	*3.7*	26.9
–15	–26.1	*0.1*	31.7
–10	–23.3	1.9	36.8
–5	–20.6	4.1	42.5
0	–17.8	6.5	48.6
5	–15.0	9.1	55.2
10	–12.2	11.9	62.3
15	–9.4	15.0	70.0
20	–6.7	18.4	78.3
25	–3.9	22.1	87.3
30	–1.1	26.1	96.8
35	1.7	30.4	107.1
40	4.4	35.0	118.0
45	7.2	40.1	129.7
50	10.0	45.4	142.2
55	12.8	51.2	155.5
60	15.6	57.4	169.6
65	18.3	64.0	184.6
70	21.1	71.1	200.6
75	23.9	78.7	217.4
80	26.7	86.7	235.3
85	29.4	95.2	254.1
90	32.2	104.3	274.1
95	35.0	114.0	295.1
100	37.8	124.2	317.2
105	40.6	135.0	340.5
110	43.3	146.4	365.0
115	46.1	158.4	390.7
120	48.9	171.2	417.7
125	51.7	184.6	445.9
130	54.4	198.7	475.6
135	57.2	213.6	506.5
140	60.0	229.2	539.0
145	62.8	245.7	572.8
150	65.6	262.9	608.1

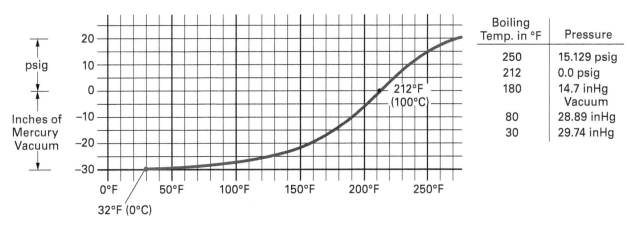

Figure 10 Pressure/temperature relationship for water.

Each of the two refrigerants has only one pressure associated with each temperature. Each combination of temperature and pressure represents the point of saturation for that refrigerant. For example, HFC-410A at 118 psig (814 kPa) has a saturation temperature of 40°F (4.4°C). At these conditions in a closed space, both liquid and vapor coexist in some proportion. If any heat is added, some of the liquid will vaporize until a new pressure/temperature balance is reached. If heat is removed, some of the vapor will begin to condense to a liquid.

The P/T chart helps you determine the state of a refrigerant inside a system. If a gauge is connected and a thermometer is attached at the same point, the readings can be compared to a P/T chart. The saturation temperature of 40°F for HFC-410A at 118 psig is used here as an example:

- If the gauge and thermometer read the same values as shown on the P/T chart, then both liquid and vapor are present in the space. A small volume of refrigerant could be in transition from one state to another. Since there is no measurable temperature change during a change of state, that cannot be confirmed. This P/T combination generally indicates that the refrigerant is stable, and no significant change of state is active.

- If the temperature reads 50°F (10°C) instead of 40°F (4.4°C) at the same pressure, then two things are apparent. First, there is no liquid refrigerant present here; all of the refrigerant is in its vapor state. Further, the refrigerant vapor has been superheated 10°F (5.6°C). Where superheat exists, liquid refrigerant cannot be present.

- If the temperature reads 30°F (–1.1°C) instead of 40°F (4.4°C) at the same pressure, it indicates there is no vapor present. All of the refrigerant is in its liquid state. In addition, the refrigerant has been subcooled 10°F (5.5°C). Where subcooling exists, refrigerant vapor isn't present.

You may not fully grasp these concepts until you have a chance to apply them. The next section provides an overview of a refrigeration cycle in action. It will further demonstrate the concepts learned thus far and how they apply to an operating system.

1.3.0 The Refrigeration Cycle

There are many variations of cooling systems used for comfort, food preservation, and industrial processes. What they have in common is a mechanical refrigeration system. *Figure 11* shows a very basic system with the following components identified:

- *Compressor* — Creates the pressure differential needed to make the refrigerant flow and allow the cycle to work.

- *Condenser* — A heat exchanger that transfers heat from the refrigerant to the outdoor air.

- *Metering, or expansion, device* — Creates a pressure drop that lowers the boiling point of the refrigerant before it enters the evaporator.
- *Evaporator* — A heat exchanger that transfers heat from the room to the refrigerant circulating within.

Also shown in *Figure 11* is the interconnecting piping—the refrigerant lines. Together, the components and lines form a closed circuit. The three primary refrigerant lines are as follows:

- *Hot gas, or discharge, line* — Carries hot refrigerant vapor from the compressor to the condenser. This line is where the highest pressure and temperature is found. It's also another place where only vapor is present.
- *Liquid line* — Carries liquid refrigerant from the condenser to the metering device.
- *Suction line* — Carries refrigerant vapor from the evaporator to the compressor. The suction line is where the lowest pressures are found, and, by design, only refrigerant vapor should be present.

The arrows in *Figure 11* show the direction of flow and the relative temperatures of the refrigerant. The purpose of the refrigerant is to transport heat from one place to another. Technically, a refrigerant is any substance that absorbs heat by evaporating at a low temperature and pressure and rejects heat by condensing at a higher temperature and pressure.

Remember that refrigerant circuit operation is basically the same in most systems. The events that take place within the system are continuous when it is running. Only the type of refrigerant used, the size and style of the components, and the locations of the components and the lines change from system to system. Various accessories are used in many systems when needed to perform special functions.

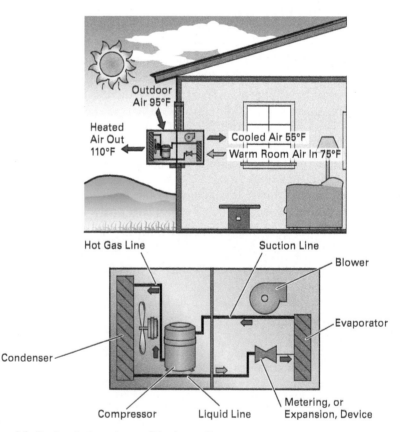

Figure 11 Basic window air conditioning unit.

1.3.1 Basic Component Functions

The refrigeration cycle is largely based on two principles already presented:

- As a liquid changes to a vapor (vaporizes), it absorbs a great deal of heat. As a vapor returns to its liquid state (condenses), it rejects a great deal of heat.
- The boiling point of a liquid can be changed by altering the pressure applied to it.

Refer to *Figure 11* once again. The refrigerant flows through the system in the direction indicated by the arrows. The compressor serves as a good starting point.

The compressor receives low-temperature, low-pressure vapor and compresses it. It creates a pressure differential, causing refrigerant to flow. The refrigerant moves toward the inlet of the compressor because it is where the lowest pressure is found. Like air, refrigerant flows from a point of higher pressure toward a lower pressure.

Once compressed, the refrigerant exits the compressor as a high-temperature, high-pressure vapor. The refrigerant vapor is now much hotter than the air outdoors. A temperature around 180°F (82°C) would be common. It travels to the condenser via the discharge line. The compression process also adds heat to the vapor. This is in addition to the heat it absorbed in the evaporator.

The condenser is an assembly of coiled tubing through which the hot vapor flows. The condenser must reject the heat that entered the system, including the **heat of compression**. Air that is cooler than the refrigerant vapor inside moves across the coil. The refrigerant vapor transfers its heat to the air. As it does so, it eventually cools to its condensing temperature for the current pressure. It then begins the change of state to its liquid form.

As the process continues, all the vapor eventually changes to liquid. The change of state is a latent process, so the refrigerant temperature didn't change until it was over. Now any additional cooling of the liquid can be measured. Since it will be cooled below the saturation temperature for the current pressure, any change represents subcooling. The high-pressure, high-temperature, subcooled liquid then travels through the liquid line to the metering device.

The metering device regulates the flow of refrigerant to the evaporator. It also causes a major pressure loss. Think of it as a restriction in the line, like placing your thumb over the end of a garden hose. It is a small hole, or *orifice*, that reduces the pressure of the liquid refrigerant.

Because of the pressure drop, the refrigerant will change into a cool or cold liquid at its saturation condition, ready to vaporize from heat absorbed in the evaporator. As it leaves the metering device, it is mostly liquid but also contains some vapor. What happens at the metering device will be discussed in greater detail later in this module.

The evaporator receives the low-temperature, low-pressure liquid refrigerant from the metering device. The evaporator is also an assembly of coiled tubing that exposes the cold liquid to the warmer air passing over. Heat is transferred through the tubing to the refrigerant. This causes the refrigerant to vaporize. For most of its journey through the evaporator, the refrigerant will continue to boil, and its temperature will remain stable as it does.

By the time the refrigerant leaves the evaporator, all of it should be vapor. Some amount of measurable superheat should be present, proving that only vapor is present. As it enters the suction line, it is a low-temperature, low-pressure vapor on its way to the compressor.

In the next section, we'll review the cycle again, but in greater detail.

1.3.2 Typical Air Conditioning System Refrigeration Cycle

A typical air conditioning system is shown in *Figure 12*. The system is divided into two sections based on pressure. The high-pressure side includes all the components in which the pressure is at or above the condensing pressure.

Heat of compression: Heat energy added to a vapor during the compression cycle. For compressors that place the motor in the vapor stream, motor heat is also added to the vapor.

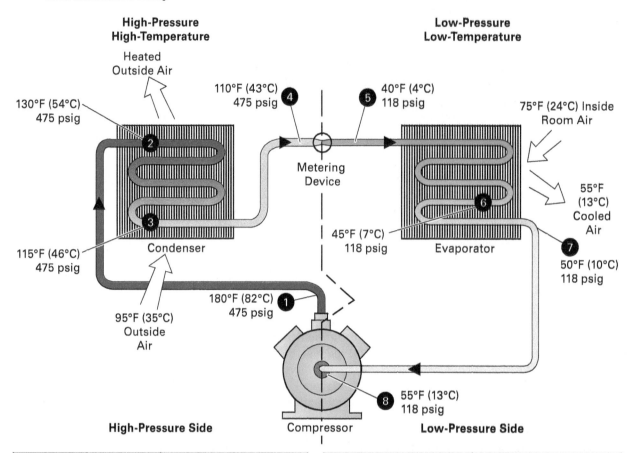

Temperature		Refrigerant/Pressure		Temperature		Refrigerant/Pressure	
°F	°C	psig	kPa	°F	°C	psig	kPa
30	–1	97	667	120	49	417	2,874
31	–1	99	681	121	49	423	2,913
32	0	101	696	122	50	428	2,952
33	1	103	709	123	51	434	2,992
34	1	105	724	124	51	440	3,031
35	2	107	738	125	52	445	3,071
36	2	109	753	126	52	451	3,112
37	3	111	768	127	53	457	3,153
38	3	114	783	128	53	463	3,194
39	4	116	798	129	54	469	3,235
40	4	118	814	130	54	475	3,278
41	5	120	829	131	55	482	3,321
42	6	123	845	132	56	488	3,363
43	6	125	862	133	56	494	3,407
44	7	127	878	134	57	501	3,451
45	7	130	895	135	57	507	3,495
46	8	132	911	136	58	513	3,540
47	8	135	929	137	58	520	3,585
48	9	137	946	138	59	527	3,631
49	9	140	963	139	59	533	3,677
50	10	142	980	140	60	540	3,724

HFC–410A Pressure/Temperature Chart

	High-Temp, High-Pressure Vapor
	High-Temp, High-Pressure Liquid
	Low-Temp, Low-Pressure Mixture
	Low-Temp, Low-Pressure Vapor

Figure 12 Typical refrigeration cycle using HFC-410A refrigerant.

The higher pressure of the system, beginning at the compressor discharge, is called the *head pressure, discharge pressure,* or *high-side pressure.* These terms are interchangeable.

The lower pressure of the system is called the *suction pressure* or *low-side pressure.* It begins at the outlet of the metering device. The dividing line between the low- and high-pressure sides of the system cuts through the compressor and the metering device. These locations are where the major pressure changes occur.

This system is typical of systems used for **comfort cooling**. It uses HFC-410A as the refrigerant. Note the P/T chart provided beneath the figure. HFC-410A reaches saturation at 40°F (4.4°C) at a pressure of 118 psig (814 kPa). The 40°F temperature is the typical target for evaporators used in comfort cooling.

The following walkthrough demonstrates the concepts and P/T relationships covered thus far. Assume a room temperature of 75°F (24°C) and an outdoor temperature of 95°F (35°C). These values, of course, vary from hour-to-hour in reality.

The numbered order of the walkthrough corresponds to the numbers shown in *Figure 12.* Thus, we begin at Point 1—the compressor discharge:

1. After compression, the hot, superheated gas from the compressor flows through the discharge line to the condenser. The gas is at 180°F (82°C) and at a pressure of 475 psig. Since the saturation temperature for 475 psig is 130°F (54°C), the refrigerant vapor has gained about 50°F (28°C) of superheat (180°F − 130°F = 50°F). All of this superheat must be removed before the vapor can condense.

2. Because the refrigerant in the condenser is significantly hotter than the outside air, it quickly rejects heat to the air passing through. The temperature falls to 130°F after a few passes through the condenser. At 475 psig, once the temperature falls to 130°F, condensation can begin. Heat continues to be rejected to the cooler air. Remember, though, that the refrigerant will remain at 130°F in the condenser until the change of state is complete. No sensible change in temperature occurs during this process.

   After the refrigerant has traveled most of the way through the condenser, all the refrigerant has condensed into a liquid. Once this happens, subcooling begins, reducing the temperature of the liquid.

3. At this point in the condenser, the liquid has lost another 15°F (8°C) of temperature to the air. In other words, the liquid refrigerant is now subcooled by 15°F (130°F condensing temperature − 115°F actual temperature = 15°F).

4. Subcooled liquid from the condenser travels through the liquid line to the metering device. Depending on its length and the environment the liquid line passes through, additional cooling may occur. In this example, the temperature has dropped another 5°F (3°C), lowering the refrigerant temperature to 110°F (43°C). Again, this extra 5°F of subcooling is completely dependent upon the environment. The temperature can also rise if the environment is very hot.

5. The metering device controls the flow of liquid refrigerant to the evaporator. The subcooled liquid from the condenser enters at 110°F and a pressure of 475 psig. Since the orifice is very small, the pressure falls to 118 psig at the exit. The result of that instant pressure drop is one of the most difficult parts of the cycle to understand.

   The pressure drop creates a very unstable condition for the refrigerant, as it departs from the carved-in-stone conditions of the P/T chart for this refrigerant. At this new low pressure, the refrigerant must rapidly cool to the 40°F (4°C) value shown on the chart. That represents a 70°F (39°C) temperature change that must happen spontaneously. The refrigerant must literally cool itself by instantly flashing some of the liquid refrigerant to a vapor. When it does, a lot of heat is instantly rejected from the liquid;

Comfort cooling: Cooling related to the comfort of building occupants. The temperature range for comfort cooling is considered to be 60°F (~15°C) to 80°F (~27°C).

NOTE

Liquid and vapor flowing through tubes and pipes always experience some pressure loss due to friction along the walls. For clarity, no pressure losses have been considered in the walkthrough that follows.

remember that the change of state requires a huge amount of heat to be transferred. All the heat required comes from the liquid, with some of it flashing to vapor until the new saturation condition is reached (40°F at 118 psig).

Once the refrigerant is through the metering device, there is a mixture of liquid and vapor. The ratio is roughly 80 percent liquid to 20 percent vapor. The vapor is referred to as *flash gas*. Although the vapor is useless for cooling purposes in the evaporator, it is a loss that must occur for the cycle to work.

The liquid-vapor mixture enters the evaporator at a temperature of 40°F and a pressure of 118 psig, boiling as heat is absorbed from the air passing over. The boiling continues until the liquid is gone and only vapor remains.

6. Since the refrigerant in the evaporator is cooler (40°F) than the room air (75°F), it absorbs heat and continues to vaporize. After traveling about 90 percent of the way through the evaporator, all the refrigerant has vaporized. The volume of vapor increased as it travelled through the evaporator, while the volume of liquid decreased. Remember that the refrigerant remains at 40°F throughout the process.

At Point 6, the temperature is 45°F (7°C). Since the temperature increased, it confirms that there is no more liquid remaining. If there was, the temperature would still be 40°F. The refrigerant now shows 5°F of superheat (45°F – 40°F = 5°F).

7. As it flows through the last few passes of the evaporator, the vapor continues to absorb heat from the warmer air, thus raising its temperature to 50°F, as shown. The vapor is now superheated 10°F (6°C). The vapor flows into the suction line and toward the compressor inlet. The room air is returned to the room at a temperature of about 55°F (13°C). The air temperature has been reduced 20°F (11°C). This is a common target for comfort cooling systems.

8. At the compressor, note that the temperature of the vapor is 5°F (3°C) warmer than it was at Point 7. The vapor can pick up additional superheat from the environment as it travels. As mentioned with subcooling, this is dependent on the environment in which the suction line is routed. The suction line is insulated, for several reasons. One is that any heat absorbed from the environment is heat the system must reject elsewhere. Thus, it robs cooling capacity from the application. It also prevents the tubing from "sweating" since the suction line temperature in operation is usually well below the dew point. If the suction line is short, there is little opportunity to absorb unwanted heat, but it will still sweat. In a packaged unit, for example, the suction line may only be 2' (0.6 m) long.

This walkthrough represents a typical system and how it is designed to operate. In reality, one or more of the conditions are constantly changing. There are many factors that can, and do, affect the cycle.

Think About It

Adding Capacity

As the liquid refrigerant passes through the metering device, some of the liquid must flash to a vapor. This cools the remaining refrigerant down to the saturation temperature for the current pressure. The flash gas is useless for cooling in the evaporator since it can't absorb much heat. But there is a way to minimize that loss. How might that be done?

1.4.0   Pressure and Temperature Measurement

Several different instruments are used to make temperature and pressure measurements. Accurate tools are important, since you will make many decisions based on what they report. Quality instruments are crucial to ensure a successful troubleshooting process.

1.4.1 Thermometers

Electronic thermometers (*Figure 13*) now dominate temperature measurement in the HVACR industry. Temperature-measurement capability has also been built into many multimeters, further increasing their versatility.

Electronic thermometers generally use either a **thermocouple** or **thermistor** type of temperature probe, or both, to sense the heat and generate a temperature indication. Thermocouple probes tend to be rugged and less expensive than thermistors. Thermistor probes use a semiconductor electronic element in which resistance changes based on temperature.

Several different probes are often used with the same electronic thermometer. For example, the long, slender probe in *Figure 13* is designed for use in air or for immersion in liquids. The other probe shown is designed to clamp onto tubing. A spring-loaded clamp keeps the sensor firmly against the tube. This probe is ideal for measuring the temperature of refrigerant lines. Refrigerant line temperature readings must be accurate, and accurate measurement requires excellent contact between the sensing device and the line.

Many electronic thermometers can accommodate two or more probes at the same time. This allows you to monitor more than one temperature at a time. These models can calculate and display the difference in temperature between the attached sensors.

Electronic thermometers are precision measuring instruments. Be sure to read and follow the manufacturer's instructions for operating and maintaining your thermometers. Calibration may occasionally be required.

Remember that thermometers, regardless of the type, are more accurate around the middle of their range. When reading temperatures that are close to the minimum and maximum values of their range, they are less accurate.

Stick Thermometers

Stick, or *pocket*, thermometers come in various forms (*Figure 14*). They are popular because they are rugged, small, and typically inexpensive. Because the scale of a dial thermometer is small, it may be hard to read accurately. A digital version, also shown in *Figure 14*, provides a digital readout.

Infrared Thermometers

Handheld infrared thermometers (*Figure 15*) are also very popular among technicians. Infrared thermometers allow the user to measure surface temperature without contacting the object. For this reason, they are also called *non-contact thermometers*. They work by detecting infrared energy emitted by the surface. Laser-sighting helps you aim them like a pistol. They are essential for taking temperature readings in hard-to-reach areas, like overhead piping. Highly reflective surfaces, however, can affect the reading. Some have an adjustable setting related to reflective surfaces. Applying paint or tape with a matte finish to the surface can also solve the problem.

It is important to understand the distance-to-spot (D:S) ratio of infrared thermometers. The D:S ratio is the ratio of the distance to the surface compared to the diameter of the temperature-measurement area (*Figure 16*). For example, if the D:S ratio is 12:1 and you measure an object 12" (~30 cm) away, it will average the temperature over a 1" (~25 mm) diameter spot. If the surface is 12' (~3.7 m) away, the spot becomes 1' (~0.3 m) in diameter. The greater the distance, the larger the averaged spot becomes. Unless the surface being measured is as large or larger than the spot, surfaces in the background will also be averaged. Be sure to know the D:S ratio for your instrument and use it properly. The ratio can vary from 1:1 to 60:1. Those with higher ratios are much better for measurements at a distance, but they are more expensive.

Thermocouple: A device comprised of wire made from two different metals joined at the end that generates a scalable voltage based on the temperature of the junction.

Thermistor: A semiconductor device that changes resistance with a change in temperature, typically used as a temperature sensor.

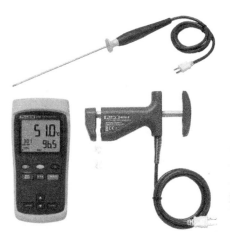

Figure 13 Electronic thermometer.
Source: Fluke Corporation

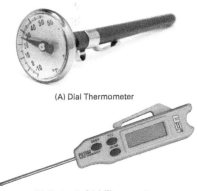

(A) Dial Thermometer

(B) Electronic Stick Thermometer

Figure 14 Dial and electronic stick thermometers.
Source: iStock@popovaphoto (14A)

Figure 15 Infrared thermometer.

Thermometer Testing

It's a good idea to check a thermometer occasionally and ensure it is accurate. While you may not be able to calibrate some, testing it lets you know if it needs attention. Simple dial thermometers often have a hex nut to turn the dial for calibration. Refer to *Figure 17*.

To check a thermometer at a high temperature, prepare a pot of boiling water. Place the probe in the pot with at least 2" (~5 cm) of it submerged. Don't allow the tip to touch or get too close to the bottom. With the water at a full boil, check the temperature reading; it should be 212°F (100°C). If not, remove it from the pot and use the hex nut to adjust it to the proper reading. After making an adjustment, retest it.

> **WARNING!**
>
> Burners and boiling water can cause serious burns. Use the appropriate PPE and handle the thermometer carefully when it is removed from the water. Ensure that no flammable materials are near the heating surface.

To check or calibrate a thermometer at a low temperature, prepare a mixture of crushed ice and only enough cool water to bring it to the top of the ice. Make sure that the ice isn't floating too far from the bottom. Place the thermometer in the mixture, again with at least 2" (~5 cm) of the probe submerged and the tip off the bottom. Wait until the temperature stabilizes and then check the reading. Adjust as necessary and retest.

All thermometers capable of measuring temperature in the ranges where water freezes and boils can be checked this way. However, not all thermometers can be field calibrated. Some may need to be returned to the manufacturer. Check the documentation that comes with the thermometer to determine how the calibration is done, or where to return it for service.

1.4.2 Gauge Manifolds and Other Pressure Measurement Devices

The gauge manifold is a frequently used HVACR service tool. It is used to charge or remove refrigerant and to measure pressures in an operating system.

Figure 18 shows a standard two-valve gauge manifold set. It consists of two pressure gauges mounted on a manifold with valves and hoses. A *compound gauge* is mounted on the left side of the manifold and a high-pressure gauge is mounted on the right. A compound gauge allows measurement of pressures both above and below atmospheric pressure. It is used to measure low-side, or suction, pressures. Most compound gauges used today can measure pressures above atmospheric pressure in a range from 0 psig to 250 psig. However, compound gauges are often equipped with a *retard range* up to 500 psig. The retard range provides an extra safety cushion for the gauge, but there is no level of accuracy in the retard range. They measure from 0 inHg to 30 inHg below atmospheric pressure.

> **NOTE**
>
> A standard compound gauge is not an accurate way of measuring vacuum levels. Other instruments, called micron gauges, are needed to accurately read vacuum levels. The gauge manifold can indicate the presence of a vacuum, however. Micron gauges are presented in NCCER Module 03205, *Leak Detection, Evacuation, Recovery, and Charging*.

> **WARNING!**
>
> Most of today's refrigerant gauge manifolds are designed to accommodate the pressures of HFC-410A. However, older manifolds, gauges, and hoses were manufactured with lower ranges that more closely match the pressures of older refrigerants. These manifolds, gauges, and hoses are not suitable for use with HFC-410A. Ensure that all gauge manifolds and accessories are rated for the appropriate pressures. The use of tools without the proper ratings may cause components to burst violently and without warning.

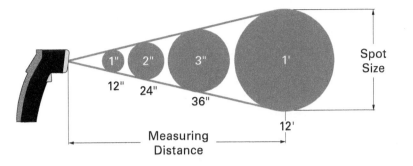

Figure 16 A 12:1 D:S ratio illustrated.

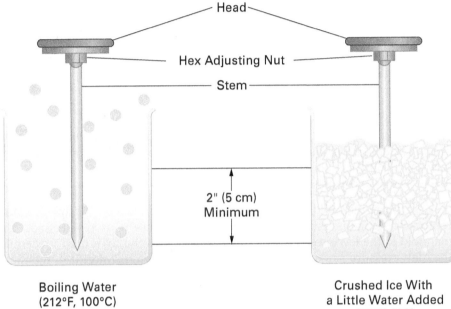

Figure 17 Calibrating a pocket thermometer.

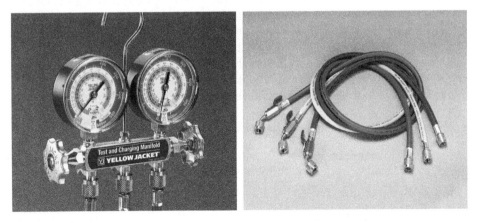

(A) Manifold with Gauges   (B) Hose Set

Figure 18 Gauge manifold set.
Source: Ritchie Engineering Company, Inc. / YELLOW JACKET

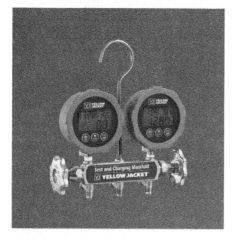

Figure 19 Manifold with electronic gauges.
Source: Rtchie Engineering Company, Inc. / YELLOW JACKET

The high-pressure gauge on the right side of the manifold is used to measure system high-side, or discharge, pressures. Most high-pressure gauges today can measure system pressures in the range of 0 psig to 800 psig. The range of both the low- and high-pressure gauges accommodates today's newer, environmentally friendly refrigerants such as HFC-410A. The hoses must also be rated for 800 psig.

Electronic versions of manifold gauges are also available (*Figure 19*). Most have the capability of reading the pressures and displaying the corresponding saturation temperature for the refrigerant in use.

A standard gauge manifold set has two hand valves and three hose ports. The hand valves are adjusted to control the flow of refrigerant to and from the system during servicing activities. To read system operating pressures, the valves remain closed. The gauge manifold is connected to the system and/or other service equipment through a set of service hoses. These hoses must be equipped with fast, self-sealing fittings that trap refrigerant in the hoses when disconnected. These fittings meet the requirements of the Clean Air Act by minimizing the release of refrigerants.

WARNING!

Once connected, refrigerant hoses are exposed to hundreds of pounds of refrigerant pressure. When disconnected, an uncontrolled hose can cause an injury, or liquid refrigerant can be released and cause burns. Wear the appropriate PPE and manage the hoses carefully when connecting to and disconnecting from a pressure source.

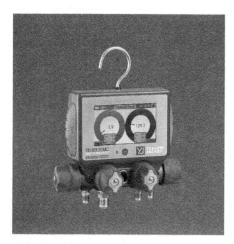

Figure 20 Four-valve electronic manifold.
Source: Ritchie Engineering Company, Inc. / YELLOW JACKET

Most gauge manifolds and service hoses are color coded. The low-pressure compound gauge is usually blue. A blue service hose is used to connect the manifold to the low-pressure side of a system. Red is the color used to identify the high-pressure gauge. A red service hose is used to connect it to the high-pressure side of the system.

The center hose port is the utility port. This port is normally connected with a yellow service hose to service equipment, such as a cylinder of refrigerant. When not in use, the utility port is best capped or plugged. Note that the hoses are the same except for their color.

Gauge manifold sets are also available with four hand valves and ports (*Figure 20*). This type of manifold offers convenience by eliminating the need to switch a single utility hose between different sources.

The gauge manifold is a precision tool. Gauge accuracy is critical. The technician must handle it with care both during use and in transport. The calibration of the gauges should be checked regularly. Without any pressure applied and open to the atmosphere, the gauges should read 0 psig. To check their accuracy, look straight ahead at the needle. It should split the 0 psig mark on center.

Standard analog gauges are easily calibrated as follows:

Step 1 Ensure that there is no pressure inside the hoses. They should be open to the atmosphere, without any residual refrigerant inside.

Step 2 Remove any protective boots around the gauge and unscrew the protective lens.

Step 3 The calibration screw is typically a small screw located on the face of the gauge. The needle position can be adjusted by turning this screw in one direction or the other with a small common screwdriver until the needle accurately splits the 0 psig mark (*Figure 21*).

Step 4 Lightly tap the gauge on the back to ensure that the needle is not under any tension that causes it to move away from the desired position.

Step 5 Reinstall the protective lens covers (do not overtighten).

Figure 21 Calibrating a refrigerant gauge.

Digital gauges can often be field calibrated as well. For some models, it is as simple as turning the gauge on, and then quickly tapping the On button two times. Other models may require a different procedure. Check the manufacturer's documentation to ensure the proper procedure is used.

Another instrument used to measure pressure is the manometer. Both electronic and non-electronic manometers (*Figure 22*) are common. Manometers are designed to measure very low pressures, usually in inches of water column or similar units. They are often used to measure the gas pressure in heating systems as well as the pressure in air distribution systems.

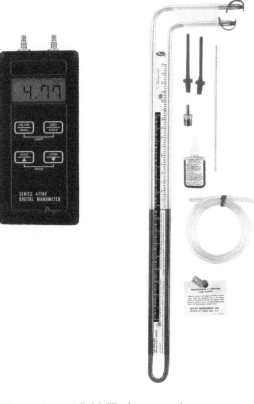

Figure 22 Electronic and fluid-filled manometers.
Source: Image courtesy Dwyer Instruments, LLC

1.0.0 Section Review

1. To convert a differential temperature in Fahrenheit (°F) to Celcius (°C), _____.
   a. divide by 1.8, then subtract 32°C
   b. multiply by 1.8
   c. multiply by 0.55
   d. divide by 0.55

2. If a refrigerant under pressure has a saturation temperature of 45°F, but the measured temperature is 55°F, what does this tell you?
   a. The refrigerant is likely contaminated with water.
   b. The refrigerant is likely mixed with another refrigerant.
   c. There is no vapor present; only liquid remains.
   d. There is no liquid present; only vapor remains.

3. Once compressed, refrigerant exits a compressor as a _____.
   a. low-temperature, high-pressure vapor
   b. high-temperature, high-pressure vapor
   c. low-temperature, high-pressure liquid
   d. high-temperature, low-pressure liquid

4. An infrared thermometer with a D:S ratio of 10:1 aimed at a surface 12'
away measures the temperature of a spot _____.
   a. 1.6' in diameter
   b. 1.2' in diameter
   c. 1' in diameter
   d. 0.83' in diameter

2.0.0 Refrigerants

Performance Task

3. Identify refrigerants using cylinder color codes.

Objective

Identify common refrigerants and their identifying characteristics.
   a. Identify fluorocarbon refrigerants.
   b. Describe the use of ammonia as a refrigerant.
   c. Identify various refrigerant containers and their safe handling requirements.

Fluorocarbon: A compound formed by replacing one or more hydrogen atoms in a hydrocarbon with fluorine atoms.

Refrigerants are used in cooling systems to transfer heat. This section briefly introduces several of them. You will learn more about their characteristics and specific uses in NCCER Module 03301, *Refrigerants and Oils*. This section is limited to fluorocarbon and ammonia refrigerants.

2.1.0 Fluorocarbon Refrigerants

Most manmade, or synthetic, refrigerants in popular use today are fluorocarbon compounds or a mixture of fluorocarbon refrigerants. Included in this group are refrigerants such as HCFC-123 (R-123), HCFC-22 (R-22), HFC-410A (R-410A), and HFC-134a (R-134a). These refrigerants all originate from one of two base molecules: methane and ethane.

Compounds: As related to refrigerants, substances formed by a union of two or more elements in definite proportions by weight; only one molecule is present.

Refrigerant and the Law

It is unlawful to knowingly release refrigerants into the atmosphere, per Sections 608 and 609 of the US Environmental Protection Agency's Clean Air Act. The punishment for knowingly and willfully releasing refrigerants into the atmosphere can be significant.

Substantial releases, even those that are accidental, must be reported. Because these substances have been linked to both global warming and the destruction of Earth's ozone layer, new refrigerants that are more environmentally friendly have emerged. However, they too must be carefully managed and most refrigerants currently in use cannot be released into the atmosphere.

Hydrocarbons: Compounds containing only hydrogen and carbon atoms in various combinations.

Halocarbon: Hydrocarbons, like methane and ethane, that have most or all their hydrogen atoms replaced with the elements fluorine, chlorine, bromine, astatine, or iodine.

Halogens: A term used to identify substances containing one or more of the elemental halogens chlorine, fluorine, bromine, astatine, or iodine.

Methane and ethane are called hydrocarbons because they are organic compounds that contain only hydrogen and carbon atoms. When most or all the hydrogen atoms in the methane or ethane molecule are replaced with elements such as chlorine, fluorine, or bromine, the new molecule is called a halocarbon, short for *halogenated hydrocarbon*. Chlorine, fluorine, and bromine are chemically related elements called halogens. To *halogenate* a substance means to cause some other element to combine with a halogen. When all the hydrogen atoms in a hydrocarbon molecule are replaced with chlorine or fluorine, the molecule is said to be *fully halogenated*.

Halocarbons in which at least one or more of the hydrogen atoms have been replaced with fluorine are called *fluorocarbons*. Fluorocarbon refrigerants fall into three groups: chlorofluorocarbons (CFCs), hydrochlorofluorocarbons (HCFCs), and hydrofluorocarbons (HFCs), based on their chemical structure. Note that fluorocarbons must contain some fluorine. Halocarbons may contain fluorine as well as other halogens, or they may not contain any fluorine at all.

2.1.1 CFCs, HCFCs, and HFCs

Traditionally, each refrigerant had a trade name or an R (refrigerant) name. Names like R-22, R-12, R-502, etc., were assigned by the American Society of Heating, Refrigerating, and Air-Conditioning Engineers (ASHRAE). These names were substituted for the true chemical names. For example, R-22 describes the refrigerant with the chemical name of *chlorodifluoromethane*. You will likely agree that numbers are much better than long chemical names.

However, the way refrigerants are named has changed. ASHRAE now uses acronyms ahead of the number, such as CFC, HCFC, and HFC. These acronyms help identify the chemical family the refrigerant comes from on sight.

The number previously assigned to a given refrigerant by ASHRAE has been retained. Both old and new names are used in the craft. Some examples of changed names for commonly used refrigerants are shown in *Table 3*.

TABLE 3 Examples of Old and New Refrigerant Names

Old Name	New Name
R-22	HCFC-22
R-123	HCFC-123
R-134a	HFC-134a
R-407C	HFC-407C
R-410A	HFC-410A

CFCs, or *chlorofluorocarbons*, contain chlorine, fluorine, and methane. These refrigerants were very popular for many years. Well-known CFCs include CFC-12, previously used in automobiles and refrigerators, and CFC-11, a low-pressure refrigerant traditionally used in centrifugal chillers. CFCs are now rarely encountered. They have been phased out due to their chlorine content.

HCFCs, or *hydrochlorofluorocarbons*, contain hydrogen as well as the other substances, as the name implies. They have far less chlorine in their structure than CFCs. Popular HCFCs include HCFC-22, which is still used in a number of comfort cooling systems. The production of new HCFC-22 has been limited for years. As of 2020, no new HCFC-22 is manufactured in the United States. However, recycled and reclaimed HCFC-22 can still be used.

HFCs, or *hydrofluorocarbons*, do not contain any chlorine. A great deal of refrigerants used today are HFCs, such as HFC-410A and HFC-134a. Since they do not contain chlorine, they do not threaten the ozone layer.

NOTE

As of model year 2021, US motor vehicle air conditioning (MVAC) systems can no longer be based on HFC-134a, due to the excessive global warming potential (GWP).

Refrigerant Compound Numbering

Each refrigerant compound has a specific chemical composition, which can be shown graphically. For example, the composition of HCFC-22 appears as shown:

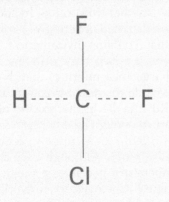

The name of the refrigerant is derived from its components. You can see from this graphic that it is an HCFC—H for hydrogen, the first C for chlorine, F for fluorine, and the second C for carbon. The numbers—22 in this case—relate to the number of atoms of each component in a single molecule of the refrigerant. Refrigerant blends have a different numbering approach.

There is evidence that the ozone layer surrounding Earth is damaged by various chemicals, most notably chlorine. The ozone layer filters out harmful radiation from the sun that would otherwise reach the surface and harm life in various ways. The chlorine in CFC and HCFC refrigerants has contributed to the problem. Since the passage of the Clean Air Act in 1990, refrigerants bearing chlorine have come under increasing government regulation and control. The US Environmental Protection Agency (EPA) is responsible for making and enforcing laws pertaining to the use of these refrigerants.

CFCs had the greatest impact on the ozone layer because they contained the greatest volume of chlorine. Since this class of refrigerants posed the greatest threat to the environment, the production of CFCs was banned years ago. Many new refrigerants have been developed to take their place. HCFC refrigerants are currently being phased out, although not as aggressively. They too contain chlorine, but in lesser amounts.

Most refrigerants in common use today are HFCs that contain no chlorine at all. They are believed to be harmless to the ozone layer. Although this helps protect the ozone layer, HFCs do contribute to the greenhouse effect and appear to accelerate global warming when released to the atmosphere. As a result, their release is also unlawful. Due to the role they appear to play in global warming, HFCs are being replaced with new refrigerants as they become available and are deemed safe for use.

2.1.2 Refrigerant Blends

Many new refrigerants have been developed by blending two or more refrigerants together. Blends can be made from combining similar HCFCs or HFCs together, or by combining the two types. Since HCFCs are being eliminated, any blend that contains an HCFC will be banned in the United States, if it hasn't already. If a refrigerant blend contains an HCFC, then its name must also begin with that acronym.

A compound is a substance formed by a union of two or more elements in definite proportions by weight. When a compound is created, only one molecule is present. HCFC-22 is a good example of a pure compound. Compounds exhibit very predictable behavior in the refrigeration circuit.

However, many blends are not pure compounds. When two or more substances are blended, they usually don't form a new compound. In fact, the molecules from the blend components remain intact and are still present in the blend. A blend that retains the properties of its components is simply a refrigerant **mixture**.

The behavior of a mixture may be predictable, but there are differences in how they behave in a system. For example, when a leak develops in a system, it is possible that one refrigerant in the mixture will escape in a larger proportion than another refrigerant. The location of the leak and the system operating status can even determine whether this occurs. In addition, the superheat and subcooling values must be calculated differently.

Consider a mixture that contains hypothetical Refrigerant 1 (R-1) and Refrigerant 2 (R-2). Assume the two have different boiling points and have their own unique P/T charts. In a mixture, they both retain their previous characteristics. Therefore, while the new refrigerant mixture 1/2 is at work in the refrigeration circuit, R-1 may be in the process of boiling and vaporizing while R-2 has not yet absorbed enough heat to begin the process.

This makes it more difficult to accurately determine the superheat or subcooling values for the mixture. Although each mixture has its own P/T chart, the saturation temperature becomes a range of temperatures, rather than one specific temperature. Chemists try to work with blend components that do not differ dramatically, but minor differences in the behavior of the two components are common. It is important to understand that all refrigerants do not behave precisely as described earlier in this module.

Mixture: As related to refrigerants, a blend of two or more components that do not have a fixed proportion to one another and that, however thoroughly blended, are conceived of as retaining a separate existence; more than one type of molecule remains present.

2.2.0 Ammonia

Anhydrous ammonia (R-717) has excellent heat-transfer qualities and is used mainly in large medium- and low-temperature applications. These include ice plants, food storage, and large food-processing plants.

Since ammonia is not in the same family as the other refrigerants discussed, the ASHRAE-assigned designation does not contain one of the CFC, HCFC, or HFC acronyms. The term *anhydrous* means that it is free of water. Ammonia is highly soluble in water. When ammonia must be discharged from a system, it is frequently discharged in water to reduce the local odor.

Though not classified as poisonous, ammonia vapors have a harsh effect on the respiratory system. Only small quantities of the vapor can be safely inhaled. Exposure for five minutes to 50 parts per million (ppm) is the maximum exposure allowed by the Occupational Safety and Health Administration (OSHA). Ammonia is hazardous to life at 5,000 ppm and is flammable at 150,000–270,000 ppm.

Ammonia has a distinct odor that can be detected at just 3–5 ppm. The odor becomes very irritating at 15 ppm. Anyone working on an ammonia system needs to be specifically trained to handle it safely and have the proper equipment. As far as the environment is concerned, ammonia is considered quite safe.

WARNING!

Ammonia (NH_3) is classified by *ASHRAE Standard 34* as a Group B2L refrigerant, due to concerns regarding toxicity and flammability. Ammonia's sharp and irritating pungent odor is self-alarming, typically causing workers to vacate an area well before dangerous thresholds of inhalation develop. This also prompts quick leak repairs. However, it is important to recognize that certain concentrations of ammonia in air can be toxic. For more information regarding the safe handling of ammonia, visit ASHRAE at **www.ashrae.org**; the National Institute for Occupational Safety and Health (NIOSH) at **www.cdc.gov/niosh**; and the International Institute of Ammonia Refrigeration (IIAR) at **www.iiar.org**.

Ammonia Applications

Ammonia (NH_3) has been used in refrigeration systems in the United States since the late 1800s. The first patents for ammonia refrigeration machinery were filed in the 1870s. It was first used in ice plants. Practically everything we eat, as well as many beverages, is processed or handled by at least one facility that uses ammonia as a refrigerant before it reaches our hands.

Technicians working with ammonia systems must use refrigerant gauge manifolds and gauges constructed of steel, as well as hoses with steel fittings.

Source: Ritchie Engineering Company, Inc. / YELLOW JACKET

Since ammonia does not represent an environmental problem and is reasonably priced in comparison to synthetic refrigerants, it remains popular for large-scale applications. Ammonia has four distinct advantages over the halogenated refrigerants:

- Ammonia is roughly 3–10 percent more efficient in transferring heat than halocarbon refrigerants. This means that less ammonia must be circulated to transfer an equal amount of heat. As a result, the operating cost for ammonia-based systems can be significantly lower.

- Since less ammonia needs to flow, pipe sizes can be smaller.

- Ammonia is substantially less expensive than manufactured refrigerants.

- Ammonia is safer for the environment, since there is no potential for ozone destruction, and it does not contribute to the greenhouse effect.

To be fair, there are also some disadvantages:

- Ammonia is toxic in very high concentrations. However, it's obvious smell is easy to detect at levels far below those that represent a hazard. Personnel who suspect a problem typically have plenty of time to react. Ammonia vapors also rise quickly, since they are lighter than air, helping to eliminate the hazard.

- Ammonia and copper are not compatible. This means that all components that ammonia encounters in an operating system must not contain copper. Piping systems for ammonia are typically made exclusively of steel. Heat transfer components such as evaporators or condensers usually contain steel or aluminum tubing.

Ammonia will likely remain a very popular refrigerant in the food-processing industry for years to come. Although it is not cost-effective for use in most small systems, it is an excellent choice for large systems.

2.3.0   Refrigerant Containers and Handling

Refrigerants come in disposable, returnable, or refillable metal containers that vary in shape and size (*Figure 23*). Low-pressure refrigerants such as HCFC-123 come in standard steel drums or cylinders. They have boiling points close to, or slightly above, common room temperatures. The pressure they exert on the container is much less than that of medium- and high-pressure refrigerants, such as HCFC-22, HFC-134a, and HFC-410A. If improperly handled, the pressurized containers can burst or leak, causing damage, injury, or even death in some cases.

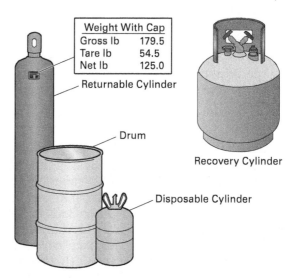

Figure 23 Refrigerant containers.

Refrigerant Safety

All refrigerants and their containers must be handled with care. Refrigerants can cause injuries several different ways.

Any refrigerant container under pressure can potentially burst or leak. A container of refrigerant with a low boiling point that bursts will expel liquid refrigerant in all directions. The refrigerant will be rapidly absorbing heat and flashing from liquid to vapor. If the refrigerant liquid contacts you, a frostbite injury is likely. Whenever handling refrigerants, technicians should wear gloves and safety glasses at a minimum.

Although most refrigerants used are non-toxic, enough refrigerant vapor in a confined space displaces the oxygen and makes breathing difficult or impossible. It is always a good idea to be aware of the volume of refrigerant contained in a room or similar space where work is going on. Consider what steps need to be taken if there is a sudden and uncontrolled release of refrigerant in the area in which you are working. Review the SDS for any chemical substances you use on the job.

2.3.1 Disposable Cylinders

Disposable cylinders are not manufactured for reuse. Since they are steel, they should be stored in dry locations to prevent rusting. They should also be transported carefully. Over time, rough handling or excessive heat could cause them to explode, especially if weakened by rust or corrosion. For added protection, keep disposable refrigerant cylinders in their original cartons when practical.

> **WARNING!**
>
> Disposable cylinders must never be refilled. It is not only dangerous but also against the law. Violators can be fined up to $25,000 and can face up to five years in jail.

For many years, most disposable cylinders were either 30 lb or 50 lb packages. However, many new refrigerants come in 24 lb, 25 lb, or 27 lb cylinders.

Disposable cylinders should be recycled as scrap once they are empty. Be sure that the cylinder pressure is reduced to 0 psig, and then render the cylinder useless by puncturing the **rupture disk** (*Figure 24*).

2.3.2 Returnable Cylinders

Returnable cylinders go back to the manufacturer to be serviced and refilled. They may also be used for recovering large refrigerant volumes. These containers should never be filled beyond 80 percent of their volume. Overfilling can result in an explosion if the contents are heated and begin to expand. The pressure increases rapidly with even slight changes in temperature if the liquid has no room for expansion. The term *full* means a vessel that is filled to 80 percent of its physical capacity. The remaining 20 percent of the volume must remain available for expansion.

Cylinders and tanks are stamped with various weight values. These values are shown in *Figure 23* and include the following:

- *Tare weight* — The empty weight of the vessel.
- *Gross weight* — The combined weight of the vessel (tare weight) plus the weight of the refrigerant when the vessel is full.
- *Net weight* — The weight of the contents in the vessel. For example, when ordering 50 lb of refrigerant from a supplier, it is the net weight that is being identified. Manufacturers design vessels so that when the full net weight is reached, 20 percent of the volume remains for liquid expansion.

The date when a cylinder was last tested is stamped on the shoulder or collar of returnable cylinders. Returnable and reusable cylinders must typically be retested every five years.

2.3.3 Recovery Cylinders

Cylinders with a 50 lb capacity are commonly used for recovery. These cylinders are very strong and can be reused many times. Their tare weight, at around 28 lb, is much heavier than a disposable cylinder. All cylinders designed for the recovery and storage of used refrigerants are painted gray with the cylinder shoulders painted yellow.

A returnable cylinder should never be substituted for a recovery cylinder. If the cylinder does not have the proper markings for a recovery cylinder, do not use it for this purpose. Like the others, recovery cylinders must never be filled beyond 80 percent of their capacity.

The US Department of Transportation (DOT) governs the construction and labeling of cylinders used for refrigerant storage and recovery. Cylinders used to recover HFC-410A must be specifically manufactured and labeled for use with this refrigerant due to the higher pressures.

2.3.4 Identifying Refrigerants

Refrigerant containers are color coded and labelled to identify the refrigerant they contain. These labels also include important safety and health information. *Table 4* lists the color for some common refrigerant containers. The color helps you readily identify refrigerants.

Rupture disk: A pressure-relief device that protects a vessel or other container from damage if pressures exceed a safe level. A rupture disk typically consists of a specific material at a precise thickness that will break or fracture when the pressure limit is reached, creating a controlled weakness, thereby protecting the rest of the container from damage.

Rupture Disk

Figure 24 Rupture disk of a disposable cylinder.

TABLE 4 Color Codes for Common Refrigerants

Refrigerant	Color Code
HCFC-22	Light green
HCFC-124	Dark green
HCFC-409A	Medium brown
HFC-410A	Rose
HFC-422B	Dark blue
HCFC-123	Light blue-gray
HCFC-408A	Medium purple
HFC-404A	Orange
HFC-407C	Brown
HFC-134a	Light blue

If the type of refrigerant in a system is unknown, it can usually be identified by using one of the following methods:

- Check the manufacturer's service literature for the equipment.
- Check the nameplate on the equipment. If a refrigerant conversion has taken place, the unit should be clearly labeled.
- Check the data marked on the metering device.

2.3.5 Refrigerant Safety Precautions

Butyl-lined gloves and safety glasses must be worn to avoid getting refrigerant on your skin or in your eyes. When released to atmospheric pressure, refrigerant can cause frostbite or burn the skin.

Refrigerants can cause suffocation. Although common refrigerants are not toxic, heavy concentrations displace oxygen, depriving the body of breathable air. Always maintain ample ventilation. Refrigerant vapor is invisible, has little or no odor, and is heavier than air. Be especially careful in low places where it might accumulate.

Equipment rooms and other areas where there are large volumes of refrigerant must have alarm systems that detect low oxygen levels and sound an alarm. A self-contained respirator must be available near areas containing large volumes of refrigerant. Use a respirator if you must enter a contaminated area. Some equipment rooms have a mechanical ventilation system that can clear contaminated air from the room quickly.

In addition to wearing protective clothing and equipment, follow these rules when handling refrigerants:

- Always double-check to be sure you are using the proper refrigerant. The containers are color coded and labeled to identify their contents. Container labels also include product, safety, and warning information.
- Refer to technical bulletins and Safety Data Sheets (SDSs) for information important to your health. They describe the flammability, toxicity, reactance, and health problems that could be caused by a refrigerant or other substance.
- Do not drop, dent, or abuse refrigerant containers. Do not tamper with their safety devices unless the intent is to disable an empty disposable cylinder.
- When some refrigerants are exposed to an open flame, a toxic gas may be formed. Avoid situations that place refrigerant vapors and an open flame, such as a brazing torch, in the same space.
- Use a proper valve wrench to open and close valves that require a tool; do *not* use adjustable wrenches.
- Replace the valve cap and hood cap to protect the cylinder valve when not in use.
- Secure containers in place during transport to protect them. Secure containers with straps or chains in an upright position.
- Do not store containers where the temperature can cause the pressure to exceed the cylinder relief valve settings.

2.0.0 Section Review

1. The difference between a halocarbon and a fluorocarbon refrigerant is that _____.
   a. the fluorocarbon always has fluorine in it, but not all halocarbons do
   b. fluorocarbon refrigerants contain fluorine, while halocarbon refrigerants still have carbon
   c. a halocarbon refrigerant never has chlorine in it while a fluorocarbon refrigerant does
   d. halocarbon refrigerants are highly toxic

2. OSHA established the *maximum* exposure to ammonia vapors for a five-minute period at how many parts per million (ppm)?
   a. 25 ppm
   b. 50 ppm
   c. 75 ppm
   d. 100 ppm

3. HFC-410A refrigerant comes in a disposable container that is painted _____.
   a. dark brown
   b. light green
   c. dark green
   d. rose

3.0.0 Cooling System Components

Objective

Identify the major components of cooling systems and their function.
   a. Identify various types of compressors.
   b. Identify different types of condensers.
   c. Identify different types of evaporators.
   d. Describe devices used to meter refrigerant flow.
   e. Recount basic refrigerant piping concepts.
   f. Identify various accessories used in refrigerant circuits.

Performance Task

4. Identify the compressor, condenser, evaporator, metering device, and accessories in a cooling system.

Primary components required for the operation of a mechanical refrigeration system include compressors condensers, evaporators, metering devices, and piping. These components are reviewed in detail in the following sections.

3.1.0 Compressors

The compressor is the heart of the refrigeration circuit. It creates the pressure difference that causes refrigerant to flow. The direction of flow is always from a higher pressure to a lower pressure, much as heat moves from a higher temperature to a lower temperature. The result is that liquids and gases can be made to move by adjusting or changing the pressure around them. Liquids cannot be compressed, and substantial amounts of liquid flowing into a compressor can easily damage or destroy it.

3.1.1 Compressor Configurations

Compressors are usually driven by an electric motor. Very large compressors can be driven by internal combustion engines or steam turbines, but they are rare. Compressors are divided into the following three groups based on the way they are joined to their motors or engines:

- Open compressors
- Hermetic compressors
- Semi-hermetic compressors

Open Compressors

An open compressor is physically separated from its motor. The drive shaft of the compressor extends through the case. A special mechanical seal is used on the rotating shaft to prevent refrigerant leakage. Open compressors can be disassembled in the field for service and repair if needed.

The motor can be connected directly to the drive shaft in a *direct-drive* arrangement, as shown in *Figure 25*. This means the compressor turns at precisely the same speed as the motor. However, the compressor can also be belt driven. The motor is then placed beside the compressor rather than behind it. Belt-driven arrangements allow the motor to rotate at one speed while the compressor rotates at a different speed. The combination of pulley sizes determines the speed.

Hermetic Compressors

The hermetic compressor and motor are sealed together in a welded steel shell. They share a common drive shaft. Hermetic compressors (*Figure 26*) are compact, less noisy, and require no maintenance since they are sealed. This also means the compressor must be replaced when it fails, regardless of the nature of the problem. The failed compressor is scrapped or returned to the manufacturer for evaluation or warranty credit.

Hermetic compressors are limited in capacity. Hermetic models with capacities of 10 tons or less are common. Larger models do exist but are rare.

Figure 25 Open, direct-drive industrial refrigerant compressors.
Source: iStock@kadmy

Figure 26 A hermetic scroll compressor.

Semi-Hermetic Compressors

In a semi-hermetic compressor (*Figure 27*), the compressor and motor share the same housing but not a drive shaft. The two shafts are mechanically connected inside. These compressors fill the space between open and hermetic models. They can be partially opened for field service and repair by removing the heads (*Figure 28*). This allows access to the valves.

If the motor fails, the entire compressor is replaced. Failures related to the crankshaft, pistons, and piston rods also result in replacement. Repairing these items in the field are very difficult without shop equipment. However, they are rarely scrapped when they fail. Instead, they are disassembled and remanufactured or rebuilt at shops designed for this purpose. They are then resold as *rebuilt* or *remanufactured*. A remanufactured compressor is generally considered a better investment, since all the components are returned to factory specifications.

Figure 27 A semi-hermetic compressor.

Figure 28 A cut-away showing some of the field-serviceable components of a semi-hermetic compressor.

The following five types of compressors are commonly used in mechanical refrigeration systems:

- Reciprocating
- Rotary
- Scroll
- Screw
- Centrifugal

3.1.2 Reciprocating Compressors

Reciprocating compressors remain very common, although scroll compressors have taken their place in many applications. They use one or more pistons moving back and forth within a cylinder or cylinders (*Figure 29*). The pistons, like the pistons of an automotive engine, compress the refrigerant vapors. The suction and discharge valves are synchronized with the piston action. They control the intake and discharge of refrigerant vapor. The compressors shown in *Figure 25* and *Figure 27* are reciprocating models.

Figure 30 shows a cut-away of a hermetic reciprocating compressor. The motor is on top, with the piston oriented sideways. Reciprocating compressors are found in open, semi-hermetic, and hermetic forms. Semi-hermetic reciprocating compressors are used in commercial air conditioning. Open reciprocating compressors are used mostly in industrial refrigeration.

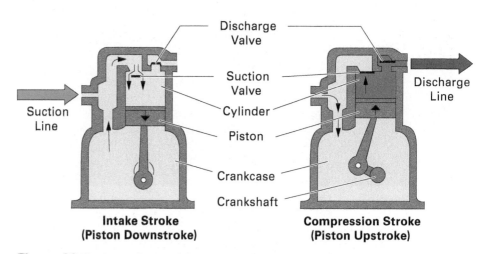

Figure 29 Reciprocating compressor operation.

3.1.3 Rotary Compressors

Rotary compressors are usually hermetic compressors. They are frequently used on appliances, room air conditioners, and central air conditioning systems with less than five tons of cooling capacity. There are two basic types of rotary compressors: *stationary vane* and *rotary vane*. The rotary-vane type is used in HVACR applications.

The rotor of rotary-vane compressors (*Figure 31*) is centered on the drive shaft. However, the drive shaft is positioned off-center in the cylinder. Two or more vanes that slide in and out to follow the shape of the cylinder are mounted on the rotor. As the rotor turns, the vanes trap refrigerant vapor and compress it against the cylinder wall. The vapor is forced out of the discharge port. The vanes also separate the high- and low-pressure areas from each other.

Figure 30 Cut-away view of a hermetic reciprocating compressor.

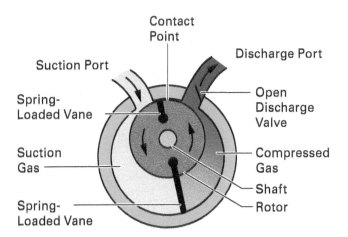

Figure 31 Rotary-vane compressor operation.

Scroll Compressor Sounds

Scroll compressors often produce unusual sounds when they start up and shut down. If you have never heard the sounds before, you might think the compressor is defective. When you have the opportunity during your training, listen closely to the sounds so you can recognize them in the field.

3.1.4 Scroll Compressors

Scroll compressors (*Figure 32*) are hermetic models. Of all the compressor types, scroll compressors have the fewest moving parts. Since their development, their use has grown substantially. They are now used in most residential and light-commercial cooling systems on the market. Durability and efficiency are the primary reasons.

Unlike reciprocating compressors, there are no suction or discharge valves. However, a check valve is used near the discharge port to help prevent reverse rotation. The scroll compressor achieves compression using two spiral-shaped parts called *scrolls* (*Figure 33*). One is fixed; the other is driven and moves in an orbiting action inside the fixed scroll. There is light contact between the two pieces.

Refrigerant gas enters the suction port at the outer edge of the scroll (*Figure 34*). The orbiting action draws gas into pockets between the two spirals. As rotation continues, the gas opening is momentarily sealed off and the gas is forced into smaller pockets. The compressed vapor is pushed out through the discharge port at the center of the stationary scroll.

3.1.5 Screw Compressors

Screw compressors (*Figure 35*) are used in large commercial and industrial applications requiring capacities from 20–750 tons. They are made in both open and hermetic styles. They are most often found on chillers.

Screw compressors use a matched set of screw-shaped rotors, one male and one female, enclosed within a cylinder (*Figure 36*). The motor drives the male rotor, which indirectly drives the female rotor. The male rotor normally turns faster than the female rotor because it has fewer lobes. Typically, the male rotor has four lobes, while the female rotor has six.

As the rotors turn, they compress the gas between them. The screw lobes form the boundaries separating several compression chambers, which move down the compressor at the same time. The gas entering the compressor is moved through a series of progressively smaller pockets until the gas exits at the compressor discharge. This characteristic makes them similar to scroll compressors.

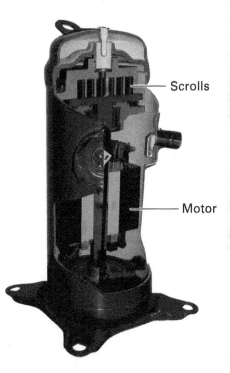

Figure 32 Scroll compressor cutaway.

Figure 33 Scroll plates.

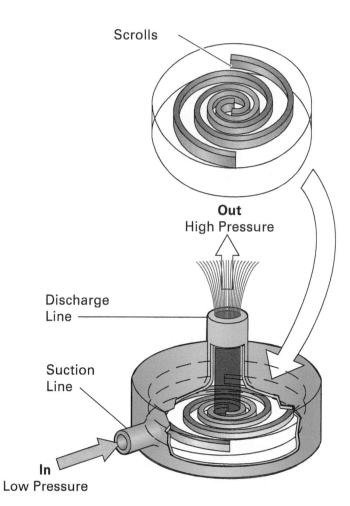

Scrolls

Out
High Pressure

Discharge
Line

Suction
Line

In
Low Pressure

Figure 34 Scroll compressor operation.

Figure 35 Cutaway of a small screw compressor.

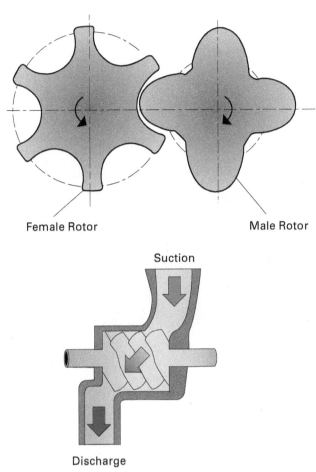

Female Rotor   Male Rotor

Suction

Discharge

Figure 36 Screw compressor operation.

3.1.6 Centrifugal Compressors

Centrifugal compressors are made in both open and semi-hermetic designs. They are typically used in commercial/industrial chilled water systems with loads exceeding 100 tons. Standard models range up to 10,000 tons of capacity, with custom models exceeding 20,000 tons.

As shown in *Figure 37*, the compressor is located on top of the condenser and evaporator assemblies. The unit chills water to be circulated for comfort cooling. The chiller is also water-cooled, with water flowing through the condenser and then outdoors to a cooling tower where the heat is rejected.

Centrifugal compressors use an impeller with many blades that rotate in a spiral-shaped housing (*Figure 38*). The impeller is rotated at very high speeds, typically around 10,000 rpm. Refrigerant vapor is fed into the housing at the center of the impeller. The impeller literally throws the incoming vapor in a circular path outward between the blades and into the compressor housing. This action, called *centrifugal force*, applies pressure to the vapor and forces it out the discharge port.

Centrifugal compressors do not typically develop pressure differentials nearly as high as other compressor types. Several impellers can be placed in series to create a greater pressure difference (*Figure 39*). The discharge from the first stage is fed into the inlet of the next stage. A compressor that uses one impeller is considered a *single-stage unit*, while one that uses two impellers is called a *two-stage unit*. If additional stages of compression are used, they are simply referred to as *multistage units*.

Most centrifugal compressors operate using low-pressure refrigerants. However, more recent designs allow the use of refrigerants that operate at higher pressures. When low-pressure refrigerants are used, the low-pressure side of the system usually operates in a vacuum. Only a small positive pressure develops on the discharge side.

Figure 37 Centrifugal compressor and chiller.
Source: Shutterstock.com/Mohd Nasri Bin Mohd Zain

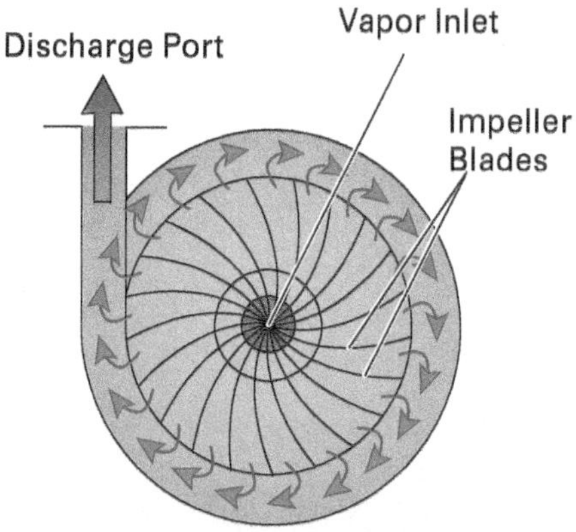

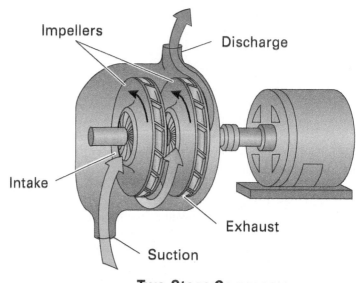

Two-Stage Compressor

Figure 38 Centrifugal compressor operation.

Figure 39 Two-stage centrifugal compressor operation.

3.2.0   Condensers

Condensers reject the heat that was absorbed in the cycle by the refrigerant. They accept high-pressure, high-temperature vapor from the compressor and condense it to a high-temperature, high-pressure liquid. The heat is generally transferred to air or water.

As the refrigerant enters the condenser, it first rejects the substantial superheat that is present. Once the saturation temperature is reached, condensation begins. Condensers must reject the heat that was absorbed in the evaporator section as well as the heat of compression. To that end, their capacity for heat rejection must be greater than the evaporator's capacity to absorb heat.

Three types of condensers will be discussed in the sections that follow:

- Air-cooled condensers
- Water-cooled condensers
- Evaporative condensers

3.2.1 Air-Cooled Condensers

Air-cooled condensers (*Figure 40*) reject the heat to the outdoor air. They come in a lot of shapes and sizes. Normally, the saturation temperature of the refrigerant flowing through the condenser is about 25°F to 35°F (14°C to 19°C) warmer than the air around it. This means that a saturation temperature of 120°F to 130°F (49°C to 54°C) is typical in the condenser when the outdoor air is 95°F (35°C). The hot refrigerant vapor that initially enters the condenser may approach 200°F (93°C).

Propeller (axial) fans are used with most air-cooled condensers to circulate air across the coil. Because air-cooled condensers require a lot of air over their surfaces, their location and the temperature of the air are very important to efficient operation. The higher the temperature of the condensing air, the more work the compressor must do to raise the refrigerant temperature to an appropriate temperature.

Air-cooled condensers are typically used in residential and commercial air conditioning systems up to about 80 tons. Water-cooled systems are often used in larger systems. However, much larger models up to several hundred tons of capacity are available when necessary (*Figure 41*).

Air-cooled condensers are generally of two types: *fin-and-tube condensers* and *plate condensers*. Both are illustrated in *Figure 42*. In the fin-and-tube condenser, the refrigerant vapor passes through the rows of tubing. The tubing is encased in metal fins. The fins provide increased surface area to the air passing by, thereby increasing heat transfer. Cooler air passing over the fins and tubing absorbs heat from the refrigerant.

Figure 40 Condensing unit with the top removed, exposing the condenser coil.
Source: iStock@CRobertson

Figure 41 Large air-cooled refrigerant condensers.
Source: iStock@DenBoma

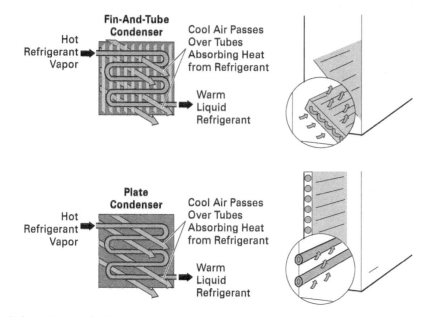

Figure 42 Basic air-cooled condenser designs.

More fins generally result in better heat transfer and a smaller coil to transfer the heat. However, more fins also result in increased resistance to airflow and a tendency to collect more debris from the air. A compromise between size and fin spacing is required.

Microchannel coils are a new way to get more from less. Coil manufacturers have developed coils that eliminate the standard round copper or aluminum tubing altogether. Instead, flat tubes with multiple small passages inside are used (*Figure 43*). The idea is to increase the refrigerant surface area in contact with the heat-transfer surface, while also minimizing the refrigerant volume. Microchannel coils have increased efficiency while maintaining or reducing the size of the coils overall. Condenser coils were first fitted with microchannel coils, but they are also now being used for evaporator coils.

A plate condenser is formed by two sheets of metal that have been pressed or stamped into the correct shape and then welded together. The plates provide a larger surface area for the transfer of heat to the surrounding air. However, this approach is generally confined to small domestic refrigeration units.

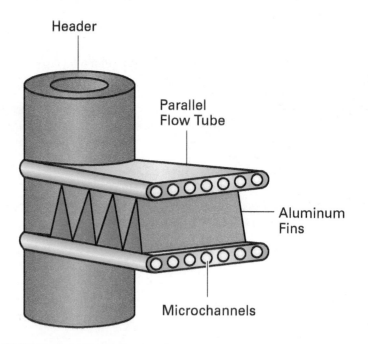

Figure 43 Microchannel coil construction.

3.2.2 Water-Cooled Condensers

Water-cooled condensers are more complex, more expensive, and require more maintenance than air-cooled condensers. However, they are also more effective and operate at much lower condensing temperatures—about 15°F (8°C) lower. This allows the system compressor to run at lower discharge pressures, requiring less power.

There are four basic water-cooled condenser designs, as shown in *Figure 44*:

- *Tube-in-tube* — In the tube-in-tube condenser, the water flows through one or more curved tubes. The high-temperature, high-pressure refrigerant supplied from the compressor flows in the opposite direction in another tube. They are also known as *coaxial heat exchangers*.

- *Shell-and-tube (both vertical and horizontal configurations)* — Shell-and-tube condensers may be either horizontal or vertical. Vertical condensers contain straight, vertical tubes encased in a metal shell. Refrigerant from the compressor enters the top of the condenser shell. The water also enters at the top and travels down through the vertical tubes where it absorbs heat from the refrigerant in the shell. As the refrigerant vapor condenses on the outside of the water tubes, the liquid drains to the bottom of the shell. Both the refrigerant and water exit from the bottom. The horizontal design allows the water tubing to be looped back and forth.

- *Shell-and-coil* — Construction of the shell-and-coil condenser is similar to the shell-and-tube condenser. The exception is that the tubing is wound around inside the shell, forming a coil. As the refrigerant vapor condenses on the cooler water tubes, liquid collects at the bottom of the shell, where it leaves the condenser.

- *Plate-and-frame* — The plate-and-frame condenser is also known as a *plate heat exchanger*. It consists of a series of metal plates held in place by a metal frame. Within the heat exchanger, the water flows through small channels, while the refrigerant flows through a separate series of channels. The plates are gasketed, welded, or brazed together to prevent the fluids from mixing. Plate-and-frame units are very effective. It is surprising to see how much heat can be transferred by a small unit. The difference is found in the amount of surface area the design provides for heat transfer.

3.2.3 Cooling Towers

Water-cooled condensers are often connected to a cooling tower (*Figure 45*). Cooling towers are not condensers since nothing is being condensed there. However, a discussion of water-cooled condensers would not be complete without mentioning cooling towers.

In a cooling tower, heat from the water in circulation is rejected into the atmosphere. The condenser water is pumped to the top. The water flows through openings in the top by gravity or is sprayed into the tower. Inside, the warm water is directly exposed to the air. A portion of the water evaporates, lowering the temperature of the remaining water. The cooled water collects at the bottom and is circulated back to the condenser.

Forced-draft cooling towers are common in commercial/industrial HVACR. Models like the one shown in *Figure 46* are called *counterflow towers* since the air travels in the opposite direction through the tower (from bottom to top). As shown, the blower assembly is located at the base of the tower. Some towers operate the fans in stages or at varying speeds.

An *induced-draft tower* (*Figure 47*) is one that pulls the air through the tower rather than pushing it through. A propeller fan in the top of the tower draws air up through the falling water droplets.

Since some water evaporates as part of the cooling process, the water lost must be replaced to maintain proper circulation. This is done using a simple float valve that adds water as needed. Electronic water-level sensors and valve controllers are also used.

Case History

Protecting Water-Cooled Condensers

Water-cooled condensers must be protected from freezing at all times. When a water-cooled condenser freezes, the ice will likely burst the tubes as it thaws. Water can then enter the refrigeration circuit in large volumes, after the refrigerant pushes its way out.

A service call to a small pork processing facility demonstrated just how much damage could be done. A refrigeration system stopped working, so plant personnel relocated the product to a different storage area. A service call was not placed for several weeks. Upon arrival, the technician connected a set of gauges to the compressor. Rusty water quickly filled the gauge manifold.

The entire system, as well as the compressor had been flooded with water. The equipment had to be replaced. The interconnecting piping was blown out with nitrogen and evacuated for days, but it was saved.

Removing any amount of water from a refrigeration circuit is expensive and time-consuming at best. Often, the damage is too extensive to make the effort practical.

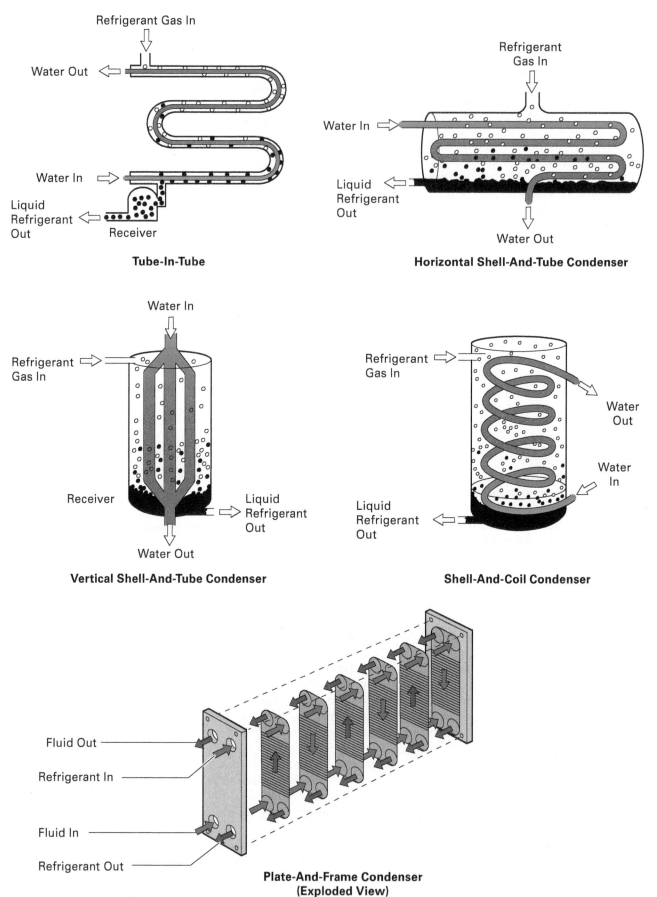

Figure 44 Common water-cooled condenser designs.

Figure 45 Cooling tower.

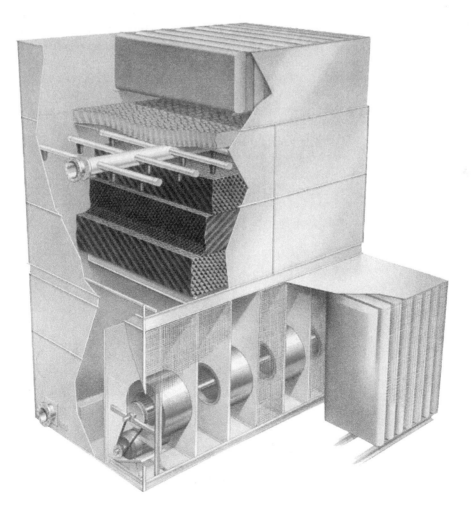

Figure 46 Forced-draft cooling tower.
Source: ©2022 SPX Cooling Technologies, Inc. All rights reserved. Images used with permission for educational purposes only.

Figure 47 An induced-draft cooling tower.
Source: ©2022 SPX Cooling Technologies, Inc. All rights reserved. Images used with permission for educational purposes only.

Did You Know?

Legionnaires' Disease

Legionnaires' disease is a very serious pneumonia-like lung infection caused by bacteria called *Legionella*. *Legionella* bacteria is often found in fresh water and rarely causes illness. Where water remains at a warmer temperature, such as in hot tubs and cooling towers, *Legionella* can grow out of control. When small droplets of water containing the bacteria get into the air, they are easily inhaled.

Cooling towers produce the water droplets that can carry the bacteria to humans. Fortunately, Legionnaires' disease is not considered contagious, but research continues. If you develop pneumonia-like symptoms and believe you have been exposed to *Legionella*, seek medical care immediately. If you have been around hot tubs or cooling towers, be sure to mention this to your healthcare provider.

The bacterium was named in 1976, when many people who attended an American Legion convention in Philadelphia became ill. About 5,000 cases of Legionnaires' disease continue to be diagnosed each year in the United States alone. It proves to be deadly for one out of ten, leading to roughly 50 deaths per year. Proper cooling tower treatment programs are essential to avoid *Legionella* growth as well as other biological growth.

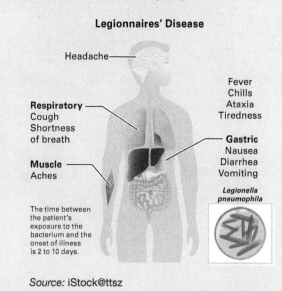

Source: iStock@ttsz

3.2.4 Evaporative Condensers

Evaporative condensers (*Figure 48*) combine the functions of a water-cooled condenser and cooling tower. They transfer heat from the refrigerant to the water, and then from the water to the air, all in the same enclosure. The condenser water evaporates directly from the tubes of the condenser. Each pound of water evaporated removes about 1,000 Btus of heat from the refrigerant.

As shown in *Figure 48*, air enters the bottom of the unit and flows upward over the refrigerant-filled coil. Water is sprayed over the coil at the same time. Both the air and the water absorb heat from the refrigerant. Mist eliminators, located above the water spray, help keep the mist from escaping the tower. Cooled by both air and water, the refrigerant in the coil condenses.

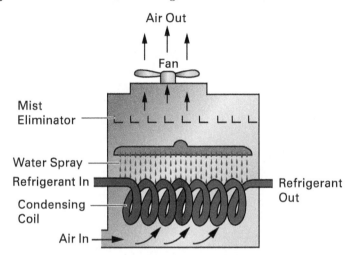

Figure 48 Typical evaporative condenser.

Did You Know?

Multiple Cooling Towers

Cooling towers are often installed in groups, or *cells*, to gain the needed capacity. The individual tower cells can be isolated from each other so that maintenance can be performed on one cell while the system continues to operate.

Source: iStock@boonsom

3.3.0 Evaporators

Evaporators (*Figure 49*) are used to extract heat from air or a substance. Low-temperature, low-pressure refrigerant from the metering device enters the evaporator and immediately begins to absorb heat. As it progresses through the evaporator, the heat causes it to boil and change to its vapor state. During this latent process, a great deal of heat is absorbed.

Figure 49 Refrigeration evaporator in a walk-in cooler.
Source: iStock@Doralin Tunas

It is important to remember that the heat absorbed by the refrigerant during the change of state is exponentially greater than what it absorbs during a sensible change in temperature. If there isn't enough liquid refrigerant provided to the evaporator, it changes to a vapor very quickly. This leaves the majority of the evaporator filled with vapor. The capacity of the evaporator is seriously reduced as a result. Vapor cannot absorb a significant amount of heat.

For evaporators to operate at peak efficiency, all the surface area inside the tubes must be exposed to flashing liquid. In a perfect world, the last remaining drop of liquid boils to a vapor as the refrigerant exits the coil. However, liquid refrigerant cannot be allowed to return to the compressor. The challenge is to strike a balance between maximizing cooling capacity while also ensuring that **refrigerant floodback** does not occur.

Most evaporators fall into one of two broad categories: *direct-expansion* and *flooded* (*Figure 50*).

3.3.1 Direct-Expansion (DX) Evaporators

Direct-expansion (DX) evaporators are the most widely used. DX evaporators have one or more continuous tubes, or coils, through which the refrigerant flows. The refrigerant, with a small amount of flash gas mixed in, enters the evaporator where it is gradually heated by the medium until it vaporizes.

The flow of refrigerant into the DX coil is controlled by the metering device. It supplies just the right amount of refrigerant so that it is all vaporized before it reaches the outlet.

Note the evaporator shown in *Figure 51*. Downstream of the metering device, a distributor is usually found. The distributor directs the refrigerant to each individual pass through the coil. The distributor is designed to ensure that each branch receives an equal share of the refrigerant.

Refrigerant floodback: A significant amount of liquid refrigerant that returns to the compressor through the suction line during operation. Refrigerant floodback can have several causes, such as a metering device that overfeeds refrigerant or a failed evaporator fan motor.

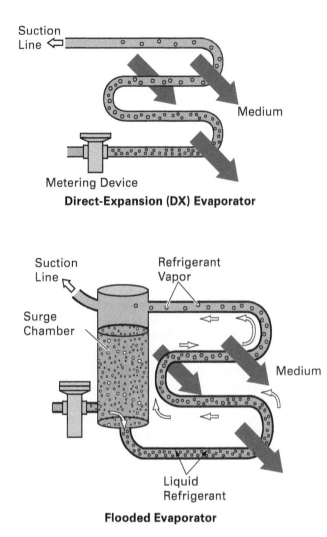

Direct-Expansion (DX) Evaporator

Flooded Evaporator

Figure 50 Direct-expansion and flooded evaporator operation.

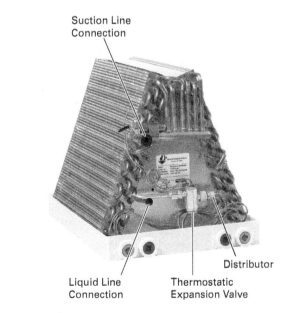

Figure 51 A direct-expansion evaporator coil.
Source: Advanced Distributor Products

Surge chamber: A vessel or container designed to hold both liquid and vapor refrigerant. The liquid is generally fed out of the bottom into an evaporator, while vapor is drawn from the top of the container by the compressor to maintain refrigerant temperature through pressure control.

3.3.2 Flooded Evaporators

In a flooded evaporator, a vessel known as a **surge chamber** is connected between the input and output of the evaporator tubing, as shown in *Figure 50*. Liquid refrigerant enters the surge chamber from the metering device. The metering device is designed to maintain a predetermined liquid level inside the chamber. Liquid flows into the evaporator coil and equalizes with the level in the surge chamber. The refrigerant vapor exits through the top of the surge chamber on its way to the compressor. The level in the surge chamber determines how much of the evaporator is flooded.

Flooded evaporators are typically used only in large food processing and similar applications. They are not generally used in comfort cooling evaporators, although some large chillers may use this approach.

3.3.3 Evaporator Construction

Evaporators can be classified by the way they are constructed. Refer to *Figure 52*. The primary classes are as follows:

- *Bare-tube evaporators* — There are two types of bare-tube evaporators: single-circuit (path) and multiple-path. Multiple-path evaporators are often used because they save space and reduce the number of metering devices needed. Each path is simply a copper, aluminum, or steel tube shaped in a way that best matches the job. The piping is the only surface used to transfer heat. Bare-tube evaporators have limited applications.

- *Finned-tube evaporators* — Thin fins of aluminum or copper are attached to the tubing, like those used with the finned-tube condenser. The fins increase the surface area exposed to the air. This increases heat transfer. This type of coil is by far the most common in the HVACR industry. Since the evaporator coil is usually wet with condensate, dirt and debris can collect on the fins quickly. Thus, the air must be properly filtered. Evaporator fin spacing is determined for each application to provide the best balance between efficiency and maintenance issues. Fin spacing is measured in fins per inch (FPI).

- *Plate evaporators* — Similar to plate condensers, plate evaporators have a length of tubing woven along and firmly bonded to a metal plate. The plate provides greater surface area for heat transfer. This type of evaporator was used as a shelf in older domestic freezers. They are still used extensively in food processing. Foods such as fish can be placed in a single layer on each plate, or between two plates, where they are frozen very quickly, or *flash frozen*, through maximum contact with the evaporator surface (*Figure 53*).

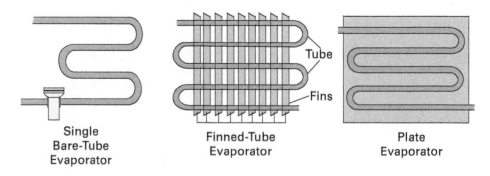

Figure 52 Evaporator construction.

Figure 53 Plate-freezing equipment for food processing.
Source: Teknotherm Refrigeration

3.3.4 Chilled-Water System Evaporators

Because of the expense and issues associated with long runs of refrigerant piping, cooling in large facilities is often done with chilled water. The water is considered a secondary refrigerant. When the temperature of the water needs to be below freezing, **brine** is used in place of water.

A refrigeration unit, called a *chiller,* is used to cool the water or brine. Technically, brine is defined as water saturated with salt. However, water that is mixed with alcohol or glycol is also referred to as brine in the craft. The salt or glycol lowers the freezing point of the water. Glycol is the primary ingredient in automotive antifreeze.

Shell-and-tube or shell-and-coil evaporators are used to transfer heat from the water. Chiller evaporators can be DX or flooded. In a DX chiller, the colder liquid refrigerant runs through the tubing while the warmer water or brine circulates within the shell around the tubes.

In a flooded chiller, the water is circulated through the tubing, which is located on the bottom of the evaporator shell. This tubing is submerged below the refrigerant level inside the shell. A float valve serves as a metering device in older models. The top portion of the evaporator shell is reserved for refrigerant vapor to collect and be drawn back to the compressor.

Brine: Water containing large amounts of salt. In the HVACR industry, the term typically describes any water-based mixture that contains salt or glycols to lower its freezing point.

3.4.0 Refrigerant Metering Devices

The metering device is located between the condenser outlet and the evaporator inlet. High-pressure, high-temperature liquid refrigerant from the condenser enters the metering device. It leaves as a low-pressure, low-temperature mixture of liquid and a small amount of vapor. The metering device performs two vital functions:

- It provides the pressure drop that lowers the boiling point of the refrigerant.
- It ideally meters the liquid refrigerant flow into the evaporator at a rate matching the rate at which the refrigerant is vaporizing.

There are several types of metering devices. They can be divided into two basic categories: *fixed* and *adjustable*. *Figure 54* shows examples of both types. This section briefly describes these devices. Metering devices are presented in greater detail in NCCER Module 03303, *Metering Devices.*

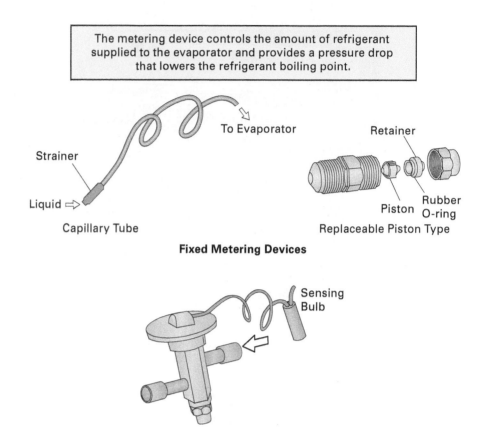

> The metering device controls the amount of refrigerant supplied to the evaporator and provides a pressure drop that lowers the refrigerant boiling point.

Fixed Metering Devices

Adjustable Metering Device (TXV)

Figure 54 Examples of common fixed and adjustable metering devices.

Fixed metering devices are used mainly in domestic refrigerators, freezers, and some residential/light-commercial AC units. Adjustable metering devices are used most often on systems with variable load requirements, and on systems that need to operate at higher efficiencies. Although fixed devices were very popular for years in residential cooling systems, the demand for higher efficiency has favored the use of adjustable metering devices in newer equipment.

The proper performance of the metering device relies heavily on the condition of the liquid refrigerant it receives. Indeed, the performance of the entire refrigeration circuit relies heavily on its condition. The condition being referred to here is the amount of subcooling.

Remember that some liquid refrigerant must flash to the vapor state as it passes through the metering device. This provides the cooling effect for the remaining liquid. Enough liquid must flash to cool it from its entering temperature (105°F, or 40°C, is common) to the saturation temperature for the new pressure. In a comfort cooling system, 40°F (4°C) is considered a normal saturation temperature. That requires a 65°F (36°C) temperature drop that must occur instantly. Any vapor that is created to provide the cooling effect is not available in the evaporator to transfer heat. Thus, a lower liquid temperature (more subcooling) entering the metering device equals more cooling capacity in the evaporator coil.

Consider how much cooling capacity is lost when the liquid leaving the condenser isn't subcooled at all. In other words, the refrigerant isn't fully condensed. A clogged condenser coil can easily cause this condition. The result is even warmer liquid being fed to the metering device to be cooled by flash gas, and less liquid refrigerant for it to work with.

3.4.1 Fixed Metering Devices

Fixed metering devices have a fixed opening, or *orifice*, size. The capillary tube, as shown in *Figure 54*, is one of the oldest metering devices. It is a fixed-length, small-diameter copper tube. Longer lengths and smaller diameters produce more pressure drop. Manufacturers determine the size and length of each capillary tube to provide a precise amount of refrigerant flow and pressure drop. Long capillary tubes are often coiled to conserve space and protect them from damage.

Another type of fixed metering device is simply known as a *fixed restrictor*. This device is a compact and rugged assembly that is installed at the evaporator inlet. The housing contains a piston that can slide back and forth. The piston has an orifice size that matches the capacity of the equipment. The correct piston size is installed at the time of system installation, per the manufacturer's guidance, to ensure it matches system capacity.

The volume of refrigerant flow is determined by the size of the orifice opening and the pressure applied. As the outdoor temperature increases, the pressure also increases, pushing more refrigerant through the fixed metering device.

The advantages of the fixed device are its simplicity and very low cost. The distinct disadvantage is their inability to adjust to changing load and operating conditions. For example, when a comfort cooling system is started up and the indoor temperature is 90°F (32°C), the evaporator is extremely efficient at transferring heat. This is due to the high temperature difference between the air and the refrigerant. As a result, all of the liquid admitted by the fixed device vaporizes quickly, leaving a great deal of the evaporator coil starved. This causes sluggish cooling performance and an above-normal superheat, slowing the process of cooling the space. A metering device that can sense the demand for more refrigerant and respond is a much better approach.

Fixed Restrictors

Fixed restrictors replaced the capillary tube in many residential air conditioning and heat pump systems. The piston is less susceptible to plugging and is easy to service in the field. In addition, they simplify the refrigerant circuit required for heat pumps. The piston is free to move inside the housing. When pressure is applied from one direction, the piston is pushed against its seat and refrigerant can only flow through the orifice. When the direction of flow is reversed in a heat pump, the piston is pushed in the opposite direction, and refrigerant can flow both through and completely around it. This results in only a slight pressure drop.

3.4.2 Adjustable Metering Devices

Adjustable metering devices differ significantly in their construction and how they are controlled. Adjustable metering devices work to regulate refrigerant flow so that the evaporator capacity matches the cooling load. There are six types of adjustable metering devices in common use:

- Hand-operated expansion valve
- Low-side float valve
- High-side float valve
- Automatic expansion valve
- Thermostatic expansion valve
- Electric and electronic expansion valves

The first three on the list are generally associated with commercial/industrial refrigeration systems. Of course, the hand-operated expansion is manually controlled, so it has few applications. The thermostatic expansion valve (TXV or TEV) is the most common type of expansion valve. However, electric or electronic expansion valves (EXVs or EEVs) are becoming more popular due to their efficiency and versatility.

A TXV controls the amount of refrigerant flowing through the evaporator by sensing the superheat in the suction line at the evaporator output. It is designed to maintain a constant superheat value. Refer to *Figure 55*.

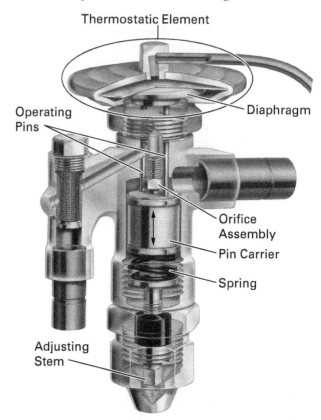

Figure 55 Internal construction of a typical TXV.
Source: Courtesy of Sporlan Division - Parker Hannifin Corporation

The TXV has a diaphragm at the top, with a valve and seat below it. The top portion containing the diaphragm is called the *power element*. The pressure exerted on the underside of the diaphragm is a combination of the evaporator refrigerant pressure and the adjustment spring pressure. The spring is adjusted to exert a pressure that represents the desired level of evaporator superheat. Pressure on top of the diaphragm is applied by refrigerant inside a remote sensing bulb. The bulb is secured to the suction line. It is important to note that the refrigerant inside the bulb and power element is not from the refrigerant circuit. It is a tiny amount of refrigerant permanently sealed inside by the manufacturer.

If the temperature of the suction line changes as the result of a changing load, a corresponding pressure change is transmitted from the bulb to the valve mechanism. This pressure change counteracts the combined evaporator and adjustment spring pressures to increase or decrease the valve opening, controlling the refrigerant flow.

The position of an EXV (*Figure 56*) is controlled by a signal such as a 4–20 mA current signal or a 0–10VDC voltage signal. The signal intensity is based on the temperature of the refrigerant leaving the evaporator, monitored by a thermistor. Many open and close in predetermined steps to accurately position the valve for the demand.

One advantage of the EXV is its increased accuracy. While a TXV is often set to maintain superheat values of 10°F to 12°F (approximately 6°C) for comfort cooling applications, an EXV can reliably maintain superheat values below 5°F (3°C), while still preventing refrigerant floodback. Since a bit more of the evaporator surface is bathed in boiling refrigerant, the system operates more efficiently.

Figure 56 Electric expansion valve (EXV).
Source: HVACGuru

TXVs

The photo in this feature shows a TXV installed in a system. The sensing bulb is wrapped in insulation to ensure that it senses only the temperature of the refrigerant leaving the evaporator. It must not be influenced by the external temperature. This TXV is *externally equalized*, as indicated by the connection from the bottom of the TXV to the suction line. This design allows the superheat to be accurately maintained independent of the pressure drop through the evaporator.

Improper installation of a TXV can prevent the system from operating correctly. The sensing bulb must be securely fastened to a clean, straight, horizontal section of the suction line close to the evaporator outlet. In addition, the bulb must be thoroughly insulated with waterproof insulation to prevent it from being influenced by the surrounding air.

Note that the short tube leaving the TXV has not been insulated. As a result, it will sweat when in operation. Remember that the outlet side will be around 40°F (4°C). That short section should be insulated.

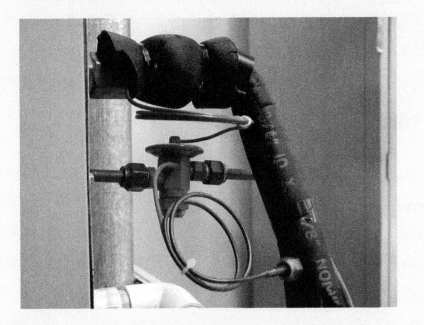

3.5.0 Refrigerant Piping

Most refrigerant piping is constructed from air conditioning and refrigeration (ACR) copper tubing. Aluminum, steel, and stainless steel may also be used in some applications. The piping layout is determined by the system designer.

There are three characteristics of a good piping layout:

- It avoids excessive pressure drop.
- It reliably returns oil to the compressor crankcase.
- It protects the compressor from floodback and slugging.

Slugging: Traditionally refers to a significant volume of oil returning to the compressor at once, primarily at startup. For example, a trap in a suction line may fill with oil during the off cycle, and then leave the trap as a *slug* at start-up. The term is incorrectly used by some to describe refrigerant floodback.

The purpose of the piping system is to provide a path for the flow of refrigerant from one component to another. Refrigerant flow must be accomplished without excessive pressure drop, such as that caused by friction, restrictions, and similar conditions.

Some compressor lubricating oil circulates throughout most refrigerant circuits. The piping layout, therefore, must allow for the return of oil to the compressor. A large slug of oil entering the compressor through the suction line can seriously damage the compressor. Oil must return, but at a manageable pace. Good piping practices help to minimize the potential for damage by slugging.

In *Figure 57*, the major refrigerant lines are identified. The *suction line* carries cold, low-pressure vapor from the evaporator to the compressor. The *hot gas line* carries hot, high-pressure vapor from the compressor to the condenser. The *liquid line* carries the liquid refrigerant from the condenser or receiver (a vessel sometimes used to store refrigerant in a system) to the metering device. Note that some of the components shown here are not necessary in every refrigerant circuit.

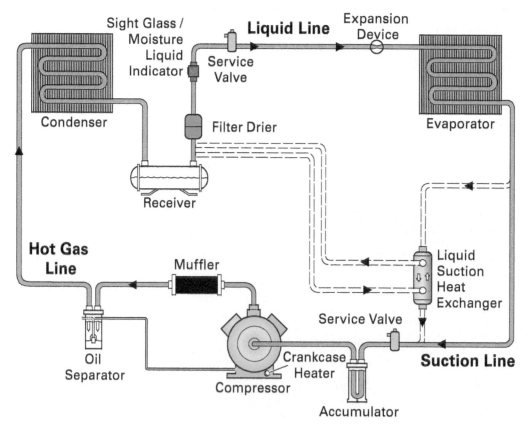

Figure 57 Major refrigerant lines.

Certain basic principles apply to all refrigerant lines:

- Keep them simple and pitch horizontal lines slightly in the direction of flow. This helps to maintain the flow of oil in the right direction.

- The pressure drop in lines should be as low as practical. The velocity of the refrigerant in the lines affects the pressure drop. The refrigerant velocity is also crucial to the movement of oil through the system. If the velocity is too low, the oil may not be encouraged to return to the compressor.

Each of the three major lines are discussed in the sections that follow.

3.5.1 Suction Line

The suction line is the most critical and challenging line from a layout and installation perspective. The pressure drop at full load must be within practical limits. Larger line sizes provide a lower pressure drop. However, oil movement and its return to the compressor must be maintained under minimum load conditions. Larger line sizes work against this because the velocity is lower. The size of the suction line must represent a balance between acceptable pressure drop and velocity. In some situations, the suction line size must change at various points in the system to ensure oil return.

The suction line design must prevent liquid refrigerant and oil slugs from draining directly to the compressor. Although very small amounts of liquid returning to the compressor are tolerable, the return of oil in slugs and refrigerant floodback must be avoided.

The suction line is larger than the liquid line because it handles vapor. The same amount of refrigerant must flow through both the liquid and suction lines, but the vapor is in an expanded state. If the suction line is undersized, capacity is lost due to pressure drop. If the suction line is oversized, it may prevent proper oil return. A larger line also increases the installation cost.

Avoid creating points where oil will collect and become trapped, other than those that have a strategic purpose. Suction piping should generally pitch down toward the compressor.

In suction risers, oil is carried up by the refrigerant gas, as shown in *Figure 58*. A minimum velocity must be maintained to keep the oil moving up. The trap at the bottom of the riser creates a point of turbulence in the line as the vapor turns the corners. This helps to mix, or *entrain*, the accumulated oil and assists in lifting it to the top of the riser.

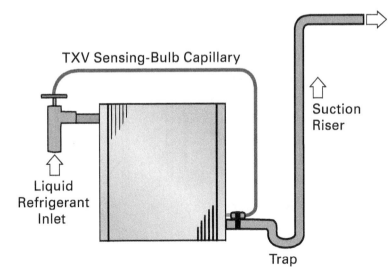

Figure 58 A suction line riser.

When a system's capacity varies because of compressor capacity control or some other arrangement, a riser may be sized smaller than the remainder of the suction line to increase the velocity (*Figure 59*). This helps ensure oil moves up the riser.

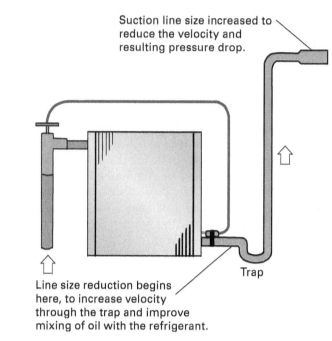

Suction line size increased to reduce the velocity and resulting pressure drop.

Line size reduction begins here, to increase velocity through the trap and improve mixing of oil with the refrigerant.

Trap

Figure 59 A reduced suction line riser.

However, it may not be possible to find a pipe size for a suction riser that ensures both oil movement at minimum capacity and an acceptable pressure drop at maximum capacity. In that case, a double suction riser (*Figure 60*) is used. When refrigerant flow is reduced, refrigerant flows up only one riser. Oil held in the trap due to insufficient velocity blocks the other riser. The trap for this purpose should be as shallow as possible, as shown in *Figure 60*. This minimizes the amount of oil it takes to block the path. At full load, when the compressor is operating on all cylinders, refrigerant vapor and oil travel up both risers.

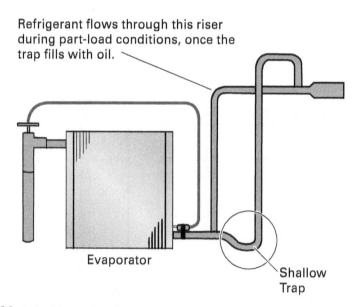

Refrigerant flows through this riser during part-load conditions, once the trap fills with oil.

Evaporator

Shallow Trap

Figure 60 A double suction riser.

When the compressor is at the same level or below the evaporator, a riser to at least the top of the evaporator coil should be placed in the suction line (*Figure 61*). This is to prevent liquid draining from the evaporator to the compressor inlet during the off cycle. The suction line loop can be ignored if the system is designed for **pump-down control**.

Pump-down control requires a solenoid valve in the liquid line, upstream of the metering device. With a pump-down control, the compressor does not shut down when the thermostat is satisfied. Only the liquid solenoid valve closes. The compressor continues to run, vaporizing and drawing out any remaining refrigerant in the evaporator, until a pressure switch turns it off. Pump-down control is most often found in refrigeration applications.

3.5.2 Hot Gas Line

Considerations for the hot gas line are similar to those for suction lines. This line carries both vapor and oil, just like the suction line. The three primary concerns related to the hot gas line are as follows:

- The pressure drop at full load must be maintained within reasonable limits.
- Oil circulation must be maintained under minimum load conditions.
- Condensed refrigerant and oil must be prevented from draining back to the compressor during shutdown.

Figure 62 shows a hot gas line in its simplest form. It should pitch down from the compressor to the condenser. If the condenser is above the compressor, a riser must be placed in the line. Like a suction riser, smaller pipe may be used to ensure upward oil movement at part-load conditions. Where the system load varies over a wide range, a double riser may be required.

Since the hot gas line connects to the compressor head, provisions must be made to prevent oil or condensed refrigerant from flowing back into the compressor during the off cycle. As shown in *Figure 62*, a loop to the floor between the compressor and the riser (a deep trap) will provide an adequate reservoir to trap and hold liquids.

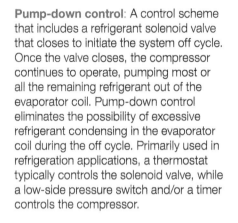

Pump-down control: A control scheme that includes a refrigerant solenoid valve that closes to initiate the system off cycle. Once the valve closes, the compressor continues to operate, pumping most or all the remaining refrigerant out of the evaporator coil. Pump-down control eliminates the possibility of excessive refrigerant condensing in the evaporator coil during the off cycle. Primarily used in refrigeration applications, a thermostat typically controls the solenoid valve, while a low-side pressure switch and/or a timer controls the compressor.

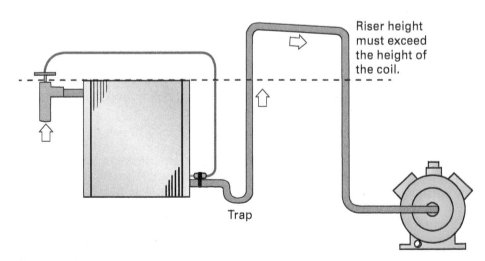

Figure 61 Routing the suction line above the top of the evaporator coil.

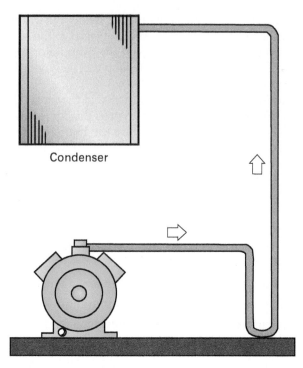

Condenser

Figure 62 Hot gas line routing.

3.5.3 Liquid Line

The layout of the liquid line is important, but its size is not as critical as the others. Oil mixes freely with many refrigerants in their liquid form. Therefore, it is not necessary to provide high velocities in liquid lines to ensure oil return. Traps in the liquid line do not create oil-return problems.

It is desirable to have a subcooled liquid reach the metering device at a sufficient pressure for proper operation. An excessive pressure drop can cause some of the refrigerant in the liquid line to flash back to vapor. In fact, a liquid line that is too small becomes an undesirable metering device, much like a capillary tube. The pressure drop in liquid lines can be due to pipe friction, long vertical rises, and/or undersized accessories.

Any vertical risers in the liquid line will experience a predictable pressure drop as the result of lifting the liquid. This is known as *static loss*, and it results from the weight of the liquid column. For common refrigerants, the pressure at the top of the riser will be lower by roughly 0.4–0.5 psig (2.7–3.4 kPa) for every foot of height. A 20' (6 m) riser will experience a pressure drop at the top of 8–10 psig (55–69 kPa) as a result. No line-sizing technique can eliminate this condition. Only a sufficient amount of subcooling in the liquid can prevent flash gas from forming in a liquid-line riser.

Accessories should be selected based on their pressure drop at a given flow rate. Line size is a secondary consideration.

3.5.4 Insulation

Liquid lines are not normally insulated unless the environment is significantly hotter than the refrigerant. The hot gas line temperature is generally well above the surrounding environment and is unaffected by it. Any heat it loses to the environment is actually a bonus.

Suction lines are always insulated to prevent condensation on the outside of the line. It also reduces the amount of heat picked up from the environment. Any heat gained from the environment still has to be rejected by the condenser. Suction line insulation can be covered with a vapor retarder and weatherproofed. Some project specifications require it, especially in industrial refrigeration applications. However, vapor retarders and weatherproofing are rarely applied on residential and light commercial systems.

3.6.0 | Refrigerant Circuit Accessories

Additional components can be added to the basic refrigerant circuit to improve safety, operation, efficiency, or servicing. A number of them are shown in *Figure 63*. Some of these components are factory-installed in equipment, while others may be installed in the field. This section briefly describes the most common accessories. They include the following:

- Service valves
- Filter driers
- Sight glass / moisture indicators
- Mufflers
- Check valves
- Receivers
- Suction line accumulators
- Heat exchangers
- Oil separators

3.6.1 Service Valves

Service valves provide access points that installers and service technicians can use to measure system pressures and perform service procedures. They can also be used to isolate a refrigerant line.

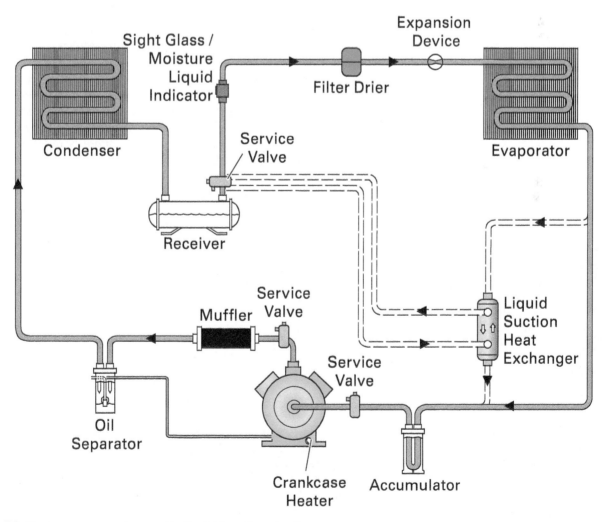

Figure 63 Common accessories used in the refrigeration circuit.

Most air conditioning systems have factory-installed service valves like the ones shown in *Figure 64*. There are two valves—one for the liquid line and one for the suction line.

Each of the valves shown has three positions. A special square wrench (*Figure 65*) is used to turn the valve stem and change the position of the valve. There are four common positions, as shown in *Figure 66*:

- *Back seated* — With the stem fully backed out, the refrigerant can flow freely through the valve. The service port is inactive. This allows you to remove the cap and connect a gauge manifold without any loss of refrigerant. This is the position the valve should be in for normal operation.

- *Front seated* — When you receive a new condensing unit that has refrigerant in it, the valves will be in this position. The main line is closed off. The service port is open to line. Once the lines are connected to the new unit, a vacuum pump can be connected to evacuate the system, while the refrigerant remains isolated in the unit.

- *Neutral* — This position is used when charging, evacuating, or recovering refrigerant from the entire system, including the condensing unit. The unit, refrigerant line, and service port are all open to each other.

- *Cracked off the back seat* — This position is used when taking pressure readings. The line and unit are open to each other; refrigerant can move freely through the valve. The service port is also active, open just enough to get a valid pressure reading. Opening it too far for this purpose would interfere with the flow through the valve.

Note that there are a few variations on the market, but this arrangement is common for residential and commercial AC condensing units.

Schrader valves (*Figure 67*) are like the valves on automobile tires, but they are not identical or interchangeable. They are spring-operated valves used to make a quick connection to a system. Manifold gauge hoses have fittings with a valve stem depressor to push down on and open the valve.

When a Schrader valve is used for charging, evacuation, recovery, etc., it is best to remove the core to provide better flow through the fitting. You can't remove it using the tool shown in *Figure 67*, since there is refrigerant under pressure. This tool can only be used when the system is at atmospheric pressure. A special tool designed for this purpose is shown in *Figure 68*. It can remove and hold the core internally while the gauge hose remains attached, without any loss of refrigerant. A built-in ball valve provides another level of control.

Although service valves are more reliable than Schrader valves, they are only necessary at primary line connections. Schrader valves do tend to leak small amounts of refrigerant, so it is very important to always keep them tightly capped when not in use.

CAUTION

Do not use common or adjustable wrenches on service valves. The stem can easily be damaged by wrenches that do not fit perfectly.

Figure 64 Service valves.

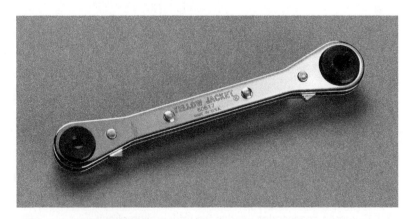

Figure 65 HVACR service ratchet.
Source: Ritchie Engineering Company, Inc. / YELLOW JACKET

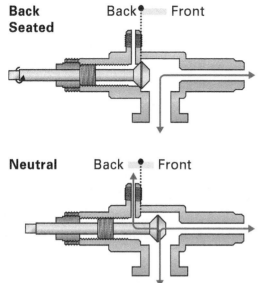
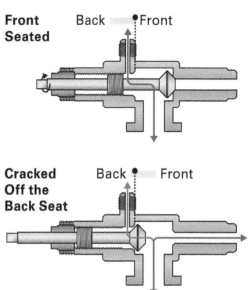

Figure 66 Service valve positions.

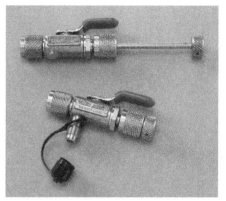

Figure 68 A 4-in-1 valve-core tool.
Source: Ritchie Engineering Company, Inc. / YELLOW JACKET

Figure 67 Schrader valve cores and tool.
Source: Ritchie Engineering Company, Inc. / YELLOW JACKET

3.6.2 Filter Driers

The filter drier combines the functions of a filter and a drier in one device (*Figure 69*). The filter protects metering devices and the compressor from foreign matter such as dirt, scale, or rust. The drier removes moisture from the system and traps it where it can do no harm. The drier portion is made of a **desiccant** material that absorbs water directly from the refrigerant (*Figure 70*).

Filter driers are best installed in the liquid line just ahead of the metering device. They are typically replaced any time the system is opened for a lengthy period to make a system repair.

3.6.3 Sight Glass and Moisture Indicator

The sight glass is a window that allows the technician to view the condition of the system refrigerant. A sight glass (*Figure 71*) shows whether the refrigerant is fully condensed into a liquid or contains vapor, appearing as bubbles. One common location for a sight glass is at the condenser outlet to view the condition and flow of refrigerant leaving the condenser. A better location is near the inlet of the metering device, since this is where the refrigerant condition is most important.

Desiccant: A material or substance, such as silica gel or calcium chloride (CaCl), that seeks to absorb and hold water from any adjacent source, including the atmosphere.

NOTE

With some refrigerant blends, due to the differing characteristics of blend components, a sight glass may not provide a reliable indication of the refrigerant condition. One or more blend components may cause bubbles when subcooling is actually sufficient and present.

A moisture indicator is a small water-sensitive spot installed in the viewing port of a sight glass. It is exposed to the passing refrigerant and changes color depending on the amount of water in the refrigerant. Even a minute amount of water in the circuit dilutes the oil, increases the chance of internal corrosion, and can create a restriction by freezing at the outlet of the metering device.

In heat pump systems, refrigerant flows in both directions. Standard filter driers are designed for flow in one direction only. *Bi-directional filter driers* are available for heat pump applications when necessary.

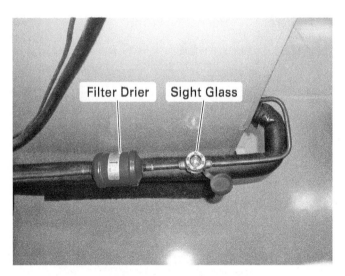

Figure 69 Filter drier and moisture-indicating sight glass.

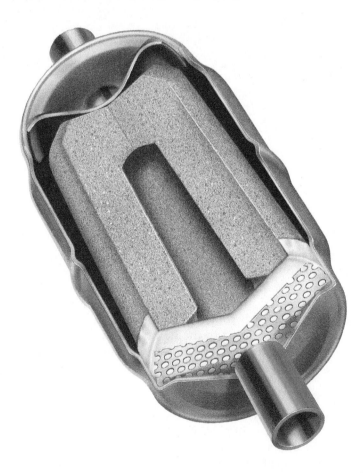

Figure 70 A cut-away of a typical filter drier, showing the filter media and desiccant material.
Source: Courtesy of Sporlan Division - Parker Hannifin Corporation

3.6.4 Compressor Mufflers

Mufflers are used in some systems with reciprocating compressors. Reciprocating compressors generate sound pulsations that can be transmitted through the piping. A muffler (*Figure 72*) installed in the discharge line, as close to the compressor as practical, dampens these pulsations. Smaller mufflers look like common filter driers from the outside.

3.6.5 Check Valves

A check valve (*Figure 73*) allows refrigerant to flow in one direction only. They are typically found in heat pumps, although new designs in heat-pump metering devices are eliminating them. Most have a simple ball that can move to either end of its housing. At one end, the curvature of the housing matches the ball, and flow in the direction is blocked. At the other end, the ball is prevented from blocking the opening by a guard. This allows refrigerant to flow freely around the ball and through the valve.

The ball inside most check valves is steel, making it magnetic. A magnet can be used from the outside to test its movement when the system is not in operation. By moving the magnet back and forth along the housing, you can hear the ball making contact at each end. This lets you know it is free to move.

Figure 71 Sight glass with moisture indicator.
Source: Ritchie Engineering Company, Inc. / YELLOW JACKET

Filter Driers

Filter driers are most often installed in the liquid line. They are often replaced whenever the refrigerant circuit is opened for service.

If a severe compressor failure has occurred, a special liquid line filter drier designed for one-time use may be installed, then replaced with a clean one after the refrigerant has circulated for a while. Filter driers used for this purpose are designed to maximize the removal of acid that results from the melting of compressor motor-winding sealants during an internal electrical failure.

A suction line filter drier may also be installed, for added protection and acid absorption. Once they have done their job and the system is clean, suction line filter driers should be completely removed from the circuit. Note the gauge port on the end. It allows you to check the pressure at the filter inlet. A second pressure measurement from a point downstream allows you to determine the pressure drop through the filter. If the filter becomes fouled with debris, the pressure drop will rise.

Figure 72 Compressor muffler.
Source: Courtesy of Sporlan Division - Parker Hannifin Corporation

Gauge Port

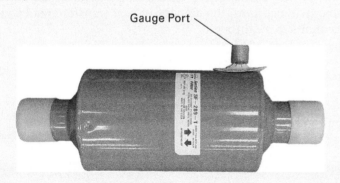

Suction Line Filter Drier

Figure 73 Refrigerant check valve.

3.6.6 Receivers

The receiver is a tank or container used to store liquid refrigerant. It is needed in some systems to accommodate sudden changes in the refrigeration load. It can also store the system charge during service procedures.

A receiver is installed in the liquid line between the condenser and the metering device. Receivers are not necessary or practical for most cooling systems. They are most often used on systems that have varying refrigeration loads, such as commercial freezers, and systems that must provide cooling even in cold weather.

3.6.7 Accumulators

An accumulator (*Figure 74*) is basically a trap designed to prevent refrigerant floodback to the compressor. If a significant amount of liquid refrigerant enters the compressor, damage or failure may result. Traces of liquid that return are not a problem, but a significant volume returning at once can damage a compressor. Accumulators are found in the suction line, usually near the compressor inlet.

Large accumulators may have heaters that help to vaporize refrigerant liquid that accumulates. Refrigerant vapor is drawn from the top of the accumulator to be returned to the compressor. A small orifice at the bottom of the internal U-tube allows tiny amounts of both liquid refrigerant and oil to return to the compressor. In larger systems, trapped oil is piped and metered back to the compressor. The small orifice shown ensures that oil can get back to the compressor, and that any liquid refrigerant that collects is controlled. Small amounts of both oil and liquid refrigerant can enter the orifice.

3.6.8 Heat Exchangers

Several types of heat exchangers can be used in refrigeration systems to improve efficiency or operation. The *liquid-to-suction* and *liquid-to-liquid heat exchangers* are two possibilities.

The liquid-to-suction heat exchanger (*Figure 75*) transfers some of the heat from the warm liquid refrigerant leaving the condenser to the cool suction vapor leaving the evaporator. This increases efficiency by further subcooling the liquid refrigerant. In addition, it serves to evaporate any small amount of liquid refrigerant that could return to the compressor from the evaporator. The amount of heat exchanged is determined by the temperature difference between the two mediums, the amount of heat transfer area inside, and how fast the mediums move through.

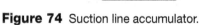

Figure 74 Suction line accumulator.

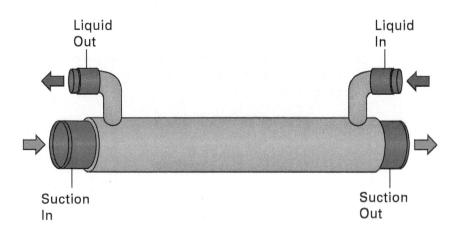

Figure 75 Liquid-to-suction heat exchanger.

A liquid-to-liquid heat exchanger can be used to desuperheat the hot gas using a water supply. They are similar in appearance to liquid-to-suction heat exchanges. Not only does the refrigeration system benefit from the extra refrigerant cooling, but the water is warmed and can be used as the cold-water supply to a water heater. For example, hotels that do a lot of laundry can benefit from recovering heat from their cooling systems. NCCER Module 03404, *Energy Conservation Equipment*, explores heat recovery and similar applications.

Case History

Proper Accumulator Installation

A radio station studio placed a service call with an HVACR service contractor regarding the cooling system for the studio. The system had been installed by others and was less than two years old. The customer reported that the system had already experienced seven compressor failures! Since the installer was unable to solve the problem, they were looking for help.

It only took a few minutes on-site for the technician to determine the real problem. The suction accumulator had been removed and reinstalled backwards. As oil returned through the suction line, all of it was being trapped in the accumulator. Once removed, it was found to be full of oil.

The Bottom Line: When installing any component in a refrigerant line, make sure the direction of flow is correct. Most components have the proper orientation stamped or marked on the body.

3.6.9 Oil Separators

Oil serves several purposes in the refrigerant circuit:

- Lubricate the compressor.
- Seal tiny passages around piston rings and similar components.
- Dampen compressor noise.
- Provide cooling for the compressor and motor.

Compressor lubrication is the most important function. Small amounts of oil from the compressor mix with the refrigerant and travel with it through the circuit.

Oil separators (*Figure 76*) minimize the amount of oil in circulation. It is unnecessary for the average system. The compressor of a residential system, for example, has enough oil to support itself while some is moving through the system. Oil separators are non-existent on residential and commercial comfort-cooling systems. In large refrigeration systems, though, the compressor may not have enough oil to compensate for the volume in circulation.

Oil coats the inside of every component through which it passes. Too much oil in circulation reduces the heat-transfer efficiency of the coils. Separating the oil from the refrigerant can improve heat transfer.

In large systems with extensive piping, there are many places where oil can be trapped. Reducing the amount of oil in circulation and controlling its movement reduces the potential for oil to be trapped in the circuit.

Supermarket refrigeration systems often use many compressors operating in stages, all connected to the same refrigerant circuit (*Figure 77*). In these applications, the oil from the compressors mixes and the supply is shared. An oil separator combined with oil management controls provides each compressor with the right amount of oil. Otherwise, one compressor may be flooded with oil while another starves.

An oil separator for a single compressor is typically installed in the hot gas line as close to the compressor discharge as practical. Separators have a reservoir, or *sump*, to collect the oil. A float valve in the sump maintains a seal between the high-pressure and low-pressure sides of the system. The valve automatically returns oil to the compressor through an orifice when the oil level rises enough to lift the float.

The oil separator is charged with oil before the system is started up to ensure an adequate supply from the beginning. Otherwise, it may require the entire oil charge from the compressor to fill it up, leaving the compressor empty. A small filter drier and a sight glass may be installed in the oil return line to the compressor. The sight glass simply helps the technician determine that oil is moving.

Figure 76 Oil separators.

Oil Separator

Figure 77 Supermarket rack system with a common oil separator.
Source: Zero Zone Parallel Refrigeration System, Courtesy of Zero Zone Inc

3.0.0 Section Review

1. Reciprocating compressors can be found in the any of the following configurations *except* _____.
    a. hermetic
    b. semi-hermetic
    c. open
    d. centrifugal

2. A cooling tower within which the airflow is forced up from the bottom as water trickles down is called a _____.
    a. counterflow tower
    b. blast tower
    c. downflow tower
    d. natural tower

3. The metering device for a flooded evaporator is designed to _____.
    a. maintain a specific superheat setting
    b. maintain a predetermined liquid level
    c. maintain a specific subcooling value
    d. ensure that there is no vapor in the surge chamber

4. The amount of refrigerant flow through a fixed-orifice metering device is dependent upon the _____.
    a. superheat at the outlet of the coil
    b. amount of subcooling in the liquid line
    c. size of the opening and the pressure applied
    d. size of the opening and the temperature of the liquid

5. Which primary refrigerant line is considered the *least* critical in its layout and installation?
    a. Liquid line
    b. Suction line
    c. Hot gas line
    d. Condensate line

6. Which refrigerant circuit accessory provides a place to store system refrigerant during service procedures and to accommodate sudden changes in the refrigeration load?
    a. Accumulator
    b. Receiver
    c. Muffler
    d. Heat exchanger

4.0.0 Cooling System Control

Objective

Identify the common controls used in cooling systems and how they function.
  a. Identify common primary controls.
  b. Identify common secondary controls.

Performance Task

4. Identify the compressor, condenser, evaporator, metering device, and accessories in a cooling system.

Controls are the devices used to start, stop, regulate, and/or protect the refrigeration system. Primary controls start or stop the refrigeration cycle, either directly or indirectly, by sensing temperature, humidity, pressure, or similar variables. Secondary controls regulate and protect the system. This section introduces you to some common control devices.

4.1.0 Primary Controls

Primary controls include, but aren't limited to, the following devices:

- Thermostat
- Pressure switches
- Time clock

4.1.1 Thermostats

Thermostats sense and respond to temperature change in a conditioned space. They switch the system on or off at a preset temperature by controlling a set of contacts in the control circuit. This may be done in several different ways.

A *cooling thermostat* makes on a temperature rise and breaks on a temperature fall. A *heating thermostat* makes on a temperature fall and breaks on a temperature rise. Of course, most comfort thermostats provide both functions.

Simple thermostats respond to temperature change by the warping of a bimetal strip (*Figure 78*). A bimetal device operates on the principle that different metals expand or contract at different rates. The electrical contact is mounted directly to the bimetal strip. The mating side of the contact is magnetized. As the two contacts become close, the magnetic energy overcomes the spring tension, and the contacts snap together. These thermostats are known as *snap-action thermostats* for that reason.

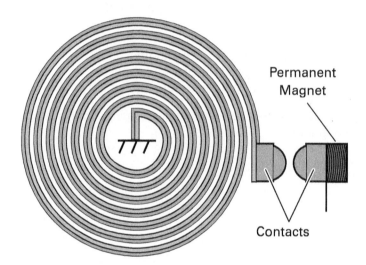

Figure 78 Snap-action thermostat operation.

A *remote-bulb thermostat* responds to pressure applied from the bellows attached to a sensing bulb (*Figure 79*). A small copper capillary tube connects the bulb to the diaphragm. The bulb is filled with a refrigerant or similar substance. The bulb pressure increases or decreases as the temperature rises or drops. An increase in bulb pressure causes the bellows to expand, mechanically moving the electrical contacts closer or farther apart. The action is the opposite when the pressure drops.

Electronic thermostats (*Figure 80*) use electronic components to sense temperature and perform switching functions. These thermostats typically use a thermistor to sense temperature.

Electronic thermostats are generally more accurate and reliable than other types. They contain microprocessors that allow programming for automatic startup, shutdown, and daily setpoint changes. Some also incorporate humidity control, eliminating the need for a separate humidistat.

Many older thermostats contained mercury. Due to the associated health hazards, thermostats and other control devices that contain mercury have largely been banned from the marketplace. The regulations regarding their use and disposal vary by state. Check the regulations in your local area to determine whether mercury-filled thermostats and controls are accepted for proper disposal of the hazardous waste.

Mercury is a hazardous metal in a liquid form. Exposure can cause severe brain and kidney damage. Mercury can either be ingested or absorbed through the skin. Metallic mercury slowly evaporates when exposed to air. Mercury vapor inhalation can also lead to serious and chronic health problems.

Handle devices containing mercury with extreme caution; do not handle mercury with your bare hands. For cleanup guidelines and related precautions, contact your state agency responsible for hazardous waste. Mercury spill kits are available and highly recommended for HVACR technicians.

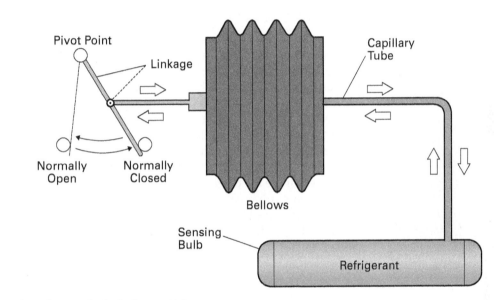

Figure 79 Operation of a remote-bulb thermostat.

Figure 80 A programmable electronic thermostat.
Source: iStock@lucadp

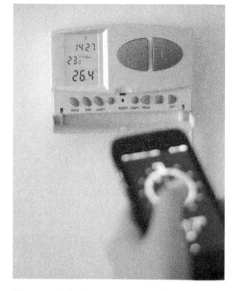

Figure 81 Wireless thermostat control through an app.
Source: iStock@adrian825

Figure 82 Non-adjustable high- and low-pressure switches.

Figure 83 Adjustable bellows-type pressure switch.

Wireless models can be connected to the local network and accessed through various apps (*Figure 81*). These thermostats allow a user to make a change from virtually anywhere.

4.1.2 Pressure Switches

Pressure switches (*Figure 82*) control based on variations in pressure. They work on both the high-pressure and low-pressure sides of the system. Adjustable pressure switches (*Figure 83*) can directly control the operation of a compressor when the low-side pressure has reached a preset value. This simple arrangement has often been used instead of a thermostat to control small refrigeration units.

A low-pressure switch is also used to control the compressor in a system designed for pump-down control. They also act as a safety device to open the control circuit if system pressure exceeds a safe level. Thus, pressure switches can be used as both primary and secondary controls.

A bellows pressure switch is connected to the refrigerant circuit through a capillary tube. As the pressure changes, the pressure in the bellows also changes. The bellows expands with a pressure increase and contracts with a pressure decrease. The movement opens or closes a control circuit.

4.1.3 Time Clocks

Time clocks (*Figure 84*) are often used to start and stop a refrigeration system or selected components within a system. Since there are many thermostats today that can provide time control of HVACR systems, time clocks like these are now used almost exclusively as defrost controls for commercial refrigeration systems.

During a defrost cycle for a walk-in freezer, for example, the timer shuts down the refrigeration equipment, including the fans on the evaporator coil. Electric heaters built into the evaporator coil are staged on, and the ice melts from the coil. If a defrost termination thermostat (discussed later in this section) doesn't terminate the defrost early, the time clock eventually ends the cycle. The refrigeration equipment then restarts and goes back to work. The evaporator fans may remain off for a short period, to avoid blowing water from the coil face onto the product below. In minutes, the droplets refreeze, and the fans can safely restart.

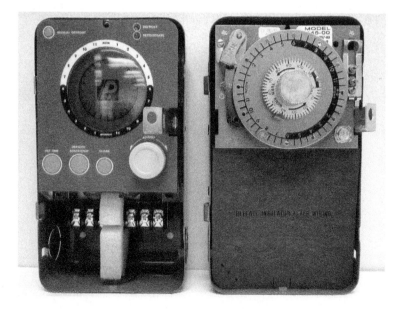

Figure 84 Electronic and mechanical time clocks.

4.2.0 Secondary Controls

Secondary controls regulate and/or protect the refrigeration system while it is in operation. They are often referred to as *operating controls* because they keep the cycle adjusted and running properly. Although there are many more, secondary controls include the following:

- Relays and contactors
- Electrical and thermal overloads
- Pressure switches
- Temperature switches
- Oil-pressure safety switches
- Flow switches

Relays, contactors, and various overloads were introduced in NCCER Module 03106, *Basic Electricity*. Pressure switches, since they can also be considered primary controls, were introduced in the previous section.

4.2.1 Temperature Switches

The term *thermostat* generally brings to mind a device on the wall used to control a system. *Temperature switches* that serve many purposes, though, can also be called thermostats. They come in many styles and can control various functions. For example, the temperature switch shown in *Figure 85* is used on heat pumps. Heat pumps must defrost themselves in the winter, and the system needs to know when the ice has been melted away to end the cycle. The thermostat shown is preset for a specific temperature. When that temperature is exceeded, its contacts close and the defrost cycle is terminated. It is designed to be clamped or strapped firmly against a round refrigerant line or coil U-bend.

Temperature switches are also used as safety switches to shut down a system when unacceptable high or low temperatures are reached.

4.2.2 Oil-Pressure Safety Switches

An oil-pressure safety switch (*Figure 86*) is a pressure-actuated safety control used to protect against a loss of oil pressure. A *differential pressure switch* is used for this purpose. A differential switch monitors two pressure sources and responds to the difference between the pressures. They are used on larger semi-hermetic and open compressors.

> **NOTE**
>
> Primary and secondary controls are presented in greater detail in NCCER Module 03314, *Control Circuit and Motor Troubleshooting*, as well as in other modules where they are relevant.

Figure 85 Defrost termination thermostat.
Source: HVACGuru

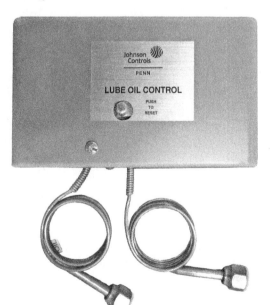

Figure 86 Oil-pressure safety switch.

The switch monitors both the pressure in the crankcase (from which the pump draws the oil) and the pressure leaving the pump. The difference between the two pressures is the net oil pressure. When the net oil pressure falls below the setpoint, the contacts open, and the compressor shuts down.

Since there is no net oil pressure when the compressor starts, oil safety switches have timing devices built in that allow a minute or two for the pressure to build before they disable the compressor. There are also electronic models of this device available.

4.2.3 Flow Switches

Flow switches are mainly used to enable or disable a system based on water flow. They prevent a system from starting when there isn't adequate flow. A water flow switch is an integral part of ensuring that a chiller has adequate water flow through the evaporator before it starts. They can also be used in air distribution systems to control devices such as electronic air cleaners.

One type used to sense air movement is called a *sail switch*. It is little more than a thin metal rod with a plastic or thin metal sail attached. The rod is attached to a mechanical switch at the other end. A more sophisticated way to sense airflow is through pressure differences in a duct. *Figure 87* shows a pressure switch used to sense the difference in duct pressure and atmospheric pressure. Pressure differential is an indirect but reliable method of confirming airflow. NCCER Module 03108, *Introduction to Heating*, discusses a similar switch used to confirm a draft through a gas furnace.

A flow switch used to monitor liquid flow is shown in *Figure 88*. The accessory paddle shown is built in sections. The pipe size determines how many sections are attached. Some paddles are made in one long piece and are cut to fit. The liquid stream pushes against the paddle, which positions a set of electrical contacts inside the enclosure. Differential pressure switches can also be used to detect liquid flow in a system. Engineers and system designers determine which method best suits their application.

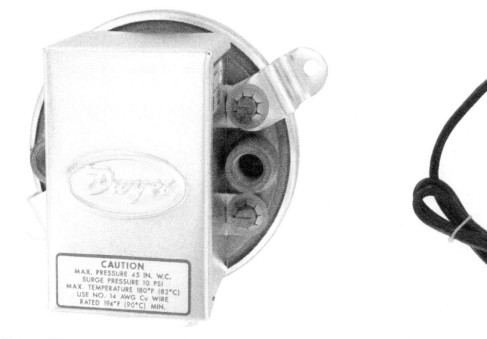

Figure 87 Air-pressure differential switch.
Source: Image courtesy Dwyer Instruments, LLC

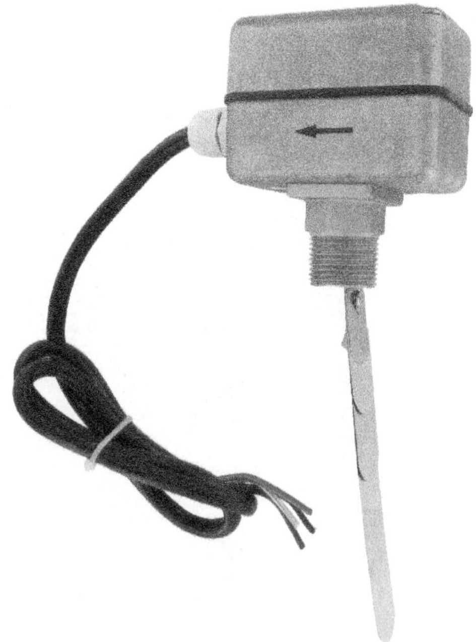

Figure 88 Liquid flow switch, with paddle attached.
Source: Image courtesy Dwyer Instruments, LLC

4.0.0 Section Review

1. What should you do after you remove a thermostat containing mercury?
    a. Throw the thermostat in a trash receptacle.
    b. Find out where mercury-filled devices can be taken for proper disposal.
    c. Collect the mercury from the thermostat in a container.
    d. Dispose of the mercury on the ground or down a drain, then put the rest in a trash receptacle.

2. Which of the following statements about oil-pressure safety switches is *true*?
    a. Oil-pressure safety switches are primary system controls.
    b. Most oil-pressure safety switches are found on hermetic compressors.
    c. Oil-pressure safety switches monitor the gross oil pressure.
    d. Oil-pressure safety switches are differential pressure switches.

Module 03107   Review Questions

1. Raising the temperature of 30 lb of water by 10°F requires _____.
    a. 3 Btus
    b. 30 Btus
    c. 150 Btus
    d. 300 Btus

2. Sensible heat is heat that _____.
    a. produces a change in state without a change in temperature
    b. is gained when changing a liquid to a vapor
    c. is gained when changing a vapor to a liquid
    d. can be measured by a thermometer

3. Superheat can be gained _____.
    a. after all of a solid changes to a liquid
    b. while changing a vapor to liquid
    c. after all liquid has been vaporized
    d. while changing a liquid to a vapor

4. When the temperature of a liquid refrigerant is below its condensing temperature, the liquid is said to be _____.
    a. superheated
    b. saturated
    c. subcooled
    d. boiling

5. If a gas is superheated 15°F (8°C) to a temperature of 87°F (31°C), we can determine that its _____.
    a. boiling point is 72°F (22°C)
    b. freezing point is 102°F (39°C)
    c. saturation point is 102°F (39°C)
    d. latent heat of fusion is 72°F (22°C)

6. A ton of refrigeration is based on the amount of heat required to melt one ton of ice in _____.
    a. 1 minute
    b. 1 hour
    c. 4 hours
    d. 24 hours

7. Zero gauge pressure (0 psig) corresponds on the absolute scale to _____.
    a. 10.4 psia
    b. 14.7 psia
    c. 27.4 psia
    d. 44.7 psia

8. A gas or liquid always flows _____.
    a. from a lower pressure to a higher pressure
    b. from a higher temperature to a lower temperature
    c. from a higher pressure to a lower pressure
    d. in a straight line

9. In a refrigeration system, the low-pressure, low-temperature vapor is converted into a high-pressure, high-temperature vapor by the _____.
    a. compressor
    b. evaporator
    c. expansion (metering) device
    d. condenser

10. The low side of the refrigeration system includes the _____.
    a. inlet of the compressor
    b. muffler
    c. receiver
    d. condenser

11. HFCs do *not* contain _____.
    a. hydrogen
    b. fluorine
    c. carbon
    d. chlorine

12. Which of the following statements about ammonia is *true*?
    a. Ammonia vapors are heavier than air.
    b. By the time you detect the odor of ammonia, the concentration is already high enough to cause injury.
    c. Ammonia is more efficient at transferring heat than halocarbon refrigerants.
    d. Ammonia is only used in large chilled-water systems.

13. What is the required color code for refrigerant recovery cylinders?
    a. Orange with gray shoulders
    b. Green with yellow shoulders
    c. Gray with yellow shoulders
    d. Yellow with gray shoulders

14. A compressor with a piston that travels back and forth in a cylinder is a _____.
    a. reciprocating compressor
    b. rotary compressor
    c. centrifugal compressor
    d. scroll compressor

15. The main purpose of a condenser is to _____.
    a. store liquid refrigerant
    b. reject heat from the refrigerant
    c. remove water from a refrigerant
    d. transfer heat to a refrigerant

16. The main purpose of an evaporator is to _____.
    a. store liquid refrigerant
    b. reject heat from a refrigerant
    c. remove water from a refrigerant
    d. transfer heat to a refrigerant

17. A TXV regulates the flow of refrigerant to the evaporator to maintain a consistent _____.
    a. subcooling
    b. superheat
    c. discharge pressure
    d. airflow

18. The operation of a double suction riser depends on _____.
    a. a sufficient volume of oil to close off the trap at the base
    b. a low velocity to get the oil to flow
    c. the compressor always operating at or near 100 percent of its capacity
    d. liquid refrigerant mixing with the oil

19. The accessory found only in the suction line is the _____.
    a. receiver
    b. accumulator
    c. muffler
    d. oil separator

20. The flow of air or liquids in duct or pipes can be monitored by flow switches or by using _____.
    a. thermometers
    b. differential pressure switches
    c. relays
    d. sight glasses

Answers to odd-numbered Module Review Questions are found in *Appendix A.*

Cornerstone of Craftsmanship

Todd Ferrara
Co-Owner and Vice President
Standard Heating and Air Conditioning

How did you choose a career in the industry?

The HVACR craft is an integral part of my family's business history. My father started an HVAC company in 1930, and I am now one of the third-generation owners. Even my grandfather and uncles worked in the HVAC industry.

Who inspired you to enter the industry?

My father, as well as some of his coworkers, in the early 1970s. I was just a teenager, but I would come into work with my dad, and then ride along on calls to carry tools and help the skilled workers.

What types of training have you been through?

I chose to go to school and attended a sheet metal program (class of 1980) at the local trade school. Today, it is known as the Dunwoody College of Technology. From 1979 to 1980, we were expected to punch a time clock as we entered and exited school each day. I have also attended many HVAC-related training programs provided by industry partners such as Bryant, York, Research Products, Honeywell, and others. The State of Minnesota sponsors code seminars that I attend, all of which teach the knowledge of the day and what we need to know as trade professionals.

How important is education and training in construction?

The training that can and should be provided, both on the job as well as in the classroom, is needed to develop your expertise. This enhances your ability to make good choices and seek the best solutions for problems encountered on the job. A proper mentor to guide you is essential! Getting along with others and understanding both generational and cultural differences are key for good communication and cooperation in today's market.

How important are NCCER credentials to your career?

I find that an accredited program is helpful to demonstrate that you and your employer are committed to investing in yourself. There are many such programs that are great, but NCCER is exceptional. NCCER checks off all the boxes related to accreditation and fresh, applicable content. These are very important characteristics in a program. To master your skills, you must have great content, qualified instructors on the job, and mentors willing to work with you and guide you toward success.

How has training/construction impacted your life?

I have been an instructor and have been involved in the training of over 150 skilled craftworkers. I thoroughly enjoy seeing people grow and become good at what they do. This is something that cannot really be measured, but it has meant a lot to me. I feel that I have made a difference and have impacted the lives of many craft professionals over the years.

What kinds of work have you done in your career?

Most of my hands-on work has been related to residential HVAC, sheet metal, and gas piping. On the business side, I've worked in various leadership roles, including personnel recruitment, hiring, and business operations. Today, I am the vice president and co-owner of Standard Heating and Air Conditioning, a Minneapolis\St. Paul Minnesota Contractor servicing the Twin Cities area for over 90 years. I oversee the deployment of new technology internally and continue to develop our training programs. I also assist in managing the fleet and facilities.

What do you enjoy most about your job?

Making lives better—watching others be successful at their jobs and making a great living in the process. I enjoy improving the quality of our work overall and providing a positive work environment.

What factors have contributed most to your success?

Hiring great people has certainly been a factor. My mentors over the years, like my dad Tony Ferrara and my brother Ted, have had a huge influence. Their example helped me develop proper craft skills, strong business ethics, and a community mentality. My current partner and niece, Claire Ferrara, has taken leadership to another level and continues to improve herself, making our business impactful in the community as well as internally.

Would you suggest construction as a career to others? Why?

Yes! The industry is always growing and ever-changing. The need for skilled people is enormous. One important aspect is the low barrier to enter this field. If you have good human qualities and are motivated to work and learn, you can find an opportunity in the construction industry.

If you really want to work and earn great wages, the opportunities today are amazing. I am involved with "Project Build Minnesota"—a program devoted to encouraging people to enter the construction industry and supporting their choice to do so. Finding a career that you love is very different than finding a job!

What advice would you give to those new to the field?

Search each day for things you can love about your work. Make that a part of who you are or who you wish to be. Never stop educating yourself and others. Be committed to lifelong learning, constantly improving your skills and knowledge. And finally—embrace change.

Interesting career-related fact or accomplishment?

The growth of our business is an accomplishment I am proud of. I've spent my career being involved in the community and the HVACR industry at the state and local level.

How do you define craftsmanship?

Craftsmanship is the ability to embrace an idea or problem and, using your skills, solve it!

Answers to Section Review Questions

Answer	Section Reference	Objective
Section 1.0.0		
1. c	1.1.1	1a
2. d	1.2.3	1b
3. b	1.3.1	1c
4. b	1.4.1	1d
Section 2.0.0		
1. a	2.1.0	2a
2. b	2.2.0	2b
3. d	2.3.4; *Table 4*	2c
Section 3.0.0		
1. d	3.1.2	3a
2. a	3.2.3	3b
3. b	3.3.2	3c
4. c	3.4.1	3d
5. a	3.5.3	3e
6. b	3.6.6	3f
Section 4.0.0		
1. b	4.1.1	4a
2. d	4.2.2	4b

Air Distribution Systems

Source: iStock@vchal

Objectives

Successful completion of this module prepares you to do the following:

1. Describe the factors related to air movement and its measurement in air distribution systems.
   a. Describe how pressure, velocity, and volume are interrelated in airflow.
   b. Describe air distribution in a typical residential system.
   c. Identify common air measurement instruments.
2. Describe the mechanical equipment and materials used to create air distribution systems.
   a. Describe various blower types and applications.
   b. Describe various fan designs and applications.
   c. Demonstrate an understanding of the fan laws.
   d. Describe common duct materials and fittings.
   e. Identify the characteristics of common grilles, registers, and dampers.
3. Identify the different approaches to air distribution system design and energy conservation.
   a. Identify various air distribution system layouts.
   b. Describe heating and cooling air movement resulting from various air distribution system designs.
   c. Explain how to maximize energy efficiency through the proper sealing and testing of air distribution systems.

Performance Tasks

Under supervision, you should be able to do the following:

1. Use a manometer to measure static pressure in a duct.
2. Use a velometer to measure the velocity of airflow at supply diffusers or registers.
3. Use a velometer to calculate the volume of airflow in a duct.

Overview

HVAC systems use ductwork to deliver conditioned air to the spaces being cooled or heated. The ductwork is made of sheet metal, fiberglass ductboard, or fabric. The performance of any HVAC system is closely related to the performance of the air distribution system. The ductwork must be of the proper size and type, and it must be correctly installed and sealed. It is important for an HVAC technician to understand how deficiencies in an air distribution system affect performance and how to recognize these deficiencies as the source of a problem.

Digital Resources for HVACR

SCAN ME

Scan this code using the camera on your phone or mobile device to view the digital resources related to this craft.

Industry Recognized Credentials

If you are training through an NCCER-accredited sponsor, you may be eligible for credentials from NCCER's Registry. The ID number for this module is 03109. Note that this module may have been used in other NCCER curricula and may apply to other level completions. Contact NCCER's Registry at 1.888.622.3720 or go to www.nccer.org for more information.

You can also show off your industry-recognized credentials online with NCCER's digital badges. Transform your knowledge, skills, and achievements into badges that you can share across social media platforms, send to your network, and add to your resume. For more information, visit www.nccer.org.

1.0.0 Air Movement and Measurement

Performance Tasks

1. Use a manometer to measure static pressure in a duct.

2. Use a velometer to measure the velocity of airflow at supply diffusers or registers.

3. Use a velometer to calculate the volume of airflow in a duct.

Objective

Describe the factors related to air movement and its measurement in air distribution systems.

a. Describe how pressure, velocity, and volume are interrelated in airflow.

b. Describe air distribution in a typical residential system.

c. Identify common air measurement instruments.

An HVAC system can't perform better than its air distribution system. Understanding the principles of air distribution is essential to accurately evaluate a system's performance. An ideal air distribution system must perform as follows:

- Supply the right volume of air to each conditioned space.
- Provide air with characteristics that don't cause discomfort.
- Operate efficiently without being noisy.
- Be reasonably accessible and require a minimum amount of maintenance.

Virtually all air distribution systems are forced-air systems. Some source of momentum is required for air to travel through a series of ducts. The major components that make up a forced-air system are as follows:

- An air handling unit (AHU) or fan coil unit (FCU). This may also be a furnace in residential and light commercial systems.
- A series of ducts that carry the conditioned, or supply, air
- A series of ducts that carry unconditioned, or return, air
- Grilles and registers to distribute the supply air and collect the return air

An air handling unit is also called an *air handler*. A smaller unit used in a residential or light commercial application that has a coil assembly in the same cabinet with the blower is commonly called a *fan coil unit (FCU)*. Regardless of which name is used, the air moving equipment must include a blower to create air movement.

Figure 1 shows the basic components of a forced-air system. In this case, the air movement is provided by a furnace blower. Systems may use different types of air handling equipment, duct materials, and air terminal devices. The location of the terminals varies as the structure requires. Additional accessories or equipment are often added to an air distribution system to achieve the desired humidity, air volume, or cleanliness.

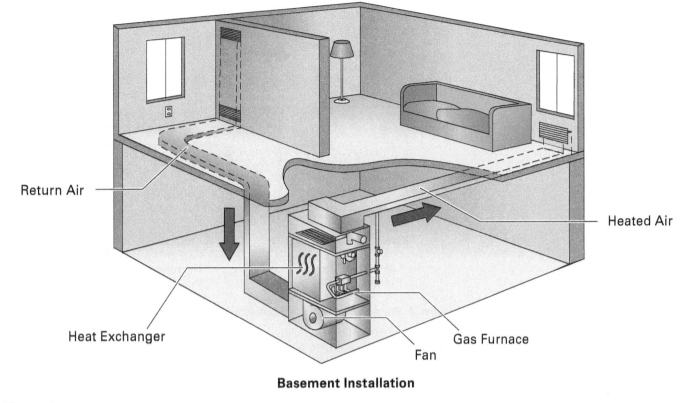

Return Air

Heated Air

Heat Exchanger

Fan

Gas Furnace

Basement Installation

Figure 1 Basic air distribution system.

Duct System Design

HVAC technicians are not generally expected to design or size the duct system. That is the job of the system designer. However, understanding the factors that go into good system design is very helpful when comfort or performance problems are encountered. Especially in the commercial environment, complaints related to the air distribution system are very common.

1.1.0　Airflow and Pressure in Ducts

Air can be moved by creating a positive pressure in one area and a negative pressure in another. Creating a pressure difference causes air to move from an area of higher pressure to an area of lower pressure. The pressure difference doesn't need to be substantial for some movement to occur. The pressure difference, however, does generally determine how fast the air moves. Wind is generated and behaves the same way.

All fans and blowers produce these changes in pressure. The inlet of a spinning blower wheel is at a negative pressure, while a positive pressure develops at the outlet. As noted earlier, air moves from a higher pressure toward a lower pressure (*Figure 2*).

At the return air grille and duct, the air pressure is lower than it is in the rest of the room. Therefore, air moves toward and then into the duct. The pressure decreases to its lowest point at the blower wheel inlet. The blower increases the pressure to its highest level at its outlet. From there, the air flows through the supply duct and into the conditioned space. The difference in the pressures causes air to move from the supply duct opening toward the return duct opening. As the air moves through the room in the general direction of the return air grille, room air mixes with the supply air.

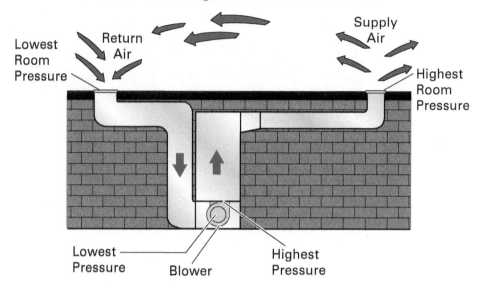

Figure 2 Pressures in an air distribution system.

Velocity: The speed at which air is moving, usually measured in feet per minute.

The pressure difference created by the blower affects the volume and **velocity** of the air. The velocity is measured in *feet per minute (fpm)* or in meters per second (m/s). The volume of air moving is measured in *cubic feet per minute (cfm)*. In the metric system, it is measured in cubic meters per hour (m³/h).

Air volume and velocity in ducts can be measured using instruments presented later in this module. The air volume in a duct is calculated by multiplying the velocity of air (in fpm) by the cross-sectional area that it is moving through (in square feet). The equations look like this:

$$\text{Volume flow rate (cfm)} = \text{duct area (ft}^2) \times \text{velocity (fpm)}$$

Or:

$$\text{Volume flow rate (m}^3/\text{h)} = \text{duct area (m}^2) \times \text{velocity (m/s} \times 3{,}600 \text{ seconds/hour)}$$

Variations of the formula can be used to find other values such as the following:

$$\text{Velocity (fpm)} = \frac{\text{Volume flow rate (cfm)}}{\text{Area (ft}^2)}$$

$$\text{Area (ft}^2) = \frac{\text{Volume flow rate (cfm)}}{\text{Area (ft}^2)}$$

As an example, consider a duct that is 14" × 12" in size. The air is moving at a velocity of 900 fpm. Begin by calculating the area of the duct in square inches:

$$\text{Area (in}^2) = 14" \times 12"$$
$$\text{Area} = 168 \text{ in}^2$$

Now convert the area in square inches to square feet by dividing by 144:

$$168 \text{ in}^2 \div 144 = 1.16 \text{ ft}^2$$

Now the air volume can be calculated using the area and velocity data:

$$\text{Volume flow rate (cfm)} = \text{Area} \times \text{Velocity}$$
$$\text{Volume flow rate} = 1.16 \text{ ft}^2 \times 900 \text{ fpm}$$
$$\text{Volume flow rate} = 1{,}044 \text{ cfm}$$

To use the metric formula, the area of the duct must be converted to square meters. The velocity is converted to, or measured in, meters per second. Multiplying by 3,600 seconds per hour changes the velocity in meters per second (m/s) to meters per hour (m/h).

Although ductwork is required to guide the air to its destination, friction on the inside surface resists the flow of air. The velocity of airflow across the duct is not uniform as a result. It varies from the lowest speed next to the duct walls to the highest speed found at the center. The rougher the duct surface, the greater the resistance to airflow. The resistance increases as the speed of the air increases. Some of the blower's energy is used in overcoming the resistance of the duct.

Additional friction and resistance can come from changes in direction, duct lining materials, and anything protruding into the airstream. *Figure 3* shows some of the joints and changes in direction found in duct systems. The blower must work against this resistance as well.

Fittings that add resistance gradually reduce the volume of air moving through as they steal energy. The size and design of the registers and grilles also affects the airflow in a system. For example, a return air grille with fins spaced ½" apart across its face resists airflow more than one with fins spaced 1" apart. The angle of the fins also makes a difference. If the air has to turn to enter, a slight friction loss occurs. These friction losses have a significant effect on sizing the blower and ducts. Friction losses may be referred to as *static pressure loss*, *static pressure drop*, or simply *pressure drop*.

1.1.1 Measuring Duct Pressure

There are three pressure values that exist in an air distribution system: **static pressure (SP)**, **velocity pressure**, and **total pressure**.

Static pressure is the pressure exerted uniformly in all directions within a duct system. In a supply air duct, think of it as the bursting or exploding pressure that acts on all internal surfaces. As shown in *Figure 4*, pressure can be applied to a manometer for measurement through openings in a **pitot tube**. Note that the static pressure inlet to the pitot tube is perpendicular to the direction of airflow.

Static pressure (SP): The pressure exerted uniformly in all directions by air within a duct system, usually measured in inches of water column (w.c.) or centimeters of water column (cm H_2O).

Velocity pressure: The pressure developed in a duct due to the linear movement of the air.

Total pressure: The sum of the static pressure and the velocity pressure in an air duct.

Pitot tube: A special tube housing a second tube used to capture pressure measurements from a moving air stream.

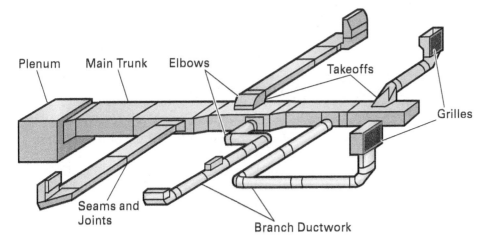

Figure 3 Joints that cause friction loss in ductwork.

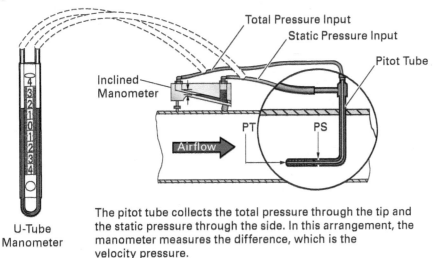

The pitot tube collects the total pressure through the tip and the static pressure through the side. In this arrangement, the manometer measures the difference, which is the velocity pressure.

Figure 4 Measuring duct pressure.

Duct Pressure Ranges

The Sheet Metal and Air Conditioning Contractors National Association (SMACNA) assigns classifications to ducts according to their pressure range. The operating pressure range is a primary factor in the required structural characteristics of the duct. Classifications are stated in inches of water column, sometimes spoken or written as *inches of water gauge*. These terms are interchangeable. The duct pressure classifications are as follows:

Pressure Class	Range
½"	Up to ½" w.c.
1"	½" to 1" w.c.
2"	1" to 2" w.c.
3"	2" to 3" w.c.
4"	3" to 4" w.c.
6"	4" to 6" w.c.
10"	6" to 10" w.c.

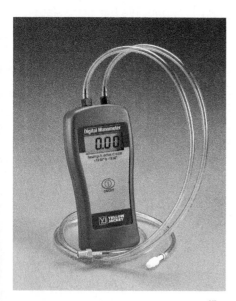

Figure 5 Digital manometer.
Source: Image Courtesy of Ritchie Engineering Inc/YELLOW JACKET

External static pressure (ESP): The cumulative resistance of all obstructions and ductwork in an air distribution system beyond the blower assembly itself.

Static pressure can also be measured by inserting a plain tube just through the wall of the duct, with the opening perpendicular to the direction of flow.

Velocity pressure is the pressure in a duct system caused by the forward movement of air. Velocity pressure acts in the direction of airflow only. It is the difference between the total pressure and the static pressure. In other words, total pressure is the sum of the static and the velocity pressures. Mathematically, their relationships look like this:

Total pressure = static pressure + velocity pressure
Static pressure = total pressure − velocity pressure
Velocity pressure = total pressure − static pressure

Duct pressures are very small when compared to other pressures you generally measure. Usually, the pressure is far less than 1 psi, so that unit of measure isn't practical. Duct pressure is typically measured in inches of water column (w.c.). This unit is based on how far the pressure can lift a column of water. You'll recall that manometers and inches of water column are used to measure gas pressure in a furnace.

Figure 4 shows a manometer connected to a pitot tube in a duct, configured to measure the velocity pressure. The total pressure is being measured through the opening pointing into the airstream. Static pressure is ported separately through the side of the tube to the opposite end of the manometer. The two pressures oppose each other, and the manometer reading represents velocity pressure.

A manometer is the perfect tool to measure duct pressure. Two types are shown in *Figure 4*. The inclined manometer has a smaller range and measures in very small increments. The U-tube manometer is not as accurate for small values. Digital manometers (*Figure 5*) are becoming increasingly popular because of their range and accuracy, and because there is no fluid inside to manage.

1.2.0 Residential Air Distribution Systems

Figure 6 shows an air distribution system for a typical single-story home. The discussion that follows demonstrates the concepts and pressure relationships that have been covered so far, along with a few new ones.

More airflow is generally needed for cooling than for heating. Therefore, the air handler and air distribution system are usually selected and designed around the cooling mode. In this example, assume that 3 tons of cooling capacity is required for a home, based on heat load calculations. That's 36,000 Btuh.

In HVAC, one rule of thumb is that comfort cooling systems require about 400 cfm of air per ton of cooling. Therefore, the system must supply at least 1,200 cfm of air (3 tons × 400 cfm). In *Figure 6*, an FCU that can provide the needed cooling capacity and air volume has been selected.

Manufacturers provide airflow performance charts for air handling equipment to show the capacity of the blower. See *Table 1* for an example. The selected unit can provide slightly more than the needed 1,200 cfm against a static pressure of 0.5" w.c. with the high speed selected. With lower resistance, more airflow is produced. Note that there is a very slight difference in blower performance based on the voltage provided to the motor. The higher voltage squeezes a little more work from the motor.

Note the use of the term total **external static pressure (ESP)** at the top of the airflow performance chart. The external static pressure represents all the resistance to airflow in the system beyond the equipment. This value usually doesn't include the pressure loss of the indoor coil. Charts from the manufacturer provides the pressure loss associated with a variety of components, including the indoor coil, at a given air volume. When designing a system, it is always necessary to confirm what components have been accounted for in the performance chart.

In *Figure 6*, the static pressure loss of the evaporator coil is shown as 0.08" w.c. If the cooling coil has not been considered in the manufacturer's chart, it must be accounted for in the design. Wet coils impose a considerable amount of resistance to airflow. Once an evaporator coil has been cooled below the **dew point** by the refrigerant, water vapor condenses on it quickly.

The system shown in *Figure 6* has a total of 12 supply air outlets, selected for airflows between 50 and 150 cfm each. The return air is taken into the system through two centrally located return air grilles.

While reviewing the drawing, remember that the structure is part of the system too. The supply-air grilles are often located around the perimeter of a home. This is where a great deal of the heat gain or loss occurs. The supply air travels through the conditioned space of the home as it moves toward the return air grilles.

A discussion of the design shown in *Figure 6* can begin at the return air grilles. The blower has created a negative pressure at the grilles. In the example, a pressure loss of −0.02" w.c. occurs at the face of each grille. The two grilles, then, produce a pressure loss of 0.04" w.c. As the air flows down the return duct towards the blower, another 0.06" w.c. is lost to friction.

Dew point: The temperature at which air becomes saturated with water vapor (100 percent relative humidity).

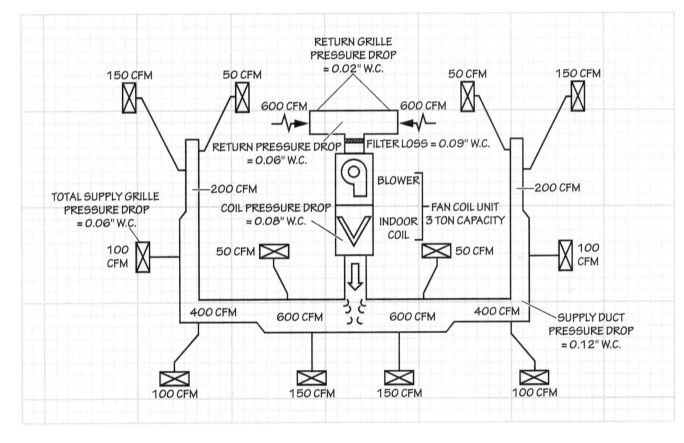

- System Cooling Capacity = 3 Tons
- At 400 CFM per Ton, Total Airflow Required = 1,200 CFM
- Total Return Duct Static Pressure Loss at 1,200 CFM = 0.19" w.c.
     0.06" w.c. Return Duct Static Pressure Loss
     0.09" w.c. Filter Static Pressure Loss
     0.04" w.c. Total Return Air Grille Loss
- Total Supply Duct System Static Pressure Loss at 1,200 CFM = 0.26" w.c.
     0.08" w.c. Evaporator Coil Static Pressure Loss
     0.12" w.c. Supply Duct Static Pressure Loss
     0.06" w.c. Total Supply Grille Static Pressure Loss
- Total External Static Pressure Loss = 0.45" w.c.

Figure 6 Typical residential air distribution system.

TABLE 1 Example Airflow Performance Chart

Model and Size	Motor Speed	Total External Static Pressure (inches w.c.)											
		0.10"		0.20"		0.30"		0.40"		0.50"		0.60"	
		208V	230V	208V	230V	208V	230V	208V	230V	208V	230V	208V	230V
FCU36	High	1495	1566	1428	1487	1365	1435	1311	1366	1232	1285	1150	1190
	Medium	1260	1388	1228	1344	1176	1281	1132	1235	1060	1166	996	1077
	Low	1080	1240	1034	1181	1004	1144	955	1094	919	1028	862	963

Transitions: Joints that accommodate a change in duct size.

Figure 7 Air damper.
Source: Image courtesy of Greenheck ©2019

Air filters can produce substantial resistance. The resistance also increases as the filter loads with debris, leaving fewer spaces for air to move through. In the example, the filter has caused a loss of 0.09" w.c.

The air at the blower outlet moves through the supply duct to a tee. The tee is used to change the direction at two 90-degree angles. That causes some pressure drop. The semicircles shown in the supply duct of *Figure 6* represent *turning vanes*. Turning vanes are simply fixed, rounded blades that help the air turn around corners in rectangular duct. The friction of the duct walls adds to the loss, as do the **transitions** protruding into the supply air duct.

Each supply duct section immediately after the tee must handle 600 cfm of air. Four branch duct outlets are connected just after the tee—two on each side of the duct, before the first transitions in duct size. This reduces the volume of air needed in the next sections of the trunk to 400 cfm on each side. Another 200 cfm of air is delivered on each side of the duct before the final transitions in duct size occur. This reduces the volume of air to the final sections of the supply duct to 200 cfm on each side. As the air exits the supply air grilles, some additional pressure loss results. In the example, it is shown to total 0.06" w.c for all the supply grilles.

At each point in the duct following a substantial change in the air volume, the duct transitions to a smaller size. The transitions play an important role in system performance. The duct is resized to maintain a relatively equal velocity and static pressure through each section. This has a self-balancing effect on the system. If the size remained the same throughout, too much air would travel down the full length of the duct and to the grilles at the end of the run. The grilles closest to the FCU would not deliver the desired airflow, while too much air would be delivered at the grilles beyond.

Dampers (*Figure 7*) are often installed in branch ducts to provide a means of balancing the volume of air supplied to each area. Dampers are used to throttle the airflow. Technically, they add resistance, discouraging airflow. Although the design may call for a specific volume of air from each grille, an occupant may prefer a warmer or cooler room. Accessible dampers provided at the grille face provide a means to make minor adjustments in the airflow.

Figure 6 shows the results of the analysis, with a total external static pressure loss of 0.45" w.c. Refer again to *Table 1* and compare the total ESP in the example to the performance of the blower. This blower should have no problem moving the desired volume of air against an ESP of 0.45" w.c. at high speed. Depending upon the voltage provided to the unit, the medium speed may provide the needed airflow. Blower speed settings are made by the technician who starts the new equipment following its installation.

Note that this example does not represent a perfect design. It simply demonstrates the effects of pressure loss and how all portions of the air distribution system contribute some level of resistance to airflow. Poor duct design results when all the losses are not considered or when accepted design standards are ignored. Poor installation techniques also create a lot of extra resistance that wasn't considered in the plan.

1.3.0 **Air Measurement Instruments**

Evaluating the performance of an air distribution system requires some measurements. These include the temperature and humidity of the air, the velocity at which it moves, and the pressure in the duct system. A variety of instruments are used to measure these values.

1.3.1 Temperature and Humidity Measurement

Several instruments are used to measure air characteristics. The following are three of the most common instruments:

- Thermometers
- Psychrometers
- Hygrometers

Thermometers

Digital thermometers (*Figure 8*) are the best choice to measure air temperatures. They are fast and accurate. They use either a thermocouple or thermistor to sense temperature. Many electronic thermometers have two or more probes so that measurements can be monitored at multiple locations. Most electronic thermometers of this type can calculate and display the difference in temperature between the two locations. Many of today's instruments are also wireless, allowing you to place a sensor somewhere and monitor changes from another place on the jobsite.

Many digital multimeters (DMMs) can also be used to measure temperature. This feature requires accessory probes that can sense temperature. Some DMMs can also use a noncontact infrared accessory to measure temperature from a distance.

Electronic thermometers are precision instruments. Be sure to follow the manufacturer's instructions for operating and calibrating them. Treat them with respect.

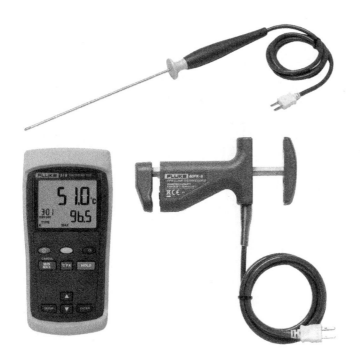

Figure 8 Digital thermometer.
Source: Reproduced with Permission, Fluke Corporation

Dry-bulb temperature: The temperature measured using a standard thermometer, representing the sensible heat present.

Wet-bulb temperature: Temperature measured with a thermometer fitted with a wetted wick wrapped around the sensing bulb. The bulb is exposed to a moving airstream and is cooled by evaporation, the rate of which depends on the amount of moisture in the airstream.

Relative humidity (RH): The ratio of moisture present in a given sample of air compared to the volume of moisture it can hold when saturated (at dew point), expressed as a percentage.

Psychrometric chart: A comprehensive chart that presents the important properties of air and water vapor as they relate to each other.

Psychrometers

A sling psychrometer (*Figure 9*) is also used to measure temperature. It has two thermometers, one to measure the **dry-bulb temperature** and the other to measure the **wet-bulb temperature**. The sensing bulb of the wet-bulb thermometer is covered with a wick saturated with water before taking a reading. To make sure the temperatures are accurate, the sling psychrometer is spun rapidly in the air. This enhances evaporation from the wet-bulb thermometer wick. The evaporation causes a lower temperature reading due to its cooling effect. This method has been used for many years and is still reliable and accurate.

The two temperatures can be used to find the **relative humidity (RH)**. This is done using either a chart on the psychrometer or a separate **psychrometric chart** (*Figure 10*). On a psychrometric chart, locate the point where the measured dry-bulb and wet-bulb temperature lines meet, then read the closest relative humidity line. Relative humidity, expressed as a percentage, compares the amount of moisture in the air to the maximum amount it can hold at a given temperature. Remember that warm air can hold more moisture than cold air.

Digital psychrometers (*Figure 11*) are easier to use, but their calibration should be checked regularly to ensure they provide accurate readings. No charts are required. The instrument can display relative humidity in real time.

1.3.2 Pressure Measurement

Several instruments and accessories are used to measure the pressures in ducts. Among these are three common devices:

- Manometers
- Differential pressure gauges
- Pitot tubes and static pressure tips

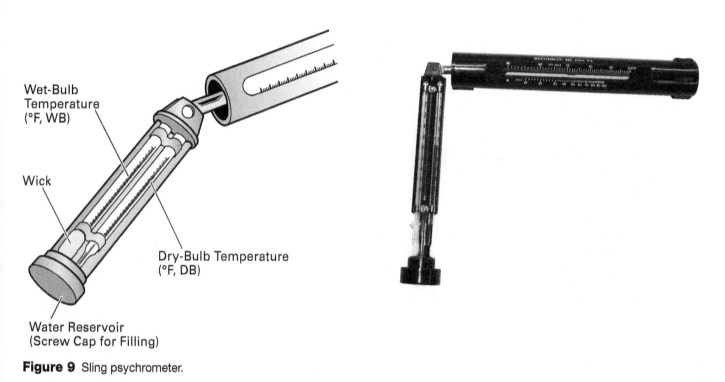

Figure 9 Sling psychrometer.

Wet-Bulb Temperature (°F, WB)

Wick

Dry-Bulb Temperature (°F, DB)

Water Reservoir (Screw Cap for Filling)

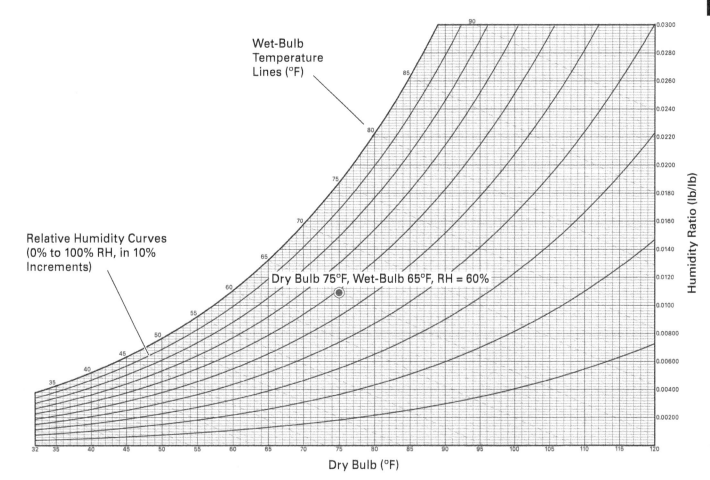

Figure 10 labels:
Wet-Bulb Temperature Lines (°F)
Relative Humidity Curves (0% to 100% RH, in 10% Increments)
Dry Bulb 75°F, Wet-Bulb 65°F, RH = 60%
Humidity Ratio (lb/lb)
Dry Bulb (°F)

Figure 10 Psychrometric chart.

Manometers

Manometers are ideal for measuring the low pressures found in air distribution systems. Manometers can be calibrated in inches or millimeters of water column, as well as a variety of other units.

Manometers use water or a special type of oil as the measuring fluid. The manufacturer of the gauge specifies the liquid to be used and substitution is not recommended. Be sure to follow the manufacturer's guidance.

Manometers come in many styles, including U-tube, inclined, and combined U-inclined styles (*Figure 12*). Electronic manometers are widely used today because of their convenience.

To check the difference in pressure between two sources, each is attached to one of the manometer ports. The column moves in relation to the difference between the two pressures. However, in many cases, one port is left open to the atmosphere. This provides a direct reading of the pressure on the connected port.

Manometers work on the principle that pressure causes a change in the level of the liquid column. When a positive or negative pressure is applied to one side, the column of liquid moves until it reaches equilibrium. If too much pressure is applied to a liquid-filled manometer, the liquid will be expelled through any open port.

U-tube and inclined manometers are available in several low-pressure ranges. Inclined manometers are usually calibrated in the lowest pressure ranges and are more sensitive than U-tube manometers. U-inclined manometers combine the sensitivity of the inclined manometer with the higher-range capability of the U-tube manometer. Inclined-vertical manometers combine an inclined section for high accuracy with a vertical manometer section for extended range. Some are permanently mounted to a piece of equipment so a reading can be taken at a glance.

Figure 11 Digital psychrometer.
Source: Reproduced with permission, Fluke Corporation

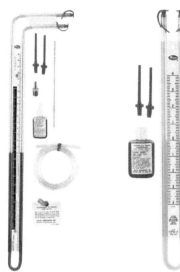

(A) Dual-Range,
Dual-Inclined

(B) U-Tube

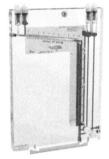

(C) Inclined-Vertical

(D) Digital

Figure 12 Manometers.
Source: Image Courtesy of Dwyer Instruments
LLC

To get accurate readings with inclined and vertical-inclined manometers, the instrument must be level. A built-in spirit level is used for this purpose. Many models have a screw-type leveling adjustment to help ensure accuracy.

Differential Pressure Gauges

Differential pressure gauges (*Figure 13*), also known as *Magnahelic®* gauges, are usually calibrated in either inches of water column or psi. Several models are available covering several pressure ranges from up to 10" w.c. They are also available with metric units of measure, such as millimeters of water, Pascals, and kilopascals.

Differential pressure gauges can accept two pressure inputs. One port can be left open to the atmosphere. This is correct when you are measuring duct pressure alone, since you want to see how it compares to atmospheric pressure. To measure the pressure drop across a set of air filters, both ports must be connected to the duct. One measures the duct pressure entering the filters, and the other measures the duct pressure after the filters. The difference is the pressure drop created by the filters. As the filters collect debris, the difference increases.

Pitot Tubes and Static Pressure Tips

Pitot tubes and static pressure tips (*Figure 14*) are probes used with manometers and pressure gauges to make measurements inside ductwork. Regardless of how accurate the instruments are, the readings will be wrong and misleading if they are not collected correctly.

The standard pitot tube used in ducts 8" across and larger has an outer tube with eight equally spaced 0.04" (1.02 mm) holes used to sense static pressure. For measurements in ducts smaller than 8", smaller pitot tubes with an outer tube and four equally spaced static pressure holes are best. Pitot tubes come in various lengths ranging from 6" to 60" (15 cm to 1.5 m), with graduation marks to show the depth of insertion in a duct.

A pitot tube consists of an impact tube fastened concentrically inside a larger tube. The inner tube receives the total pressure input. The outer tube receives static pressure inputs from the sensing holes that are at a 90-degree angle to the airflow. The tubes transfer pressure from the sensing holes to the ports of the manometer.

Figure 13 Differential pressure gauge.
Source: Image Courtesy of Dwyer Instruments
LLC

Flow Hood

A flow hood, also known as a *balometer*, is a special instrument used in balancing air distribution systems. It is held over a diffuser where it can sense the total volume of air flowing through and shows the air volume on an integral display. The information is then documented on appropriate forms where it can be compared to design criteria.

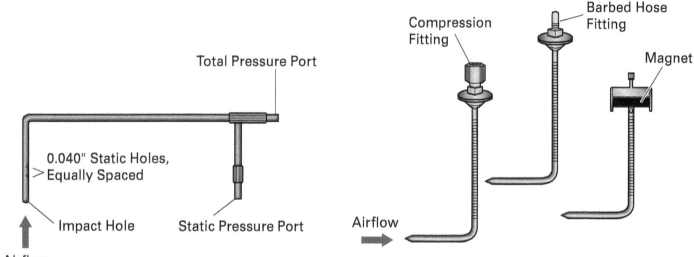

Figure 14 Pitot tube and static pressure tips.

When the total pressure tube is connected to the high-pressure port of the manometer alone, velocity pressure is indicated directly. The other manometer port remains open to the atmosphere. To be sure of accurate velocity pressure readings, the pitot tube tip must be pointed directly into the duct air stream. Velocity pressure, though, isn't of as much interest to HVAC technicians as the static pressure. Static pressure tips are typically L-shaped and have four of the equally spaced sensing holes.

1.3.3 Air Velocity Measurement

Velometers and anemometers, such as the one shown in *Figure 15*, are used to measure the velocity of airflow. The measurement of air velocity is done to check the operation of an air distribution system. It is also done when balancing system airflow.

Most velometers provide a direct air velocity reading in feet per minute or meters per second. Some can provide direct readings in cfm or m³/h with the input of duct size so that internal calculations can be made. Anemometers collect velocity data in several different ways.

Some anemometers, like the one in *Figure 15*, use a rotating vane (propeller) or balanced-swing vane to sense the air movement. When it is positioned to make a measurement, the vane rotates in the air stream. The rotational speed is converted into a velocity reading for display. In the swinging-vane velometer, the air stream causes the vane to tilt at different angles in response to the measured air velocity. The position of the vane is converted into a velocity reading.

Another type is the *hot-wire anemometer* (*Figure 16*). It uses a sensing probe that contains a small resistance-heating element. When the probe is held perpendicular to the air stream being measured, the temperature of the element changes due to the airflow. This causes its resistance to change, which alters the volume of current flow to the meter circuitry. Hot-wire anemometers can be used for very low velocity readings, such as a very light draft.

Depending on the probe or attachment used, velometers can measure air velocities up to 10,000 fpm (~50 m/s). Some electronic models can average up to 250 individual readings to calculate an average velocity. Wireless sensors are available for remote monitoring and system control.

The average velocity in a duct is used to determine the total airflow passing through. The average must be calculated by taking multiple velocity readings across a cross section of duct. For rectangular ducts with dimensions less than 30" × 30" (76 cm × 76 cm), 25 readings are typically taken, equally spaced across the height and width of the duct. The readings are then averaged. Multiplying the area of the duct by the average velocity provides a reasonably accurate volume.

Figure 15 A rotating-vane anemometer.
Source: Photo Courtesy TSI Incorporated

Revolutions per minute (rpm): The common unit of measure for rotational speed.

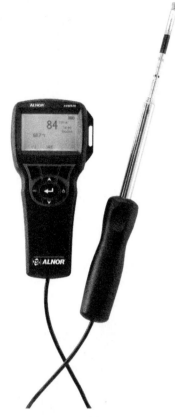

Figure 16 Hot-wire anemometer.
Source: Photo Courtesy TSI Incorporated

NOTE

If the entire surface of the pulley or wheel is reflective, the surface may need to be dulled to provide an accurate measurement.

1.3.4 Rotational Speed Measurement

The rotational speed of motors and blowers must often be checked when working with air distribution systems. This is especially true when working with belt-driven blowers. The speed at which a shaft is rotating is measured in **revolutions per minute (rpm)**. The easiest way to determine the fan rpm is to measure it directly with a tachometer (*Figure 17*).

There are two types of tachometers: contact and noncontact. Using a noncontact type is safer and more convenient when the motor is in a hard-to-reach place. Some manufacturers make a combination contact/noncontact model tachometer that can be used to make rpm measurements both ways, such as the one shown in *Figure 17*. To measure the rpm of a shaft with a contact tachometer, do the following:

WARNING!

Working near a rotating shaft is hazardous. Keep your hands and body away from the shaft. Ensure that no part of your clothing can be caught and wrapped around the shaft. Use a noncontact tachometer whenever practical.

Step 1 Turn the drive motor on.
Step 2 Contact the end of the motor or blower shaft with the tachometer sensor tip.
Step 3 Allow the reading to stabilize, and then read the rpm.

Noncontact tachometers are also well suited for use with pulleys or blower wheels that are much larger than a drive shaft. To measure the rpm with a noncontact tachometer, use the following steps:

Step 1 Ensure the drive motor is off and properly disabled.
Step 2 Place a reflective mark or a piece of reflective tape on the device to be measured.
Step 3 Turn the motor back on.
Step 4 Point the tachometer light beam at the pulley or blower wheel and hold it steady. The optical eye of the tachometer counts the rotations of the reflective surface.

Figure 17 Combination contact/noncontact tachometer.
Source: Courtesy Checkline

1.0.0 Section Review

1. Multiplying the cross-sectional area of a duct by the air velocity is used to calculate _____.
   a. static pressure
   b. velocity pressure
   c. air volume
   d. air density

2. The supply-air grilles in a residential system are often placed around the perimeter of the structure because _____.
   a. they are easier to install there
   b. the perimeter is where most of the heat gain or loss occurs
   c. the return-air grilles are also located there
   d. it allows the air distribution ducts to be shorter in length

3. A manometer is used to measure _____.
   a. pressure
   b. velocity
   c. dry-bulb temperature
   d. relative humidity

2.0.0 Air Distribution Equipment and Materials

Objective

Describe the mechanical equipment and materials used to create air distribution systems.

a. Describe various blower types and applications.
b. Describe various fan designs and applications.
c. Demonstrate an understanding of the fan laws.
d. Describe common duct materials and fittings.
e. Identify the characteristics of common grilles, registers, and dampers.

Performance Tasks

There are no Performance Tasks in this section.

An air distribution system consists of the fan or blower that moves the air, the ductwork, and the grilles or registers that are used as air entry and exit points. In addition, dampers are used in some systems to control airflow. These items are the focus of this section.

2.1.0 Blowers

The blower provides the pressure difference required to move air. It must overcome the pressure losses caused by both the supply and return ductwork and deliver the desired volume of air.

The terms *fan* and *blower* are often used interchangeably, but the correct term often depends on the application. For example, a *blower* generally describes a device used to move air against the resistance of a duct system. A *fan* describes something used to move air when there is little or no significant resistance to airflow.

2.1.1 Belt- and Direct-Drive Blowers and Fans

Two types of blowers are commonly used in air distribution systems: belt-driven and direct-drive types (*Figure 18*). In a belt-driven blower, the blower motor is connected to the blower wheel by a belt and pulley. The blower speed is adjusted mechanically by a change in the pulley diameter. Belt-driven blowers are most common in commercial HVAC equipment.

(A) Belt-Drive Blower

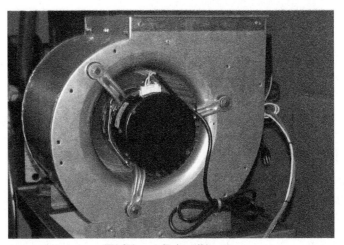

(B) Direct-Drive Blower

Figure 18 Belt-drive and direct-drive blowers.

The blower wheel of a direct-drive blower is mounted directly on the motor shaft. Most residential and light commercial equipment uses multispeed or variable-speed motors with direct-drive blowers. The blower speed is adjusted electrically by changing the blower motor wiring connections, or by changing the program settings of an electronically commutated motor (ECM). This enables the speed of the motor to be adjusted to match the requirements of each system.

2.1.2 Centrifugal Blowers

Centrifugal blowers (*Figure 19*) are used with forced-air systems because they work best against resistance. They can be used in very large systems and those that are classified as high-pressure systems. As is true of most blowers, the airflow is at a right angle (perpendicular) to the shaft on which the wheel is mounted.

The wheel is mounted in a scroll-shaped housing, which helps to develop the pressure change. Basic centrifugal blowers include the following designs:

- *Forward-curved* — Forward-curved centrifugal blowers are the most common in both residential and commercial systems. As shown in *Figure 20*, the tips of the blades in a forward-curved blower are inclined in the direction of rotation.

- *Backward-inclined* — The blades of the backward-inclined centrifugal blower (*Figure 21*) are inclined away from the direction of rotation. Visually, they may appear to be installed backwards. Typically, these blowers are used in industrial systems that require heavy-duty blower construction and stable air delivery against a significant resistance. The individual blades are often welded in place. Backward-inclined blowers operate at a higher efficiency than forward-curved blowers. However, they also operate at higher speeds, and therefore tend to be noisier.

- Smaller backward-inclined wheels are usually supplied with flat blades, while larger wheels are supplied with airfoil-shaped (curved) blades to improve efficiency. Blowers using airfoil blades generally run more quietly than others.

• *Radial* — Radial blower wheels (*Figure 22*) have straight blades that are, to an extent, self-cleaning. This makes them more suitable for air systems that have particles or grease in the air stream. They are also used in other applications such as pneumatic conveying systems. The wheels of radial blowers are simple in construction, with narrow blades that often resemble paddles. They can withstand the high speeds needed to operate at higher levels of resistance. However, for high static pressures and high-speed operation, the blades are welded in place.

Figure 19 Industrial centrifugal blower.
Source: iStock@Warut1

Figure 20 Forward-curved centrifugal blower wheel.

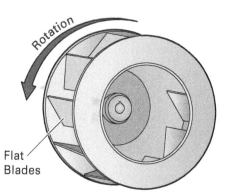

Backward-Inclined Fan Wheel

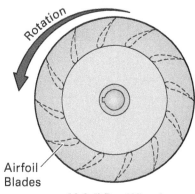

Airfoil Fan Wheel

Figure 21 Backward-inclined centrifugal blower wheels.

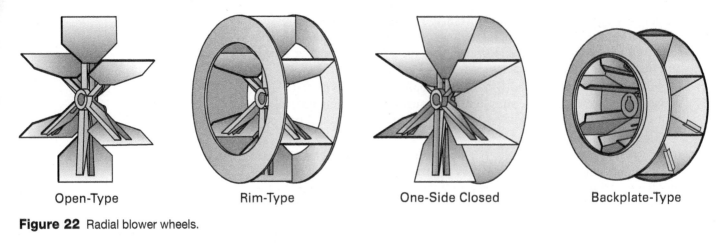

Open-Type   Rim-Type   One-Side Closed   Backplate-Type

Figure 22 Radial blower wheels.

2.2.0 Fans

Fans are typically used in applications where there is little resistance to airflow. Examples include through-wall exhaust fans and condenser fans. Although there are many variations, most are either propeller fans (axial) or duct fans (tube-axial). The airflow is normally parallel to the drive shaft of a fan, as opposed to the perpendicular relationship found in blowers.

2.2.1 Propeller Fans

Propeller, or axial, fans (*Figure 23*) are efficient at *free air delivery* and are commonly used on condenser fans or exhaust fans in HVACR applications. Free air delivery is the condition when there is little or no restriction to airflow.

Propeller fans are usually mounted in a short **venturi** to cause the air to flow in a straight line through the fan. A venturi is a ring surrounding the blades on a propeller fan. To achieve the best performance, the blade must be properly positioned in the venturi. If the position is not what the manufacturer specified, the performance will drop and the noise might be louder. Propeller fans are louder than centrifugal blowers or fans, so they are normally used where noise is not a concern.

2.2.2 Duct Fans

The airflow from a duct fan is also parallel to the shaft on which the wheel is mounted. However, duct fans are housed in a cylindrical tube. The tube acts as a long venturi and allows them to operate at higher static pressures than propeller fans.

Duct fans are commonly used in ducted exhaust systems that move a lot of air against significant resistance. A fan is considered *ducted* if the attached duct length is greater than the distance between the inlet and outlet of the fan blades.

The two types of duct fans commonly used are *tube-axial fans* and *vane-axial fans (Figure 24)*. A tube-axial fan discharges air in a helical or screw-like pattern. A vane-axial fan has vanes on the discharge side of the propeller that help the air discharge in a straight line. This reduces turbulence, thereby improving efficiency and capacity.

Venturi: A ring or panel surrounding the blades on a propeller fan to improve fan performance.

2.3.0 The Fan Laws

The performance of all fans and blowers is governed by rules commonly known as the *fan laws*. The volume of air movement, fan speed, external static pressure, and horsepower (hp) are all closely related in air distribution. There are three fan laws:

- *Fan Law 1* — The volume of air delivered by a fan varies directly with the speed of the fan. If you are considering a change to increase or decrease air volume, these equations are used:

$$\text{New volume flow rate} = \frac{\text{New rpm} \times \text{Existing volume flow rate}}{\text{Existing rpm}}$$

Or:

$$\text{New rpm} = \frac{\text{New volume flow rate} \times \text{Existing rpm}}{\text{Existing volume flow rate}}$$

- *Fan Law 2* — The static pressure of a system varies directly with the square of the ratio of the fan speeds:

$$\text{New static pressure} = \text{Existing static pressure} \times \left(\frac{\text{New rpm}}{\text{Existing rpm}}\right)^2$$

- *Fan Law 3* — The horsepower varies directly with the cube of the ratio of the fan speeds:

$$\text{New hp} = \text{Existing hp} \times \left(\frac{\text{New rpm}}{\text{Existing rpm}}\right)^3$$

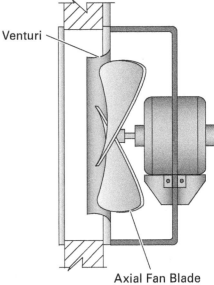

Figure 23 Propeller (axial) fan.

Example:

Assume the current conditions for an air-moving system are as follows:

- 5,000 cfm flow rate
- 1,000 rpm blower speed
- 0.5" w.c. ESP
- 0.5 hp motor

If the airflow is to be increased to 6,000 cfm, what new rotational speed is needed, and what will the new static pressure and horsepower requirement be?

Solution:

Use the first fan law to calculate the new rpm needed:

$$\text{New rpm} = \frac{\text{New volume flow rate} \times \text{Existing rpm}}{\text{Existing volume flow rate}}$$

$$\text{New rpm} = \frac{6,000 \text{ cfm} \times 1,000 \text{ rpm}}{5,000 \text{ cfm}}$$

$$\text{New rpm} = \frac{6,000,000}{5,000 \text{ cfm}}$$

$$\text{New rpm} = 1,200 \text{ rpm}$$

Next, use the second fan law to calculate the new static pressure:

$$\text{New static pressure} = \text{Existing static pressure} \times \left(\frac{\text{New rpm}}{\text{Existing rpm}}\right)^2$$

$$\text{New static pressure} = 0.5" \text{ w.c.} \times \left(\frac{1,200 \text{ rpm}}{1,000 \text{ rpm}}\right)^2$$

$$\text{New static pressure} = 0.5" \text{ w.c.} \times 1.2^2$$

$$\text{New static pressure} = 0.5" \text{ w.c.} \times 1.44$$

$$\text{New static pressure} = 0.72" \text{ w.c.}$$

Finally, use the third fan law to calculate the new horsepower requirement:

$$\text{New hp} = \text{Existing hp} \times \left(\frac{\text{New rpm}}{\text{Existing rpm}}\right)^3$$

$$\text{New hp} = 0.5 \text{ hp} \times \left(\frac{1,200 \text{ rpm}}{1,000 \text{ rpm}}\right)^3$$

$$\text{New hp} = 0.5 \text{ hp} \times 1.2^3$$

$$\text{New hp} = 0.5 \text{ hp} \times 1.728$$

$$\text{New hp} = 0.864 \text{ hp}$$

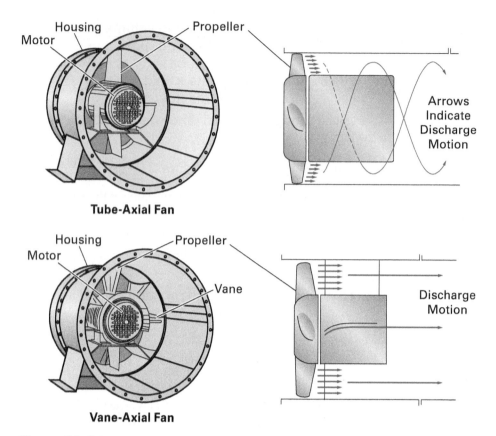

Figure 24 Tube-axial and vane-axial duct fans.

What these figures reveal is that the current motor will be overloaded if the air volume is increased to 6,000 cfm. A 1 hp motor will be required to handle the additional load.

Although these calculations aren't needed by an HVAC technician regularly, there are times when they are helpful to analyze airflow characteristics. It is important to understand how one factor can significantly affect another value. This fact alone should be remembered when troubleshooting airflow problems.

2.3.1 Fan Curves

Manufacturer's fan curves show the performance of a fan or blower at different conditions. Refer to *Figure 25*. If the values for two of the three characteristics shown on the chart are known, the missing value can be found.

For example, assume that the static pressure of a system is 1.4" w.c. and the blower is running at 900 rpm. The intersection of these two lines is represented by the black dot in *Figure 25*. To find the air volume, drop down vertically from the intersection and read the value of 7,500 cfm. This means that this particular fan can produce 7,500 cfm at these conditions. As the external static pressure and speed change, so does the air volume.

2.4.0 Duct Materials and Fittings

Building codes are not the same across the nation. Local governments have codes that determine the type of materials and methods that can be used in air distribution systems. Although they may be similar from place to place, a technician must become familiar with and follow the local codes for each job.

Load calculations determine the heating and cooling loads for a building. This means determining how many Btus of cooling and heating are required to maintain comfort, based on the local weather conditions. From the calculated

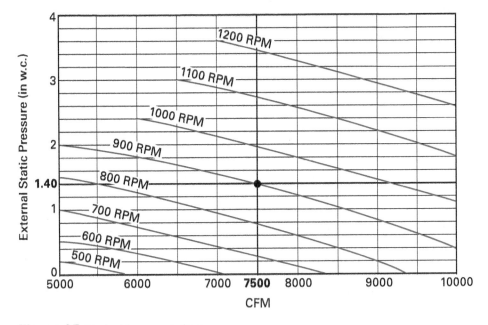

Figure 25 Typical fan curve chart.

load, the needed air volume can be determined for each area of a building. The size of the ducts is then based on the air volume needed to satisfy those requirements. A variety of charts and calculators can be used to find the correct duct sizes.

This section describes the primary components of residential air distribution systems and presents some of the considerations. The basic components of a duct system include the following:

- Main trunk and branch ducts
- Fittings and transitions
- Air diffusers, registers, and grilles
- Dampers

A wide variety of materials are used in HVAC duct designs. *Table 2* shows a few of the common materials used to construct duct systems, and where the material would be applied. As the table shows, the most common material used is galvanized steel. The other materials are used for more specific applications such as kitchen exhausts, moist air, and so on. Air shafts, typically found in multi-story buildings, are usually made of concrete or gypsum board.

Duct systems are installed in basements, crawl spaces, attics, and even within concrete floors. Galvanized sheet metal, fiberglass, and flexible duct materials are typically used for comfort systems. When installed in a concrete slab, ducts are usually made of metal or plastic. Where light weight is needed, aluminum or fabric duct can be used.

TABLE 2 Duct Materials and Their Applications

Material	Applications
Galvanized steel	Widely used for most HVAC systems
Aluminum	For systems with high moisture-laden air or special exhaust systems
Stainless steel	For kitchen exhaust, fume exhaust, or high moisture-laden air
Concrete	Underground ducts and air shafts
Rigid fibrous glass	Interior, low-pressure HVAC systems
Gypsum board	Ceiling plenums, corridor ducts, and air shafts

2.4.1 Galvanized Steel Duct

Galvanized steel ducts can be round, oval, or rectangular. All three shapes may be used in the same duct system. Popular sizes of round and rectangular steel duct, along with an assortment of standard fittings, can be acquired from HVACR supply houses, with some assembly required. Many HVACR shops can fabricate most of the components needed to construct a system.

One reason that metal remains popular for ductwork is that it offers the least amount of resistance to airflow. Other common products, such as fiberglass ductboard and flexible duct, offer more resistance.

Because sheet metal ducts are rigid, the layout must be well planned. An appropriate thickness of metal must be used for a given duct size to maintain its structural integrity. The thickness of sheet metal is expressed as a *gauge* (*Figure 26*). Galvanized sheet metal is typically available from 8 to 30 gauge (ga). Higher gauges indicate thinner metal, as shown in *Table 3*. Galvanized sheet metal thicker than 8 ga is called out by its actual thickness, in inches. Note that the gauges of metals such as stainless steel and aluminum are different from that of galvanized steel.

Larger ducts are made from heavier metal than smaller ducts. This helps reduce *drumming*—popping noises heard when the system blower starts and stops. Some refer to this as *oil-canning*, which refers to the sound metal oil cans from days gone by would make when squeezed. This happens as the duct inflates and deflates. Heavier metal also improves the structural integrity. Lines or ridges, called *cross-breaks* (*Figure 26*), are often applied to the ducts to help them resist drumming. A cross-break is shown in *Figure 26*. *Table 4A* and *Table 4B* show common metal gauges used, based on duct size and application.

Sections of square or rectangular duct are assembled using any one of several methods. S-slip and drive connectors (*Figure 27*) are commonly used for small rectangular duct found in homes. Standing seams are quite waterproof, and

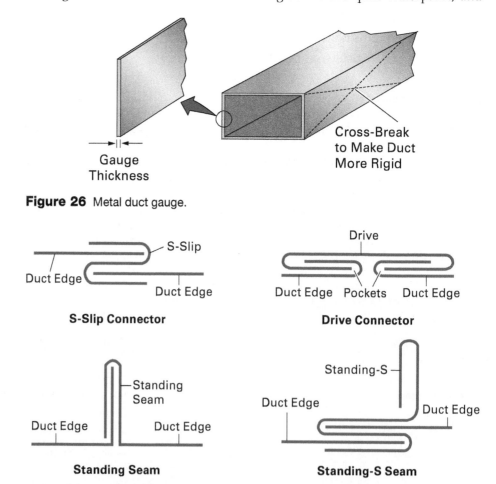

Figure 26 Metal duct gauge.

Figure 27 Common rectangular metal-duct connection methods.

TABLE 3 Standard Sheet Metal Gauge Thicknesses (Inches)

Gauge	Galvanized Steel	Aluminum	Stainless Steel
3	—	0.2294	0.2500
4	—	0.2043	0.2344
5	—	0.1819	0.2187
6	—	0.1620	0.2031
7	—	0.1443	0.1875
8	0.1680	0.1285	0.1650
9	0.1532	0.1144	0.1562
10	0.1382	0.1019	0.1406
11	0.1233	0.0907	0.1250
12	0.1084	0.0808	0.1094
13	0.0934	0.0720	0.0937
14	0.0785	0.0641	0.0781
15	0.0710	0.0571	0.0703
16	0.0635	0.0508	0.0625
17	0.0575	0.0453	0.0562
18	0.0516	0.0403	0.0500
19	0.0456	0.0359	0.0437
20	0.0396	0.0320	0.0375
21	0.0366	0.0285	0.0344
22	0.0336	0.0253	0.0312
23	0.0306	0.0226	0.0281
24	0.0276	0.0201	0.0250
25	0.0247	0.0179	0.0219
26	0.0217	0.0159	0.0187
27	0.0202	0.0142	0.0172
28	0.0187	0.0126	0.0156
29	0.0172	0.0113	0.0141
30	0.0157	0.0100	0.0125
31	0.0142	0.0089	0.0109
32	0.0134	0.0080	0.0102
33	—	0.0071	0.0094
34	—	0.0063	0.0086
35	—	0.0056	0.0078
36	—	—	0.0070

standing-S seams add additional strength and rigidity to the joint. These seams are used on large and heavy duct, especially when it is exposed to weather.

Some prefabricated ducts are assembled using a snap-lock approach (*Figure 28*). In addition to the connection methods shown here, there are other styles used for large, heavy ducts and unique applications.

Round duct sections are normally fastened together with self-tapping sheet metal screws. One end of the joint is *crimped* to allow it to slip inside the other piece (*Figure 29*). The joint can be sealed with a flexible duct-sealing compound applied with a paint brush or caulking gun. Leaking duct joints reduce the amount of air available to the structure and waste energy.

A ductwork system must be well supported so that it does not move significantly or fall. In earthquake-prone areas, there are additional codes that govern duct installation and security. Long runs of duct can also develop stress from thermal expansion and contraction as it heats and cools. This type of movement can be accommodated by using flexible or fabric joints at various points in the system.

TABLE 4A Typical Metal Duct Gauge Thickness and Aspect Ratios (1 of 2)

RECTANGULAR DUCT

Rectangular Duct Width (Inches)	Commercial		Residential
	Sheet Metal (Galvanized)	Aluminum	Sheet Metal Galvanized
Up To 12	26	0.020	28
13–23	24	0.025	26
24–30	24	0.025	24
31–42	22	0.032	–
43–54	22	0.032	–
55–60	20	0.040	–
61–84	20	0.040	–
85–96	18	0.050	–
Over 96	18	0.050	–

TABLE 4B Typical Metal Duct Gauge Thickness and Aspect Ratios (2 of 2)

ROUND DUCT

Round Duct Diameter (Inches)	Commercial Sheet Steel Galvanized Gauge	Residential Sheet Steel Galvanized Gauge
Up to 12	26	30
13–18	24	28
19–28	22	–
27–36	20	–
35–53	18	–

Plenum: A chamber at the inlet or outlet of an air handling unit. Ducts typically attach to a plenum.

Flexible connections also help eliminate vibration transmission. Metal duct systems can easily transmit vibration from the air handling equipment. A flexible connector is common where ducts connect to the equipment or its **plenum**. *Figure 30* shows how flex connectors are applied.

2.4.2 Fiberglass Ductboard

Fiberglass ductboard can be used instead of metal duct in some applications. Its inner surface creates more friction loss than does metal duct, but it is quieter because it absorbs noise. It is also self-insulating. Fiberglass ductboard is often used to insulate metal duct in many commercial and industrial applications.

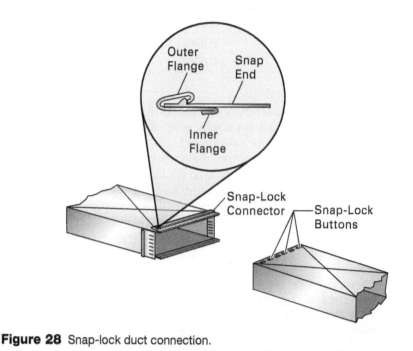

Figure 28 Snap-lock duct connection.

Proprietary Connectors

Proprietary connectors are unique connection systems invented and patented by their manufacturer. These systems are often used with large ductwork and ductwork that requires a very tight seal. The illustrations show examples of a connection system that is very popular for commercial and industrial duct applications.

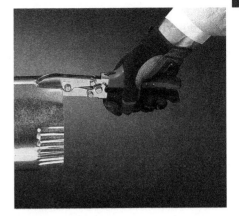

Figure 29 Crimping round duct.
Source: Malco Products, Inc.

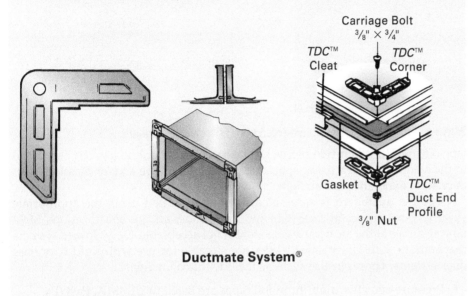

Ductmate System®

Source: Ductmate Industries

Ductboard is available in flat sheets for fabrication, or in round prefabricated sections. It is commonly sold in 1″, 1.5″, and 2″ (25 mm, 38 mm, and 51 mm) thicknesses with a foil backing. The backing is reinforced with fibers to make it stronger. The inside surface of the ductboard is coated to prevent the erosion of fibers. An abundance of fiberglass particles released into the air can be harmful to health.

Duct is fabricated from the sheets of fiberglass using special knives or programmable cutting machines. The material cuts very easily and neatly. The knives can cut away the fiberglass while leaving the foil backing intact to form

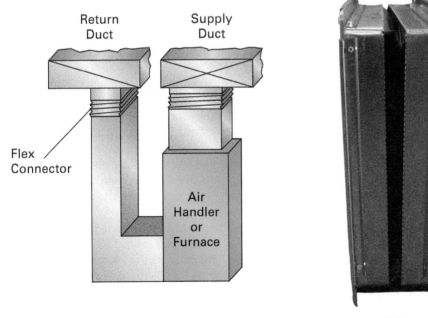

(A)   **(B)**

Figure 30 Duct flex connectors.

Health Issues

At one time, there were three major concerns related to ductboard. The first concern was related to the growth of mold and fungus resulting from the porous material absorbing moisture. This could cause serious health problems for building occupants. The second concern was the difficulty in cleaning ductboard since its surface tends to collect and hold dust and dirt. The third concern was the potential for the glass fibers to loosen from the duct and be swept into the air stream. These concerns have been largely eliminated by biocides in the fiberglass to prevent the growth of microorganisms, and by the application of an interior surface sealant.

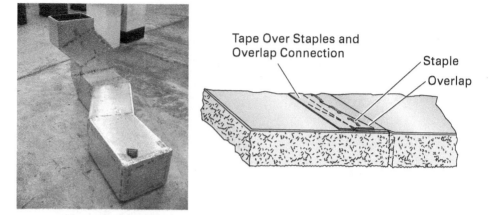

Figure 31 Practice joints constructed from fiberglass ductboard.

laps (*Figure 31*). When two pieces are fastened together, the lap can be stapled to the adjoining board using special staples. Special tape and/or mastic is then applied to make the joint airtight.

Once the ductboard is assembled, it must be sealed using the appropriate closure system, in accordance with applicable codes and job specifications. Most tape available at the local hardware store is not acceptable. Only closure systems that comply with UL Standard 181A are suitable for use with rigid fiberglass duct systems. These closure systems include the following:

- Pressure-sensitive aluminum foil tapes are listed in UL 181A, Part (P).
- Heat-activated aluminum foil or scrim tapes are listed in UL 181A, Part (H). This is the preferred method, but it does require heat and extra time.
- Mastic and glass fabric tape closure systems are listed in UL 181A, Part (M).

Fiberglass duct must be very well supported, or it will sag. In this case, the extra support needed has nothing to do with weight. Special hangers with a wider surface area must be used to avoid cutting into the duct. The requirements for hanger spacing are also different from that required for metal duct.

One disadvantage of fiberglass duct is that it is not as sturdy as sheet metal. Therefore, it cannot be installed in areas where it might be subject to physical damage. It also cannot withstand high pressures.

2.4.3 Flexible Duct

Flexible round duct (*Figure 32*) comes in sizes up to 24" (~60 cm) in diameter. It is available with and without insulation and is protected by a **vapor retarder** made of fiber-reinforced vinyl or foil backing. Insulated material is generally required.

Vapor retarder: A barrier placed over insulation to stop water vapor from passing through the insulation and condensing on a cold surface.

NOTE

Vapor retarders were formerly referred to as *vapor barriers*. The industry reconsidered that term in recent years, finding it misleading. These materials do often allow water to pass through, even though the rate may be insignificant. The word *barrier* insinuates that water cannot penetrate the material at all.

Flexible duct is typically used in spaces where obstructions make rigid duct difficult or impossible to install. One common use is to connect a branch line to the main duct. Flexible duct is easy to route around corners and other obstacles.

Flexible duct runs should be kept as short and as straight as possible. Gradual bends should be used since tight turns can significantly reduce the airflow. If a connection to a ceiling diffuser needs to turn sharply at the end, it is better to use an insulated metal elbow at the input to the diffuser than to bend flexible ducts too tightly to form the connection. This is because the metal elbow is smoother inside and will not collapse.

Even when properly installed, most flex duct creates two to five times as much resistance as an equal metal duct. The material should be stretched when installed in a straight run, but not tightly. This reduces the friction loss. To avoid sags in a run, it must be amply supported with 1" (2.5 cm) wide or fibrous material to keep the duct from collapsing. Some flexible duct comes with built-in eyelet holes along the seam for hangers.

Flex duct branches should never exceed 12' (3.7 m) in length; 6' (1.8 m) is a more desirable goal. Local codes often dictate the maximum length of flexible duct connections.

2.4.4 Fittings and Transitions

Duct fittings such as elbows, **takeoffs**, and **boots** change the direction of airflow, or change the size or shape of a duct. Transitions are used to change from one size or shape of duct to another. *Figure 33* shows some examples of these fittings.

Air moving in a duct has inertia that makes it want to flow in a straight line. Changing the direction of the air increases the friction loss, since the air uses a duct's surface to encourage the turn, creating a scrubbing effect against the duct wall.

Every fitting in a duct run adds friction to some degree. Energy is consumed to overcome the added resistance of the fittings. Because of this, the number of fittings used in a duct system must be minimized to reduce the total friction in a system.

A system is needed to determine how much resistance a given fitting creates. Thus, the resistance of fittings is related to duct length. This means that each fitting creates resistance equal to a certain number of feet of straight duct of the same size. For example, a 90-degree elbow of a given size may offer the same resistance as 10' (3 m) of straight duct. With the resistance of fittings converted to feet of straight duct, the total resistance of a system is very easy to calculate. The number of feet of straight duct is added to the total equivalent length of the fittings. The sum is then found on a chart that shows the resistance of a given duct size per foot, or per 100'.

The friction loss for various fittings is available in a set of charts available in American Society of Heating, Refrigerating, and Air Conditioning Engineers (ASHRAE) and Sheet Metal and Air Conditioning Contractors' National Association (SMACNA) publications. Duct designers must often refer to the proper tables as they work.

The total equivalent length for a duct run is calculated by adding all the equivalent lengths for fittings to the actual length of straight duct used. *Figure 34* shows several types of elbows as an example of how their equivalent lengths change. Also shown is an elbow with an equivalent length of 30' (9.1 m) added to two sections of duct totaling 100' (30.5 m) in length. The resulting pressure drop is the same as the pressure drop of a straight duct 130' (39.6 m) long, due solely to the change in direction.

Takeoffs: Connection points installed on a trunk duct that allow the connection of branch ducts.

Boots: Sheet metal fittings designed to transition from the branch duct to the receptacle for the register, grille, or diffuser to be installed.

Figure 32 Examples of flexible duct.
Source: iStock@Tatabrada

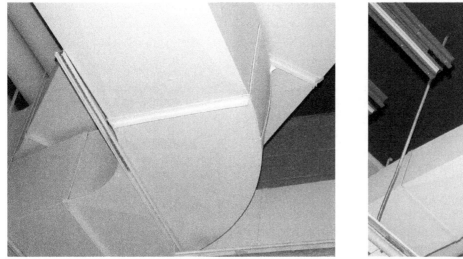

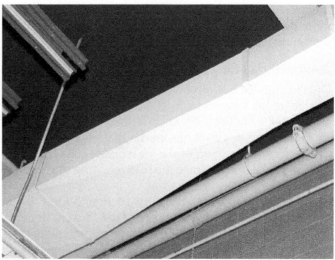

Elbow

Transition

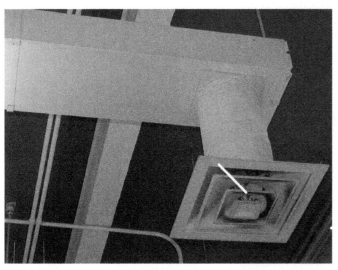

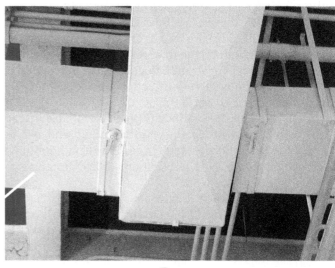

Takeoff

Tee

Figure 33 Examples of duct fittings.

2.5.0 Diffusers, Registers, Grilles, and Dampers

Air terminal devices distribute supply air into the conditioned space. When properly applied, they blend the supply air with the room air so that the room is comfortable without excess noise or drafts. The terms *diffuser*, *register*, and *grille* are used to describe different kinds of terminals. Although they are often used interchangeably in conversation, they each have their own specific definition:

- *Diffuser* — An outlet that discharges air into a room in a widespread, fan-shaped pattern. The pattern may be fixed or adjustable.

- *Register* — An outlet that discharges air into a room in a more concentrated pattern. Some have adjustable pattern deflectors, and they often have a volume control damper built in as well.

- *Grille* — A louvered covering on an opening created for the passage of air into, or out of, a room. Grilles have many different face designs, although most are simple bars or blades. One major purpose is to conceal the opening from view. Some are fixed and can direct air in one direction only, while others are adjustable and can redirect the air. Grilles are not equipped with dampers like a register.

Figure 35 shows common registers, grilles, and diffusers. The floor register provides distribution for both heating and cooling when used on the perimeter of a home. It is installed parallel to the outside wall, about 6" to 8" (15 to 20 cm) away from the wall. Typically, fixed vanes are installed crosswise at an angle to spread the air stream, but there are more decorative versions as shown in *Figure 35*. Floor registers normally have an integral damper for volume control.

Baseboard diffusers are mounted on the floor with the back against the wall. Like floor registers, they normally have a built-in volume damper. The choice between these or floor registers usually depends on the construction characteristics of the home.

Sidewall registers (not shown) can be used low or high on a wall. Low installations are best for heating. They are fed from the back and mounted flush with the wall just above the baseboard. Some are made to discharge air in multiple directions. Low sidewall registers normally include a volume damper.

High-sidewall registers provide poor heating performance since hot air naturally rises. However, they are adequate for heating in warmer climates. They provide good performance in cooling. When used with a return air grille in each room, the cooling performance is even better. High-sidewall registers are mounted flush with the top edge about 6" to 12" (15 to 30 cm) from the ceiling.

Ceiling diffusers can be round, square, or rectangular. The vast majority of them are found in commercial buildings. Some mount flush to a hard ceiling, while others are surface-mounted. They are easily installed in acoustic ceilings, with their weight supported by the T-bars. The supply air is fed from above. Ceiling diffusers can distribute supply air equally in all directions, or be designed for one-, two-, or three-sided discharge. Accessory volume dampers are commonly added to the back. They perform very well in cooling and are adequate for heating. However, they do allow warm air to **stratify** in the upper part of a room, like most systems that deliver warm air from above.

Stratify: To form or arrange into layers. In air distribution, layers of air at different temperatures will tend to stratify unless they are forced to move and mix. Warmer air stratifies above cooler air.

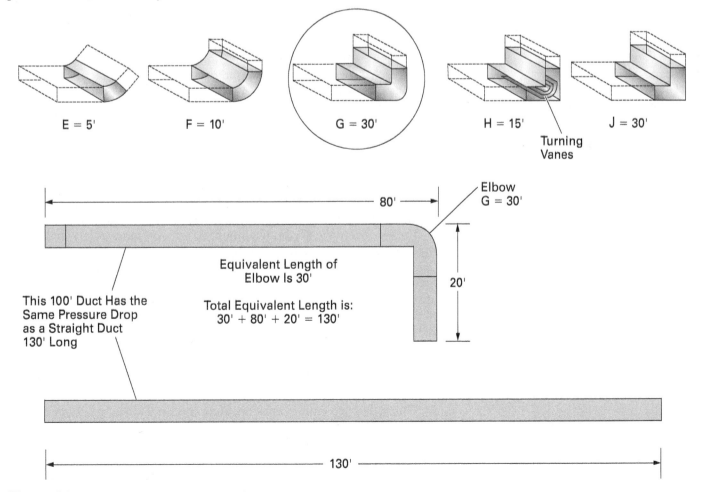

Figure 34 Example of equivalent length values.

(A) Floor Register

(B) Baseboard Register

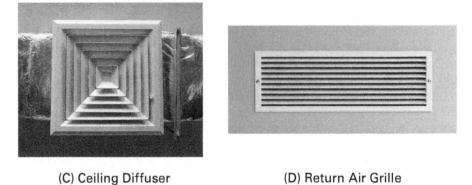

(C) Ceiling Diffuser

(D) Return Air Grille

Figure 35 Registers, grilles, and diffusers.
Sources: Shutterstock.com/B Calkins (33A), Shutterstock.com/Claude Huot (33B), Shutterstock.com/Nuntaporn Keeratibunharn (33C), iStock@KangeStudio (33D)

Manufacturers provide tables that show the performance of their grilles, registers, and diffusers. *Table 5* shows an example of the performance data for a square ceiling diffuser. The table provides the designer with the pressure loss for a given velocity of air exiting the diffuser. This is known as the *face velocity*. In addition, the air volume is provided, along with the *throw*. Throw refers to the distance the air travels when it exits the diffuser at the face velocity shown. An air velocity of 50 fpm (0.254 m/s) is considered the point where the air is no longer moving forward with any real effect.

Note that each diffuser also has an **A_k factor** assigned. The A_k factor represents the free area of the grille in square feet. The A_k factor is used to determine the air volume by multiplying it times the average velocity of the air. Air balancing is presented in detail in another module.

Designers reference the chart with a desired air volume in mind. For example, assume that a copier room needs 400 cfm of conditioned air. According to *Table 5*, a 14" (36 cm) neck size will provide 410 cfm at face velocity of 500 fpm, a pressure loss of 0.16" w.c., and a throw of 11.5'. The neck size refers to the size of the duct connection—round, in this case. A 16" (41 cm) diffuser can provide the same airflow at a face velocity of 400 fpm and a throw of 10'. In other words, the 16" diffuser will move the same amount of air a bit more gently and more quietly. More pressure loss at the opening means more noise too.

A face velocity of 500 fpm from a ceiling diffuser is usually considered reasonable for comfort. Higher velocities and longer throws allow for greater mixing of the room air, but with higher pressure losses, noise, and a risk of the room feeling drafty. The final selection is made based on the designer's knowledge of the priorities.

2.5.1 Air Dampers

Dampers are used to direct and control airflow in duct systems. Without dampers, some areas would receive too much air while others do not receive enough. This is due to common imperfections and limitations in system design and installation.

Some dampers are adjusted manually by an occupant. Others are motorized or have actuators that change their position in response to external controls. Fire and smoke dampers, for example, help protect building occupants in emergency situations by closing tightly on command.

A_k factor: An area factor used with registers, grilles, and diffusers that reflects the free area, relative to a square foot, for airflow.

TABLE 5 Performance Data for a Square Diffuser

Face Velocity (fpm)		300	400	500	600	700	800	900	1,000
Pressure Loss (inches w.c.)		0.006	0.011	0.016	0.024	0.031	0.043	0.050	0.066
Neck Size 6" $A_k = 0.165$	CFM	52	66	88	104	120	136	159	176
	*Throw (ft)	3.5	4.5	5.5	6.5	8.0	9.0	10.0	11.0
Neck Size 8" $A_k = 0.280$	CFM	85	110	140	170	195	225	250	280
	Throw (ft)	4.5	5.5	7.0	8.5	10.0	11.0	12.0	14.0
Neck Size 10" $A_k = 0.420$	CFM	125	170	210	250	295	335	380	420
	Throw (ft)	5.1	6.6	8.2	9.7	11.8	13.3	15.4	16.4
Neck Size 12" $A_k = 0.595$	CFM	180	240	300	355	415	475	535	595
	Throw (ft)	6.0	8.0	10.0	11.5	13.5	15.5	17.5	19.0
Neck Size 14" $A_k = 0.820$	CFM	245	330	410	490	575	655	740	820
	Throw (ft)	7.0	9.0	11.5	13.5	16.0	18.0	20.0	22.5
Neck Size 16" $A_k = 1.030$	CFM	310	410	515	620	720	825	925	1,030
	Throw (ft)	7.5	10.0	12.5	15.0	18.0	20.0	22.0	25.0
Neck Size 18" $A_k = 1.330$	CFM	404	535	672	806	938	1,075	1,212	1,345
	Throw (ft)	8.5	11.0	14.0	17.0	20.0	23.0	26.0	28.0
Neck Size 20" $A_k = 1.600$	CFM	480	640	800	960	1,120	1,280	1,440	1,600
	Throw (ft)	9.5	12.0	16.0	18.0	22.0	25.0	28.0	31.0
Neck Size 22" $A_k = 1.900$	CFM	570	760	950	1,140	1,330	1,520	1,710	1,900
	Throw (ft)	10.5	13.5	17.0	19.0	24.0	27.0	30.0	33.0
Neck Size 24" $A_k = 2.300$	CFM	690	920	1,150	1,380	1,610	1,840	2,070	2,300
	Throw (ft)	11.0	14.5	18.5	22.0	26.0	30.0	33.0	36.0

*Terminal velocity of 50 fpm

A balancing damper for a branch duct should be installed in an accessible place. The closer the damper is to the point of connection for the branch, the better. Branch takeoffs at the main duct are the ideal location.

Balancing dampers should be tight fitting with minimum leakage around the blade. *Figure 36* shows a butterfly damper that would be installed at a branch takeoff. Their leak resistance influences their cost; low-leakage dampers are more expensive. The accessory dampers found at supply diffusers and registers should not be used to balance a system. When partially closed, they disrupt the performance of the diffuser or register and may increase noise significantly. These dampers should only be used by room occupants to make minor adjustments in the airflow.

Larger dampers used in rectangular ducts and at outside air intakes have multiple blades. The two types are *opposed blade dampers* and *parallel blade dampers* (*Figure 37*). Well-constructed multi-blade dampers can close very tightly.

2.5.2 Fire and Smoke Dampers

Fire dampers are used to maintain the fire-resistance ratings of walls, partitions, and floors penetrated by HVAC ducts. Fire dampers prevent the spread of fire through the air distribution system. The dampers are normally held open with a **fusible link** (*Figure 38*) that is usually set to melt at 165°F (74°C). Fusible links and dampers with ratings of 212°F (100°C) and 286°F (141°C) are also available. If a fire occurs, the link melts and the damper closes forcibly by spring action.

Fusible link: Mechanically, a device that holds a fire damper in the open position and is designed to melt at a specific temperature. Once melted, the damper opens by spring pressure. Fusible links are not reusable and must be replaced.

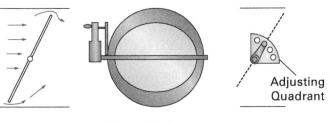

Adjusting Quadrant

Butterfly Damper

Figure 36 Round butterfly damper.

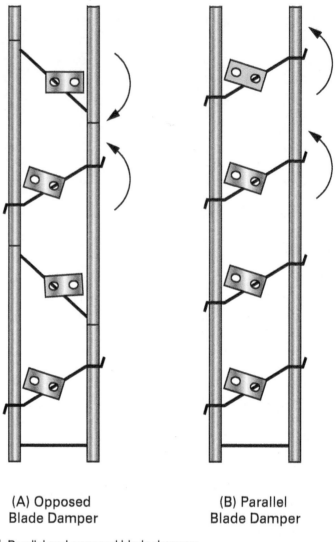

(A) Opposed
Blade Damper

(B) Parallel
Blade Damper

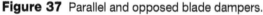

Figure 37 Parallel and opposed blade dampers.

Fire dampers usually have a fire resistance rating of either $1\frac{1}{2}$ or 3 hours. Building walls with a 3-hour fire rating require a 3-hour rated damper in ducts that pass through. Partitions with fire ratings of less than 3 hours generally require $1\frac{1}{2}$-hour rated dampers. The rating for the fire damper must be at least 75 percent of the fire rating for the wall, floor, or partition. Therefore, a fire damper rated for $1\frac{1}{2}$ hours can be used for a barrier rated for up to 2 hours—a common fire rating for walls. Likewise, a damper rated for 3 hours can be used in barriers with fire ratings up to 4 hours.

In addition to the fire-resistance rating, fire dampers have a *static* or *dynamic* air-closure rating. Static-rated dampers can only be used in HVAC systems where the equipment is automatically shut down when a fire is sensed. In that case, no air should be flowing in the ducts when the damper closes. Dynamic-rated dampers are designed to close even if the HVAC system remains running and air is moving through the ducts. They are capable of positive closure despite the added air pressure.

Smoke can be as deadly as fire, so controlling the spread of smoke in a building is critical. Smoke dampers help deal with smoke management. Smoke dampers can be passive in their function, where they close to isolate a section of duct. They also can be part of an engineered smoke control system that directs smoke outdoors once the fire has been extinguished.

Most smoke dampers are operated electrically, controlled by a smoke or heat detector, fire alarm, or automated building control system. Smoke dampers are rated for leakage in accordance with *UL Standard 555S, Leakage Rated Dampers for Use in Smoke Control Systems*. Class 1 has the lowest leakage rating, and Classes 2 and 3 have higher leakage ratings. Smoke dampers have temperature, air velocity, and air pressure ratings that show how they will perform in various conditions. The proper selection of life safety accessories for a system is critical.

Combination fire and smoke dampers (*Figure 39*) are also available. These dampers must conform to the rating agency requirements for both fire and smoke dampers. The use of combination dampers is guided by NFPA 90, *Installation of Air Conditioning and Ventilation Systems*.

Fusible Link

Figure 38 Fire damper fusible link.

Figure 39 Combination fire and smoke dampers at a wall penetration.
Source: Certified Life Safety Inspection Services

Case History

Air Loss at the Airport

A service call from frustrated maintenance personnel working at a small airport resulted from a lack of airflow to a number of passenger areas. The problem seemed minor at first but became progressively worse over several years. After looking over the equipment and controls, the visiting service technician asked for a set of plans. With plans in hand that identified the location of fire dampers in the system, a systematic inspection of each one revealed four with broken fusible links that were tightly closed, as designed. Replacing the links restored the airflow quickly.

The Bottom Line: Although HVAC technicians often find themselves focusing on equipment, it is essential to remember that the equipment is just one part of a larger picture. Regardless of how well it runs, the equipment is no better than its air distribution component. Don't forget that when troubleshooting an unfamiliar system.

2.0.0 Section Review

1. Which type of blower is commonly used in residential systems?
    a. Backward-inclined centrifugal
    b. Forward-curved centrifugal
    c. Radial
    d. Vane-axial

2. The type of fan usually housed in a short venturi is the _____.
    a. duct fan
    b. tube-axial fan
    c. vane-axial fan
    d. propeller fan

3. An air distribution system is equipped with a $\frac{1}{2}$ hp motor and a belt-driven blower. The system is moving 1,200 cfm of air with the blower turning at 950 rpm. The system static pressure is 0.38" w.c. If the belt loosens and the blower speed slows to 790 rpm, what will the new air volume be? The answer is rounded to the nearest whole number.
    a. 625 cfm
    b. 998 cfm
    c. 1,332 cfm
    d. 1,443 cfm

4. A homeowner wants to add air conditioning to an existing warm-air heating system. The existing system moves 870 cfm of air at a static pressure of 0.37" w.c., and it has a $\frac{1}{3}$ hp motor driving a blower turning at 1,050 rpm. If the system needs to move 1,000 cfm of air to support the new air conditioning system, what motor horsepower will be required? The answer has been rounded to the nearest hundredth.
    a. 0.29 hp
    b. 0.37 hp
    c. 0.51 hp
    d. 0.61 hp

5. Round metal duct sections are normally connected using _____.
    a. snap-lock fasteners
    b. duct tape
    c. sheet metal screws
    d. drive clips

6. An air terminal device that directs air in a relatively concentrated, non-spreading stream is a _____.
    a. diffuser
    b. register
    c. louver
    d. damper

3.0.0 Air Distribution System and Energy Conservation

Objective

Identify the different approaches to air distribution system design and energy conservation.

a. Identify various air distribution system layouts.

b. Describe heating and cooling air movement resulting from various air distribution system designs.

c. Explain how to maximize energy efficiency through the proper sealing and testing of air distribution systems.

Performance Tasks

There are no Performance Tasks in this section.

Air distribution systems typically consist of a supply duct system and a return duct system. The supply duct is usually more complex in its layout. The design and layout depend on the structure and its intended use. Large and specialized systems can be quite complex. Since residential systems are more uniform and relatively simple, they are used as the basis for learning.

To size the system, a designer begins with a heat load analysis. This determines the heat gain or loss for the structure and the individual areas. To create the analysis, the designer divides heat gain and heat loss factors into two categories: *external* and *internal*.

External factors include the outside air temperature range, the relative humidity, and the position of the building relative to the sun. Internal factors include lighting, mechanical equipment, office machines, the number of people in the building, and the operating schedule. These factors vary based on the type and size of the building, how it is used, and where it is located. Internal factors typically produce a heat gain, working against the cooling system but helping to keep it warm in the winter.

Major factors affecting the design approach are building construction and climate. The designer must also consider other factors such as the applicable building codes and overall cost. The best system design is one that balances performance with cost while ensuring a safe, effective, and reliable installation.

3.1.0 Air Distribution System Layouts

The type of duct system used is determined by climate and the building construction features. Most homes use perimeter duct systems where the construction features allow it. Perimeter systems have floor registers or ceiling diffusers near the outside walls. They can be applied in homes with crawl spaces or attics, and even those built on concrete slabs. Floor registers generally provide better combined heating and cooling performance than ceiling diffusers.

Air handlers and furnaces in the attic may be suspended from the rafters with vibration isolators (*Figure 40*). This minimizes vibration transmission to the structure. They may also be laid on the attic floor. The unit is usually placed in a horizontal position, but an attic with enough height may allow it to be installed in an upflow or counterflow position (*Figure 41*).

The main ducts can be made from fiberglass ductboard or insulated sheet metal with a vapor retarder. Branch ducts are usually round flexible duct or sheet metal covered with foil-faced insulation. Remember that the excessive use and poor installation of flexible ducts diminishes airflow. Branch lines longer than 6' (~2 m) should start at the main trunk as metal, and then transition to flexible duct within 6' of the terminals. The return grille is generally in a central hallway ceiling or wall, but larger installations require more than one for best performance.

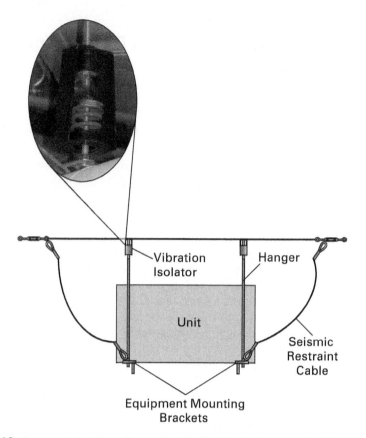

Figure 40 Suspended air handling unit with vibration isolators and seismic restraints (earthquake-prone areas).

Perimeter systems can be created using several schemes. The four common layouts that are presented here included the following:

- Perimeter loop
- Radial
- Extended plenum
- Reducing trunk

3.1.1 Perimeter-Loop Duct System

Perimeter-loop systems (*Figure 42*) are often used in homes built on concrete slabs. The perimeter loop is a continuous duct of consistent size embedded in the slab. It runs close to the outside walls, with the outlets located in the floor directly above the loop. The loop is fed by several branches from the plenum, maintaining an equal pressure throughout.

When the blower is running, warm air in the ducts helps keep the entire slab warmer in winter. Heat loss to the outside is reduced by insulation placed around the edges of the slab and beneath the duct.

3.1.2 Radial Duct System

In a radial duct system, also shown in *Figure 42*, each perimeter outlet is fed directly from the supply air plenum. The loop is eliminated. Also known as *spider systems* or *plenum systems*, radial duct systems are used in small homes or additions. The ductwork can be installed in a concrete slab, crawl space, or attic.

Volume control dampers should be used in all branch duct runs to facilitate balancing the system airflow. Unlike the loop system, the pressure throughout the system is not consistent. That means dampers are needed to balance the volume, throttling the damper on branches receiving too much air. The air will favor the branches with the least resistance to flow. A single return air grille is typically used.

Figure 41 Upflow furnace in an attic.

Radial duct systems are economical to install but not very common due to size limitations, noise, and questionable performance unless the structure is very small.

3.1.3 Extended-Plenum Duct System

The extended-plenum system (*Figure 43*) uses round or rectangular trunks for the main supply and return. The supply duct remains the same size over its entire length, leading to the name. These systems can be installed in a crawl space or attic, but they are not recommended for slab floors. This is another situation where a slab floor does not provide access to balancing dampers. Like radial systems, dampers for balancing this system are essential.

Separate branch ducts run from the main trunk to each supply outlet. The extended-plenum system works best when the air handler is in the center of the main duct's length. However, the supply duct can also be routed in one direction only.

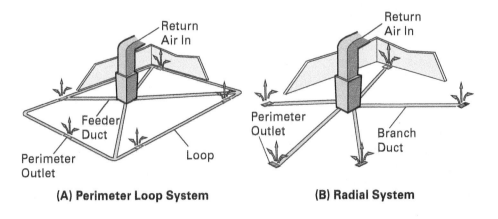

(A) Perimeter Loop System   **(B) Radial System**

Figure 42 Two types of perimeter duct systems.

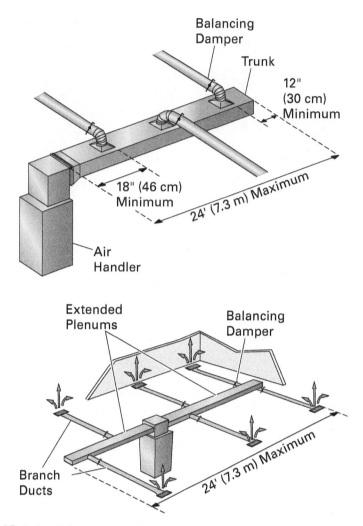

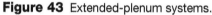

Figure 43 Extended-plenum systems.

A volume damper should be installed in each branch, at the takeoff. This allows the airflow to be balanced since the trunk duct has no transitions in size to help balance the airflow.

An extended-plenum system requires careful duct sizing and balancing dampers. But it can provide reasonable performance in small homes. In a larger structure that requires a significant amount of duct to reach supply outlets, it is a poor choice. The following are some recommended practices for laying out an extended-plenum duct system:

- The supply and return ducts should extend no more than 24' (7.3 m) from the air handler.
- The first branch should be at least 18" (46 cm) from the beginning of the main duct.
- The main trunk should extend at least 12" (30 cm) past the last branch takeoff.
- Balancing dampers should be installed at each branch takeoff.

3.1.4 Reducing-Trunk Duct System

A reducing-trunk system (*Figure 44*) provides the best performance but at a slightly higher cost. It works well in larger homes that require longer duct runs. Remember that extended-plenum systems are limited to 24' (7.3 m) in length. The reducing-trunk system is a good choice for systems regardless of where the air handling equipment is positioned.

When properly sized, the same pressure loss is maintained from one end of the duct system to the other. This allows all branches to operate at roughly the same pressure. When it is properly designed and installed, it requires limited

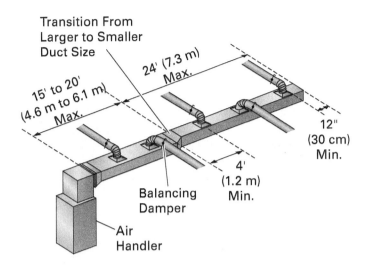

Figure 44 Reducing-trunk duct system.

effort to balance. The following are some recommended practices for laying out a reducing trunk duct system:

- The first main duct section should be no longer than 20' before a transition occurs.
- The length of each reduced section should not exceed 24'.
- The first branch duct connection downstream of a transition should be several feet from the transition fitting. This allows the air turbulence caused by the fitting to subside before reaching the branch duct.
- The trunk duct should extend at least 1' beyond the last branch takeoff.

3.1.5 Furred-In Duct Systems

Many multifamily facilities, such as apartments, have neither a crawl space nor an attic. When independent systems are provided for each apartment, the air distribution system is often *furred-in* (*Figure 45*). That means it is installed *on* the finished ceiling, but then concealed with the same ceiling material. The finished ceiling is lowered in that area as a result. This is usually done in a hallway. With the registers and diffusers in the high sidewall of the adjacent rooms, the ceiling in these areas can remain at a normal height.

Small fan coil units are made specifically for installation in a furred-in space (*Figure 46*). For small systems, the return air grille is installed immediately beneath the unit. The supply duct may only consist of a few inches of duct to connect to a sidewall register, as shown in *Figure 46*.

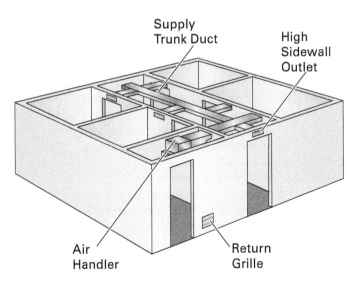

Figure 45 Furred-in duct system.

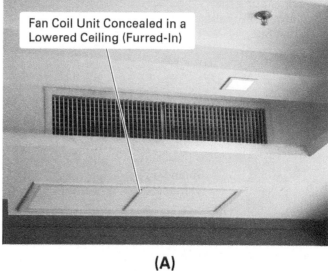

(A)

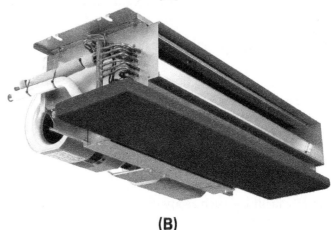

(B)

Figure 46 Fan coil unit installation.
Sources: iStock@VacharapongW (46A), Carrier Corporation (46B)

The registers are placed on the high sidewall of the adjoining interior walls. As a result, the branch line connections are very short, extending only to the walls of the furred area. The furred-in approach can make use of both extended-plenum and reducing trunk systems. If the supply duct length extends more than 24' (7.3 m), a reducing trunk should be used.

3.2.0 Heating and Cooling Room Airflow

Understanding airflow patterns helps to determine which layout is best for an application. It also helps when you must investigate comfort-related complaints.

The airflow from a perimeter system sweeps the outside walls and windows. This helps offset the heat losses and gains at the source. The return air grilles are located on the interior walls or ceilings in a central location. For improved performance, individual returns can be installed in each room. However, individual room returns increase the installation cost of a system significantly and are rarely chosen as a result.

The centralized location of the return grilles helps to draw air from the perimeter toward the center of the building. To allow air to leave the rooms at the same rate as it enters, doors must remain open or be *undercut* to leave space beneath them for air to pass. Alternatively, a **transfer grille** can be placed in the wall or door. However, transfer grilles have a negative effect on acoustics—voices can easily be carried from room to room. Air must have a way to escape from the room. Otherwise, the supply air volume drops sharply once the room is pressurized.

Transfer grille: A grille usually installed on walls or doors, with one mounted on each side of an opening, that allows air to pass freely into or out of an enclosed space.

Figure 47 shows the room airflow for a perimeter system with floor registers. During the heating mode, the heated air moves up the wall to the ceiling. Because it is warmer and lighter than the room air, it spreads across the ceiling, and then down the inside wall as it mixes and cools. Some room air is induced from the floor into the flow of warm air and mixes with it. A stratified zone of cooler air is formed at the floor.

During the cooling mode, the supply air travels up the wall, with some striking the ceiling. Because it is cooler and heavier than room air, the air travels a shorter distance and then drops back down into the room as shown. The cool air mixes with the room air well, leaving only a small, stratified layer of warm air near the ceiling. High return grilles can minimize this problem but doing so has a negative effect on heating performance.

Figure 48 shows the room airflow with high sidewall outlets. This pattern is typical of furred-in duct systems. In the cooling mode, cool air moves across the ceiling and down the far wall. The room air mixes well with the supply air and almost no stratification occurs. In the middle of the room, the air has the tendency to tumble gently. Since the cool air falls, some is drawn down from the ceiling quickly as it exits through or under the door.

In the heating mode, the warm air tends to stay closer to the ceiling. Again, the negative pressure at the doorway will draw some air down, but the warm air will generally travel farther across the room before turning down. The airflow

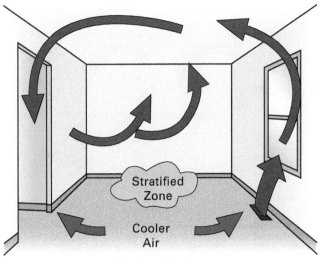

Heating Mode

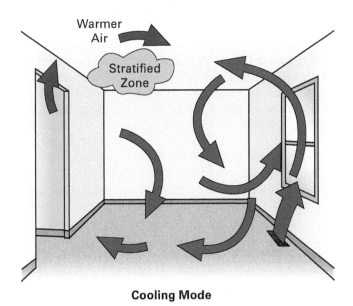

Cooling Mode

Figure 47 Room air distribution patterns for a perimeter duct system.

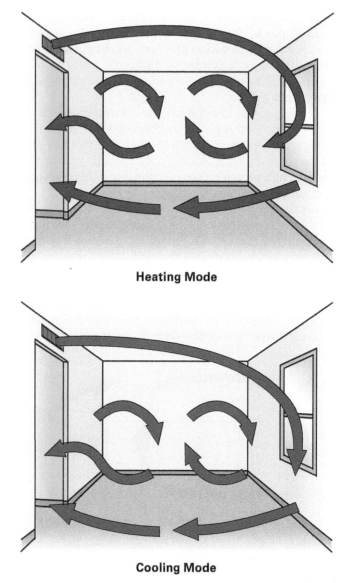

Heating Mode

Cooling Mode

Figure 48 Room air distribution patterns for high sidewall outlets.

pattern in the room is not significantly different than in cooling, but all air movement and mixing takes place higher in the room. High sidewall outlets do perform better for cooling than heating. It is difficult to warm the lower portion of the room using high sidewall outlets, especially when the return grilles are installed high.

Ceiling registers and diffusers perform best for cooling but perform poorly for heating. High ceilings make heating performance worse. In the cooling mode, supply air mixes well with the room air (*Figure 49*). Air motion in the room is good with few stagnant areas. In the heating mode, the warm air clings to the ceiling without reaching the occupied space until it cools. When the return grille is also in the ceiling, the air is not encouraged to move lower. Using a return grille mounted low on the wall encourages better air movement and mixing to some degree. However, since air is being delivered to the room from above, return air often must be drawn from above as well due to structural limitations. Virtually all commercial and industrial facilities must live with this arrangement.

3.3.0 Energy Efficiency in Air Distribution Systems

Duct systems must be properly installed, sealed, and insulated to be energy efficient. Ductwork in unconditioned spaces must be insulated to prevent it from losing or gaining heat, and to prevent condensation. Poorly sealed duct

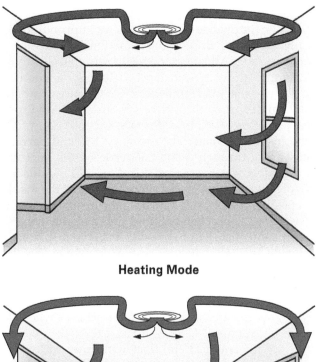

Heating Mode

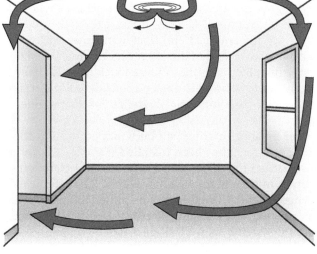

Cooling Mode

Figure 49 Room air distribution patterns for ceiling diffusers.

joints are also a significant cause of wasted energy. Leaking supply ducts lose conditioned air, usually to unconditioned spaces. Leaking return ducts draw unconditioned air from attics and crawl spaces. They may also draw in foul odors.

Pressure testing can determine if ductwork is properly sealed. This type of testing might be done as part of a whole-house energy efficiency analysis. Such testing is often performed by HVAC contractors who have incorporated building weatherization into their list of services. Duct testing is usually a specified part of the start-up and commissioning process for commercial buildings.

3.3.1 Insulation and Vapor Retarders

When ductwork passes through an unconditioned space, heat transfer takes place between the duct walls and the surrounding air. In the cooling mode, an astounding amount of water from the surrounding air can condense on metal duct. Since the supply air temperature is around 55°F (13°C), the duct surface is below the dew point of air in unconditioned spaces. As a result, insulation with a reliable vapor retarder must be applied to the duct. When fiberglass ductboard is used, the material serves as the insulation and the foil on the outside acts as a vapor retarder. However, if the temperature difference and humidity in the space is great enough, even fiberglass ductboard will sweat on the outside.

Figure 50 Condensation on a supply air duct.

The importance of a good vapor retarder cannot be overstated. Condensation forms very quickly on the outside of a cold duct in an unconditioned space (*Figure 50*). This typically happens in unseen and sometimes inaccessible locations. The results can be devastating. Allowing an area to stay consistently moist prompts mold and mildew growth. Water constantly dripping on wood encourages rot. Entire structures have been made uninhabitable because of sweating ducts and pipes. Mold and mildew growth present serious short- and long-term health concerns. These problems are all in addition to wasted energy.

Although the insulating material provides a thermal barrier, water vapor can easily find its way through it and condense on the duct wall. The vapor retarder must provide a water-resistant shield around the duct and insulation. Even a small flaw in it can allow water vapor to enter. If the insulating material becomes wet beneath the vapor retarder, both the insulation and vapor retarder are compromised, and the problem becomes progressively worse.

Metal duct can be insulated in two ways. It can be insulated on the outside and/or on the inside. Insulation inside the duct, known as *duct liner*, is installed by the duct fabricator. It is either glued or fastened to tabs mounted on the inside duct wall as shown in *Figure 51*.

Duct liner also helps to make the system quieter. However, if liner is to be installed, the designer must consider the liner thickness when sizing the duct. Since duct liner is commonly 1" (2.5 cm) thick, a 12" × 12" (30 cm × 30 cm) duct becomes a 10" × 10" (25 cm × 25 cm) duct with liner added. For a duct that size, the loss of area is roughly 30 percent. Increasing the size of the duct to accommodate the liner increases the job cost. Duct liner also has a higher friction loss than smooth sheet metal. This must be considered in fan sizing. Duct liner is most often applied in commercial installations to provide insulation and noise attenuation.

Insulation and a vapor retarder can be wrapped around the outside of the ductwork before or after it has been installed (*Figure 52*). Most duct insulation is a foil or vinyl-backed fiberglass batt. The material is commonly referred to as *duct wrap*. It comes in several thicknesses, with 2" being typical. The insulation must be taut when installed, but it cannot be pulled too tightly. As the fiberglass material is compressed, it loses some of its insulating value. Insulation must be applied to round metal duct as well, as shown in *Figure 53*.

Once applied, all seams must be secured and sealed with an approved tape. To avoid condensation, any punctures in the vapor retarder must be well sealed.

Figure 51 Duct liner.
Source: Special Thank You to Ray's Metal Works

In many cases, the taped areas are painted with a sealant to prevent water vapor from passing through.

ASHRAE Standard 90.1, *Energy Standard for Buildings Except Low-Rise Residential Buildings*, specifies the minimum acceptable **R-value** of insulation that must be used for covered applications. The R-value must be calculated for both heating and cooling operation. The required R-value is determined by which operating mode requires the thickest insulation to meet the standard. Of course, local codes can always exceed a given standard.

Even though there is a smaller temperature difference between return air duct and the surrounding air, it should also be insulated. A common practice is to line return-air plenums with duct liner to reduce fan noise, especially if the return grille is not far from the air handling equipment.

R-value: A number, such as R-19, that indicates the ability of a material to resist the flow of heat. The higher the R-value, the better the insulating ability.

3.3.2 Duct Sealing

All air distribution systems should be free of air leaks. Leaks allow conditioned air to be lost, or air from outside the conditioned space to be drawn into the return duct. Leaks result in higher energy costs.

In a residential setting, duct leakage can be tested by pressurizing the duct system with a small portable fan with a variable-speed motor (*Figure 54*). To find air leaks, all the grilles and registers are tightly sealed with a special plastic or similar material. The fan speed is adjusted to maintain a specific pressure in the duct. Then the flow rate of air into the duct is measured with instruments. With all the openings in the system tightly closed and the equipment off, it is assumed that any air volume moving into the duct is escaping through leaks.

Another method involves a **blower door** attached to the building entry (*Figure 55*). The fan speed is controlled to exhaust air from the building and maintain a consistent negative pressure throughout. With the home depressurized, a

Blower door: An assembly that fits a door frame or other opening and contains a fan used to depressurize a building by drawing the air out.

Figure 52 Insulated duct sections.
Source: Shutterstock.com/DESIGNFACTS

Figure 53 Insulating round metal duct.
Source: iStock@MyrKu

Figure 54 Duct pressurization fan kit.

Figure 55 A blower door to depressurize a home.
Source: "1000 Blower Door.jpg" is licensed under the Creative Commons Attribution 3.0 Unported license.

pressure measurement pan is connected to a manometer and placed over each register or grille to obtain a pressure reading. Since the entire structure is under a negative pressure, the duct system should be at the same negative pressure. Leaks that allow air to enter the duct result in the duct pressure being higher than the pressure in the room. A perfectly sealed duct should show no pressure difference. A substantial pressure difference indicates a large leak.

Discovering leaks exist is one step. Locating and repairing them can be a challenging and time-consuming task. A flashlight and inspection mirror can be used to inspect hard-to-see areas. Since it is unlikely that metal duct will leak along its surface, joints and seams are the focus of the inspection. Poorly supported and sagging duct sections tend to open joints. If ductwork is sagging, make sure it is properly supported before attempting to seal any leaks. Repositioning the duct after sealing will likely destroy the repair. Listening when the system is running helps locate leaks too. Visible dirty streaks near joints and seams act as tattletales.

WARNING!

Always wear the proper personal protective equipment (PPE) including safety glasses and a dust mask or filter when inspecting duct systems outside the conditioned space.

WARNING!

Vermiculite was once widely used as loose attic insulation. The product looks like small light-brown or gold-flecked pebbles. During mining, some of this product was contaminated with asbestos. If you find vermiculite attic insulation, take care not to disturb it and use appropriate breathing protection around it.

It can be assumed that older duct systems leak somewhat. Sheet metal duct in new homes is less likely to leak because improved codes require them to be properly sealed at the time of installation. When sealing a duct system, either of these two approaches can be used:

- Spend the time to find each leak and seal only the leaking areas.
- Seal all accessible seams and joints in the system.

Finding individual leaks can take a lot of time. It is often more cost effective to simply seal all seams and joints.

Mastic duct sealant (*Figure 56*) is non-toxic and easy to apply. It comes in pails for brush application, or in tubes for use in a caulking gun.

Follow the manufacturer's instructions regarding the technique and the proper PPE when applying mastic. All joints and seams should be sealed (*Figure 57*). That includes seams on plenums, air handling unit connections, joints between duct sections, register boots, takeoffs, and branch line connections. Longitudinal seams in snap-lock round and rectangular ducts must also be sealed.

Sealing Leaks with Duct Tape

It seems that duct tape would be a natural choice for sealing duct leaks. In reality, it is a poor choice, and temporary at best. Scientists at the Lawrence Berkeley National Laboratory tested over a dozen duct tape products. The tests simulated actual system operating conditions. They found that some duct tapes held up better than others, but all the products failed over time. The products used today to seal ducts are designed for that purpose and must meet specific standards. What we have longed referred to as *duct tape* doesn't get the job done.

Going Green

Sealing Duct Leaks

The US Department of Energy estimates that 20 to 30 percent of the air passing through a duct system is lost to leakage. For that reason, it is important to find and seal leaks in forced-air duct systems. Visit the Energy Star website for more information about duct leakage, energy costs, and problem resolution at
https://www.energystar.gov/campaign/heating_cooling/duct_sealing.

Before sealing a joint, remove any dust or dirt. If a joint has a gap less than $\frac{1}{4}$" (~6 mm), use a brush to apply a thick coat of mastic that is at least 2" (~5 cm) wide. If the joint has a gap greater than $\frac{1}{4}$", embed fiberglass mesh tape in the first coat of mastic (*Figure 58*). Apply a second coat of mastic to cover the mesh. Do not turn on the air handler or disturb the joint until the mastic has cured.

Minor slits, punctures, and gaps in fiberglass ductboard are repaired using reinforced tap or fiberglass mesh tape embedded in mastic. The scope of the repair is based on the size and shape of the damage.

Flexible ducts that are damaged can be replaced or repaired. To repair a damaged area, try the following procedure:

- Cut the vapor retarder and insulation from around the inner core at point of the damage.
- Use wire cutters to cut the wire reinforcement and remove the damaged portion. The duct section is now cut in two.
- Insert a sheet metal sleeve at least 4" (~10 cm) long into the ends of each duct section to serve as a splice.

Figure 56 Duct sealing mastic.

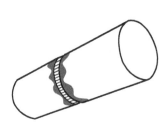

Straight Joints

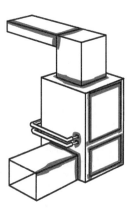

Plenums and
Air Handlers

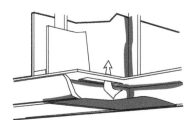

Building Cavities
Used as Duct

Boots

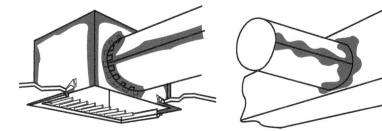

Tees, Wyes, and Ells

Figure 57 Duct sealing locations.

Figure 58 Fiberglass mesh tape.

- Secure the inner portion of the flexible duct tightly to the splice with nylon ties.
- Apply mastic and tape as necessary over the sliced area to ensure a reliable seal.
- Pull the insulation and vapor retarders of the two ducts together over the splice and seal them in the same manner.

Case History

High-Tech Duct Sealing

A process is now available that allows air duct leaks to be sealed from the inside. A special machine injects a mist containing the sealant into the duct. The sealant seeks out the leaks and plugs them from the inside. This high-tech procedure can be expensive and requires special equipment. Workers also require special training. However, the results have apparently been good.

In one case study, a contractor building a new home called on Aeroseal, a company specializing in the field of duct sealing. Before the home could be considered done, the builder had to prove that duct leakage did not exceed 70 cfm. The first test showed the leak rate to be over 500 cfm—roughly 26 percent of the total airflow. After the duct sealant application, the test revealed a leak rate less than 25 cfm. The inspectors were pleased, and the transfer of ownership took place on schedule.

3.0.0 Section Review

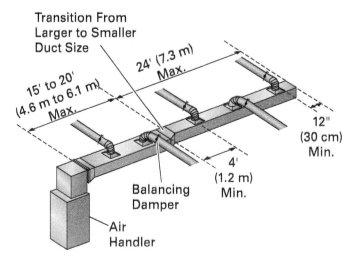

Figure SR01

1. What type of duct system is shown in *Figure SR01*?
    a. Extended plenum
    b. Perimeter loop
    c. Reducing trunk
    d. Radial

2. Which supply air diffuser and register location offers the *best* cooling performance?
    a. Floor
    b. Low sidewall
    c. Ceiling
    d. Baseboard

3. A blower door may be used during residential leak testing to _____.
   a. disperse a sealant throughout the ducts
   b. establish a positive pressure in the duct system
   c. establish a negative pressure in the home
   d. establish a positive pressure in the home

Module 03109   Review Questions

1. Within an air distribution system, the *highest* pressure is found at the _____.
   a. conditioned space
   b. inlet to the return duct
   c. inlet to the blower
   d. outlet of the blower

2. The static pressure, velocity pressure, and total pressure measured in a duct system are typically measured in _____.
   a. pounds per square inch or kPa
   b. inches of mercury or centimeters of mercury
   c. inches of water column (w.c.) or centimeters of water column (cm H_2O)
   d. cubic feet per minute or cubic meters per hour

3. An air distribution system for a light commercial building is designed to work with a 5-ton cooling system. About how much airflow must the system blower be capable of supplying?
   a. 1,200 cfm
   b. 1,650 cfm
   c. 2,000 cfm
   d. 2,350 cfm

4. External static pressure loss is the _____.
   a. total pressure loss in the air handler and its component parts
   b. total pressure loss in an air distribution system, beyond the equipment
   c. velocity pressure loss across the air handler
   d. difference between the total pressure and the static pressure

5. A psychrometer is used to measure _____.
   a. wet-bulb and dry-bulb temperature
   b. static pressure and velocity pressure
   c. grains of moisture in the air
   d. wet-bulb temperature, dry-bulb temperature, and static pressure

6. Pitot tubes are used for measuring _____.
   a. air temperature
   b. air density
   c. relative humidity
   d. air pressure and velocity

7. The type of blower or fan that is well suited to applications with a high resistance to airflow and may appear to be installed backwards is the _____.
   a. forward-curved centrifugal blower
   b. backward-inclined centrifugal blower
   c. propeller fan
   d. radial centrifugal blower

8. An existing air distribution system has 6,000 cfm of airflow created by a blower operating at a speed of 1,200 rpm. To reduce the airflow to 5,000 cfm, what new blower speed is required?
   a. 833 rpm
   b. 1,000 rpm
   c. 1,150 rpm
   d. 1,300 rpm

9. The duct material listed here that has the *least* resistance is _____.
   a. sheet metal
   b. fiberglass ductboard
   c. flexible duct
   d. internally insulated sheet metal

10. Fittings and transitions used in a duct system have _____.
    a. the same friction loss per foot as the same size straight duct
    b. less friction loss per foot as the same size straight duct
    c. a friction loss equal to some length of the same size straight duct
    d. twice the friction loss as the same size straight duct

11. Dampers built into supply diffusers and registers are *best* used to _____.
    a. add moisture to the air
    b. reduce duct noise
    c. allow minor airflow adjustments by the occupant
    d. balance airflow throughout a system

12. The ideal location for a balancing damper in a branch duct is at the _____.
    a. takeoff
    b. boot
    c. transitions
    d. air handling unit

13. A fire damper will close when _____.
    a. a fire has burned for at least 3 hours
    b. the fusible link melts
    c. the HVAC equipment shuts off
    d. smoke is detected

14. A duct system installed in a concrete slab that helps keep the entire floor warm is the _____.
    a. perimeter-loop system
    b. reducing trunk system
    c. perimeter radial system
    d. overhead trunk system

15. Under what conditions would it be necessary to seal a gap in a duct with fiberglass tape embedded in mastic?
    a. It is always done that way.
    b. When the opening exceeds $\frac{1}{4}$" (~6 mm).
    c. It is only done to seal fiberglass ductboard.
    d. When the opening is less than $\frac{1}{4}$" (~6 mm).

Answers to odd-numbered Module Review Questions are found in *Appendix A*.

Cornerstone of Craftsmanship

Thomas Egan
HVACR Field Technician
CMS Mechanical

How did you choose a career in the industry?

I never wanted to be in the same place day after day and I liked trouble shooting problems. I had a summer job with an HVAC company, and that prompted me to examine all the opportunities that the craft offered.

What types of training have you been through?

I initially completed an HVACR certification program through Lincoln Tech. I've also attended CAT chiller training programs in Tampa, FL and Eindhoven, Holland. Since there are many new refrigerants on the horizon, I've also attended refrigerant-specific training for R-290 refrigerant, which is essentially refrigerant-grade propane.

How important is education and training in construction?

It is extremely important! The HVACR industry is constantly changing. Learning about new techniques and equipment is important. Staying up to date with the industry is essential if you want to be a good technician.

How has training/construction impacted your life?

As a Lincoln Tech HVACR graduate, many opportunities were made available. Those opportunities have taken me to many places across the world. But most importantly, the training prepared me to troubleshoot and tackle any problem I face in the craft.

What kinds of work have you done in your career?

I've worked with residential systems, but a great deal of my experience is in commercial and specialized systems. I've installed ice floors for Disney on Ice, along with my wife, who is also HVACR certified. I've worked with large chillers, ice machines, reach-in and walk-in coolers and freezers, and even generators.

Tell us about your current job.

For the last three years, I've worked for CMS Mechanical as a commercial field technician. Our company installs and maintains commercial refrigeration equipment. My days usually include conducting periodic maintenance, troubleshooting problems, and replacing failing and outdated equipment.

What do you enjoy most about your job?

I truly enjoy working with my hands, and every job is different! Even though some tasks can be repetitive, they are done in different environments.

What factors have contributed most to your success?

Hard work combined with patience when challenges arise. I've also continued to learn and I always try to leave a jobsite better than I found it.

Would you suggest construction as a career to others? Why?

Yes! If you enjoy working with your hands and like to be challenged by each job.

What advice would you give to those new to the field?

Tools are important—buy high-quality tools and treat them well. Make an effort to learn something from every task, and always be willing to own your mistakes.

Interesting career-related fact or accomplishment?

My previous position, related to ice rinks, took me to 6 different continents and 32 countries to install systems and supervise the crews. I have friends all over the world!

How do you define craftmanship?

Craftsmanship is taking pride in the work you do and in the condition in which you leave a jobsite.

Answers to Section Review Questions

Answer	Section Reference	Objective
Section 1.0.0		
1. c	1.1.0	1a
2. b	1.2.0	1b
3. a	1.3.2	1c
Section 2.0.0		
1. b	2.1.2	2a
2. d	2.2.1	2b
3. b	2.3.0	2c
4. c	2.3.0	2c
5. c	2.4.1	2d
6. b	2.5.0	2e
Section 3.0.0		
1. c	3.1.4; Figure 44	3a
2. c	3.2.0	3b
3. c	3.3.2	3c

Section Review Calculations

Section 2.0.0

Question 3

Use Fan Law 1 to determine the new volume flow rate:

$$\text{New volume flow rate} = \frac{\text{New rpm} \times \text{Existing volume flow rate}}{\text{Existing rpm}}$$

$$\text{New volume flow rate} = \frac{790 \text{ rpm} \times 1{,}200 \text{ cfm}}{950 \text{ rpm}}$$

$$\text{New volume flow rate} = \frac{940{,}000}{950 \text{ rpm}}$$

$$\text{New volume flow rate} = \textbf{998 cfm}$$

Question 4

First, use Fan Law 1 to determine the new rpm:

$$\text{New rpm} = \frac{\text{New volume flow rate} \times \text{Existing rpm}}{\text{Existing volume flow rate}}$$

$$\text{New rpm} = \frac{1{,}000 \text{ cfm} \times 1{,}050 \text{ rpm}}{870 \text{ cfm}}$$

$$\text{New rpm} = \frac{1{,}050{,}000}{870 \text{ cfm}}$$

$$\text{New rpm} = 1{,}207 \text{ rpm}$$

Then, use Fan Law 3 to determine the new horsepower:

$$\text{New hp} = \text{Existing hp} \times \left(\frac{\text{New rpm}}{\text{Existing rpm}} \right)^{3}$$

$$\text{New hp} = 0.333 \text{ hp} \times \left(\frac{1{,}207 \text{ rpm}}{1{,}050 \text{ rpm}} \right)^{3}$$

$$\text{New hp} = 0.333 \text{ hp} \times 1.15^{3}$$

$$\text{New hp} = 0.333 \text{ hp} \times 1.52$$

The horsepower is **0.51 hp**

Basic Copper and Plastic Piping Practices

Source: Mueller Streamline Co.

Objectives

Successful completion of this module prepares you to do the following:

1. Recognize different types of copper tubing and the related fittings.
   a. Identify and describe the characteristics of copper tubing products.
   b. Identify various copper fittings.
2. Explain how to mechanically join copper tubing.
   a. Explain how to measure, cut, bend, and swage copper tubing.
   b. Describe how to join copper tubing using flare and compression joints.
   c. Describe how to join copper tubing using press-to-connect and push-to-connect fittings.
   d. Explain how pressure testing on refrigerant lines is conducted.
   e. Identify common hangers and supports associated with copper tubing installations.
3. Recognize different types of plastic piping and explain how they are joined.
   a. Recognize and identify different types of plastic piping.
   b. Explain how to join various types of plastic piping.

Performance Tasks

Under supervision, you should be able to do the following:

1. Cut and bend copper tubing.
2. Join copper tubing using a flared connection.
3. Join copper tubing using a compression fitting and ferrule.
4. Assemble press-to-connect joints in copper tubing according to the manufacturer's instructions.
5. Cut and join PVC pipe and fittings.

Overview

Copper tubing is used extensively in HVACR work. Copper is the base material for virtually all common refrigerant piping. Plastic piping is also used in heating and cooling systems for a variety of purposes. Plastic piping may be used to carry chilled water for large cooling systems, condenser water for water-cooled mechanical systems, or even to vent high-efficiency gas furnaces. This module will introduce copper and plastic piping materials and explain how these piping systems are applied and assembled.

Digital Resources for HVACR

SCAN ME

Scan this code using the camera on your phone or mobile device to view the digital resources related to this craft.

Industry Recognized Credentials

If you are training through an NCCER-accredited sponsor, you may be eligible for credentials from NCCER's Registry. The ID number for this module is 03103. Note that this module may have been used in other NCCER curricula and may apply to other level completions. Contact NCCER's Registry at 1.888.622.3720 or go to **www.nccer.org** for more information.

You can also show off your industry recognized credentials online with NCCER's digital badges. Transform your knowledge, skills, and achievements into badges that you can share across social media platforms, send to your network, and add to your resume. For more information, visit **www.nccer.org**.

1.0.0 Copper Tubing and Fittings

Performance Tasks

1. Cut and bend copper tubing.
2. Join copper tubing using a flared connection.
3. Join copper tubing using a compression fitting and ferrule.
4. Assemble press-to-connect joints in copper tubing according to the manufacturer's instructions.

Objective

Recognize different types of copper tubing and the related fittings.

a. Identify and describe the characteristics of copper tubing products.
b. Identify various copper fittings.

Going Green

Copper

The cost of copper has skyrocketed in recent years due to excessive worldwide demand. Regardless of its cost and value, copper is 100 percent recyclable and should never be disposed of in a landfill or other permanent disposal site. If a material can be recycled, it should be recycled. Collect copper scrap and waste from piping installations and demolished facilities, and then turn it over to a responsible party for recycling. To deter copper thieves and prevent them from selling their ill-gotten gains, documentation of the exchange is now required in many areas.

Halocarbon: Any compound containing carbon and one or more halogens (chlorine, fluorine, bromine, astatine, and iodine). For example, HCFC-22 contains chlorine, making it a halocarbon. Refrigerants bearing chlorine have been or are being phased out for environmental reasons.

Malleable: A characteristic of metal that allows it to be pressed or formed to some degree without breaking or cracking. For example, copper is malleable, while cast iron is not.

An HVACR technician must be able to work with a variety of piping and tubing materials. Copper tubing is used to transport refrigerant in virtually all residential and commercial air conditioning systems. Steel is also used with a few refrigerants, such as ammonia. Copper, steel, and plastic are all used to transport water in HVACR systems for various purposes. Steel, copper, and a few special types of tubing are used for fuel gases. The final choice of a piping material for an application is often controlled by building codes.

This module focuses on handling, cutting, bending, and joining copper tubing and plastic piping. Mastering these skills is essential to an HVACR technician's success.

1.1.0 Characteristics of Copper Tubing

HVACR technicians and installers work with copper tubing and fittings consistently. It is the primary material used for refrigerant piping. Therefore, it is essential for you to learn all there is to know about copper tubing and how to make reliable connections.

Air conditioning and refrigeration systems using **halocarbon** refrigerants use copper tubing. Some aluminum is used to construct coils, but its use is limited due to its durability and workability. Steel pipe and fittings, as well as aluminum, are used with some non-halocarbon refrigerants such as ammonia. Despite the higher cost of copper, it significantly outperforms all other materials in most HVACR applications.

Copper tubing of all types must meet the specifications set forth by ASTM International (formerly known as the American Society for Testing and Materials). It must be constructed of 99.9 percent copper. Copper is relatively soft and **malleable**, with high thermal and electrical conductivity. The surface molecules of copper tend to combine with oxygen in the air, causing its natural, bright orange or yellow-orange color to change to a dull brown or even black color. The copper reacts with the oxygen in the air to form copper oxides, commonly called *tarnish*. The process is referred to as **oxidation**.

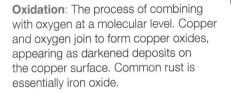

In the presence of heat, such as that from a torch when **soldering** or **brazing**, the oxidation process accelerates dramatically. During the brazing process, the copper oxides form so quickly that they often form loose, black flakes that readily fall away from the tubing. The surface of the tubing also remains discolored.

In craft work, there are two primary categories of copper tubing. The two categories are based on the temper, or hardness, of the material. **Annealed** copper tubing is often referred to as *soft copper*. It is generally sold in manageable rolls of various lengths. Both 25' and 50' rolls are common. Annealed tubing is very easy to bend and form. As long as it is done with the proper tools and technique, it can be formed without significant damage.

Hard-drawn tubing, commonly referred to as *hard copper*, is fabricated by drawing, or pulling, the copper through a series of dies to decrease its size to the desired diameter. This process hardens the tubing, making it more rigid. Bending hard-drawn tubing in the field is not generally recommended; instead it can be done with shop equipment.

Both annealed and hard-drawn tubing are shown together in *Figure 1*. The working pressure rating of hard copper is roughly 60 percent higher than that of soft copper. Both types, however, are quite capable of handling common refrigerant circuit pressures. Hard copper tubing is sold in straight lengths, usually 20' long. However, they are often cut into 10' lengths by the supplier for easier handling and transportation by technicians. Both soft and hard copper can be joined and installed using a variety of methods and fittings.

Soft copper tubing is available in very small sizes as *capillary tubes*. Soft copper tubing is most often used for smaller refrigerant lines in HVACR work. Although larger sizes are available, it is rare for sizes larger than $1\frac{1}{8}$" outside diameter (OD) to be used in the craft. Hard copper tubing is commonly available from $\frac{1}{4}$" OD to $6\frac{1}{8}$" OD. More information about sizing conventions is provided in the next section.

Oxidation: The process of combining with oxygen at a molecular level. Copper and oxygen join to form copper oxides, appearing as darkened deposits on the copper surface. Common rust is essentially iron oxide.

Soldering: A heat-based method of joining metals using another alloy with a melting point lower than the metal(s) being joined. The alloy is used as a filler metal to bond and fill any gaps between the pieces by capillary action. Soldering uses filler metals that have a melting point below 842°F (450°C).

Brazing: A heat-based method of joining metals using another alloy with a melting point lower than the metal(s) being joined. The alloy serves as a filler metal to bond and fill any gaps between the pieces by capillary action. Brazing uses filler metals that have a melting point above 842°F (450°C).

Annealed: Describes a metal that has been heat-treated, making it more formable yet tougher after it cools.

Hard-drawn: Describes a material, such as hard-drawn copper tubing, that has been drawn through dies to form its shape and dimensions. Each die is progressively smaller than the previous until the desired size is reached.

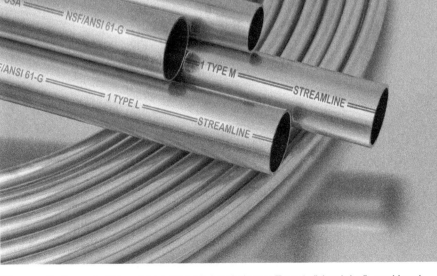

Figure 1 Annealed Type M (red label), hard-drawn Type L (blue label), and hard-drawn Type K (green label) copper tubing.
Source: Mueller Streamline Co.

Tubing? Pipe? Which Is It?

The terms *pipe* and *tube* are often used interchangeably. Some say there is one characteristic that separates the two. Pipe is typically sized by its inside diameter, while tubing is sized by its outside diameter. However, this is not really the case.

For example, a common 1" steel pipe would be expected to have an *inside* diameter of 1". However, the actual inside diameter is roughly 0.050" greater. The copper industry refers to its tubular products as tubing. So, a 1" copper tube would be expected to have an *outside* diameter of 1". Again, this is not the case. In fact, most copper tubing products have an outside diameter $\frac{1}{8}$" larger than their spoken size. Therefore, a 1" copper tube, regardless of its wall thickness, has an outside diameter of $1\frac{1}{8}$". The size of any pipe or tube, as spoken, is usually based on the approximate, or *nominal*, size. That specific dimension is rarely found on the material, inside or outside. Further, it is usually based on the inside diameter, except for one particular type of copper tubing.

Another important fact does nothing to clarify the difference. Both pipes and tubes come in different wall thicknesses. As the wall thickness increases, the outside diameter does not change. The added wall thickness shrinks the inside diameter, rather than increasing the outside diameter. This is true for both pipes and tubes. However, the nominal size remains the same.

The issue is undoubtedly confusing. Regardless of the discrepancies and valid arguments about the differences in pipes and tubes, it is the tendency of experienced HVACR technicians and installers to consider tubes to be relatively thin-walled materials, often with some level of flexibility, while the walls of a pipe are relatively thick.

1.1.1 Copper Tube Types

Different types of copper tubing are identified with one or more letters. The primary difference in each type is the wall thickness. A copper tubing size of 1" will be used in this section to demonstrate the variations in wall thickness.

Type M copper tubing has the thinnest wall of the four common products. It is also the lightest in weight and lowest in cost. Type M is typically used in domestic plumbing systems for water, but it is suitable for many other low-pressure and drainage applications. A 1" Type M copper tube has a wall thickness of 0.035" and an OD of $1\frac{1}{8}$".

Type L copper tubing has a thicker wall and can be used for higher pressures than Type M. A 1" Type L copper tube has a wall thickness of 0.050" but still has an OD of $1\frac{1}{8}$".

Note that, as the wall thickness of copper tubing varies from type to type, the OD for a given size does not change. In other words, 1" Type M and 1" Type L copper tubing have the same OD. As the wall thickness increases, only the inside diameter of the tubing decreases.

Type K tubing has the thickest wall and, of course, is the heaviest in weight per foot. Type K is typically used in the most challenging and critical applications. It is rarely specified in the HVACR industry for field use, but it may be chosen for underground use. Specifications usually identify the minimum type and weight of tubing to be used, but not a maximum value. A heavier weight is almost never substituted for a lighter one by contractors due to the excessive cost.

Type DWV has the thinnest walls, but it is not commonly used today due to the cost. The acronym *DWV* stands for *drain, waste,* and *vent*. It is not designed for pressurized applications. Since it is primarily designed for DWV service, the smallest size available is $1\frac{1}{4}$". This size has a wall thickness of 0.040". Due to the cost of the material compared to plastic options, it is typically used in DWV applications where an attractive appearance is required. There is little incentive to use it where the installation is hidden from public view. HVACR personnel very rarely work with Type DWV, if ever.

Type ACR tubing is the most important to HVACR technicians. ACR is an acronym for *air conditioning and refrigeration*. This tubing was specifically developed for refrigerant piping. However, it can be used in any application where Type L has been specified. This type is the same as Type L tubing, with the same dimensions.

What sets it apart is that ACR is purged of air, then filled with nitrogen and capped after fabrication. This eliminates oxygen inside the tube, thereby preventing the formation of copper oxides. When a new length of Type ACR is uncapped, it is extremely clean and shiny inside. This level of cleanliness is very important to the refrigeration circuit. The caps or plugs should always be replaced on the unused portion of ACR tubing once it is cut. Although all the nitrogen has escaped and air has entered, replacing the plugs prevents the constant entry of fresh oxygen as well as dirt. Type ACR is available in both hard and soft versions.

One other important feature is that ACR is called out by size differently. While a length of Type L tubing would be called out as 1" (its nominal inside diameter), Type ACR of the same size would be listed and purchased as $1\frac{1}{8}$" tubing (the actual OD).

This is a key point to remember. HVACR personnel refer to copper tube sizes by their actual OD because they are generally working with Type ACR. Plumbers, on the other hand, call out copper tubing by the nominal size (inside diameter) since they are generally working with types other than ACR. This can cause some confusion between craftworkers and suppliers. Craftworkers must be specific about the copper tubing type and size they need.

There are also types of copper tubing designed specifically for medical gas service. They are typically identified as Medical Gas Type K and Medical Gas Type L. The dimensions remain the same as their common counterparts, but like ACR, the difference is in the preparation and cleanliness required.

1.1.2 Identifying Markings

The labeling on copper tubing contains very important information. Manufacturers must permanently etch or stamp each product to show the tube type, the name or trademark of the manufacturer, and the country of origin. This same information must be printed on hard-drawn tubes in a specific color that corresponds to the tube type. Type ACR does not require these same

TABLE 1 Copper Tubing Color Codes

Type	Color Code
K	Green
L	Blue
M	Red
DWV	Yellow
ACR	Blue

Figure 2 Identifying markings on Type ACR tubing.

permanent markings, but they may be present. Hard-drawn ACR tubing is best identified by the printed markings (*Figure 2*). Soft ACR tubing may not have printed or permanent markings at all. *Table 1* shows five types of copper tubing and the corresponding color code for labeling. Note that the Type ACR color code is the same as Type L.

1.1.3 Copper Tube Sizing

The sizing of copper tubing other than ACR is based on the nominal, or standard, size of the inside diameter (ID). However, remember that various types have different wall thicknesses, and the ID of the tubing changes as the wall thickness changes. The OD remains the same. This means that the OD of Type M, Type L, and Type K tubes is the same for any given size. The ID of each one, however, is different.

Figure 3 provides a comparison of dimensions for each type of copper tubing, based on a nominal size of 1". Remember that Type ACR is referred to by its OD. In the example shown then, all types would be identified as 1" except for Type ACR. It is identified as $1\frac{1}{8}$" tubing. The dimensions shown are based on the ASTM standards.

Table 2 provides some detailed information about Type ACR tubing. Charts for all tubing types are readily available from a number of sources, such as the Copper Development Association.

Thankfully, one feature of copper tube sizes holds true. Tubing sizes stated as nominal have an actual OD that is $\frac{1}{8}$" larger than the stated size. For example, 1" Types K, L, and M all have an OD of $1\frac{1}{8}$". Type ACR of the same size would be called $1\frac{1}{8}$" OD. As a result, the same fittings can be used on, and will properly fit, any of the four types of copper tube.

It should be noted that copper tubing is made to metric standards as well as US standards. ASTM Standard B88 covers US-standard copper tubing requirements. ASTM Standard B88M is a companion publication covering the metric sizes. Metric copper tubing has several different types as well, identified as Types A, B, and C. Type A has the thickest wall, while Type C has the thinnest wall. There are a few other standards to which copper tubing is manufactured across the world.

It is important to remember that the metric and US-standard tubing sizes and fittings are not interchangeable. Unlike metric hardware, such as nuts and bolts, metric copper tubing is virtually nonexistent in American products or installations. You should only expect to encounter it in products manufactured abroad unless they were built to an American specification. Remember that metric fittings must be used with metric tubing.

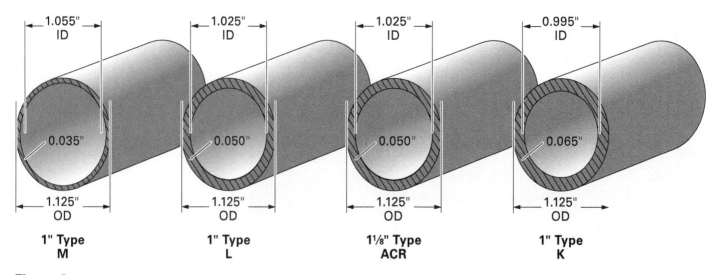

Figure 3 Tubing dimensions for a nominal 1" copper tube size, for each type of copper tubing.

TABLE 2 Dimensions and Physical Characteristics of Type ACR Copper Tubing

Nominal OD (Inches)	Form (Annealed or Hard-Drawn)	Dimensions (Inches)			Calculated Values		
		OD	ID	Wall Thickness	Bore Area (Square Inches)	Tube Weight (Pounds per Linear Foot)	Holding Capacity (Cubic Feet per Linear Foot)
1/8	A	0.125	0.065	0.30	0.00332	0.0347	0.00002
1/4	A	0.250	0.190	0.030	0.0284	0.0804	0.00020
	H	0.250	0.200	0.025	0.0314	0.0680	0.00022
5/16	A	0.312	0.248	0.032	0.0483	0.109	0.00034
3/8	A	0.375	0.311	0.032	0.076	0.134	0.00053
	H	0.375	0.315	0.030	0.078	0.126	0.00054
1/2	A	0.500	0.436	0.032	0.149	0.182	0.00103
	H	0.500	0.430	0.035	0.145	0.198	0.00101
5/8	A	0.625	0.555	0.035	0.242	0.251	0.00168
	H	0.625	0.545	0.040	0.233	0.285	0.00162
3/4	A	0.750	0.680	0.035	0.363	0.305	0.00252
	H	0.750	0.666	0.042	0.348	0.362	0.00242
7/8	A	0.875	0.785	0.045	0.484	0.455	0.00336
	H	0.875	0.785	0.045	0.484	0.455	0.00336
1 1/8	A	1.125	1.025	0.050	0.825	0.655	0.00573
	H	1.125	1.025	0.050	0.825	0.655	0.00573
1 3/8	A	1.375	1.265	0.055	1.26	0.884	0.00875
	H	1.375	1.265	0.055	1.26	0.884	0.00875
1 5/8	A	1.625	1.505	0.060	1.78	1.14	0.0124
	H	1.625	1.505	0.060	1.78	1.14	0.0124
2 1/8	H	2.125	1.985	0.070	3.09	1.75	0.0215

HVACR technicians should also be aware that the copper tubing used by manufacturers to construct finned coils for air conditioning and refrigeration systems does not meet the same specifications as commercially available copper tubing. The walls are typically very thin, allowing for maximum heat transfer. This also makes them more prone to physical and environmental damage. All other copper tubing in a typical HVACR unit, other than that contained within the coil assemblies, is usually Type ACR or Type L.

1.2.0   Copper Tube Fittings

There are many types of copper tube fittings. Each type is used for a specific joining method. The fittings introduced here include flare fittings, compression fittings, soldered and brazed fittings, and both press-to-connect and push-to-connect fittings.

Regardless of the type of fitting used, virtually any deviation in the wall of a pipe or tube creates turbulence and **pressure drop**. All fittings create some friction that increases pressure drop. In addition, abrupt changes in the direction of flow also generate friction and increase pressure drop. For this reason, it is important to plan the route of tubing and piping carefully and minimize the number of fittings. More fittings in any piping system also increase the possibility of leaks.

Pressure drop: The reduction in pressure between two points in a pipe or tube. Pressure drop results from friction in piping systems, which robs energy from the flowing fluid or vapor.

Tubing Sets

Interconnecting tubing packages are available to connect the evaporator and condensing units of split-system air conditioners and heat pumps for residential and light commercial use. The copper tubing in the packages is Type ACR, factory-charged with nitrogen and fitted with plugs. Other sets may even be pre-charged with refrigerant. The suction line is usually pre-insulated. When using a tubing package, leave the plugs in place until just before making connections. This will minimize the entry of contaminants.

Male Flare Fitting

Figure 4 Pressure control designed for a flared connection.

1.2.1 Flare Fittings

The flared joint is a common method of making a connection. It is rarely used to couple runs of tubing though. The most common application is the connection of various pressure controls to system components (*Figure 4*).

Flare fittings are not as reliable as heat-bonded joints on copper tubing. However, mechanical joints like flare fittings can be taken apart and reassembled far more easily when component service is required.

The flare fittings used in HVACR work include elbows, tees, unions, adapters, and the flare nuts that match the size and angle of the flare (*Figure 5*). They are drop-forged brass and are accurately machined to form a 45-degree flare face. Although flare fittings used in other industries may have different face angles, the 45-degree flare angle is used in the HVACR industry.

Fitting sizes are based on the size of tubing. Flare nuts are hexagon-shaped, like most nuts. Flare couplers and adapters also have a hexagon-shaped body to accommodate a wrench. Flare tees and elbows typically offer two flat sides for a wrench.

A flare nut wrench (*Figure 6*) is best used to assemble these fittings. However, a correctly sized open-end wrench, as well as an adjustable wrench when used properly, is acceptable. However, since the material is generally brass, a poorly fitted wrench can permanently damage the fitting.

Flare-to-pipe adapters are often needed to connect a copper line to equipment such as a compressor. Since the point of connection on the component is often a pipe thread, it must be adapted to allow for the connection of a flared copper line. The flare-to-pipe adapter, called a *flare adapter* and shown in *Figure 5*, accomplishes this task. Such adapters are also available as elbows and tees, and in both male and female forms.

1.2.2 Compression Fittings

Compression fittings (*Figure 7*) are also used to connect tubing to various components. Common compression fittings, however, are not generally used on refrigerant tubing in the HVACR field. Most professionals prefer flare fittings to compression fittings in most situations due to reliability.

Compression fittings take less time to assemble than flared or heat-bonded connections. A seal is obtained by connecting the tubing to the threaded fitting with a coupling nut and a compression ring. The compression ring is known as a **ferrule** (*Figure 8*). It is referred to as an *olive* in the United Kingdom. The joint is formed by compressing the ferrule between the nut and the threaded fitting, forming a tight seal.

Ferrule: A ring or bushing placed around a tube that squeezes or bites into the tube when compressed, forming a seal.

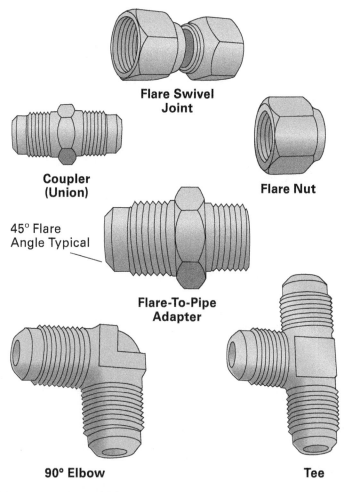

Flare Swivel Joint

Coupler (Union)

Flare Nut

45° Flare Angle Typical

Flare-To-Pipe Adapter

90° Elbow

Tee

Figure 5 Examples of flare fittings.

Figure 6 Flare nut wrench.

Compression couplings are available in types similar to those for flare fittings. They are available for tubing sizes from ¼" to 1" nominal. You will rarely encounter sizes larger than ⅜", however. Adapters are available to join compression fittings to other styles.

Tee Orientation

The orientation of tee fittings in a system where significant flow occurs can be very important. Tee fittings have *branch* and *run* connections, as shown here. Improperly connecting a tee can lead to a condition known as *bull-heading*. Bull-heading occurs when tees are installed in a position such that the flow input slams into the back of the tee. Bull-heading results in turbulence, significantly increasing the pressure drop.

Tees should be installed so that the pressure input is applied to the run. If more than one tee is installed in the line, a straight piece of pipe at least 10 tube-diameters in length between tees is recommended to avoid additional turbulence whenever possible.

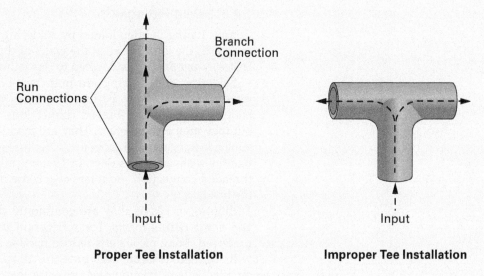

Branch Connection

Run Connections

Input

Proper Tee Installation

Input

Improper Tee Installation

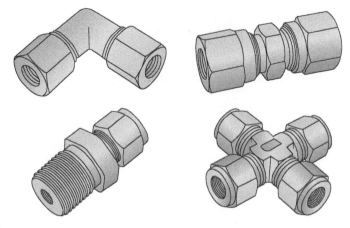

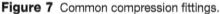

Figure 7 Common compression fittings.

Figure 8 Compression ring, or ferrule.

Ferrules come in several shapes and materials depending on the intended use. Compression fittings are used with plastic tubing as well as copper. Ferrules used with plastic tubing may be plastic or copper. Some types of compression fittings may also have a metal or plastic sleeve that slides on the tubing and passes through the compression nut before assembly.

Common compression fittings are often used on small water lines to connect humidification equipment, grease lines for large bearings, and pilot gas lines for gas heating equipment. However, the use of common compression fittings in the HVACR industry overall is limited.

1.2.3 Sweat Fittings

Copper tubing is often joined by soldering or brazing in which a soft alloy is melted in the joint between the walls of the two parts. NCCER Module 03104, *Soldering and Brazing*, is devoted to this joining process.

Soldering and brazing are both heat-bonding techniques. Both are often referred to as *sweating*. Fittings known as *sweat fittings* (*Figure 9*) are made for making soldered and brazed connections. Sweat fittings are usually copper, but they may also be brass. They are made slightly larger than the tubes to be joined, leaving enough room for solder to flow into the joint. Some sweat fittings, like the male adapter shown in *Figure 9*, allow copper tubing to be joined to a threaded connection. Adapters also come in female form, with the threads on the inside.

Elbows, or *ells* as they are commonly called, are available in long radius and short radius forms. For refrigerant piping, long radius ells are always preferred. Long radius ells reduce friction and pressure drop. Smooth piping with limited friction and pressure drop is very important in refrigerant circuits. Where friction and pressure loss are less of a concern, short radius ells can be used.

Reducing couplings are designed to connect different sizes of tubing. Each end has a socket that slips over the tubes to be connected. A *reducing bushing*, however, is designed so that one end slips over the smaller tubing while the opposite end slips inside the socket of a fitting. The bushing end has the same OD as an equally sized piece of tubing. *Figure 10* shows the two types of fittings together for clarity.

1.2.4 Press-To-Connect Fittings

Press-to-connect fittings (*Figure 11*) are designed to connect pipe and tubing using innovative pressing tools and compatible fittings. Press fitting systems have now been developed for use with copper, stainless steel, black iron pipe, and plastic tubing. With press technology, connections can be made quickly with great consistency. Pressed fittings are available for a wide variety of applications, ranging from potable water to natural gas and corrosive chemicals. Their use for refrigerant piping and other applications in the HVACR craft is expanding quickly.

Special pressing tools (*Figure 12*) and jaws for each size of fitting are required. A manually operated tool is available to assemble smaller sizes if desired, but repetitive joint assembly is much faster using a power tool. Press-to-connect joint assembly is covered in *Section Two* of this module.

Reducing Bushing

Reducing Coupling

Coupling

Male Adapter

Tee

45° Elbow

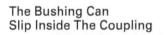
Short Radius
90° Elbow

Long Radius
90° Ell

Figure 9 Sweat fittings.

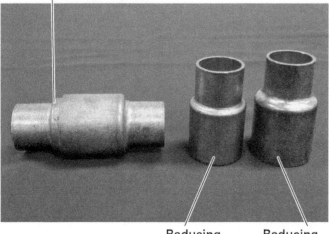

The Bushing Can
Slip Inside The Coupling

Reducing
Bushing

Reducing
Coupling

Figure 10 A reducing coupling and a reducing bushing.

The fittings shown in *Figure 11* are rated for 700 psi (48 bar) service. They are typically available in sizes from $\frac{1}{4}$" through $1\frac{3}{8}$" OD. Fittings to accommodate tubing sizes up to $4\frac{1}{8}$" OD are also on the market, but these fittings are rarely chosen for HVACR tubing sizes greater than $1\frac{3}{8}$" OD.

1.2.5 Push-To-Connect Fittings

Push-to-connect fittings (*Figure 13*) are the most recent introduction to refrigerant piping. Various designs have been developed over the years for low-pressure and less critical applications, such as water. But push-to-connect fittings have become more reliable and new designs can accommodate the higher pressures found in refrigerant circuits.

The fittings shown in *Figure 13* are rated for 870 psig (60 bar) service and hold a vacuum down to 200 microns. Most products are also designed for easy disassembly. The size range available is $\frac{1}{4}$" through $\frac{7}{8}$" OD. Since they do rely on elastomeric parts inside the brass housing, it is important to ensure that any fittings used are compatible with both the oil and refrigerant in use.

Figure 11 Press-to-connect fittings.

Figure 12 Press-to-connect power tool with $\frac{3}{4}$" jaws installed.

Figure 13 Push-to-connect fitting assortment.

1.0.0 Section Review

1. Which of the following types of copper tubing has the *thickest* wall?
   a. Type K
   b. Type L
   c. Type M
   d. Type ACR

2. Common compression joints _____.
   a. take more time to create than flare joints
   b. cannot be used for water
   c. are more reliable than flare joints
   d. are not typically used for refrigerant lines

2.0.0 Joining Copper Tubing and Fittings

Objective

Explain how to mechanically join copper tubing.

a. Explain how to measure, cut, bend, and swage copper tubing.
b. Describe how to join copper tubing using flare and compression joints.
c. Describe how to join copper tubing using press-to-connect and push-to-connect fittings.
d. Explain how pressure testing on refrigerant lines is conducted.
e. Identify common hangers and supports associated with copper tubing installations.

Performance Tasks

1. Cut and bend copper tubing.
2. Join copper tubing using a flared connection.
3. Join copper tubing using a compression fitting and ferrule.
4. Assemble press-to-connect joints in copper tubing according to the manufacturer's instructions.

With copper tubing materials now introduced, methods to join the tubing can be presented in detail. Mechanical joining methods are presented here, along with **swaging** of copper tubing. Measuring, cutting, and bending are skills needed regardless of what type of joint is being made. These techniques are presented before moving on to joint assembly.

2.1.0 Preparing Copper Tubing for Assembly

Before a joint can be made, the material must be cut to the proper length. Of course, if a very long piping run is being assembled, full lengths of tubing will simply be coupled together until obstructions are encountered or the equipment is reached. Soft copper tubing can also be bent, eliminating some fittings when it is used.

> **NOTE**
>
> Soldering and brazing skills are presented in NCCER Module 03104, *Soldering and Brazing*.

Swaging: The process of using a tool to shape metal. In context, it describes the process of forming a socket at the end of a copper tube that is the correct size to accept another piece of tubing, acting as a coupling.

2.1.1 Measuring Tubing

It is extremely important to measure tubing accurately. It is relatively easy to measure between two known points and cut a piece of tubing to that length. However, when fittings are involved, the dimensions of the fitting must be considered. The tubing must be cut to allow room for the fittings between the points of connection. Tubing that is too long or too short presents a problem. To arrive at the proper cut length, the additional length added by fittings must be taken into consideration. This is true of both mechanical and sweat fittings.

To establish the length of tubing needed to connect two fittings, two of three dimensions must be considered. Refer to *Figure 14*. There are several ways to determine the correct cut length. For sweat fittings, a measurement can first be taken between the faces of the two fittings—a face-to-face measurement. Next, the socket depth of the fitting, multiplied by two (one fitting on each end), must be added to arrive at the proper length.

Alternatively, a measurement can be made from the centers of the fittings (center-to-center distance), or to the backs of the fittings (back-to-back distance). This approach requires you to subtract the length that the fittings add to the run to arrive at the proper cut length. The installation situation typically determines the simplest and most accurate approach. Regardless of the method chosen, the dimensions of the fittings must be considered before the tubing is cut. Fitting dimensions can be field measured or derived from available charts.

Compression fittings have a socket much like sweat fittings, but it is generally shallow (*Figure 15*). However, the socket depth must still be considered. For flare fittings (*Figure 16*), the tube must be long enough to reach the top of the beveled edge, after the flare is applied. The flaring process does shorten the length of the tube, although the difference is very slight and not normally worthy of consideration. Flare joints are typically used on small sizes of soft copper tubing that can be formed to fit. Greater accuracy is needed when working with hard-drawn copper.

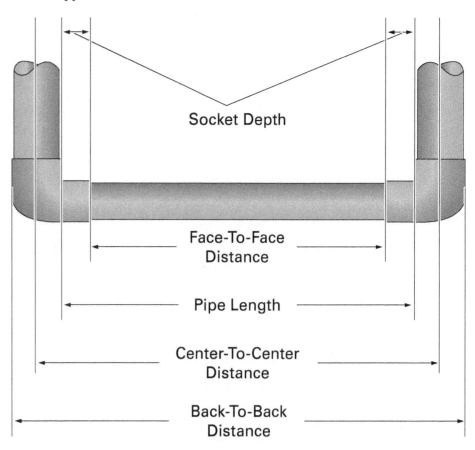

Figure 14 Typical tubing measurements to accommodate sweat fittings.

Think Like a Carpenter

When working with any kind of pipe or tubing, it's important to remember this guiding principle with which all carpenters are familiar.

Think Like a Carpenter

Measure Twice, Cut Once

Managing Soft Copper Rolls

Soft copper purchased in a roll must be straightened out before it can be measured and cut. To unroll soft copper tubing reasonably straight and minimize distortion of the tube, stand the roll up vertically and place the bottom on a flat surface. Hold the end of the tubing firmly to the surface with one hand, or lightly with a foot, and roll the tubing away as shown in *Figure 17*. This is especially important for larger sizes. Unrolling it incorrectly results in tubing that is oval instead of round.

Copper tubing that is not round is a significant problem for all types of joints. The tubing must be consistently round to create a leak-free joint. *Figure 18* shows two tools that can be used to reform the tubing. The lever-action tool is best for larger tubing and higher production. Otherwise, the hammered tools work fine.

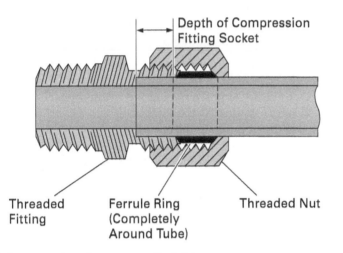

Figure 15 Cross section of a compression joint.

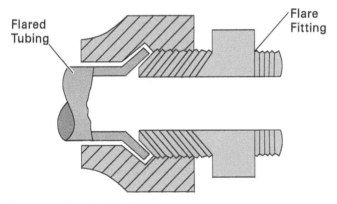

Figure 16 Cross section of a flare joint.

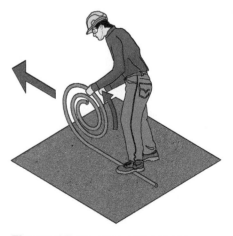

Figure 17 Unrolling soft copper tubing.

(A) Lever-Action
Rerounding Tool

(B) Hammer-Driven
Rerounding Tool

Figure 18 Tools to reform round tubing.

2.1.2 Cutting Tubing

Copper tubing is typically cut with a handheld tubing cutter (*Figure 19*). As shown, tubing cutters come in a variety of sizes to accommodate different tube sizes and for use in close quarters. The tubing cutter is always preferred to sawing because it produces a square end and creates no filings. Particles from the cutting action of a saw can be very harmful in refrigerant lines and other piping systems.

Only use a hacksaw to make the cut if there is no alternative. But if a cut is made in a stationary line, be aware that filings can be hard to collect and remove. If the tubing is in a vertical position, filings will fall in and be inaccessible. Even when held horizontally, some filings will find their way into the tube and out of reach.

Well-fitted gloves, as well as safety glasses, should always be worn whenever working with copper tubing and the related tools. Don the proper PPE, and follow these steps to use a tubing cutter:

Step 1 Place the cutter on the tube at the desired cutting point, as shown in *Figure 20*.
Step 2 Tighten the knob slightly, forcing the cutting wheel against the tube. It should not be tightened too tight, as it will distort the tube as well as make it difficult to rotate the cutter.
Step 3 Rotate the cutter around the tube. After one to three turns, the cutter will begin to feel loose. Tighten it again, but not too tight, and continue rotating. Work incrementally through the tubing. Overtightening, especially when the blade is close to penetrating the inside wall, causes the sharp edge inside the lip of the tube to be more pronounced. As the cutting wheel gets closer to the inside wall, keep the pressure light.
Step 4 The two pieces will separate cleanly once the wheel penetrates the inside wall. Be prepared to catch or support the pieces.

WARNING!

The edge inside a cut tube is extremely sharp. Avoid running your fingers across the cut edge, even when gloved.

Reaming: Using a tool to remove the sharp lip and burrs left inside a pipe or tube after cutting.

Copper plating: The bonding of a thin layer of copper to another metal surface. When plating is done intentionally, it is an electro-chemical process using an electric current to encourage and strengthen the bond. In a refrigerant circuit, it is a problem that occurs when copper molecules are in circulation, primarily at locations where there is a friction fit between two other metal surfaces.

The cut ends must be *deburred* by **reaming**. Deburring simply means to remove the sharp burrs that often result from a cut, especially when the material is sawn. Not only is this edge sharp and dangerous, but it also creates turbulence and resists flow. Further, the surface can erode over time and transport microscopic particles of copper throughout the system. The copper molecules then cling to other metal surfaces, resulting in **copper plating**. Places such as compressor piston rings, where there is friction between two pieces of metal, are common locations where copper plating forms.

Reaming can be done using a reamer built into the tubing cutter body, as shown in *Figure 20*. With the reamer extended, rotate it multiple times inside the tube to carve out the sharp lip around the inside edge. A better reamer is shown

This Design Allows for Quick
Adjustment to the Tube Size

Fold-Out
Reamer

Figure 19 Handheld tubing cutters.

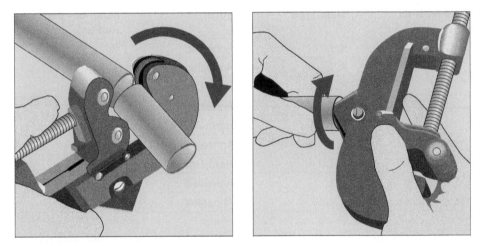

Figure 20 Cutting with a handheld tubing cutter.

in *Figure 21*. The visible end of the reamer is used to ream the outside edge
of the tube when needed. The blades in the opposite end are inverted and are
designed to ream the inside edge. The outer edge typically needs little attention,
while the inside edge always requires attention and effort to remove. Use the
tool to ream the one edge, then reverse it and ream the remaining edge.

While reaming, always point the cut end of the tubing down whenever
possible. This will prevent any burrs removed from falling inside the tubing. To
ensure any particles are removed, it is best to run a brush or cloth through the
pipe after deburring. If you are not sure that all the particles can be removed,
it may be best to skip deburring. Although the sharp edge is certainly not
desirable, it may still be better than leaving large burrs and shavings in the
piping system. Careful planning of the workflow can often help you avoid these
situations.

When using a tubing cutter is not an option, a hacksaw and sawing fixture
such as a miter box may be used. The miter box helps keep the end square and
enables a more accurate cut. The blade of the hacksaw should have at least 32
teeth per inch (tpi). A saw blade that is too coarse will chatter and snag on the
tubing walls as the blade reaches the mid-point of the tube.

After saw cutting, use a reaming tool like the one shown in *Figure 21* to ream
both the inner and outer edge. Deburr the inside edge and carefully clean the
inside of the tubing with a cloth. Turn the tube end down and lightly tap it to
dislodge any remaining particles.

Figure 21 Reaming tool.

Again, wear gloves when working with copper to avoid serious cuts. Sawing creates burrs that are more irregular and perhaps more dangerous than the edge created by a tubing cutter.

2.1.3 Bending Tubing

Soft copper tubing can be bent around obstructions. Smaller tubing can easily be bent for wide turns by hand, but care must be taken to avoid flattening the tube with a bend that is too sharp. When bending is done properly, the tubing will not collapse or buckle, but slight distortion is expected. The tubing often becomes harder due to **work hardening**. Work carefully to get the bend right on the first try, rather than trying to bend it several times.

Table 3 provides the minimum bending radius, in inches, for various sizes of tubing. Although hard-drawn tubing can be bent, only small sizes can be bent on the job. There is no advantage to using hard-drawn tubing when bending is expected, so it is best to avoid it and use soft copper instead. However, manufacturers often bend hard-drawn tubing of all sizes in the factory environment. Bending large hard-drawn tubing generally requires shop-class equipment. Note that *Table 3* is based on tubing ID rather than the OD. Add $\frac{1}{8}$" to each size listed for the equivalent size of ACR tubing.

Work hardening: A permanent hardening caused by a change in the crystalline structure of metal, caused by repetitive bending or forming.

TABLE 3 Minimum Bend Radius for Copper Tubing

Nominal ID	Tube Type	Form	Minimum Bend Radius
$\frac{1}{4}$"	K, L	Annealed	$\frac{3}{4}$"
$\frac{3}{8}$"	K, L	Annealed	$1\frac{1}{2}$"
	K, L, M	Hard-Drawn	$1\frac{3}{4}$"
$\frac{1}{2}$"	K, L	Annealed	$2\frac{1}{4}$"
	K, L, M	Hard-Drawn	$2\frac{1}{2}$"
$\frac{3}{4}$"	K, L	Annealed	3"
	K, L	Hard-Drawn	3"
1	K, L	Annealed	4"
$1\frac{1}{4}$"	K, L	Annealed	9"

Handheld tubing benders (*Figure 22*) are used to make accurate and reliable bends in both soft and hard tubing. The tool provides the necessary leverage to make a smooth bend. Handheld benders provide consistent bends with a safe radius, and the bends are usually more visually appealing than anything formed by hand.

Various sizes of benders are available to bend tubing. The bender should be designed for the tubing size to avoid damage to the tubing. For tubing greater than 1" OD, larger, manually operated or powered equipment is necessary. To ensure bends are made without damage to the tubing and have the proper radius, follow the manufacturer's guidance for use.

The common parts of a tubing bender are shown in *Figure 23*. Follow these basic steps to bend tubing with a manual bender:

Step 1 Measure as necessary and place a mark on the tubing. This mark represents where the center of the bent portion will land, as shown in *Figure 24*.

Step 2 Raise the tool handle and place the tube in the groove of the forming die.

Step 3 Refer to *Figure 25*. Align the 0° index mark on the forming die with the long index line near the end of the sliding die.

Step 4 Move the tube left or right as needed for alignment. If the measured portion of the tube is to your right, align your mark on the tubing with the R mark. If the measured portion of the tube is to your left, align your mark on the tubing with the L mark.

Step 5 For a 90-degree bend, carefully draw the sliding die handle down and along the tubing until the long index line aligns with the 90° mark. You can bend any angle you like through 180 degrees.

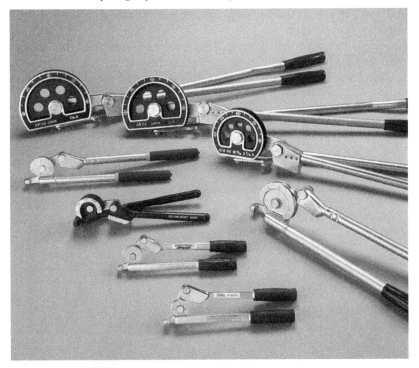

Figure 22 Handheld tubing benders.

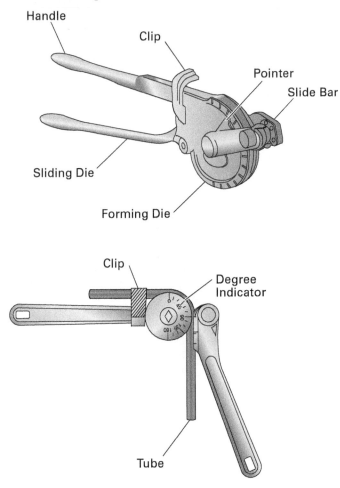

Figure 23 Tubing bender construction and components.

2.1.4 Swaging Tubing

Swaged joints are made without a fitting. Swaging refers to the use of a tool to expand the end of a piece of tubing. The process forms a female socket that is the proper size to accept the end of another piece of tubing (*Figure 26*). The joint is then soldered or brazed.

A handheld swaging tool (*Figure 27*) is one way to expand the tubing. It is expanded to match the size of a fitting socket (*Figure 28*) for the same size of tubing. Once the joint is soldered or brazed, a strong and reliable joint is formed.

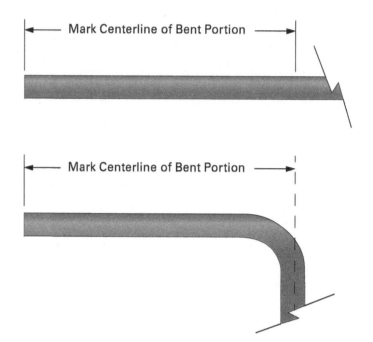

Figure 24 Marking the tubing for bending.

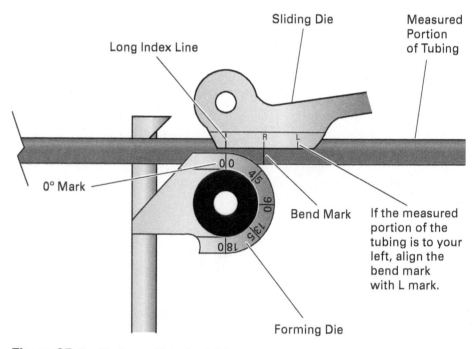

Figure 25 Positioning and bending tubing.

Hammer-driven swaging tools are simple to use. Each tool is generally made to fit a single size of tubing. Choose some appropriate gloves and insert the tool into the end of the tube. Hammer the end until the tip of the tool is fully seated inside the tubing, expanding the inside diameter as it is forced in.

Tubing is often swaged in the factory environment using automated equipment to reduce manufacturing costs. However, swaging is not often used by field technicians today, since there are other choices that are faster and equally reliable.

If many joints need to be field-swaged, it is best to use a tool that makes them easier and faster. Tools like the one shown in *Figure 29* allow a swage to be made with a single stroke of the handle. An even easier tool to use, a hydraulic swaging tool, is shown in *Figure 30*. It is lightweight, compact, and easy to handle and operate.

> **NOTE**
>
> Hard-drawn copper tubing should not be swaged in the field, although it can be done with shop-based equipment.

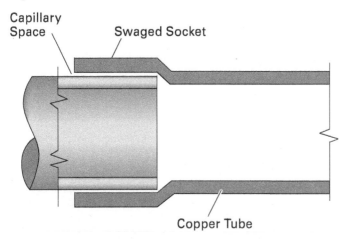

Figure 26 Assembled swaged joint.

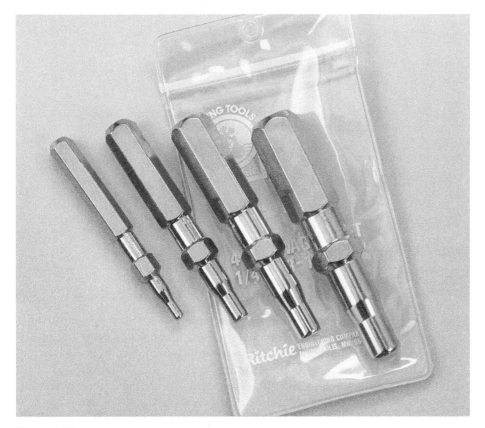

Figure 27 Hammer-driven swaging tools.

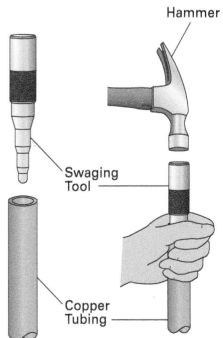

Figure 28 Swaging tubing with a hammer-driven tool.

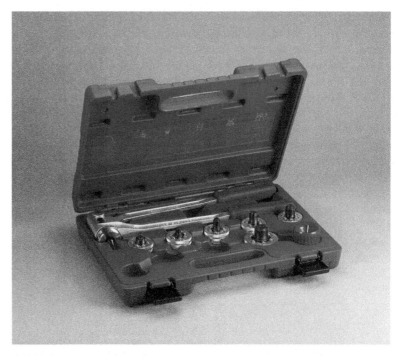

Figure 29 Tubing expander kit.

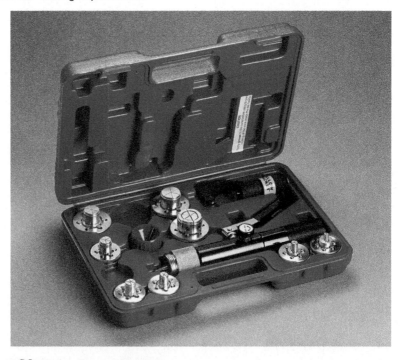

Figure 30 Hydraulic swaging tool.

2.2.0 Flared and Compression Joints

Because copper tubing is too thin for threading, mechanical joining or heat-bonding methods are used. Mechanical joining methods include flared and compression fittings that can be taken apart repeatedly if required.

Regardless of the type of joint being made, it is crucial that the end of the tubing be properly shaped, smooth, clean, and free of burrs inside and out. The shape of the tubing is a significant concern with soft copper tubing. As the tubing is unrolled, it often takes on a slight oval shape. Always unroll the tubing carefully to avoid significant distortion. Tubing that is out of round makes a poor flare, and both press-to-connect and push-to-connect fittings require perfectly round tubing to be reliable.

2.2.1 Flare Joints

Flared joints (*Figure 31*), when formed and assembled correctly, will not leak, and they can be disassembled for service as necessary. Flared copper tubing joints are acceptable in some jurisdictions for fuel gas piping as well. Most flared joints used in HVACR work are applied to $\frac{3}{8}$" or smaller tubing. A $\frac{1}{2}$" or $\frac{3}{4}$" flare joint will occasionally be encountered.

A flaring tool (*Figure 32*) spreads the end of the tube into a cone shape. The traditional tool consists of a flaring bar with openings for various tube sizes, and a yoke with a flaring cone. The yoke also incorporates a clamp to attach it to the flaring bars. Flaring bars for large sizes of tubing, such as $1\frac{1}{8}$" OD, may be usable with only one size of tubing to limit the size of the tool.

There are several flaring tool designs, some of which have accessories used to swage tubing. Better flaring tools also *burnish* the face of the flare, meaning that they smooth and polish the face as it is formed.

Better tools also allow the formation of a flare with longer shoulders, making the flare face a little wider. HFC-410A refrigerant operates at higher pressures than many other refrigerants, so additional security offered by a wider flare face is needed. *Figure 33* shows the common dimensional requirements for flares used with HFC-410A. These dimensions and the extra security they provide work equally well for other flare joints. No special flare nut is required. In *Figure 32* (B), the deluxe tool shown is designed to make the wider flare face. The traditional tool [*Figure 32* (A)] can, but you must manually compensate for the wider face by repositioning the tubing in the flaring bar. The deluxe flaring tool eliminates the guesswork. Regardless of which tool is used, it is important that the flare not be too wide to fit the flare nut without interference.

Flared End
of Tube

Flare
Fitting

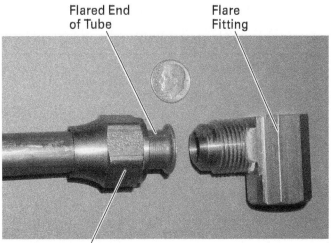

Flare Nut

Figure 31 Flare joint components.

(A) Traditional Flaring Tool   (B) Deluxe Flaring Tool

Figure 32 Flaring tools.

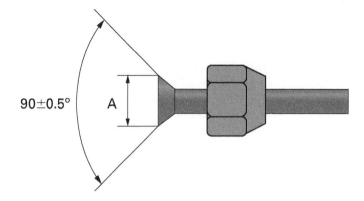

90±0.5°   A

Flare Size	"A" Dimension – Standard SAE Flare	"A" Dimension – HFC-410A Flare*
1/4"	0.327" (8.3 mm)	0.358" (9.1 mm)
3/8"	0.488" (12.4 mm)	0.520" (13.2 mm)
1/2"	0.622" (15.8 mm)	0.654" (16.6 mm)
5/8"	0.748" (19 mm)	0.776" (19.7 mm)
3/4"	0.917" (23.2 mm)	0.945" (24 mm)

*Not required by all manufacturers

Figure 33 SAE and HFC-410A flare dimensions.

Flares can be made at different angles, and the flaring cone must match the desired angle. The required flare angle for most uses in the HVACR trade is 45 degrees. This is the standard provided by the Society of Automotive Engineers (SAE). Other standards—not associated with HVACR work—require a 37-degree flare angle. Ensure your tools are designed to form a 45-degree flare.

Two kinds of flare fittings are used:

- The single-thickness flare forms a 45-degree cone that fits up against the face of a flare fitting. Single-thickness flares are the most common, but they are not as tolerant of repeated assembly and disassembly or physical stress.

- A double-thickness flare is shown in *Figure 34*. The single-thickness flare is formed first, and then the lip is folded back onto itself and compressed again. Double-thickness flare connections can be taken apart and reassembled more often without damage. Flaring kits with the required adapters can create both single- and double-thickness flares. It does take longer, but the added strength may be worth the effort.

Forming a Flare

To make a flare, first measure and cut the tubing to the desired length. Check to ensure the tube has a proper round shape. Ream the end inside and out to remove burrs. Complete removal of the inner burr is important to the quality of the flare. The surface of the tubing just inside the opening forms the sealing surface of the joint. Avoid gouging the tubing while reaming; deep scratches will allow leakage across the face. Next, remove oxidation on the outer surface using fine sand cloth or a nylon abrasive pad (*Figure 35*). If using sand cloth, be sure it is a fine grit to avoid deep scratches. A coarser cloth can be used when preparing for soldering or brazing, but a flare joint works best with a very smooth surface.

In the Groove

The groove shown here in the roller of a tubing cutter allows a damaged flare to be cut off close to the end of the tube.

Groove to
Accommodate
Flared End

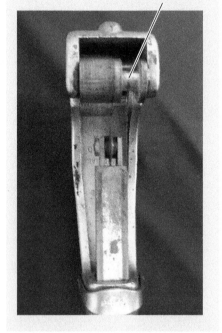

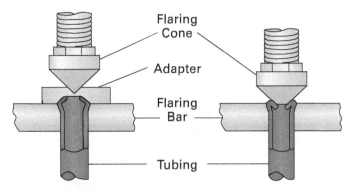

Figure 34 Forming a double-thickness flare.

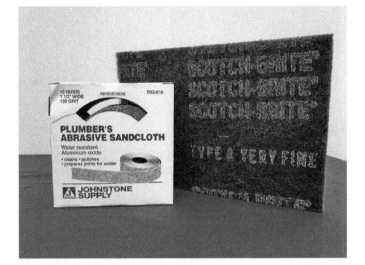

Figure 35 Nylon abrasive pads and sand cloth.

Follow these general steps to form a flare using a tool like the one shown in *Figure 32* (B). Always follow the instructions provided with the tool in use:

Step 1 Retract the cone by turning the sliding handle counterclockwise. Inspect it to be sure it is in good condition, then clean it and apply a small amount of a light lubricant (not grease).

Step 2 Loosen the wing nut on the clamp bolt and swing the die-head clamp away. This separates the two dies so the tubing can be inserted. Read the tubing sizes stamped into the dies and rotate the die assembly until the proper size is beneath the cone.

Step 3 When the dies are separated, an insertion gauge automatically swings out above the tube opening (*Figure 36*). Insert the tubing until it bumps the gauge.

Step 4 Hold the tubing in place against the insertion gauge and close the die-head clamp, bringing the two dies together. This also moves the insertion gauge out of the way. Move the clamp bolt into position and tighten the wing nut to secure the dies and tubing.

Step 5 Begin flaring the tube by turning the sliding handle clockwise. You will feel consistent resistance as the cone descends and forms the flare. As the cone meets the dies with copper between the two faces, you will feel major resistance. The flare is complete—do not overtighten the cone.

Step 6 Turn the sliding handle counterclockwise to raise the cone. Loosen the wing nut and swing the die-head clamp around to separate the dies. With the dies separated, the flared tubing end will easily fall free of the tool.

The result should be a smooth, consistent flare wide enough to cover the mating surfaces. Splits or visible flaws are reasons to reject the flare and try again.

NOTE

When flaring a long piece of tubing or a section that is secured at the opposite end, be sure to slide a flare nut over the end of the tubing before inserting it into the tool.

Flaring Cone   Insertion Gauge

Die-Head Clamp   Wing Nut

Figure 36 Inserting the tubing into the flaring tool.

Maintain your flaring tool by cleaning and lightly lubricating the moving parts occasionally. A properly maintained flaring tool will likely provide a lifetime of service.

A flare joint must be assembled using two wrenches. One is used to tighten the nut while the other holds the fitting. No sealant is generally required on the face of the joint. Adding a drop of light oil to the back of the flare, where it contacts the flare nut, can relieve friction as the joint is tightened. Thread sealant is also unnecessary. The primary sealing point of the joint is the flare face; the threads serve only to bring the fitting faces together.

2.2.2 Compression Joints

Like flare joints, compression fittings can be taken apart and reassembled. However, compression joints are not generally as reliable or as durable as flare joints under stress. They are rarely used for refrigerant piping.

The tubing is prepared in the same manner as it is for a flare joint. The tubing must be cut square, reamed, and cleaned of oxidation. Make sure that the tubing is not distorted and is free from any nicks or gouges in the area where the ferrule will land. If the compression nut and ferrule slip freely onto the tubing for roughly 1" (~2.5 cm), the tubing is probably straight and round enough for a compression connection. If the ferrule binds and will not move freely, the tubing is deformed.

Once the tube end is prepared, begin by sliding the compression nut onto the tube, with the open end toward the fitting. Then slide the ferrule onto the end of the tube. Insert the end of the tubing into the fitting socket, and push the compression nut up to the fitting. Tighten the nut finger-tight first, all the while keeping the tubing firmly seated inside the fitting socket.

One of the keys to a successful joint is to keep the tubing tight against the bottom of the socket while tightening the nut. Since two wrenches are often required—one to hold the nut and one to hold the fitting—this can be challenging. Once the nut has been tightened down roughly one-half turn, the tube will likely stay in position while the tightening is completed.

One full turn after the fitting is finger-tight is usually sufficient to complete a leak-free joint. It is tempting and easy to overtighten compression joints, especially if the wrench is longer than necessary. If the joint leaks, the fitting can still be tightened a little further. If the joint leaks due to overtightening, the tube end must be cut off and a new joint started. The ferrule cannot be removed once it is compressed onto the tube.

If a compression joint is disassembled, the end of the tubing may appear slightly bulged beyond the ferrule. This is normal, indicating that the ferrule has been compressed onto the tubing. Assembling compression joints is easily mastered with a little practice.

2.3.0 Press-To-Connect and Push-To-Connect Fittings

Press-to-connect and push-to-connect fittings are rapidly gaining in market share. Today's products are reliable and rated for the pressures found in the refrigerant circuit. Although the products are not inexpensive, the cost of labor is also high, so saving time on the job is essential. The time and materials required to complete a brazed joint makes that approach expensive as well. Press- and push-to-connect fittings can significantly increase productivity and reduce labor costs.

It is important to note that many product manufacturers require these fittings to be installed by a technician certified through a manufacturer-approved training program. The assembly instructions provided here are based on the Zoomlock® MAX product provided by the Sporlan Division of Parker Industries. Other products are similar, but you must follow the assembly guidance for the product in use. No product-specific certification can be offered herein.

2.3.1 Press-To-Connect Fitting Assembly

Before assembling a press-to-connect joint (*Figure 37*), the tubing must be properly prepared. The tubing must be cut square, reamed, and cleaned of oxidation. Make sure that the tubing is not distorted and is free from any nicks or gouges in the fitting area. Even a deep scratch can interfere with the seal.

Before preparing a joint for assembly, make sure the tubes being connected are properly supported and aligned. The fitting should not be used to force the pieces into alignment. Also ensure the selected fitting is compatible with the refrigerant and lubricating oil in use. Do not assume that all fittings are compatible with all other substances.

Follow these basic instructions to assemble a press-to-connect joint on copper tubing:

Figure 37 Press-to-connect elbows.

Step 1 Ensure that the tubing, fitting, and compression jaw set for the tool are of the proper size. Use only jaw sets approved by the manufacturer for use with their products. The jaw set must also be compatible with the power tool used.

Step 2 Cut the tubing to the desired length with a tubing cutter. Ensure the cut is square and the tubing is not distorted. Thoroughly deburr both the inside and outside edges. Clean and remove oxidation from the tube using an abrasive pad or fine sand cloth.

Step 3 Inspect the end of the tubing for flaws such as gouges or deep scratches. Ensure the fitting is clean.

Step 4 Use a manufacturer-provided depth gauge to mark the tubing at the full insertion depth (*Figure 38*). Alternatively, firmly seat the end of the tube into the fitting and mark the tube. Then remove and lay the end against the fitting to check the mark for accuracy. This mark, often referred to as a **witness mark**, is your indicator to show that the tube is fully inserted into the fitting.

Step 5 Inspect the fitting. Ensure that the seal inside is present, properly positioned, and undamaged. Apply a small amount of oil or silicone, if desired, to allow the tube to slide in smoothly.

Step 6 Insert the tubing firmly into the fitting (*Figure 39*). Rotate the tubing as it passes across the seal. Ensure the witness mark is just visible. Monitor the tube throughout the process to ensure it does not slide out.

Step 7 Align the tool jaws on the fitting, ensuring that the tool and jaws remain square to the fitting.

Witness mark: A mark made as a means of determining the proper positioning of two pieces of pipe or tubing when joining. A witness mark is typically in addition to a first mark, made in a location that will not be obscured by the joining process; the first mark is often obscured during the process.

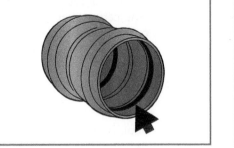

(A) Inspect Fitting

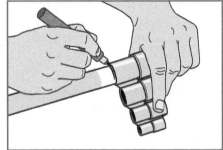

(B) Mark with Depth Gauge

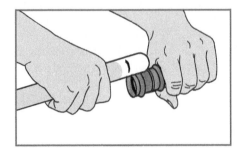

(C) Confirm Insertion Mark Is Accurate

Figure 38 Press-to-connect fitting assembly (1 of 2).

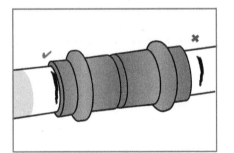

(D) Insert the Tube Firmly

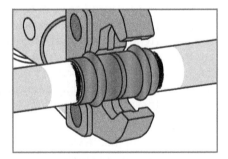

(E) Align and Square the Jaws to the Fitting

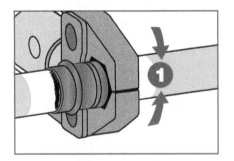

(F) Cycle the Tool Through One Complete Cycle

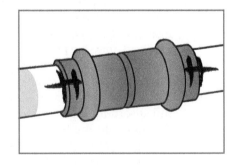

(G) Mark Completed Joints

Figure 39 Press-to-connect fitting assembly (2 of 2).

CAUTION

Do not compress a fitting more than once. If the cycle does not run as expected or the tubing moves out of the fitting, discard the fitting and begin again.

Step 8 Press and hold the button on the power tool to initiate and complete one full compression cycle.

Step 9 Mark the completed joints as shown in *Figure 39* to indicate they are fully assembled.

2.3.2 Push-To-Connect Fitting Assembly

Push-to-connect fittings (*Figure 40*) require the same tubing preparation as press-to-connect fittings. The tubing must be square, completely free of burrs and deep scratches, round, and clean. If a scratch can be felt with a fingernail, it is too deep. The coupled area of the tubing must also be free of *incise marks*—information or specifications that are stamped into the wall of the tube by its manufacturer. Labeling using inks or dyes is not a problem.

With push-to-connect fittings, you don't have the luxury of inserting the tube into the fitting to make a witness mark. Once it is fully inserted, it is assembled. For that reason, you must use the manufacturer-provided depth gauge to place the witness mark.

Assembly is as simple as it sounds. Place a witness mark with the fitting manufacturer's gauge. Then push the tube firmly into the fitting until it stops. Check the witness mark to see if the tubing is fully inserted.

Some fittings can be unlocked and removed using a special tool, such as the C-shaped tool shown in *Figure 41*. The fittings are not designed for reuse, however.

> **WARNING!**
>
> Before releasing a push-to-connect joint, ensure the refrigerant circuit is not pressurized.

Figure 40 Push-to-connect female adapters.

Figure 41 Zoomlock® PUSH removal tool.

2.4.0 Pressure Testing

Once all the joints in refrigerant piping have been assembled, they may be inspected by others to make sure that the work has been done in accordance with the job specifications and applicable codes. The refrigerant piping must also be leak tested to be sure that the joints are sound.

Pressure testing is the preferred method. Most pressure testing in the HVACR trade is done using nitrogen gas. Nitrogen is inexpensive and, if there is a leak, the technician does not risk losing expensive refrigerant and violating laws by releasing refrigerants into the atmosphere. Nitrogen is also inert and safe to handle. Although nitrogen is the most common choice, some manufacturers and codes permit the use of carbon dioxide (CO_2) for HVACR pressure testing.

> **NOTE**
>
> Refrigerant circuit leak testing is covered in greater detail in NCCER Module 03205, *Leak Detection, Evacuation, Recovery, and Charging*.

> **WARNING!**
>
> Oxygen, acetylene, and other combustible gases should never be used to pressure test a system. Compressed air is also an unacceptable test gas for refrigerant circuits, due to the moisture content.

The test pressure can vary based on what is being tested. However, the test pressure must never exceed the maximum pressure stated on the nameplate of any component exposed to the pressure. The required test pressure may be specified in construction documents, but testing is routinely done at pressures from 150 psi to 300 psi (1,030 kPa to 2,070 kPa). Components located on the low-pressure side of a refrigerant circuit typically have lower maximum pressure ratings than those on the high-pressure side.

A typical setup used to pressure test refrigerant piping is shown in *Figure 42*. There should be a hand shutoff valve, pressure regulator, pressure gauge, and pressure-relief valve in the charging line. The relief valve should be set to open at just 1 or 2 psi above the test pressure.

A refrigerant gauge manifold is often substituted for the hand valve shown in *Figure 42*. Connecting the nitrogen supply to a gauge manifold provides full control over the flow through multiple points into a system, and it provides an accurate means of measuring and monitoring the test pressure applied.

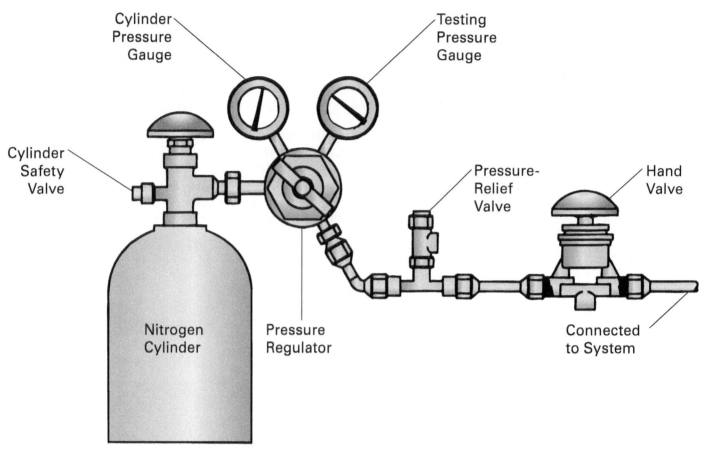

Figure 42 Nitrogen cylinder and accessories for pressure testing.

WARNING!

Make sure nitrogen tanks are secured in a cart or other stabilizing device to prevent the tanks from tipping or being knocked over. Nitrogen tanks are routinely charged to pressures exceeding 2,000 psig. If the tank valve is broken off suddenly, the tank cylinder can be propelled like a rocket by the escaping nitrogen gas.

A soap solution or commercial leak-test solution can be used to check for leaks on any pressurized circuit. This is done by wiping the joints with the solution and watching for any bubbles or foam to form. If no leaks are noted, the test pressure is often maintained for 24 hours or longer as a standing pressure test. Project specifications may require a specific time period. At the end, the pressure is evaluated to determine if any pressure has been lost. Note, however, that the pressure may rise and fall in response to temperature change.

Refrigerant-detecting instruments are of no value for leak testing when nitrogen alone is charged into a system. Since our atmosphere contains a great deal of nitrogen, a nitrogen detector would be useless. The United States Environmental Protection Agency (EPA) does allow for a small percentage of HCFC-22 to be added to the nitrogen charge, providing a trace gas that an appropriate leak-detecting device can sense.

2.5.0   Hangers and Supports

Hangers and supports provide horizontal and vertical support for piping. *Hangers* generally describe a means of support that is anchored above the piping. *Supports* carry the piping load from below, or from walls and other sturdy structural components. The principal purpose of hangers and supports is to keep the piping in alignment and to prevent it from bending or distorting. Physical stress on the piping often leads to joint failure and refrigerant leaks.

Refrigerant Leak Detectors

Portable electronic leak detectors are designed to detect minute amounts of refrigerant vapor. While some units are designed to detect a specific refrigerant compound or blend, others may be designed to detect a family of refrigerants. The device sounds an alarm and/or displays a result when refrigerant is detected. Today's detectors are typically battery powered. Many have selectable and adjustable sensitivity ranges.

Some leak detectors are designed to detect the sound of leaking vapors. However, they can be unreliable in noisy or windy environments. Similar devices are designed to detect combustible gases for work in and around heating equipment.

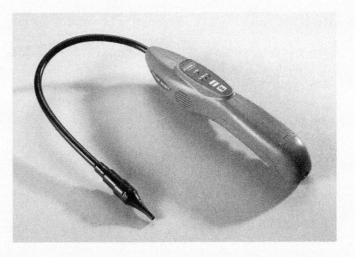

Hangers that are in direct contact with copper should be **copper-clad**. It is best to keep copper and steel isolated from each other in piping systems. Under some conditions, corrosion can occur where there is contact between the two metals. Hangers of pure copper would be unnecessarily expensive and not as strong as steel, but some small hangers and pipe straps are made from copper.

To fasten a hanger assembly to beams and other metal structures, various styles of *beam clamps* are a common choice (*Figure 43*). A piece of threaded rod or similar material is connected to the beam clamp, and the hanger is connected to the rod.

Figure 44 provides examples of two popular hanger styles. Clevis hangers are very popular for all sizes of pipe and tubing. The split ring hanger can be used to suspend piping, but they are more often used to support it from below or horizontally from a wall.

A wide variety of straps are available to secure pipe. Some simple examples of straps used for supports are shown in *Figure 45*. Riser clamps (*Figure 46*) support pipe where the pipe penetrates floors vertically.

There are many kinds of straps that are designed to hold tubing in place against a special type of U-shaped channel (*Figure 47*). Copper or copper-clad straps can be used with the channel. Steel straps are also used with copper tubing, using a plastic insert to isolate the copper from the steel (*Figure 48*). The insert also provides a firmer grip around the tube and helps to isolate any vibration. In addition, the plastic strap provides some insulation without compromising grip. Any insulation needed for the tubing can often be butted against the clamp, rather than passed through it.

Copper-clad: Components that have been coated or covered with a thin copper layer.

Adjustable I-Beam Clamp

C-Style Beam Clamp

I-Beam Clamp

Flange Clamp

Side Beam Clamp

Figure 43 Various beam clamp styles.

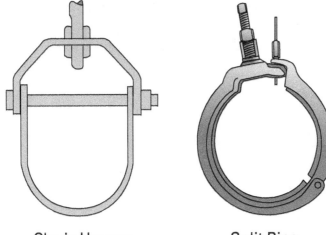

Clevis Hanger　　　**Split Ring**

Figure 44 Clevis and split ring hangers.

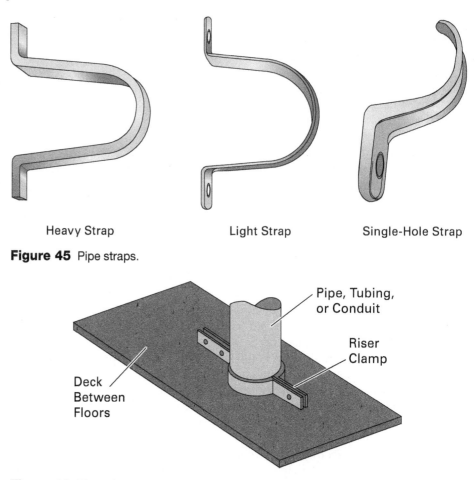

Heavy Strap   Light Strap   Single-Hole Strap

Figure 45 Pipe straps.

Pipe, Tubing,
or Conduit

Riser
Clamp

Deck
Between
Floors

Figure 46 Riser clamp.

Pipe Insulation

Even when common temperatures and humidity levels are present, condensation will form on refrigerant and cold-water piping. The condensate may drip into equipment, walls, and other places where it can cause damage. Refrigeration piping can also pick up heat from the surrounding air, causing a small efficiency loss. Similarly, heat can escape from hot-water piping. To prevent these problems, various types of piping must be insulated.

Insulation is a material that slows down the transfer of heat. Cork, glass fibers, mineral wool, and polyurethane foams are examples of insulating materials. Insulation is ideally fire-resistant, moisture-resistant, and vermin-proof. It must also provide a vapor retarder to prevent moisture from gaining access to the piping beneath. The most popular material for insulating refrigerant lines is shown here. It is a closed-cell, elastomeric material.

Type ACR soft copper tubing for refrigerant lines can be purchased in rolls with factory-applied insulation. If it is necessary to install pipe insulation at the jobsite, as much as possible should be added before the joints are assembled. That way, the insulation can be slid onto the tubing intact instead of split along its length for application. Sections that must be split are then applied to straight lengths instead of around elbows where it is more difficult to seal effectively.

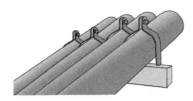

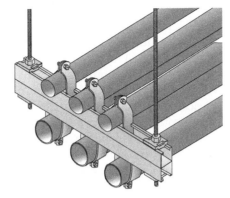

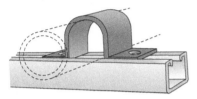

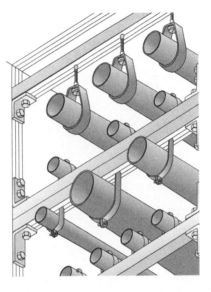

Figure 47 Pipe straps and channel.

Figure 48 Pipe straps with plastic inserts (cushion clamps).

The channel material is available in a number of depths for different applications. The most popular depths are nominally $\frac{3}{4}$" and $1\frac{1}{2}$". The channel can be mounted on walls, ceilings, or almost any other structurally reliable surface. When a number of copper tubes must be routed together, this method of support is very sound and provides a neat, consistent appearance.

When a pipe or tube must be insulated, the hangers may be selected to accommodate the insulation thickness. It is best, and often required by specifications, to ensure that the insulation is continuous through the hanger or support rather than being spliced at each one. *Figure 49* shows some straps selected to accommodate pipe insulation. However, with the insulation also inside the clamp, the clamp exerts very little grip on the pipe. Thus, they work well on a horizontal run but less so on a vertical run.

Figure 49 Pipe straps designed to accommodate insulation.

Clevis hangers are often fitted with a sheet metal saddle (*Figure 50*) to spread the weight of the load over a larger surface, keeping the insulation from being crushed and losing its effectiveness. The insulated line simply lies on the saddle.

HVACR installations often include plans and specifications that completely describe the hanger system. Project specifications are based on codes or ordinances and require strict adherence. Pipe hangers and supports are part of the project specifications. A specification for pipe hanger spacing, for example, may read as follows: *"All piping shall be supported with hangers spaced not more than 10' apart (on center). Hangers shall be the malleable iron split-ring type and shall be as manufactured by XYZ Hangers, Inc. or other approved vendor."* However, since the spacing of hangers and supports is a function of the type and size of piping used, a table to show the required hanger spacing for various piping and tubing materials is often provided.

Table 4 shows the recommended hanger spacing, as well as the hanger rod size, for copper tubing. The spacing shown in the table is valid for copper tubes filled with liquid; the added weight of liquid has already been considered. The listed hanger rod size indicates the rod diameter, followed by the number of threads per inch. Hangers for steel piping can be spaced farther apart, due to the added strength and structural integrity of the material.

When it comes to hangers and supports, more is always better than less.

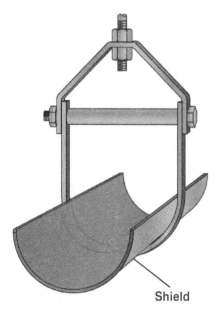

Shield

Figure 50 A clevis hanger fitted with an insulation saddle.

TABLE 4 Recommended Hanger Spacing for Copper Tubing

Nominal Tubing Size		Maximum Span		Recommended Hanger Rod Size
Inches	Millimeters	Feet	Meters	
$\frac{1}{2}$	15	5	1.52	$\frac{3}{8}$"–16
$\frac{3}{4}$	20	5	1.52	$\frac{3}{8}$"–16
1	25	6	1.83	$\frac{3}{8}$"–16
$1\frac{1}{4}$	32	7	2.13	$\frac{3}{8}$"–16
$1\frac{1}{2}$	40	8	2.44	$\frac{3}{8}$"–16
2	50	8	2.44	$\frac{3}{8}$"–16
$2\frac{1}{2}$	65	9	2.74	$\frac{1}{2}$"–13
3	80	10	3.05	$\frac{1}{2}$"–13
$3\frac{1}{2}$	90	11	3.35	$\frac{1}{2}$"–13
4	100	12	3.66	$\frac{1}{2}$"–13
5	125	13	3.96	$\frac{1}{2}$"–13
6	150	14	4.27	$\frac{5}{8}$"–11
8	200	16	4.87	$\frac{3}{4}$"–10

2.0.0 Section Review

1. After it is bent as desired, soft copper tubing usually becomes softer and more pliable.
   a. True
   b. False

2. What happens when a flare formed on a piece of tubing is too wide?
   a. It will not cover the complete face of the fitting.
   b. The flare nut will separate from the joint at higher pressures.
   c. The flare will interfere with the flare nut.
   d. The fitting and the flare face will not be at the same angle.

3. Which of the following statements about push-to-connect fittings is *correct*?
   a. Push-to-connect fittings are designed to be used many times.
   b. Push-to-connect fittings work well with tubing that is distorted.
   c. One advantage of push-to-connect fittings is their ability to withstand stress and abuse.
   d. Some push-to-connect fittings can be removed using a manufacturer-provided tool.

4. Nitrogen tanks used for pressure testing are routinely charged to pressures exceeding _____.
   a. 1,500 psig
   b. 2,000 psig
   c. 3,000 psig
   d. 4,000 psig

5. The spacing between hangers and supports for copper tubing is based primarily on the _____.
   a. tubing size
   b. length of the tubing run
   c. use of the piping system
   d. tubing type, such as Type L or Type M

3.0.0 Plastic Piping

Performance Task	Objective
5. Cut and join PVC pipe and fittings.	Recognize different types of plastic piping and explain how they are joined. a. Recognize and identify different types of plastic piping. b. Explain how to join various types of plastic piping.

Plastic pipe is used to drain condensate from furnaces and air conditioners, transport water in water-cooled HVACR systems, and even to vent combustion byproducts from condensing furnaces. Some plastic piping can be threaded, and manufactured threaded adapters are available to connect plastic piping to threaded fittings and components.

There are several types of plastic pipe used in the HVACR craft. *Figure 51* shows how plastic pipe was used in a typical condensing furnace installation. The primary advantage that plastic piping offers over metal pipe is its resistance to corrosion or rot. Many types are highly resistant to chemical agents that would destroy metal piping systems in a relatively short period of time.

3.1.0 Types of Plastic Tubing and Pipe

There are many types of plastic tubing and pipe on the market. However, some types are encountered far more often than others are. You will likely work with the following products at some point in your career, and you may work with some of them daily.

Identifying markings on plastic piping products include the manufacturer and trademarks, the standard(s) to which they were manufactured, and the nominal size. Pressure and temperature ratings are usually present as well. If the tubing can be used for potable water service, it must bear a laboratory marking to prove it.

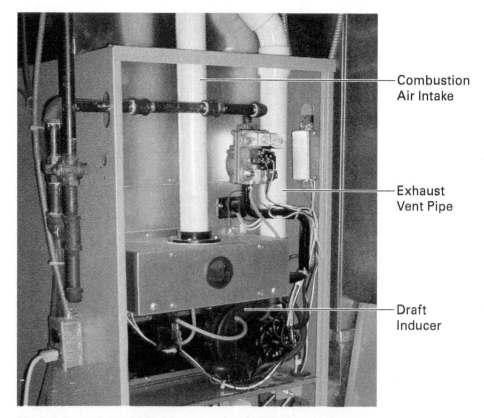

Figure 51 Plastic piping used in a condensing furnace installation.

Regardless of purchase orders and receipts, pay attention to the tubing you have in hand and be sure it is the correct product for the job. Incorrect product deliveries to a jobsite are not uncommon.

3.1.1 ABS Pipe

Acrylonitrile butadiene styrene (ABS) pipe (*Figure 52*) was originally developed in the 1950s for use in the chemical industry and in oil fields. It is rigid, smooth, and has good impact strength at low temperatures. It is typically used for drain, waste, and vent (DWV) applications. It performs reliably in the –40°F to 180°F (–40°C to 82°C) temperature range.

Although it is generally used in applications that are at or near atmospheric pressure, ABS pipe is also available for pressurized systems in industrial environments. ABS pipe is sold in 10' and 20' lengths. Most ABS pipe is black in color, but other colors can be requested.

Figure 52 Plumbing vent stack constructed from ABS pipe and fittings.
Source: Shutterstock.com/VDB Photos

3.1.2 PE and PEX Tubing

Polyethylene (PE) tubing (*Figure 53*) is a **thermoplastic** material originating from ethylene. PE is available in rolled coils for sizes less than 6" diameter. Rigid, semi-rigid, and corrugated lengths up to 40' are available for the larger sizes. Diameters up to 60" (152 cm) are available. PE tubing and pipe is available in many different colors and wall thicknesses. It can be used for DWV service, domestic water service, and even fuel gases in some locations.

Thermoplastic: Describes plastic materials that become more plastic (elastic) when heated and harden again when cooled and can generally do so repeatedly.

Figure 53 PE tubing joined with plastic compression fittings.
Source: Shutterstock.com/RosnaniMusa

Cross-linked polyethylene (PEX) represents an improvement on a good material. PEX is a **thermoset** material, rather than a thermoplastic like PE. It is more resistant to impact at low temperatures and abrasion, while the material is slightly softer and less rigid than PE.

PEX (*Figure 54*) is extremely popular in new construction for domestic hot- and cold-water systems. It is currently the most widely used flexible pipe product in the plumbing industry. It is also used in radiant floor-heating systems. A version of PEX tubing, identified as hePEX, is sealed with a special polymer barrier that prevents oxygen diffusion through the sides of the tubing. This type is specifically designed for radiant heat applications.

PEX tubing is rated for 200°F (93°C) at 80 psig, 180°F (82°C) at 100 psig, or 73°F (23°C) at 160 psig. It can be purchased in coils up to 1,000' in length, or in straight lengths of pipe.

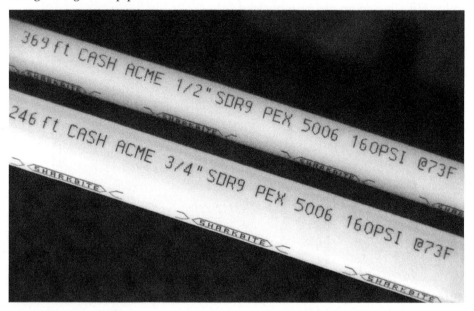

Figure 54 PEX tubing.

Unlike some plastic materials, PEX cannot be joined using solvent cements. However, PEX can be joined in a variety of other ways:

- *Clamping* — A stainless steel ring is applied over an inserted fitting, and a special tool is used to set the ring (*Figure 55*). One tool works for all fitting sizes.

- *Crimping* — A copper ring is applied over an inserted fitting, and a special tool is used to set the ring (*Figure 56*). Each size of tubing requires a different set of jaws to crimp the ring.

- *Expansion* — A heavy plastic collar is placed on the tubing, then a special tool is used to expand the end of the tubing (*Figure 57*). Then the fitting is simply inserted. This connection largely relies on the "memory" of PEX tubing since it is a thermoset. Once the fitting is inserted, the tubing slowly begins to return to its original size, creating a tight fit.

- *Expansion with compression* — This is a combination of the expansion and crimping methods. This method is slow, as it requires the use of two special tools. Each size of tubing requires a different set of jaws to crimp the ring.

- *Push-to-connect* — The fitting is simply pushed onto properly prepared tubing.

The advantage to expanded joints is that the fitting can have a larger bore. With all the others, the fitting ID must be smaller than the tubing ID, which restricts flow somewhat. Due to the speed of assembly and the need for only one tool, the clamping method is very popular.

Figure 55 Clamped PEX joints.
Source: iStock@JJ Gouin

Figure 56 Crimped PEX joint.
Source: iStock@John_Brueske

Figure 57 Expanding a PEX tubing joint.
Source: DEWALT Industrial Tool Co.

3.1.3 PVC Pipe

Polyvinyl chloride (PVC) is a rigid pipe with high impact strength, and it is used extensively in HVACR applications (*Figure 58*). It can be used in high-pressure systems at low temperatures, and as ductwork for various applications. It is also used to vent gas-fired condensing furnaces. The maximum recommended temperature for PVC is 140°F (60°C).

PVC pipe is usually joined with a solvent cement. PVC pipe can also be thermo-welded (*Figure 59*). This method is often used for larger pipe sizes that are difficult to assemble with cement. It is done using a PVC welding rod and a tool to melt the rod around the pipe joint.

Figure 58 PVC pipe and fittings.
Source: Charlotte Pipe and Foundry Company

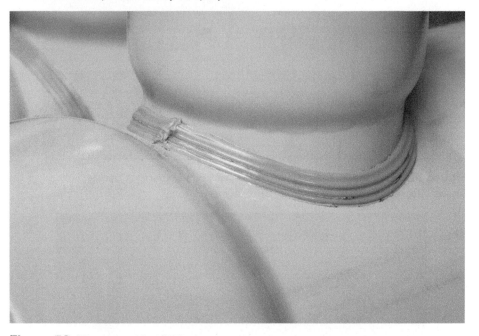

Figure 59 Thermo-welded PVC pipe.
Source: Shutterstock.com/Yuthtana artkla

PVC comes in various wall thicknesses, some of which are thick enough for threading. It is used in both pressurized and non-pressurized applications, and as electrical conduit. An informal color code does exist, but it is not universal:

- Green is generally used in sewage applications.
- Dark gray and blue, as well as white, are used for cold-water piping.
- Dark gray is also used for industrial applications, and often for electrical conduit.
- White is used in many applications, for both pressurized and non-pressurized service.

The identifying markings (*Figure 60*) are much the same as those required for PE piping. PVC is generally sold in 10' (3 m) or 20' (6 m) lengths. Note that PVC and CPVC plastic pipe have the same OD as the equivalent size of steel pipe. For example, the actual OD of both 1" PVC pipe and 1" steel pipe is 1.315".

3.1.4 CPVC Pipe

The applications and joining methods for chlorinated polyvinyl chloride (CPVC) are very similar to those for PVC (*Figure 61*). However, CPVC can be used with fluid temperatures up to 180°F (82°C) at a pressure of 100 psig. CPVC pipe and fittings typically have a yellow tint, rather than being bright white.

Since CPVC pipe does not support combustion, it may be accepted for use in fire sprinkler systems. This property also allows it to be used in some areas where combustible materials are not allowed. CPVC pipe is joined in much the same way as PVC, using a solvent cement.

¾" *Brandname* PVC-1120 SCH.40 PR 480 PSI @ 73°F ASTM D-1785 | NSF-pw

Figure 60 PVC pipe markings.

Figure 61 CPVC pipe and fittings.
Source: Charlotte Pipe and Foundry Company

3.1.5 Plastic Piping Schedules

Some pipe, including PVC, CPVC, and even steel, are made to different *schedules*. Schedules for pipe refer to their wall thickness and rated strength. The schedule system for pipe can be compared to the system of assigning letters to copper tubing (M, L, and K) with different wall thicknesses.

As the schedule number increases, the wall thickness increases. Like copper tubing, as the wall thickness of plastic (and steel) pipe increases, the ID becomes smaller. The OD does not change.

The OD is not the only thing that PVC and CPVC pipe have in common with steel piping. The wall thickness of each schedule group is also the same. For example, 1" Schedule 40 PVC and 1" Schedule 40 steel pipe both have an OD of 1.315" and a wall thickness of 0.133". Obviously, this results in PVC/CPVC pipe and steel pipe of a given schedule having the same ID as well.

PVC and CPVC pipe are primarily manufactured in Schedules 40, 80, and 120. *Table 5* shows the dimensions of Schedule 40 pipe, and *Table 6* shows the same information for Schedule 80 pipe. Remember that these same dimensions apply to steel pipe of the same schedule. Steel piping, however, is available in more schedules than plastic. Schedule 80 PVC and CPVC are used for applications with higher pressures and where greater durability is needed. Schedule 80 PVC can even be threaded like steel pipe, due to the increased wall thickness.

3.2.0 Joining Plastic Pipe

Plastic pipe may be cut with a saw (*Figure 62*) or with a ratcheting plastic pipe cutter (*Figure 63*). Saws are the best choice for larger sizes, but the ratcheting cutter is quick, clean, and easy to use for pipe sizes up to around 2" (5 cm). Saws leave burrs that must be reamed or filed away.

Ratcheting cutters can cut soft, flexible tubing with a single stroke. Rigid plastic pipe, like PVC, requires several quick strokes. Square cuts with saws may require a miter box or similar cutting guide, as shown in *Figure 62*. When many cuts are needed, a power tool, such as a chop saw or power miter saw, can be used to increase production.

A pipe wrap (*Figure 64*) provides a simple method of marking a straight line around any type of pipe to ensure a square cut. A wrap-around also has additional information and lines printed on it to help fitters with unusual cuts that may be needed. Wrap it around the pipe tightly so that the two ends meet squarely. Then use a marker to mark the pipe completely around its circumference. Once the pipe is cut, it must be deburred inside and out using a reamer or file.

TABLE 5 PVC/CPVC Schedule 40 Pipe Dimensions

Nominal Pipe Size	OD	Avg ID	Min. Wall Thickness	Nominal Weight (lb/ft)	Max. Water Pressure (psi)
⅛"	0.405"	0.249"	0.068"	0.051	810
¼"	0.540"	0.344"	0.088"	0.086	780
⅜"	0.675"	0.473"	0.091"	0.115	620
½"	0.840"	0.602"	0.109"	0.170	600
¾"	1.050"	0.804"	0.113"	0.226	480
1"	1.315"	1.029"	0.133"	0.333	450
1¼"	1.660"	1.360"	0.140"	0.450	370
1½"	1.900"	1.590"	0.145"	0.537	330
2"	2.375"	2.047"	0.154"	0.720	280
2½"	2.875"	2.445"	0.203"	1.136	300
3"	3.500"	3.042"	0.216"	1.488	260
3½"	4.000"	3.521"	0.226"	1.789	240
4"	4.500"	3.998"	0.237"	2.118	220
5"	5.563"	5.016"	0.258"	2.874	190
6"	6.625"	6.031"	0.280"	3.733	180
8"	8.625"	7.942"	0.322"	5.619	160
10"	10.750"	9.976"	0.365"	7.966	140
12"	12.750"	11.889"	0.406"	10.534	130
14"	14.000"	13.073"	0.437"	12.462	130
16"	16.000"	14.940"	0.500"	16.286	130
18"	18.000"	16.809"	0.562"	20.587	130
20"	20.000"	18.743"	0.593"	24.183	120
24"	24.000"	22.544"	0.687"	33.652	120

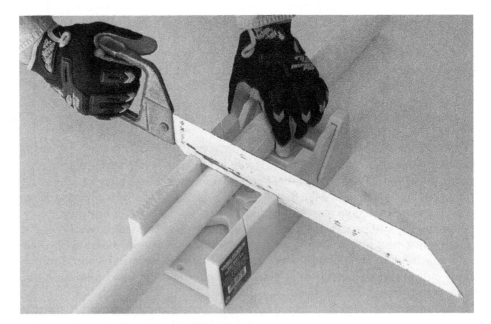

Figure 62 PVC hand saw and miter box.

3.2.1 Solvent-Cementing Products

Although there are several ways to join rigid plastic pipe, solvent cementing is the most common method. This technique is used daily by many HVACR craftworkers. PVC and CPVC pipe materials used in HVACR applications are most often joined this way.

TABLE 6 PVC/CPVC Schedule 80 Pipe Dimensions

Nominal Pipe Size	OD	Avg ID	Min. Wall Thickness	Nominal Weight (lb/ft)	Max. Water Pressure (psi)
1/8"	0.405"	0.195"	0.095"	0.063	1,230
1/4"	0.540"	0.282"	0.119"	0.105	1,130
3/8"	0.675"	0.403"	0.126"	0.146	920
1/2"	0.840"	0.526"	0.147"	0.213	850
3/4"	1.050"	0.722"	0.154"	0.289	690
1"	1.315"	0.936"	0.179"	0.424	630
1 1/4"	1.660"	1.255"	0.191"	0.586	520
1 1/2"	1.900"	1.476"	0.200"	0.711	470
2"	2.375"	1.913"	0.218"	0.984	400
2 1/2"	2.875"	2.290"	0.276"	1.500	420
3"	3.500"	2.864"	0.300"	2.010	370
3 1/2"	4.000"	3.326"	0.318"	2.452	350
4"	4.500"	3.786"	0.337"	2.938	320
5"	5.563"	4.768"	0.375"	4.078	290
6"	6.625"	5.709"	0.432"	5.610	280
8"	8.625"	7.565"	0.500"	8.522	250
10"	10.750"	9.493"	0.593"	12.635	230
12"	12.750"	11.294"	0.687"	17.384	230
14"	14.000"	12.410"	0.750"	20.852	220
16"	16.000"	14.213"	0.843"	26.810	220
18"	18.000"	16.014"	0.937"	33.544	220
20"	20.000"	17.814"	1.031"	41.047	220
24"	24.000"	21.418"	1.218"	58.233	210

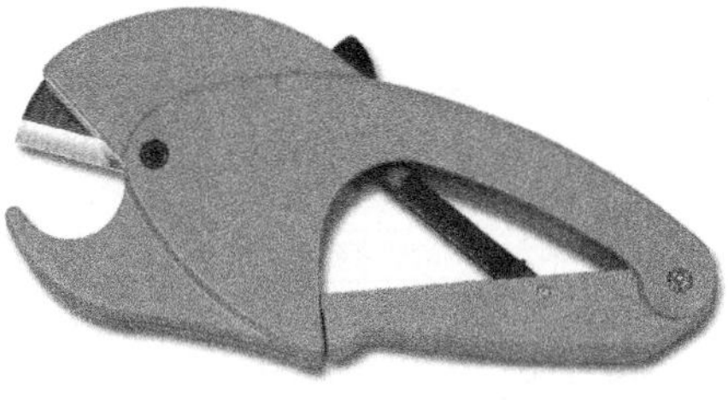

Figure 63 Plastic tubing and pipe cutter.

Figure 64 Pipe wrap for measuring, marking straight cut lines, and laying out complex cuts.

A wide variety of products are available that are related to solvent cementing PVC and CPVC pipe. They include cleaners, primers, and cements.

Both PVC and CPVC pipe surfaces must be thoroughly cleaned with an appropriate cleaner (*Figure 65*). The cleaner removes debris and any oil or grease that may be on the pipe from manufacturing or handling.

Figure 65 PVC/CPVC clear pipe cleaner.

Figure 66 PVC/CPVC primer/cleaner with a purple color.

Primers (*Figure 66*) contain an aggressive solvent that softens the surface of the material. This prepares the surface for the application of the cement, helping the cement penetrate deeper for a strong, leak-free joint. The primer shown in *Figure 66* has had a purple colorant added and also has been combined with a cleaner to eliminate the cleaning step. The use of a primer may be required by codes to ensure joints have the proper integrity. As a result, primers often have a color added to visually confirm to an inspector that a particular primer was applied.

Cements (*Figure 67*) are available in a much wider variety than cleaners and primers. Most cement products are not designed for use with both PVC and CPVC. The cleaners and primers can often be used on several different kinds of plastic piping. However, workers must ensure that the correct cement for the material is applied. Cements can be chosen for their body (thickness), color, curing time, ability to work in water, ability to perform better in hot environments, or for their safe use in potable water systems. On some projects, the characteristics of the cement to be used may be specified in the construction documents.

3.2.2 Solvent Cementing Plastic Pipe

To solvent cement PVC, CPVC, and ABS pipe, the pipe end is first deburred and examined to ensure it is square. If alignment of the fitting in a particular plane is required, clearly mark the pipe and fitting with a single line so that the two can be realigned as the fitting is slipped in place. Remember that cleaners and solvents may destroy or remove the line, so make it bold and long or use a scratch awl to lightly scratch the surface instead.

WARNING!

PVC cleaners, primers, and cements are toxic and flammable. Do not use them near an open flame, and do not breathe the vapors. Do not use PVC cement in confined spaces without the appropriate ventilation or breathing apparatus. Avoid getting all solvent-cementing products on the skin. Refer to the applicable Safety Data Sheet (SDS) for the product in use.

Use special care when solvent cementing plastic pipe in temperatures below 40°F (4°C) or above 100°F (38°C). In hot environments, make sure both surfaces to be joined are still wet with cement when putting them together. Cements set up faster in higher temperatures. In cold temperatures, longer curing times are required.

Table 7 provides some general guidelines for handling and set-up times at different temperatures, based on the pipe size. Once the joint has been made, the cement must be allowed to cure before it can withstand pressure testing. *Table 8* provides the average cure times for both PVC and CPVC pipe. Note that the times shown for curing should be extended 50 percent in wet or extremely humid environments. If this presents a problem, another cement that cures faster in humid conditions can be chosen. Note that these tables are average only. Always follow the guidelines provided by the product manufacturer.

CAUTION

Table 7 and *Table 8* serve as an example only and are not intended for field use. These values, as well as the advice for assembly in wet environments, often differ from product to product. Follow the assembly and curing guidance for the specific product in use.

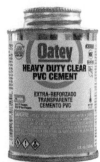
(A) Heavy-Duty Clear

(B) Heavy-Duty Gray

(C) CPVC Cement

(D) Black ABS Cement

Figure 67 A variety of PVC, CPVC, and ABS cement products.

TABLE 7 Average Handling and Set-Up Times for PVC/CPVC Cements

Temperature During Assembly	Pipe Diameter					
	½" to 1¼"	1½" to 3"	4" to 5"	6" to 8"	10" to 16"	18"+
60°F to 100°F (16°C to 38°C)	2 minutes	5 minutes	15 minutes	30 minutes	2 hours	4 hours
40°F to 60°F (5°C to 16°C)	5 minutes	10 minutes	30 minutes	90 minutes	8 hours	16 hours
20°F to 40°F (−7°C to 5°C)	8 minutes	12 minutes	96 minutes	3 hours	12 hours	24 hours
0°F to 20°F (−18°C to −7°C)	10 minutes	15 minutes	2 hours	6 hours	24 hours	48 hours

TABLE 8 Average Cure Times for PVC/CPVC Cements

Temperature During Assembly and Cure Periods	Pipe Diameter									
	½" to 1¼"		1½" to 3"		4" to 5"		6" to 8"		10" to 16"	18"+
	Up to 180 psi	180 psi+	Up to 180 psi	180 psi+	Up to 180 psi	180 psi+	Up to 180 psi	180 psi+	Up to 100 psi	Up to 100 psi
60°F to 100°F (16°C to 38°C)	1 hour	6 hours	2 hours	12 hours	6 hours	18 hours	8 hours	24 hours	24 hours	36 hours
40°F to 60°F (5°C to 16°C)	2 hours	12 hours	4 hours	24 hours	12 hours	36 hours	16 hours	48 hours	48 hours	72 hours
20°F to 40°F (−7°C to 5°C)	6 hours	36 hours	12 hours	72 hours	36 hours	4 days	3 days	9 days	8 days	12 days
0°F to 20°F (−18°C to −7°C)	8 hours	48 hours	16 hours	96 hours	48 hours	8 days	4 days	12 days	10 days	14 days

Note: Joint cure time is the time required before pressure testing the system. Chart assumes a relative humidity of 60% or less. In wetter conditions, allow 50% additional cure time.

To join CPVC or PVC pipe and fittings with solvent cement, refer to *Figure 68*. It shows a logical sequence of steps to fabricate a solvent-cemented joint. These steps illustrate typical assembly. When joining plastic pipe, always follow the cement manufacturer's instructions for the environment and materials being joined.

Before beginning, don the appropriate PPE. Safety glasses should be worn at all times when working with pipe and solvent-cementing products. Rubber or latex gloves should also be worn during the joining process. Making a proper joint requires some timing and preparation when the handling time is short. Make sure all the required materials are readily available and within easy reach.

Step 1 Cut the pipe square to the required length. Then ream the inside edge of the pipe with a reamer. **Chamfer** the outside edge of the pipe with a suitable reamer, single-cut file, or sand cloth. Examine the pipe end; if cracks are noted, cut the pipe back at least 2" from the end of the crack and examine it again. Then clean the pipe with a rag (*Figure 68* [A]).

Step 2 Test fit the pipe and fitting (*Figure 68* [B]). The pipe should fit more tightly as it reaches about two-thirds of the socket depth. Once the primer and cement are applied, the surfaces will be mildly dissolved and slippery, and the pipe and fitting will fit together fully.

Step 3 Mark the pipe and fitting with a felt-tipped pen or scratch awl to show the proper position for alignment, if necessary. Then make a witness mark for the depth of the fitting socket on the pipe. Since the mark will be removed by cleaners and primers, make the witness mark 2" away. Once the fitting is assembled, a measurement can quickly be made from the witness mark to the face of the fitting to make sure that the pipe bottomed-out in the socket. During the first few moments of assembly, the pipe is sometimes pushed out of the socket somewhat by natural processes. The witness mark helps to ensure the joint stayed together during the cure.

Step 4 Apply a cleaner/primer to the inside of the fitting with the provided dauber (*Figure 68* [C]). Rotate the dauber around the inside of the fitting several times, working it into the surface.

Chamfer: Breaking what would usually be a 90-degree angle to create a symmetrical angled surface on an edge. Chamfer is also used as a noun, as a name for the angled edge.

(A)

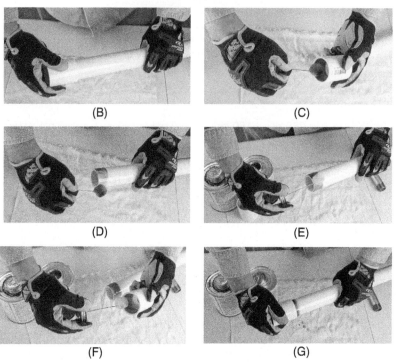

(B)   (C)

(D)   (E)

(F)   (G)

Figure 68 Fabricating a solvent-cemented joint.

Plastic Pipe Failures

Plastic piping has many applications. However, PVC, for example, should not be used at any time for above-ground compressed air installations. Incidents of material failure in compressed air systems have caused a number of injuries over the years. While it is easy to install, PVC and other types of plastic pipe are not the answer to every need.

Proper support is essential for plastic piping. Numerous incidents of failure over the years have resulted from improper support and from using the material as a means of structural support. For example, supporting other loads from even a properly supported PVC pipe is not an acceptable practice.

It is also important not to rush the pressure testing of cemented joints. Follow the manufacturer's recommendations for curing time under the current environmental conditions before applying any pressure.

Step 5   Apply the cleaner/primer to the end of the pipe (remember that some types of plastic pipe, such as ABS, do not require a primer). Apply the primer to the pipe about $\frac{1}{2}$" to 1" (~13 mm to ~25 mm) past the socket depth (*Figure 68* [D]). You may also apply the primer to the socket of the fitting a second time.

Step 6   Quickly work in a generous, even coat of cement to the pipe end while the primer is still wet (*Figure 68* [E]). Apply the cement up to the depth of the socket only. Use the same applicator (without adding more cement) to apply a thin coat of cement inside the fitting socket (*Figure 68* [F]). Do not allow excess cement to puddle in the fitting socket, as it will simply be pushed back inside the fitting during assembly.

Step 7   Apply a light second layer of cement to the pipe end.

Step 8   Immediately insert the pipe into the fitting socket, rotating the tube one-quarter to one-half turn while pushing them together (*Figure 68* [G]). This motion ensures an even distribution of cement in the joint. Properly align the fitting if an alignment mark was placed.

Step 9   Hold the assembly together firmly for about 30 seconds. This helps prevent the pipe from pushing out of the socket as the curing process begins. An even bead of cement should be visible around the joint. If this bead does not appear all the way around the socket edge, the cement was improperly applied. In this case, remake the joint to avoid the possibility of leaks. Wipe excess cement from the tubing and fitting surfaces; the small bead at the root of the joint should not be wiped away.

The procedure for joining ABS pipe with cement is the same as for PVC or CPVC, except that no primer is required on ABS pipe. Unless the pipe and joint are being used for DWV service, pressure test the piping system after the last joint assembled has reached the recommended curing time for the conditions.

3.2.3 Plastic Pipe Support Spacing

Since even rigid plastic piping products are somewhat flexible and are more sensitive to heat than metals, support must be provided at closer intervals. *Table 9* provides the recommended support spacing intervals for Schedule 40 and Schedule 80 PVC pipe.

Note that some building codes require that support be provided at a maximum of 4' (1.2 m) intervals, regardless of the pipe size. As shown in *Table 9*, smaller pipe sizes require closer intervals. It is also important to note the operating temperature in the table. As the temperature increases, the support interval becomes closer.

NOTE

Manufacturer's tables and charts for pipe support intervals do not override local codes or project specifications. With plastic piping, more support is always better than less, especially in hot environments.

TABLE 9 General Guidelines for Horizontal Support Spacing for PVC

| Nominal Pipe Size | Schedule 40 | | | | | Schedule 80 | | | | |
| | Operating Temp (°F) | | | | | Operating Temp (°F) | | | | |
	60°F	80°F	100°F	120°F	140°F	60°F	80°F	100°F	120°F	140°F
½"	4½'	4½'	4'	2½'	2½'	5'	4½'	4½'	3'	2½'
¾"	5'	4½'	4'	2½'	2½'	5½'	5'	4½'	3'	2½'
1"	5½'	5'	4½'	3'	2½'	6'	5½'	5'	3½'	3'
1¼"	6'	5½'	5'	3'	3'	6'	6'	5½'	3½'	3'
1½"	6'	5½'	5'	3½'	3'	6½'	6'	5½'	3½'	3½'
2"	6'	5½'	5'	3½'	3'	7'	6½'	6'	4'	3½'
2½"	7'	6½'	6'	4'	3½'	7½'	7½'	6½'	4½'	4'
3"	7'	7'	6'	4'	3½'	8'	7½'	7'	4½'	4'
4"	7½'	7'	6½'	4½'	4'	9'	8½'	7½'	5'	4½'
6"	8½'	8'	7½'	5'	4½'	10'	9½'	9'	6'	5'
8"	9'	8½'	8'	5'	4½'	11'	10½'	9½'	6½'	5½'
10"	10'	9'	8½'	5½'	5'	12'	11'	10'	7'	6'
12"	11½'	10½'	9½'	6½'	5½'	13'	12'	10½'	7½'	6½'
14"	12'	11'	10'	7'	6'	13½'	13'	11'	8'	7'
16"	12½'	11½'	10½'	7½'	6½'	14'	13½'	11½'	8½'	7½'

3.0.0 Section Review

1. The schedule system for plastic pipe _____.
   a. is based on the materials of construction
   b. can be used to determine the outside diameter
   c. identifies the type of cement that should be used
   d. identifies the wall thickness

2. The cure time required for PVC cements may be much shorter in moist or very humid environments.
   a. True
   b. False

1. The organization that establishes manufacturing standards for copper tubing is _____.
    a. SAE
    b. CDA
    c. ASTM International
    d. ASHRAE

2. The working pressure rating of hard-drawn copper tubing is typically _____.
    a. 30 percent higher than soft copper tubing
    b. 40 percent higher than soft copper tubing
    c. 50 percent higher than soft copper tubing
    d. 60 percent higher than soft copper tubing

3. As the wall thickness of copper tubing increases, the inside diameter of the tube decreases.
    a. True
    b. False

4. One of the main uses of flare fittings in HVACR work is to _____.
    a. connect various pressure controls
    b. connect runs of tubing greater that $1\frac{1}{8}$" OD
    c. provide a tubing socket for heat-bonded joints
    d. provide the most reliable type of joint wherever needed

5. One reason to use a tubing cutter instead of a hacksaw to cut copper tubing is that _____.
    a. stronger joints will result
    b. tubing cutters do not create loose filings
    c. the tubing can be cut off at any angle
    d. it does not create a ridge inside the tube edge

6. Swaging a copper tube prepares it for a sound, reliable fit with a compression ferrule.
    a. True
    b. False

7. Which of the following statements about push-to-connect fittings for copper tubing is *correct*?
    a. Push-to-connect fittings are designed to be reused many times.
    b. Push-to-connect fittings are not rated for the common pressures found in a refrigerant circuit.
    c. Some push-to-connect fittings come with a release tool for easy removal.
    d. Push-to-connect fittings must be made from steel.

8. The pressure-relief valve on a nitrogen pressure test set should be set to relieve pressure at _____.
    a. 1 or 2 psig above the required test pressure
    b. at a pressure 50 percent higher than the required test pressure
    c. the maximum working pressure of the nitrogen regulator
    d. the maximum safe working pressure of the tubing being tested

9. Which type of tubing or pipe would be expected to require the *closest* hanger spacing?
    a. $\frac{1}{2}$" copper tubing
    b. $1\frac{1}{2}$" copper tubing
    c. $\frac{3}{4}$" steel pipe
    d. 2" steel pipe

10. Which plastic product is the *most* widely used flexible product in the plumbing industry?

    a. ABS
    b. PVC
    c. PEX
    d. CPVC

11. The type of plastic pipe that does *not* support combustion is _____.

    a. ABS
    b. PEX
    c. PVC
    d. CPVC

12. Which type of plastic piping can be threaded?

    a. PEX
    b. Schedule 40 PVC
    c. Schedule 80 PVC
    d. Schedule 40 CPVC

13. The *most* common way to join rigid plastic pipe is _____.

    a. solvent cementing
    b. heat-bonding
    c. threading
    d. clamps

14. Plastic pipe cements take longer to cure in hot weather.

    a. True
    b. False

15. In wet environments, the listed curing times for PVC solvent-cement joints should be extended _____.

    a. 20 percent
    b. 35 percent
    c. 50 percent
    d. 65 percent

Answers to odd-numbered Module Review Questions are found in *Appendix A*.

Answers to Section Review Questions

Answer	Section Reference	Objective
Section 1.0.0		
1. a	1.1.1; 1.1.3	1a
2. d	1.2.2	1b
Section 2.0.0		
1. b	2.1.3	2a
2. c	2.2.1	2b
3. d	2.3.2	2c
4. b	2.4.0	2d
5. a	2.5.0; Table 4	2e
Section 3.0.0		
1. d	3.1.5	3a
2. b	3.2.2	3b

User Update

Did you find an error? Submit a correction by visiting **https://www.nccer.org/olf** or by scanning the QR code using your mobile device.

Soldering and Brazing

Objectives

Successful completion of this module prepares you to do the following:

1. Describe and demonstrate the safe process of soldering copper tubing.
   a. Describe and demonstrate the use of the PPE, tools, and materials needed to solder copper tubing.
   b. Describe and demonstrate the preparation required for soldering.
   c. Describe and demonstrate the soldering process.
2. Describe and demonstrate the safe process of brazing copper tubing.
   a. Describe and demonstrate the use of the PPE, tools, and materials needed to braze copper tubing.
   b. Describe and demonstrate the preparation used for brazing.
   c. Describe and demonstrate the brazing process.
   d. Describe and demonstrate the process of brazing copper tubing to dissimilar metals.

Performance Tasks

Under supervision, you should be able to do the following:

1. Properly set up and shut down an air-acetylene torch.
2. Properly prep and safely solder copper tubing in various planes, using various fittings.
3. Properly set up and shut down oxyacetylene equipment.
4. Properly prep and safely braze copper tubing in various planes, using various fittings, with a nitrogen purge.
5. Braze copper tubing to either steel or brass components.

Overview

Soldering is used primarily to join copper water lines and condensate lines in the HVACR craft. Brazing is used when mechanically strong, pressure-resistant joints are needed for refrigerant lines. Both soldering and brazing demand careful attention to safety due to the hazards associated with flammable gases and open flames. With some practice, HVACR technicians quickly master soldering and brazing techniques.

Digital Resources for HVACR

SCAN ME

Scan this code using the camera on your phone or mobile device to view the digital resources related to this craft.

1.0.0 Soldering

Performance Tasks

1. Properly set up and shut down an air-acetylene torch.
2. Properly prep and safely solder copper tubing in various planes, using various fittings.

Objective

Describe and demonstrate the safe process of soldering copper tubing.

a. Describe and demonstrate the use of the PPE, tools, and materials needed to solder copper tubing.
b. Describe and demonstrate the preparation required for soldering.
c. Describe and demonstrate the soldering process.

Nonferrous: A term used to describe metals or metal alloys that do not contain iron.

Capillary action: The movement of a liquid along the surface of a solid in a kind of spreading action.

Solder: A fusible alloy used to join metals, with melting points below 842°F (450°C).

Fillet: A rounded internal corner or shoulder of filler metal often appearing at the meeting point of a piece of tubing and a fitting when the joint is soldered or brazed.

The two methods used for joining copper tubing and fittings are soldering and brazing. Both methods fasten the metals together using a **nonferrous** filler metal that adheres to the surfaces being joined. The filler metal is drawn into and distributed between the closely fitted surfaces by **capillary action**.

The primary difference between soldering and brazing is the temperature needed to melt the filler metal. Soldering uses filler metals that melt at temperatures below 842°F (450°C), usually in the 375°F to 500°F range (191°C to 260°C). The filler metals used for brazing melt at temperatures above 842°F. This temperature dividing line between the two processes has been determined by the American Welding Society (AWS).

Soldering, also called *sweating* or *soft soldering*, is the common method of joining copper tubing and fittings for water lines and other low-pressure applications. Sweat fittings are sometimes used for low-pressure refrigerant lines, but brazing is generally preferred for all refrigerant line work. Soldered joints are used in piping systems that carry liquids at temperatures of 250°F (121°C) or below. Soldered joints are typically used in situations where specs call for copper condensate lines. They are also used in some cases to join piping in low-pressure hydronic heating systems.

Soldering involves joining two metal surfaces by using heat and a nonferrous filler metal. A nonferrous filler metal is a metal that contains no iron and is therefore nonmagnetic. The melting point of the filler metal must be lower than that of the two metals that are being joined.

Soldered joints depend on capillary action to pull and distribute the melted **solder** into the small gap between the fitting and the tubing. When the joint is filled, the solder will form a tiny bead around the joint called a **fillet**. Capillary action is the flow of liquid, in this case solder, into a small space between two surfaces. Capillary action is most effective when the space between the tubing and the fitting is between 0.002" and 0.005". To maintain the proper spacing, it is important to check the joint for correct alignment before soldering. *Figure 1* shows how capillary action works.

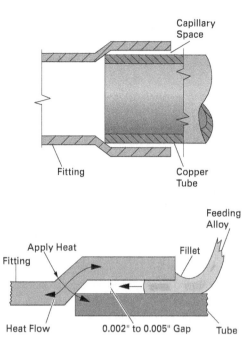

Figure 1 Capillary action.

1.1.0   PPE, Tools, and Materials Used for Soldering

Soldering does not require a lot of tools or sophisticated equipment. The equipment is relatively simple and easy to operate as long as safe work habits are maintained. The process does require some practice to master, however.

1.1.1 Soldering PPE and Safety Guidelines

Soldering involves the use of a flammable gas under pressure, open flame, and possible exposure to molten metal. To avoid injury, workers performing soldering must wear appropriate PPE and follow correct safety procedures. The following safety guidelines apply:

- Wear clothing made of nonflammable fabric such as cotton. Avoid synthetic fabrics such as nylon, which can melt and adhere to the skin.
- Wear a long-sleeve shirt to help prevent burns from sparks or molten metal. Pants should not have cuffs, as they can catch and hold hot materials.
- Wear eye and face protection.
- Wear flame-resistant gloves and high-top work boots.
- Make sure the gas cylinder is securely turned off when not in use. Gases used in soldering are flammable and explosive if exposed to a spark or flame.
- Keep a fire extinguisher handy while soldering in case material near the work ignites. When working close to flammable material such as wood, use a fire blanket or other flame/heat blocking material to protect it.
- Always solder in a well-ventilated area because fumes from the **flux** can irritate your eyes, nose, throat, and lungs. Wear respiratory equipment if working in an area with poor ventilation.
- Specific procedures are often in place for the use of an open flame. When working at field sites, be sure to acquire and follow all necessary guidelines and safety requirements related to the site.

Flux: A chemical substance that prevents oxides from forming on the surface of metals as they are heated for soldering, brazing, or welding.

Figure 2 Handheld fuel gas cylinder.

Acetone: A colorless organic solvent that is volatile and extremely flammable. In the HVACR trade, it is used as a carrier for acetylene gas in cylinders.

Figure 3 B and MC acetylene tanks.

1.1.2 Soldering Tools and Equipment

The heat used for soldering is provided by a torch that is attached to a cylinder of flammable gas. Flammable gases such as propane, MAP-Pro (propylene), and acetylene are commonly used to make sweat joints in copper tubing. Propane and MAP-Pro, which is also known as *MAPP*, are sold in small cylinders such as the one shown in *Figure 2* for handheld use. Such cylinders are fine for small projects, such as occasionally joining fittings to water lines or condensate lines. When larger, more repetitive projects are involved, the use of acetylene is far more common. It is generally sold by welding supply companies and trade-related distributors. When portability is important, acetylene can be carried in small, easily transportable cylinders (*Figure 3*). The size MC tank holds 10 ft³, while the larger size B tank holds 40 ft³. Larger acetylene cylinders are used when portability is less important.

Acetylene cylinders are packed with a porous material of some type, such as wood fibers or charcoal, that is partially saturated with **acetone**. The packing helps prevent high-pressure gas pockets from forming in the cylinder. Acetone is a colorless, flammable liquid. It is added to the cylinder until about 40 percent of the porous material becomes saturated. The acetone can hold large quantities of dissolved acetylene under pressure—as much as 25 times its own volume. Being a liquid, acetone can be drawn from an acetylene cylinder when it is not upright, or when the usage rate of the acetylene gas is excessive. The sudden appearance of flaming liquid at the end of the torch while working can be alarming.

In addition to the fuel, soldering requires a torch to provide the heat, such as the self-igniting soldering torch head shown in *Figure 4*. A press of the red button ignites the air/gas mixture. This particular torch is designed to be connected directly to a handheld propylene or propane gas cylinder. Propylene burns slightly hotter than propane, making it better suited for larger joints. An air-acetylene torch kit is shown in *Figure 5*. Note that a connecting hose and bottle fitting is used with acetylene. Since acetylene tanks are larger and heavier, the torch assembly with its fuel tank cannot be comfortably handheld.

Air-acetylene torches can usually accommodate different tips for a smaller or larger flame. The tips are numbered by the manufacturer, with larger numbers typically identifying larger tips. The tip specifications include the actual opening size, the fuel consumption at a given pressure (often rated at 14 psig), and the tubing sizes for which it is recommended. *Table 1* provides an example of performance data for air-acetylene torch tips and how it is documented.

Because soldering requires relatively low heat, torches that mix the flammable gas directly with air is all that is needed. Oxygen is required for combustion, but the air in the environment provides a sufficient amount of oxygen to produce the needed heat for soldering. Air-fuel torches are designed to draw

Figure 4 Self-igniting soldering torch head.

Figure 5 Air-acetylene torch kit.
Source: Courtesy of Uniweld Products, Inc.

TABLE 1 Air-Acetylene Torch Tip Performance Data

Tip Model	Tip Size		Gas Flow (SCFH) @ 14 psig	Recommended Tube Size (Inches)	
	in	mm		Up to 175 psi	175 psi+
TAA-0.5	$\frac{1}{16}$	4.8	2.1	$\frac{1}{8}$–$\frac{1}{2}$	$\frac{1}{8}$–$\frac{1}{4}$
TAA-1	$\frac{1}{8}$	6.4	3.4	$\frac{1}{4}$–1	$\frac{1}{8}$–$\frac{1}{2}$
TAA-2	$\frac{1}{4}$	7.9	5.8	$\frac{1}{4}$–1$\frac{1}{2}$	$\frac{1}{4}$–$\frac{3}{4}$
TAA-2.5	$\frac{5}{16}$	9.5	8.2	1–2	$\frac{1}{2}$–1
TAA-3.5	$\frac{7}{16}$	11.1	11	1$\frac{1}{2}$–3	$\frac{7}{8}$–1$\frac{5}{8}$
TAB-1	$\frac{1}{2}$	12.7	14.5	2–3$\frac{1}{2}$	1–2

Note: Data shown is an example only and not associated with a specific product; it is not for field use. Always refer to the product literature for the specific equipment in use.

in the correct amount of air for combustion from the atmosphere. Since air is roughly 80 percent nitrogen, the flame produced by an air-fuel torch burns at a lower temperature than the flame produced by an oxygen-fuel torch.

For torches that are not self-igniting, a cup-type striker (*Figure 6*) is used to light the torch. Other similar styles that produce a spark are also available. A handheld pocket lighter or common cigarette lighter should never be used for this purpose. Strikers keep the hands a safe distance from the flame and do not contain a flammable substance.

WARNING!

Do not use pocket or cigarette lighters to light any type of torch. Use only a proper striker to avoid serious burns.

Figure 6 A cup-type striker.

1.1.3 Solders and Soldering Fluxes

Solder is a nonferrous metal or metal **alloy** with a melting point below 842°F (450°C). An alloy is any substance made up of two or more metals. Solder selections are based mainly on the operating temperature and pressure of the piping system. If the system experiences dramatic temperature changes, then thermal stress may also be a selection factor. Another important point to consider is whether the system will carry potable water. Note that solders containing more than 0.2 percent lead were banned for use on potable water systems by a 1986 amendment to the Federal Safe Drinking Water Act.

The most common solder used on water lines and other low-pressure applications is an alloy made of 95 percent tin and 5 percent antimony. *Table 2* provides a comparison of pressure-temperature ratings for two different solder alloys. The 50-50 alloy of tin and lead provides very limited strength and is only suitable for low- or non-pressure applications. It also cannot be used on potable water lines. Solders composed of tin-antimony, tin-copper-silver, or tin-silver are usually recommended for applications requiring greater joint strength. The tin-antimony solder typically melts between 430°F and 480°F (221°C and 249°C) and solidifies rapidly. Generally, these solders, in the form of thick yet soft wire, are supplied on spools for easier use (*Figure 7*). The job specifications often identify the type of solder that is to be used for a specific project.

Alloy: Any substance made up of two or more metals.

Figure 7 95%-5% tin-antimony solder.

TABLE 2 Pressure-Temperature Ratings of Soldered Joints

Joining Material	Service Temperature	Fitting Type	Maximum Working Gauge Pressure (psi) for Standard Water Tube Sizes				
			Nominal Tube Size (Inches)				
			⅛–1	1¼–2	2½–4	5–8	10–12
50%-50% Tin-Lead Solder*	100°F	Pressure	200	175	150	135	100
	150°F	Pressure	150	125	100	90	70
	200°F	Pressure	100	90	75	70	50
	250°F	Pressure	85	75	50	45	40
	Saturated Steam	Pressure	15	15	15	15	15
95%-5% Tin-Antimony Solder	100°F	Pressure	1,090	850	705	660	500
	150°F	Pressure	625	485	405	375	285
	200°F	Pressure	505	395	325	305	230
	250°F	Pressure	270	210	175	165	125
	Saturated Steam	Pressure	15	15	15	15	15

Notes: This data is not for field use. Always refer to the project specifications, and to the current *ASME B16, Standards for Pipes and Fittings* for accurate information.
*The Safe Drinking Water Act Amendment of 1986 prohibits the use of any solder in potable water systems having a lead content exceeding 0.2%.

A flux (*Figure 8*) should be applied to all tube ends and fittings immediately after they are cleaned. However, since the flux is rather tacky, care must be taken to keep grit and debris off the tube and fittings once it is applied. Choosing a proper flux is very important. Soldering flux performs many functions, and the wrong flux can ruin the soldered joint. Flux performs the following functions:

- It chemically cleans and protects the surfaces of the tubing and fitting from oxidation. Oxidation occurs when the oxygen in the air combines with the recently cleaned metal. Oxidation produces tarnish or rust in metal and prevents solder from adhering. The oxidizing process is accelerated many times over when the tube and fitting is heated.

- It allows the soldering alloy or filler metal to flow more easily into the joint by promoting **wetting** of the metals.

- It floats out remaining oxides ahead of the molten filler metal.

Fluxes can be classified into three general groups: highly corrosive, less corrosive, and noncorrosive. The soldering process must render the flux inert; that is, lacking any chemical action. If not rendered inert, the flux gradually destroys the soldered joint over time. The fluxes used for joining copper tubing and copper fittings should be noncorrosive fluxes. Although corrosive types can do a better job of removing oxides and cleaning the copper, they represent a long-term risk to the joint. One noncorrosive type is composed of water and white rosin dissolved in an organic or benzoic acid base. The best noncorrosive fluxes for joining copper pipe and fittings are compounds of mild concentrations of zinc and ammonium chloride with petroleum bases. Like solders, job specifications often identify the type of flux that is to be used on a specific project. Otherwise, it is best to make use of the manufacturer's recommendations.

An oxide film begins forming on copper immediately after it has been mechanically cleaned. Therefore, it is important to apply flux immediately after cleaning copper fittings and tubing. Flux should be applied to the clean metal with a brush or swab, never with your fingers. Not only is there a chance of causing infection if you have a cut, but there is also a chance that flux could be carried to the eyes or mouth. In addition, body contact with the cleaned fittings and tubing adds to unwanted contamination of the metal.

Wetting: A process that reduces the surface tension so that molten (liquid) solder flows evenly throughout the joint.

Figure 8 Soldering flux and a flux brush.

CAUTION

Brazing flux and soldering flux are not the same products. Do not allow these fluxes to become mixed or interchanged. Carelessness can ruin the soldering or brazing task.

Flux must be stirred before each use. If a can of flux is not closed immediately after use, or if it sits unused for a long time, the chlorides separate from the petroleum base.

1.2.0 Preparing Tubing and Fittings for Soldering

To prepare tubing and fittings for soldering, the tubing must be measured, cut, and reamed, and the tubing and fittings must be cleaned. It is critical to use proper cleaning techniques to produce a solid, leakproof joint. Use the following procedure to prepare the tubing and fittings for soldering:

Step 1 If using the face-to-face method, measure the distance between the faces of the two fittings (*Figure 9*).

Step 2 Determine the **cup depth** engagement of each of the fittings. The cup depth engagement is the distance that the tubing penetrates the fitting. This distance can be found by measuring the fitting or by using a manufacturer's makeup chart. *Table 3* shows an example of a fitting makeup chart. Note that the tube size is based on inside diameter (ID), not outside diameter (OD) as is used for Type ACR copper. For $1\frac{1}{8}$" ACR tubing, for example, the nominal ID is 1".

Step 3 Add the cup depth engagement of both fittings to the measurement found in *Step 1* to find the length of tubing needed.

Step 4 Cut the copper tube to the correct length using a tubing cutter.

Step 5 Ream the inside of both ends of the copper tube using a reamer.

Step 6 Clean the tubing and fitting to a bright finish using No. 00 steel wool, an abrasive pad, an emery cloth, or a tubing brush (*Figure 10* [A–C]). Examine the fitting and tube to ensure they are well cleaned, and no significant flaws exist in the finish. Flaws in the tubing can usually be seen more easily after cleaning since oxidation may remain in a dent or gouge. The tube should be cleaned back from the end slightly farther than the cup depth of the fitting.

Step 7 Stir the flux before applying it. Using a brush or swab, apply flux to the copper tubing and to the inside of the copper fitting socket immediately after cleaning the parts (*Figure 10* [D–E]).

Step 8 Insert the tube into the socket, while turning the tube, until the tube touches the inside shoulder of the fitting.

Step 9 Wipe away any excess flux from the joint (*Figure 10* [F]).

Step 10 Check the tube and fitting for proper alignment before soldering.

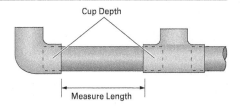

Figure 9 Face-to-face method of measuring cut length.

Cup depth: The distance that a tube inserts into a fitting, usually determined by a stop inside the fitting.

NOTE

The pipe preparation instructions in *Steps 6–9* are illustrated in *Figure 10*.

CAUTION

When cleaning copper tubing and fittings, take care to remove all the abrasions on the copper without removing a large amount of metal. Abrasions can weaken or ruin a copper joint. Do not handle the cleaned area with your bare hands because the oils in your fingers will contaminate the freshly cleaned metal.

TABLE 3 Fitting Makeup Chart

Tube Size	Depth of Cup	Tube Size	Depth of Cup
$\frac{1}{4}$"	$\frac{5}{16}$"	2"	$1\frac{11}{32}$"
$\frac{3}{8}$"	$\frac{3}{8}$"	$2\frac{1}{2}$"	$1\frac{15}{32}$"
$\frac{1}{2}$"	$\frac{1}{2}$"	3"	$1\frac{21}{32}$"
$\frac{5}{8}$"	$\frac{5}{8}$"	$3\frac{1}{2}$"	$1\frac{29}{32}$"
$\frac{3}{4}$"	$\frac{3}{4}$"	4"	$2\frac{5}{32}$"
1"	$\frac{29}{32}$"	5"	$2\frac{21}{32}$"
$1\frac{1}{4}$"	$\frac{31}{32}$"	6"	$3\frac{3}{32}$"
$1\frac{1}{2}$"	$1\frac{3}{32}$"	—	—

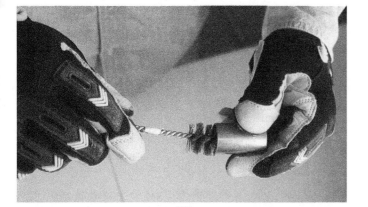

(A)

(B)

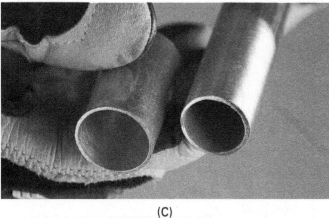

(C)

(D)

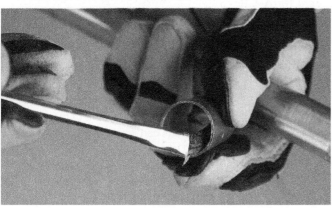

(E)

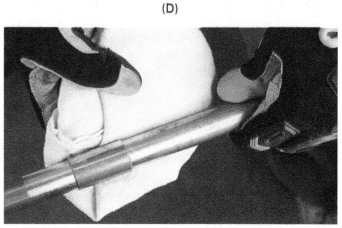

(F)

Figure 10 Preparing copper for soldering.

Acetylene

Acetylene has a lower heat content per cubic foot than any of the other fuel gases except methane or natural gas. Butane, propane, MAPP gases, and propylene all have twice as much heat content per cubic foot. This high-heat content allows much lower fuel consumption for the same amount of heating from the other gases. However, the major advantage of a properly adjusted air-acetylene torch is that higher combustion temperatures can be reached using acetylene than with any of the other gases.

1.3.0   Soldering Joints

With the tubing and fitting properly prepared, the soldering process can begin. Joints that have been prepared should be soldered within a few hours. Of course, it is best to have the soldering equipment prepared and ready so that the soldering can be done as soon as the joint has been prepared. Prepared joints should not be left for long periods of time or overnight, for example.

It is very important to ensure the tubing and fittings to be soldered are well supported. Once the heating process begins and the copper begins to expand, joints tend to sag. Remember that excess heat can damage certain components, such as valves with seals and other heat-sensitive materials. If a heat-sensitive component cannot be temporarily removed, then it should be protected by wrapping it with a very wet cloth. Heat absorbing pastes can also be used to protect components from excess heat.

Heat Sink Products

Overheating air conditioning and refrigeration piping components when soldering or brazing has always been an issue. Many components that must be soldered or brazed into the piping system, such as reversing valves and thermostatic expansion valves, are also sensitive to the extreme heat applied during the process. Some components can be disassembled before the installation, leaving only the metal body exposed to the heat. However, this is not always possible.

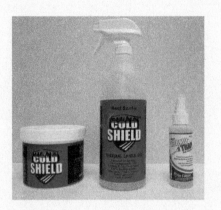

Heat sinks are materials that have a high capacity for absorbing heat. A variety of products are available to help absorb the heat that would otherwise be transferred to the component. Rather than bathe the component in the paste or gel, apply it in a ring around the tubing between the heat source and the component. If the component itself is being brazed, try to apply it on the component body away from the joint. The product should be applied with good contact and without leaving any voids where the heat can simply pass it by.

WARNING!

Always solder in a well-ventilated area because fumes from the flux can irritate your eyes, nose, throat, and lungs. Wear a respirator if required.

Step 1   Prepare the work area and protect any components in the piping system that can be damaged by the soldering process. Remove any loose, flammable materials in the area and ensure there is adequate room for physical movement around the joint. Place an appropriate fire extinguisher within easy reach.

WARNING!

Whenever soldering or brazing with an open flame, always keep a fire extinguisher of the correct type nearby. Shield any combustible materials near the work area with a fire blanket placed a safe distance away from the material. Many fires are started accidentally when using a torch through the ignition of loose materials or wood.

Step 2   Obtain either an acetylene tank and the related equipment (*Figure 11*) or a propane bottle and torch assembly. Also obtain the required solder. The instructions provided here are related to an air-acetylene torch, but the equipment manufacturer's instructions should always take precedence. For all fuel gases, follow the manufacturer's instructions carefully.

> **WARNING!**
>
> Acetylene is a relatively unstable gas. It can become volatile when released from the storage tank at pressures exceeding 15 psig or if released at a rate greater than $\frac{1}{10}$ of the tank's contents per hour. Always keep the cylinder in an upright position and do not exceed the recommended usage rate. Open the cylinder valve only one half to one turn and leave the valve wrench in place on the tank at all times. Failure to follow the proper procedures can cause a fire or explosion resulting in serious personal injury.

Step 3 Purge (clear out) the acetylene hose.

- Crack the acetylene cylinder valve open slowly until a small amount of pressure registers on the acetylene high-pressure gauge. Then open it about one half turn.

- Turn the acetylene regulator adjusting screw clockwise until a small amount of pressure shows on the acetylene working-pressure gauge. Allow a small volume of acetylene through to purge the hose, clearing it of any loose debris.

- Turn the acetylene regulator adjusting screw counterclockwise until it is loose. This shuts off the regulator output.

Step 4 Install the brazing torch handle on the ends of the hoses and close the valves on the torch.

> **WARNING!**
>
> Never adjust the acetylene regulator higher than 15 psig because acetylene becomes volatile above this pressure.

Step 5 Check the equipment for leaks.

- Adjust the acetylene regulator adjusting screw for 10 psig on the working-pressure gauge.

- Close the acetylene cylinder valve and check for leaks. If the working-pressure gauge remains at 10 psig for several minutes, there should be no significant leaks in the system. If the reading drops, there is a leak. Use a soap solution or commercial leak detection fluid to check the acetylene connections for leaks. Never use hoses that are in questionable condition, dried or cracked, or appear to have a leak. Such hoses should be destroyed and replaced.

> **WARNING!**
>
> Before lighting the torch, be sure to don gloves and goggles. Point the torch away from your body when lighting it. Use only an approved lighter such as those shown in *Figure 6* to light the torch unless it is a self-igniting model. Do not use a match, pocket cigarette lighter, or smoking materials to light the torch, because this could result in severe burns or cause the lighter to explode.

Step 6 Light the heating equipment according to the type of equipment you are using and the manufacturer's instructions.

- Open the torch handle valve and then set the acetylene regulator to the required pressure while the gas is flowing. Under no circumstances should the outlet pressure exceed 15 psig.

- Using an appropriate device for ignition, light the torch.

- Note that the torch handle valve is not generally used to control the size of the flame. It should be opened fully. The size of the flame is controlled by the torch size and the acetylene pressure regulator setting.

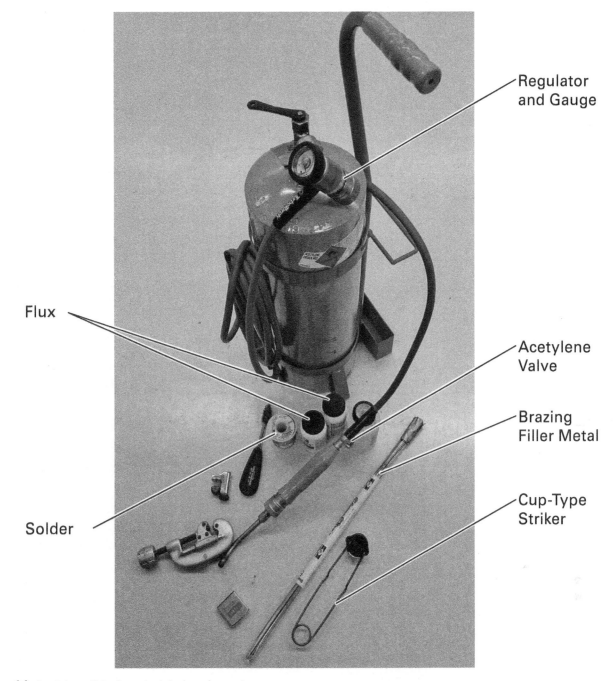

Figure 11 Acetylene B tank and related equipment.

Step 7 Heat the tubing first, with the flame perpendicular (at a right angle) to the tube. Then move the flame onto the fitting cup (*Figure 12*). Make sure not to overheat the joint. The size of the tubing determines how much time this step requires.

Step 8 Momentarily move the flame back to the tubing, and then to the back shoulder of the cup. As the flame is moved back to the cup, also back the flame away some to avoid overheating.

Step 9 Touch the end of the solder to the area between the fitting and the tube. The solder will be drawn into the joint by capillary action. The solder can be fed upward or downward into the joint. If the solder does not melt quickly on contact with the joint, move the solder away and heat the joint further. Do not melt the solder with the flame; the solder must be melted by the copper.

CAUTION

The inner cone of the flame should barely touch the metal being heated. Do not direct the flame into the socket because this will burn the flux. Be sure to keep the flame moving instead of holding the flame in one place.

Step 10 Continue to feed the solder into the joint until a ring of solder appears around the joint, indicating that the joint is filled. On nominal $\frac{3}{4}$" diameter tubing and smaller, the solder can generally be fed into the joint from one point. On larger tubing, the solder should be fed around the entire circumference of the joint.

Step 11 Allow the joint to cool until the solder solidifies. While the joint is still warm but not hot, wipe it with a clean damp cloth to remove any excess flux (*Figure 13*). Shocking the joint with water while it is very hot may cause unnecessary stress and eventual failure.

The amount of solder generally needed is equal to the diameter of the tubing for sizes 1" and below. For example, with $\frac{3}{4}$" tubing, $\frac{3}{4}$" of solder wire should properly fill the joint. As the size increases above 1", more solder than this is necessary. For example, a 2" joint with an average joint clearance of 0.005" will require about $3\frac{1}{2}$" of solder. However, in all cases, the joint clearance significantly affects the amount of solder that is needed.

Figure 12 Applying heat to the fitting.

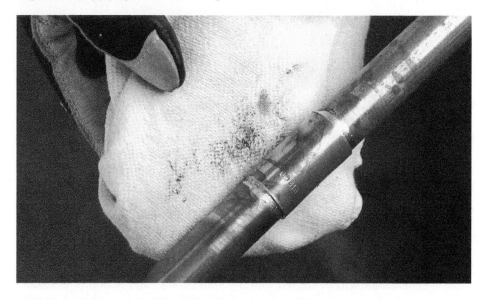

Figure 13 Wiping away excess flux.

If the joint is being done with the tubing horizontal, the top of the joint will likely heat up on its own and will not need preheating. Concentrate on the bottom and sides to avoid burning the flux at the top. Start feeding the solder near the bottom of the joint and work up towards the top. Then bring the solder back to the starting point and work up the other side toward the top. Remember that, although gravity does affect the solder, it is capillary action rather than gravity that pulls the solder into the joint.

Soldering is an art that requires practice. Watching how the solder and flux behave during practice provides the necessary information to move and position the torch flame. With practice, soldering can be mastered by anyone.

Cup-Type Striker

When using a cup-type striker to ignite a torch, hold the cup of the striker slightly below and to the side of the tip, parallel with the fuel gas stream. This helps prevent the ignited gas from deflecting back toward you from the cup. It also reduces the amount of soot that accumulates in the cup. The flint in most cup-type strikers can be replaced when it is worn out.

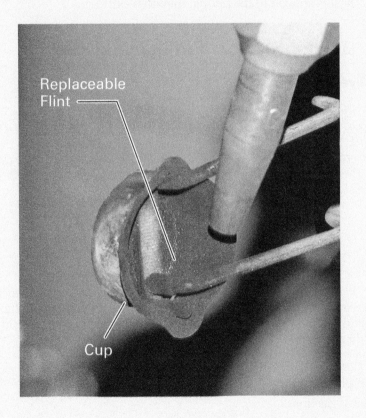

Copper Fittings

Two kinds of solder fittings are available with copper tubing. The first is a wrought copper fitting, which is made from copper tubing that is shaped into different types of fittings. Wrought copper fittings are generally lightweight, smooth on the outside, and have thin walls.

The second type is a cast fitting. This type of fitting is made using a mold. The first cast fittings had holes in the sockets to put solder in. Today, the heated copper is poured into the mold and allowed to cool. Cast copper fittings have a rougher exterior and come in a wide variety of shapes. They are typically heavier than wrought copper fittings. Wrought copper fittings are the type used in HVACR work.

1.0.0 Section Review

1. Clothing made of synthetic fabrics should be worn whenever soldering or brazing.
   a. True
   b. False

2. When applying soldering flux, _____.
   a. brazing flux can be used if no soldering flux is available
   b. the flux is applied only to the tube end using a brush
   c. the flux is applied to the tube end and fitting socket using a brush
   d. the flux can be applied using the fingers or a brush

3. When soldering a joint, _____.
   a. heat is applied directly to the solder
   b. only the tubing is heated
   c. only the fitting is heated
   d. heat is applied to the tubing, then to the fitting

2.0.0 Brazing Copper Fittings and Tubing

Performance Tasks

3. Properly set up and shut down oxyacetylene equipment.

4. Properly prep and safely braze copper tubing in various planes, using various fittings, with a nitrogen purge.

5. Braze copper tubing to either steel or brass components.

Objective

Describe and demonstrate the safe process of brazing copper tubing.

a. Describe and demonstrate the use of the PPE, tools, and materials needed to braze copper tubing.

b. Describe and demonstrate the preparation used for brazing.

c. Describe and demonstrate the brazing process.

d. Describe and demonstrate the process of brazing copper tubing to dissimilar metals.

Brazing, like soldering, uses nonferrous filler metals to join base metals that have a melting point above that of the filler metals. Brazing uses filler metals that melt at temperatures above 842°F (450°C). Brazed tubing and fittings are used in the following:

- Low-pressure steam lines
- High-pressure refrigeration lines
- Medical gas lines
- Compressed air lines
- Vacuum lines
- Fuel lines
- Other chemical lines that need extra corrosion resistance in the piping joints

Brazing, sometimes incorrectly referred to as *hard soldering*, produces mechanically strong, pressure-resistant joints. The strength of a brazed joint results from the ability of the filler metal to adhere to the base metal. However, reliable adhesion can occur only if the base metals are properly cleaned, the proper flux and filler metal are selected, and the clearance gap between the outside of the tubing and the inside of the fitting is appropriate. Ideally, the clearance is approximately 0.003" to 0.005" (0.08 mm to 0.13 mm).

2.1.0 PPE, Tools, and Materials Used for Brazing

To properly braze copper tubing and fittings, it is necessary to understand the following:

- Required PPE and safety practices
- Brazing equipment and accessories
- Filler metals and fluxes
- Tube and fitting preparations
- Setting up and lighting a torch
- Nitrogen purging equipment and use
- Brazing techniques

2.1.1 Brazing Safety

Brazing is often done with a combination of oxygen and acetylene. Although the hazards are similar to those encountered in soldering, they are compounded by the use of two flammable and explosive gases. In addition, when fuel gases are directly combined with oxygen instead of air, the temperature of the flame rises significantly. When brazing, the following safety guidelines apply:

- Wear clothing made of nonflammable fabric such as cotton. Avoid synthetic fabrics such as nylon, which can melt and adhere to the skin.
- Wear a long-sleeve shirt to prevent burns from sparks or molten metal. Pants should not have cuffs.
- Wear eye and face protection.
- Wear flame-resistant gloves and high-top work boots.
- Make sure the gas bottle is securely turned off when not in use. Gases used in brazing are flammable and explosive if exposed to a spark or flame.
- Keep a fire extinguisher handy while brazing in case material near the work ignites. When working close to flammable material such as wood, use a fire blanket or other flame/heat-blocking material to protect it.
- Always braze in a well-ventilated area because fumes from the process can irritate your eyes, nose, throat, and lungs. Wear respiratory equipment if necessary.

There are specific safety concerns associated with the use of oxyacetylene equipment that must also be considered:

- Acetylene cylinders must be stored, transported, and used in the upright position. Otherwise, liquid acetone could be drawn from the tank.
- Cylinders must be secured to a cart or structure using a safety chain or approved strap.
- When not in use, cylinders must be stored in the upright position with a safety cap installed (*Figure 14*). Note that some small acetylene cylinders, such as the MC tank, cannot be capped. When in transit, the regulator must be removed from all cylinders and the safety cap securely installed. In many states, it is against the law to transport cylinders with the regulators in place.
- Oxygen and fuel gas cylinders must be stored at least 20' (6.1 m) apart or separated by a wall at least 5' (1.5 m) high with a 30-minute minimum fire rating.
- Oxygen can cause ignition even when there is no obvious flame or spark around to set it off, especially when it comes in contact with oil or grease. Never handle oxygen cylinders with oily hands or gloves. Keep grease away from the cylinders and do not use oil or grease on cylinder attachments or valves. Never use an oxygen regulator for any other gas or try to use a regulator with oxygen that has been used with other gases.

Figure 14 Acetylene cylinder with a valve safety cap.

- A pressure-reducing regulator set for no more than 15 psig must be used with acetylene. Acetylene becomes unstable and volatile above 15 psig. An acetylene cylinder that has been laid down should be placed in the upright position for at least 20 minutes before using.

- Do not allow anyone to stand in front of the oxygen cylinder valve when opening because the oxygen is under high pressure (about 2,000 psig) and could cause severe injury when released.

- Acetylene cylinder valves should only be opened one half to one turn, and the wrench should never be removed from the valve during use. The valve on oxygen cylinders, however, should always be opened fully.

2.1.2 Brazing Equipment

Brazing is typically done using either an acetylene B-tank with an air-acetylene torch or an oxyacetylene setup (*Figure 15*). A complete B-tank setup was previously shown in *Figure 11*. The air-acetylene setup is generally used on smaller refrigerant lines, since air-acetylene produces less heat than the oxyacetylene equipment. For lines greater than 1" in diameter, installers usually select the oxyacetylene method, which provides both greater heat and better control over the movement of the filler metal. Oxyacetylene equipment mixes acetylene, as a fuel gas, with pure oxygen to produce a hotter flame.

The equipment shown in *Figure 15* is usually mounted on a hand truck and is very heavy. This type of equipment is typically used in a shop or on a jobsite when extensive brazing must be accomplished in a limited area. For small installations and service-related brazing, portable equipment that can be hand-carried by one person is generally chosen (*Figure 16*).

Transporting Fuel Gas Cylinders

Although OSHA has specific standards for the handling of fuel gas cylinders, found in *OSHA Standard 1910.253*, the Department of Transportation has primary jurisdiction over the cylinders themselves and their transportation requirements. These requirements are contained in the Code of Federal Regulations, *49 CFR, Parts 171 through 177*.

Oxygen and acetylene are compressed and shipped at medium to high pressures in cylinders. Oxygen is supplied in cylinders at pressures of about 2,000 psig. Acetylene cylinders are pressurized to about 250 psig. These cylinders must not be moved unless the protective caps are in place. Dropping a cylinder without the cap installed may result in breaking the valve off the cylinder. This allows the pressure inside to escape, propelling the cylinder like a rocket.

The oxygen and acetylene tanks require regulators to be installed before they can be used. The regulator reduces the high pressure in the tank to the desired pressure for use. Regulators (*Figure 17*) must be selected for the correct gas. The connecting fittings are typically designed to prevent a regulator for one gas from being installed on the wrong gas tank. Different tank sizes of the same gas may also have different regulator connections. Regulators must be handled carefully; avoid dropping or otherwise shocking them.

WARNING!

A regulator should always be installed on a tank before the valve is opened. The high pressure inside the tank, if released suddenly, can cause the tank to become a dangerous projectile. Although it is a generally accepted practice to slightly open (crack) the valve of a cylinder to blow out any debris before attaching a regulator, it must be done with great care to ensure the cylinder does not tip over as a reaction to the release of gas.

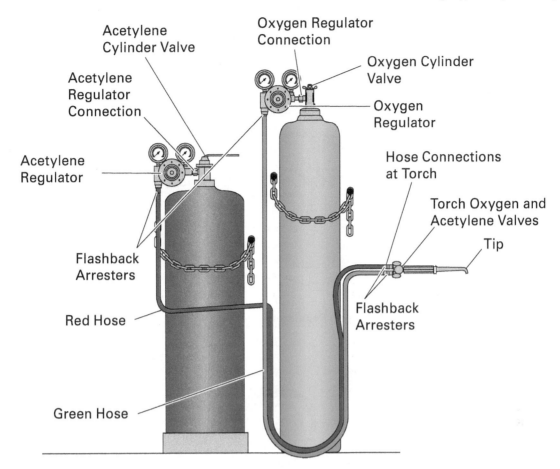

Acetylene
Cylinder Valve

Oxygen Regulator
Connection

Acetylene
Regulator
Connection

Oxygen Cylinder
Valve

Oxygen
Regulator

Acetylene
Regulator

Hose Connections
at Torch

Torch Oxygen and
Acetylene Valves

Tip

Flashback
Arresters

Red Hose

Green Hose

Flashback
Arresters

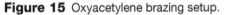

Figure 15 Oxyacetylene brazing setup.

Many regulators have two stages of pressure regulation. The first stage reduces the tank pressure to an intermediate level and has a fixed pressure setpoint. The second stage of the regulator reduces the intermediate pressure to the outlet pressure desired by the user and is adjustable from 0 psig to the maximum outlet pressure specified by the manufacturer. The final adjustment of gas pressure at the nozzle is set using the valves on the torch handle.

A regulator usually has two pressure gauges. One shows the pressure in the cylinder, while the other shows the outlet pressure setpoint. The outlet gauge is monitored as the technician sets the desired outlet pressure. The regulator is adjusted to a higher pressure by turning the adjusting screw clockwise.

Some simpler or cheaper regulators, such as those used on small acetylene tanks, provide only one stage of regulation. A single-stage regulator needs adjustment more often, as the outlet pressure tends to fall as the pressure in the bottle changes. Two-stage regulators generally prevent this from happening. Small single-stage acetylene regulators often do not show the pressure in the cylinder, but instead have a less accurate gauge that shows the volume that remains.

Hoses designed specifically for brazing and welding work must be used to connect the regulator outlets to the torch handles. The hoses are color coded for visual identification. In the United States, the oxygen hose is green, and the fuel gas hose is red. Other countries may have different standards. This is an important feature, as the hose fittings are designed so that they can only be connected to the proper gas regulator and torch handle port. Oxygen hoses have a right-handed thread, while fuel gas hoses use a left-handed thread. The female fittings with left-handed threads have a groove cut into the nut for further visual identification. New hoses should be purged with high-pressure air before use, as they often contain a white powder.

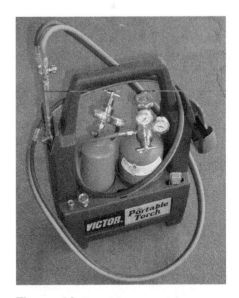

Figure 16 Portable oxyacetylene equipment.

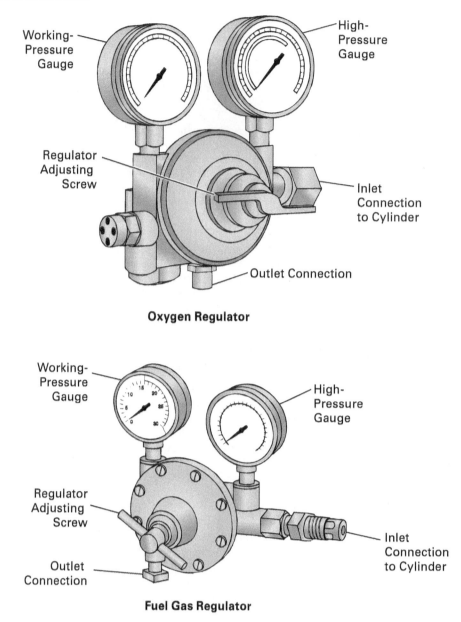

Figure 17 Oxygen and fuel gas regulators.

Flashback arrestors: Valves that prevent a flame from traveling back from the tip and into the hoses.

Figure 18 Torch wrench.

Figure 18 shows a universal torch wrench designed for regulators, hose connections, check valves, **flashback arrestors**, torches, and torch tips. The fittings for torch equipment are usually brass or bronze and certain components are often fitted with soft, flexible, O-ring seals. The seal surfaces of the fittings or O-rings can be easily damaged by overtightening with standard mechanics wrenches. The length of a torch wrench is limited to reduce the chances of damage to fittings from application of excessive torque. In some cases, manufacturers specify only hand-tightening certain fitting connections of a torch set (tips or cutting/welding attachments). In any event, follow the manufacturer's specific instructions when connecting the components of a torch set. The torch wrench can be tethered to the torch outfit so that it is always available when needed and not easily misplaced.

Oxyacetylene brazing equipment offers a wide array of torch handles and tips from which to choose. The style and size of the tip is determined by the size of the tubing and, to a lesser degree, the type of joint to be made. Brazing torch handles are offered in heavy-, medium-, and light-duty models. For most HVACR brazing applications, medium- and light-duty handles are sufficient and generally preferred since they are more maneuverable (*Figure 19*). The handle typically provides a mixing chamber for the two fuel gases before they reach the nozzle.

Note that the two handles shown in *Figure 19* offer different configurations that can make a difference to technicians. The medium-duty handle places the fuel gas adjusting valves below the technician's hand. With the light-duty handle, the adjusting valves are placed above the hand. This affects the balance of the torch to some degree. Having the adjusting valves above the hand can be more convenient, but they can also get in the way in tight spots.

Many torch handles today have built-in *flashback arrestors* (*Figure 20*). Flashback arrestors are one-way valves that prevent a pressurized flame from traveling back up the tip and into the hoses and regulators. A flashback can sometimes happen if either the oxygen or the fuel gas flow rate is inadequate, or if both the oxygen and fuel gas are turned on and then ignited at the brazing tip. If a torch handle is not already equipped with them, flashback arrestors should be installed where the hoses connect to the torch handle. Although regulators can be equipped with them at their outlet, torch-mounted models prevent a flashback from entering the hoses. To be effective, flashback arrestors must be installed with the flow arrow pointing toward the torch handle.

It is very important to note that the brazing tip must be designed to properly fit the handle. Most brazing tips from a given manufacturer are designed to fit only that manufacturer's handles. Further, tips do not fit every handle that a manufacturer offers. When selecting and purchasing tips, always be sure that the tips are compatible with the torch handle in use.

Several different brazing tips are shown in *Figure 21*. Brazing nozzle and tip sizes are indicated by a number corresponding to the size of the nozzle opening. Larger numbers indicate larger openings. *Table 4* provides a guide for the selection of tips based on the size of the tubing to be brazed. Much smaller tips are available but are not typically used for HVACR work.

TABLE 4 Tip Sizes Used for Common Tubing Sizes

Tip Size (No.)	Rod Size	Tubing Diameter
4	$3/32$"	$1/4$" to $3/8$"
5	$1/8$"	$1/2$" to $3/4$"
6	$3/16$"	1" to $1\,1/4$"
7	$1/4$"	$1\,1/2$" to 2"
8	$3/8$"	2" to $2\,1/2$"
9	$3/8$"	3" to $3\,1/2$"
10	$7/16$"	4" to 6"

The multi-flame, or rosebud, torch tip is used for larger tubing and to heat larger surfaces. Instead of a single opening in the nozzle, several smaller openings are provided. The multi-flame tip is an excellent choice for larger tubing and fittings, to broaden the area of the flame. The universal tip accepts different nozzles so that one tip can be used with many different nozzle sizes.

The brazing tip does occasionally get contaminated with solids from the combustion process. This is especially true when the torch is not set up to burn cleanly. The relatively soft brass can also be damaged at the tip, causing the flame to be erratic. Torch tip cleaners (*Figure 22*) are used to clean the tip. Tip cleaners are like small, round files that clear the material out of the way. Each tip cleaner fits a specific nozzle size. As shown here, tip cleaners typically come as a set in folding enclosures for protection. Tip drills are used for major cleaning and for holes that are plugged. Tip drills are tiny drill bits that are sized to match the diameters of the tip openings. A drill bit is mounted in a small handle for manual use.

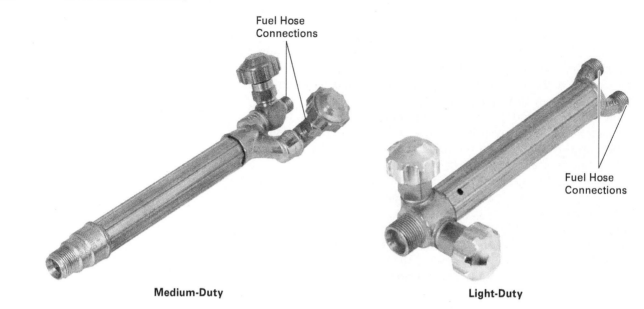

Fuel Hose
Connections

Fuel Hose
Connections

Medium-Duty

Light-Duty

Figure 19 Medium- and light-duty brazing torch handles.

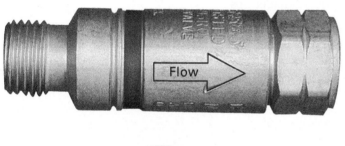

Flow

Figure 20 Flashback arrestor.

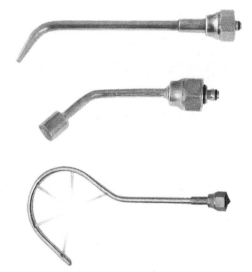

Figure 21 Various brazing tips.
Source: Courtesy of Uniweld Products, Inc.

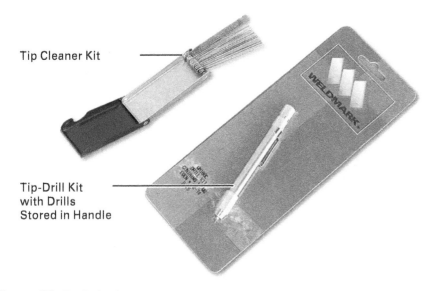

Tip Cleaner Kit

Tip-Drill Kit
with Drills
Stored in Handle

Figure 22 Torch tip cleaners.

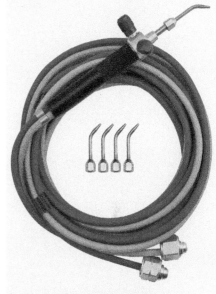

Source: © Miller Electric Mfg. LLC

2.1.3 Filler Metals and Fluxes

Filler metals used to join copper tubing are of two groups: alloys that contain 8 to 60 percent silver (the BAg series), and copper alloys that contain phosphorus (the BCuP series). *Table 5* lists brazing filler materials according to their American Welding Society (AWS) classification and principal elements. *Figure 23* shows a package of BCuP-5 brazing filler rods that contain 15 percent silver. They have been an industry standard for many years, although other alloys have dramatically increased in popularity as silver prices have increased. Unlike soldering filler metals, brazing filler metals are typically in the form of a stick.

TABLE 5 Brazing Filler Metals

Percent of Principal Element						
AWS Classification	Silver	Phosphorus	Zinc	Cadmium	Tin	Copper Classification
BCuP-2	—	7–7.5	—	—	—	Balance
BCuP-3	4.75–5.25	5.75–6.25	—	—	—	Balance
BCuP-4	5.75–6.25	7–7.5	—	—	—	Balance
BCuP-5	14.5–15.5	4.75–5.25	—	—	—	Balance
BAg-1	44–46	—	14–18	23–25	—	14–16
BAg-2	34–36	—	19–23	17–19	—	25–27
BAg-5	44–46	—	23–27	—	—	29–31
BAg-7	55–57	—	15–19	—	4.5–5.5	21–23

Figure 23 BCuP-5 brazing rods comprised of 15 percent silver, 80 percent copper, and 5 percent phosphorus.
Source: The Harris Products Group

> **WARNING!**

BAg-1 and BAg-2 contain cadmium. Heating when brazing can produce highly toxic fumes. Use adequate ventilation and avoid breathing the fumes.

The two major groups of filler metals differ in their melting, fluxing, and flowing characteristics. These characteristics should be considered when selecting a filler metal. When joining copper tubing, any of the filler metals in *Table 5* can be used. However, the filler metals used most often for close tolerances and reasonable economy are BCuP-3 and BCuP-4. BCuP-5 is better where close tolerances cannot be held in the joint, but it is generally a more expensive product.

As with soldering, fluxes are also used in brazing. Brazing fluxes must have significantly different properties due to the temperatures involved. However, a flux is not always needed in brazing. Alloys that contain phosphorus, such as BCuP-3 through BCuP-5, are self-fluxing when used in clean copper-to-copper applications. When they are used with brass, bronze, or ferrous metals, a flux is required.

Brazing fluxes (*Figure 24*) are applied using the same methods as soldering fluxes. However, high-temperature fluxes used in brazing are usually applied only to the end of the tube and not on the inside of the fitting. Brazing fluxes are more corrosive than soldering fluxes, so care must be taken to select the correct flux. The two types must never be mixed. For best results, use the flux recommended by the manufacturer of the brazing filler metals in use.

Since brazing temperatures are significantly higher than soldering temperatures, oxides form even more rapidly. Without the proper use of flux when needed, the brazed joint will not reach an acceptable level of quality or strength. Brazed joints often must be disassembled in the field for service purposes. Whenever a hermetic compressor is replaced, for example, a number of brazed joints must be taken apart. Flux should be applied to these joints as well, before they are heated. Apply the flux on top of the visible filler metal and on the adjacent areas of the tube and fitting.

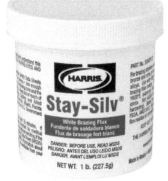

Figure 24 A high-performance brazing flux used for both nonferrous and ferrous metals.
Source: The Harris Products Group

2.2.0 Preparing for Brazing

To prepare tubing and fittings for brazing, you must follow the same procedures as you would to prepare tubing and fittings for soldering. It is critical that proper cutting and cleaning techniques be used to produce a reliable joint. Use the following procedure to prepare the tubing and fittings for brazing:

Step 1 Measure the distance between the faces of the two fittings.

Step 2 Determine the cup depth engagement of each of the fittings.

Step 3 Add the cup depth engagement of both fittings to the measurement found in *Step 1* to find the length of tubing needed.

Step 4 Cut the copper tubing to the correct length using a tubing cutter.

Step 5 Ream the inside and outside of both ends of the copper tubing using a reamer.

Step 6 Clean the tubing and fitting using No. 00 steel wool, a piece of emery cloth, an abrasive pad, or a special copper-cleaning tool.

Step 7 Apply flux (if required) to the end of the copper tubing before inserting it into the fitting. Flux does not need to be applied to the fitting cup.

Step 8 Insert the tube into the fitting socket, and push and turn the tube into the socket until the tube touches the inside shoulder of the fitting.

Step 9 Wipe away any excess flux from the outside of the joint.

Step 10 Check the tube and fitting for proper alignment before brazing.

2.2.1 Brazing Equipment Setup

The brazing heating procedure differs from soldering in that different equipment may be required to raise the temperature of the metals to be joined well above 842°F (450°C). Actually, most brazed joints are made at temperatures between 1,200°F and 1,550°F (649°C and 843°C). This is the range of temperatures where the popular brazing alloys are fluid enough to flow, without becoming too fluid. The torch flame must be significantly higher than this to achieve good control over the heating process. Because of the higher temperatures needed, oxyacetylene equipment is typically used for brazing and is the primary approach covered here.

Oxygen and acetylene cylinders must be secured with a stout cable or chain in the upright position during use, transportation, and storage to prevent them from falling and injuring people or damaging equipment. When stored at the jobsite, oxygen and acetylene cylinders must be stored separately with at least 20' (6.1 m) between them, or with a 5' (1.5 m) high, half-hour minimum fire wall separating them. Cylinders should be stored without a regulator attached, and the regulator should be removed when a cylinder is moved. With a regulator attached, the chances of shearing off the cylinder valve stem are increased, since it protrudes well beyond the body of the cylinder. Store empty cylinders away from partially full or full cylinders and make sure they are properly marked to clearly show that they are empty (*Figure 25*).

CAUTION

Take care when cleaning the tubing and fittings to remove all the abrasions on the copper without removing a large amount of metal. Abrasions can weaken or ruin a copper joint. Do not touch or brush away filings from the tube or fitting with your fingers because your fingers will also contaminate the freshly cleaned metal. Filings and burrs can also easily cause an injury.

Figure 25 Typical empty cylinder marking.

Alternate High-Pressure Cylinder Valve Cap

High-pressure cylinders can also be equipped with a clamshell cap that can be closed to protect the cylinder valve with or without a regulator installed on the valve. This enables safe movement of the cylinder after the cylinder valve is closed. This type of cap is usually secured to the cylinder body cap threads when it is installed so that it cannot be removed. When the clamshell is closed, it can also be padlocked to prevent unauthorized operation of the cylinder valve.

Latch Pin
(or Padlock)

Clamshell Open to Allow
Cylinder Valve Operation

Clamshell Closed for
Movement or Padlocked
to Prevent Operation of
Cylinder Valve

Clamshell Closed
for Transport

Alternate Acetylene Cylinder Safety Cap

Acetylene cylinders can be equipped with a ring guard cap that protects the cylinder valve with or without a regulator installed on the valve. This enables safe movement of the cylinder after the cylinder valve is closed. This type of cap is usually secured to the cylinder body cap threads when it is installed so that it cannot be removed.

Follow this basic procedure to prepare oxyacetylene brazing equipment for use:

> **WARNING!**
>
> Do not handle acetylene and oxygen cylinders with oily hands or gloves. Keep grease away from the cylinders and do not use oil or grease on cylinder attachments or valves. The mixture of oil and oxygen could cause an explosion. Make sure that the protective caps are in place on the cylinders before transporting or storing the cylinders.

Step 1   Install and securely fasten the oxygen and acetylene cylinders in a bottle cart or in an upright position against a wall or other substantial structure.

> **WARNING!**
>
> Do not allow anyone to stand in front of the oxygen cylinder valve when opening (cracking) it, because the oxygen is under high pressure (about 2,000 psig) and could blow debris into the eyes or face. In addition, the bottle reaction could cause severe injury if the valve is opened too far. Ensure that the bottle is properly secured before cracking the valve open.

Step 2   Install the oxygen regulator on the oxygen cylinder.

- Remove the cylinder protective cap.
- Open (crack) the oxygen cylinder valve just long enough to allow a small amount of oxygen to pass through and blow out any debris, then close it.
- Carefully inspect the regulator attaching threads, cylinder threads, and mating surfaces for damage (*Figure 26*).
- Turn the adjusting screw on the oxygen regulator counterclockwise until it is loose. This shuts off the regulator output and prevents the application of excessive pressure to the hose and torch when the cylinder is opened.
- Using a torch wrench, install the oxygen regulator on the cylinder (*Figure 27*). Oxygen cylinders and regulators have right-hand threads. Tighten the nut snugly. Be careful not to overtighten the nut because this may strip the threads.

Check That Fittings are Clean

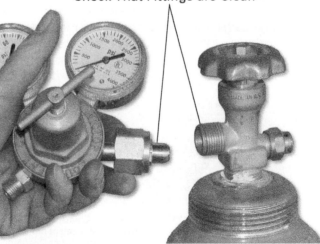

Figure 26 Inspecting the fittings.

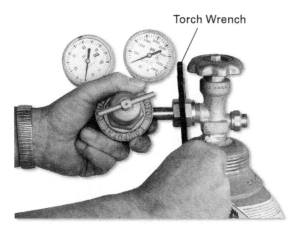

Torch Wrench

Figure 27 Installing the regulator.

WARNING!

Acetylene gas is highly flammable. Ensure there are no sources of ignition nearby when working with an acetylene cylinder.

Step 3 Install the acetylene regulator on the acetylene cylinder.

- Remove the cylinder's protective cap.
- Open (crack) the acetylene cylinder valve, using the valve wrench, just long enough to allow a small amount of acetylene to pass through the valve, then close it.
- Turn the adjusting screw on the acetylene regulator counterclockwise until it is loose. This shuts off the regulator output and prevent accidental over-pressurizing of the hose and torch during hookup.
- Using a torch wrench, install the acetylene regulator on the cylinder. Acetylene cylinders and regulators have left-hand threads. Tighten the nut snugly. Be careful not to overtighten the nut because this may strip the threads.

Step 4 Connect the hoses and brazing torch.

- If they are not already in place or a part of the torch handle, install flashback arrestors on the oxygen and acetylene hoses at the torch handle inlets.
- Connect the green hose to the oxygen regulator and the red hose to the acetylene regulator. Tighten the hoses snugly. Be careful not to overtighten the fittings because this may strip the threads.

WARNING!

Open the oxygen cylinder valve slowly. Ensure that you and others are not standing directly in front of or behind the oxygen regulator body when opening the cylinder valve in case the regulator fails suddenly.

Step 5 Purge (clean) the oxygen hose.

- Crack the oxygen cylinder valve slowly until the pressure stops rising on the cylinder gauge. This now indicates the pressure in the cylinder. Then slowly open the valve completely.
- Turn the oxygen regulator adjusting screw clockwise until a small amount of pressure (3–5 psig) shows on the oxygen working-pressure gauge. Allow a small amount of pressure to build up and purge the oxygen hose, clearing it of any loose debris.
- Turn the oxygen regulator adjusting screw counterclockwise until it is again loose. This will shut off the regulator output.

Step 6 Purge (clean) the acetylene hose.

- Crack the acetylene cylinder valve open slowly until the cylinder pressure registers on the acetylene cylinder, or high-pressure, gauge. Then open it about $\frac{1}{2}$ to $\frac{3}{4}$ of a turn.
- Turn the acetylene regulator adjusting screw clockwise until a small amount of pressure shows on the acetylene working-pressure gauge. Allow a small amount of pressure to build up and purge the acetylene hose, clearing it of any loose debris.
- Turn the acetylene regulator adjusting screw counterclockwise until it is again loose. This will shut off the regulator output.

Step 7 If not already in place, install a flashback arrestor on the acetylene inlet of the torch.

Step 8 Install the brazing torch on the ends of the hoses and close the valves on the torch. Remember that the oxygen hose has normal right-hand threads while the acetylene hose has left-hand threads.

WARNING!

Never adjust the acetylene regulator for a pressure higher than 15 psig because acetylene becomes unstable and volatile at this pressure.

Step 9 Check the oxyacetylene equipment for leaks.

- Adjust the acetylene regulator adjusting screw for 10 psig on the working-pressure gauge.
- Adjust the oxygen regulator adjusting screw for 20 psig on the working-pressure gauge.

WARNING!

Do not use an oil-based soap for leak testing because the mixture of oil and oxygen may cause an explosion.

- Close the oxygen and acetylene cylinder valves and check for leaks while the pressure remains in the hoses and torch handle. If the working-pressure gauges remain at 10 psig and 20 psig for several minutes, there should be no significant leaks in the system. If the readings drop, there is a leak. Use a soap solution or commercial leak detection fluid to check the oxygen or acetylene connections for leaks. Note that the leak could also be on the regulator assembly. If the regulator assembly is found to be leaking, stop using it immediately. Do not attempt to repair it in the field.
- Open both valves on the torch handle to release the pressure in the hoses. Watch the working-pressure gauges until they register zero, then close the valves on the torch handle.
- Turn the oxygen and acetylene regulator valves counterclockwise until they are loose. This releases the spring pressure on the regulator diaphragms and completely closes the regulators.

Step 10 Coil the hoses and hang them on the hose holder.

2.2.2 Lighting the Oxyacetylene Torch

After the oxyacetylene brazing equipment has been properly prepared, the torch can be lit, and the flame adjusted for brazing. Before beginning, it is important to know the correct oxygen and acetylene operating pressures for the tip in use. *Table 6* provides one example of a table with this data for a series of brazing tips. It is important that the information for the actual tip in use be consulted; this table should not be considered a general guide for all tips.

For hose lengths greater than 25', the pressure at the regulator is generally increased 2–3 psig to compensate for pressure drop. However, remember that the acetylene pressure should never exceed 15 psig while the gas is flowing.

There are three types of flames: neutral, carburizing (or reducing), and oxidizing. *Figure 28* shows the differences in the types of flames. The neutral flame burns equal amounts of oxygen and acetylene. The inner cone is bright blue in color, surrounded by a fainter blue outer flame envelope that results when the oxygen in the air combines with the superheated gases from the inner cone. A neutral flame is used for almost all fusion welding or heavy brazing applications.

A carburizing flame has a white feather created by excess fuel. The length of the feather depends on the amount of excess fuel in the flame. The outer flame

TABLE 6 Brazing Tip Operational Data

Tip #	Acetylene Consumption Rate (SCFH)	Acetylene Pressure (psig)	Oxygen Pressure (psig)	Welding — Metal Thickness	Brazing — Copper Tubing Nominal Size
TOA-0	1–2	5	5	$\frac{1}{32}$"	$\frac{1}{8}$"
TOA-1	2–3	5	5	$\frac{3}{64}$"	$\frac{1}{4}$"
TOA-2	2–4	5	5	$\frac{5}{64}$"	$\frac{1}{2}$"
TOA-3	3–7	5	5	$\frac{3}{32}$"	$\frac{3}{4}$"
TOA-4	5–10	5	5	$\frac{1}{8}$"	1"
TOA-5	8–18	6	7	$\frac{3}{16}$"	$1\frac{1}{2}$"
TOA-6	11–25	7	10	$\frac{1}{4}$"	2"
TOA-7	15–35	8	12	$\frac{1}{4}$" to $\frac{1}{2}$"	3"
TOA-8	25–45	9	14	$\frac{1}{4}$" to $\frac{3}{4}$"	4"
TOA-9	30–60	10	16	$\frac{3}{4}$" to $1\frac{1}{4}$"	6"

Note: This data is an example only and does not reflect the performance of any specific product. Always refer to the product literature for the specific equipment in use.

Carburizing Flame

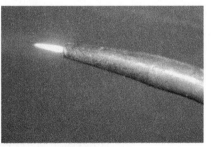

Neutral Flame

Oxidizing Flame

Figure 28 Types of flames.

envelope is brighter than that of a neutral flame and is much lighter in color. The excess fuel in the carburizing flame produces large amounts of carbon. The carburizing flame is cooler than the neutral flame and is used for light brazing to prevent melting of the base metal.

An oxidizing flame has an excess amount of oxygen. Its inner cone is shorter, with a bright blue edge and a lighter center. The cone is also more pointed than the cone of a neutral flame. The outer flame envelope is very short and often fans out at the ends. The hottest flame, it is sometimes used for brazing cast iron or other metals.

Use the following procedure to light an oxyacetylene torch.

Step 1   Set up the oxyacetylene torch as discussed previously, heeding all warnings. Make sure the desired tip is installed before lighting the torch (refer back to *Table 4* for recommended tip sizes). Be aware of the needed gas pressure settings for the tip in use.

Step 2   Adjust the torch oxygen flow.

- Open the oxygen cylinder valve slightly until pressure registers on the oxygen cylinder gauge and then stops rising; then open the valve fully.
- Turn the oxygen regulator adjusting screw clockwise until pressure shows on the oxygen working-pressure gauge. Adjust the regulator for the manufacturer's recommended pressure setting.
- Open the oxygen valve on the torch handle.
- Check the regulator and ensure that the pressure remains at the desired setting. Adjust as needed.
- Close the oxygen valve on the torch handle.

Step 3   Adjust the torch acetylene flow.

- Open the acetylene cylinder valve slightly until the pressure registers and stabilizes on the acetylene high-pressure gauge; then open the valve about one half turn.
- Turn the acetylene regulator adjusting screw clockwise until the desired pressure shows on the acetylene working-pressure gauge.
- Open the acetylene valve on the torch handle.
- Check the regulator and ensure that the pressure remains at the desired setting. Adjust as needed.
- First close, and then open the torch acetylene valve about one half turn to prepare for ignition.

> **NOTE**
>
> Always fine-tune the pressure setting with the torch valve open. When it is closed, the pressure may register slightly higher. Regulators are incapable of making pressure adjustments without gas actively flowing through.

> **CAUTION**
>
> Be sure to leave the valve wrench on the acetylene cylinder valve while the cylinder is in use so that the valve can be closed quickly in case of an emergency.

> **WARNING!**
>
> When lighting the torch, be sure to wear flame-resistant gloves and goggles. Use an appropriate device, such as a cup-type striker, to light the torch. Point the torch away from you. Always light the fuel gas first, and then open the oxygen valve on the torch handle.

Step 4   Light the torch.

- Hold the striker in one hand and the torch in your other hand. Strike a spark in front of the escaping acetylene gas.
- Open the acetylene valve on the torch until the flame jumps away from the tip about $\frac{1}{16}$". Then slowly close the valve until the flame just returns to the tip (*Figure 29*). This sets the proper fuel gas flow rate for the size tip being used.
- Open the oxygen valve on the torch slowly to add oxygen to the burning acetylene. Refer again to *Figure 28* and *Figure 29*. Observe the luminous cone at the tip of the nozzle and the long, greenish envelope around the flame, which is excess acetylene that represents a carburizing flame. As you continue to add oxygen, the envelope

> **NOTE**
>
> It is not unusual for visible soot to form in strings and float away from the acetylene-only flame. However, they can fall on an otherwise clean surface and be somewhat messy. Take the surroundings into account as you light the torch and be aware that soot is likely to form from the initial flame.

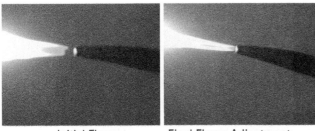

Initial Flame   Final Flame Adjustment

Figure 29 Reduce the acetylene flow until the base of the flame returns to the tip.

Tank Identification Transducers and Safety Plugs

Many gas suppliers mount transducers on their tanks so that they can readily identify the tanks. The transducer is electronically scanned, and a coded number is matched against the supplier's records for identification. This aids in quickly determining the purchaser or user of the tank as well as the required retesting date. Acetylene cylinders are also equipped with safety plugs that will release if the temperature exceeds 220°F (104°C). The safety plugs release the gas in the event of a fire to prevent the cylinder from exploding.

Cylinder Top Safety   Valve Handwheel
Plugs (1 of 2)

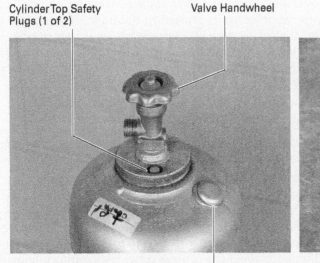

Cylinder Bottom Safety Plugs

If Present, Gas Supplier
Transducer for Cylinder
Identification

of acetylene should disappear. The inner cone will appear soft and luminous, and the torch will make a soft, even, blowing sound. This indicates a neutral flame, which is the ideal flame for welding. If too much oxygen is added, the flame becomes smaller, more pointed, and white in color, and the torch makes a sharp snapping or whistling sound. For brazing thin materials, a cooler, carburizing flame can be used to help prevent melting the base metal accidentally.

Step 5 Shut off the torch when finished brazing.

- Shut off the oxygen valve on the torch handle first.
- Shut off the acetylene valve on the torch handle. Close it quickly to avoid excess carbon buildup.
- Shut off both the oxygen and acetylene cylinder valves completely.
- Open both valves on the torch handle to release the pressure in the hoses. Watch the gauges on both regulators. With the cylinder valves closed, the pressure should fall on all regulator gauges until they register zero. Then close the valves on the torch handle.
- Turn the oxygen and acetylene regulator adjusting knobs counterclockwise until loose. This releases the spring pressure on the diaphragms in the regulators.

Step 6 Properly stow the hoses while any final preparations to braze are made.

2.2.3 Air-Acetylene Equipment Setup

In many instances, the heat provided by an air-acetylene torch alone is suitable for brazing refrigerant lines. This is especially true for smaller lines. The procedure for setting up and using the air-acetylene torch is the same as that used in the soldering process.

Air-acetylene torch tips are available in different sizes (*Figure 30*). However, it should be noted that a larger tip does not increase the temperature of the flame. It does, however, provide more heat by burning more fuel. Follow the manufacturer's guidelines for the selection of torch tips.

2.2.4 Purging Refrigerant Lines

Oil inside the tubing can vaporize when the heat of the brazing torch is applied. Oil vapor mixed with air can ignite or explode if a source of ignition is available. In addition, when copper is heated during brazing, it reacts rapidly with the oxygen in the air to form copper oxide. The oxidation occurs in such abundance, thin, flaky deposits are formed that then separate from the tubing as they cool. If oxygen from the atmosphere is present in the tubing during brazing, these deposits form on the inside as well. Refrigerant flow later washes away the coarse particles, which can plug orifices and generally pollute the system. As a precaution, all the air should be removed from the tubing being brazed. This can best be done by purging the tubing with nitrogen.

> **WARNING!**
>
> Never use oxygen to purge or pressure test tubing or any component. An explosion could result when oil and oxygen are mixed.

The pressure in a full nitrogen cylinder is 2,000 psig or more. An accurate nitrogen pressure regulator and an adjustable pressure relief valve should always be used when purging (*Figure 31*). The relief valve should be adjusted to open at a relatively low pressure, such as 7–10 psig, to keep the pressure low.

Some careful thought about the process provides the information needed to properly purge a system for brazing. First, the goal is to simply push the air

Figure 30 Air-acetylene torch tips.

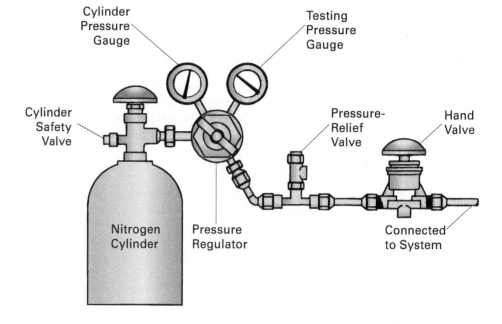

Figure 31 Nitrogen purging setup.

(and the oxygen it contains) out of the tubing and keep it out while brazing is in progress. However, nitrogen pressure cannot be allowed to build in the tubing while brazing proceeds. There is a real risk that leaks will develop as nitrogen escapes through a joint while the filler metal is in its liquid state. For this reason, the tubing circuit must remain freely open to the atmosphere at one or more points, allowing the nitrogen to escape without a pressure build-up. Remember that there is no need to use any significant pressure. Once the air is initially pushed out, the goal is to simply create and maintain a nitrogen atmosphere inside the tubing. The nitrogen pressure inside the tubing need only be slightly higher than the atmospheric pressure.

The best approach is to connect the nitrogen source to a refrigerant gauge manifold. The manifold valves provide much finer control over the nitrogen pressure applied than is possible with the regulator. The manifold hoses provide multiple outlets to connect the hoses to the system if necessary. When brazing independent tubing assemblies or circuits that have no gauge ports, ports can be installed, or cone valve plugs (*Figure 32*) can be used in a tube of the assembly. They are pushed into the end of the tubing or fitting to provide a point of connection. Note that these same valve plugs are very useful in clearing clogged condensate drains as well.

Once the connections are made and the regulator is installed, slowly open the nitrogen cylinder valve. With the refrigerant gauge manifold valves closed, set the regulator for roughly 3–5 psig. Then open the gauge manifold valve(s) and allow the nitrogen to begin circulating through the circuit. Once the air has been pushed out of the area, use the gauge manifold valve to reduce the flow to a very low volume—just enough to keep the air from being able to flow back in. The flow is sufficient when it can be felt gently in the palm of your hand. Allow the flow to continue at this pace throughout the brazing process. It should also continue to flow as the joint(s) cools.

If a refrigerant circuit is being closed and the final joints are being brazed, it is essential that a port remain open nearby the work to relieve any pressure. As the last few joints are completed, it is often best to turn off the nitrogen flow altogether. Heat from the process typically maintains enough of a draft leaving the circuit to prevent a significant amount of air from entering.

If it is necessary to braze tubing located in a confined space, make sure to attach a hose, pipe, etc., to the tubing being brazed so that the purge gas leaving the tubing is vented into the atmosphere outside. Otherwise, a hazardous atmosphere can be created within the space as the oxygen is displaced, resulting in an oxygen deficiency. OSHA regulations require that the internal atmosphere of a permit-required confined space be tested for hazardous atmosphere with a calibrated direct-reading instrument before an employee is allowed to enter the space. In addition, the atmosphere must be continuously monitored while work is in progress to detect the presence of hazardous gases or an oxygen deficiency.

Nitrogen purging is a valuable aid in maintaining the cleanliness of refrigerant circuits and creating a safer working environment. Although it is all too often omitted in field work, its value cannot be overstated, and it should be applied whenever possible.

Figure 32 Cone valve plug.
Source: Courtesy of Uniweld Products, Inc.

2.3.0 Brazing Joints

Once the flame is properly adjusted, the torch is ready to braze joints. Use the following procedure to braze a joint.

Step 1 Prepare the work area and protect any components in the piping system that can be damaged by the heat of the brazing process. Remove any loose, flammable materials in the area and ensure there is adequate room for physical movement around the joint. Place an appropriate fire extinguisher within easy reach.

Step 2 Set up the nitrogen gas to purge the tubing following the guidelines and precautions described previously.

Step 3 Put on welding goggles with a No. 4 or 5 tint.

Step 4 Set up and light the oxyacetylene brazing equipment as described previously.

- Place the chosen filler metal within easy reach and be certain that the material is freely accessible while the process is in progress.
- Light and adjust the torch to produce a neutral flame.

Step 5 Apply the heat to the tubing first, in the area adjacent to the fitting. Watch the flux (if used). It will first bubble and turn white and then melt into a clear liquid, becoming calm. At this time, shift the flame to the fitting and hold it there until any flux on the fitting turns clear.

Step 6 Continue to move the heat back and forth over the tubing and the fitting. Allow the fitting to receive more heat than the tubing by briefly pausing at the fitting while continuing to move the flame back and forth. Pause at the back of the fitting cup. For larger tubing, it may be difficult to bring the whole joint up to temperature at one time with a single-orifice tip. It will often be desirable to use a multi-flame (rosebud) heating tip, such as the one shown in *Figure 33*, to maintain a more uniform temperature over a larger area. A mild preheating of the entire fitting and adjacent tubing is recommended for larger sizes, and the use of a second torch to retain a uniform preheating of the entire fitting assembly may be necessary with the largest diameters.

Step 7 Touch the filler metal rod to the joint. If the filler metal does not melt on contact, continue to heat and test the joint until the filler metal melts. Be careful to avoid melting the base metal.

Step 8 Hold the filler metal rod to the joint and allow the filler metal to enter the joint while holding the torch slightly ahead of the filler metal and directing most of the heat to the shoulder of the fitting. When the filler metal is applied to the top of the fitting, it should run down to the bottom and fill in the joint through capillary action. Applying heat at the bottom of the joint should help to make sure the filler metal penetrates. Make sure the filler metal is visible all around the shoulder of the joint. If it is not, apply additional filler metal. However, building an excessive fillet does not improve joint quality and wastes material.

Step 9 For larger joints, small sections of the joint can be heated and brazed. Be sure to overlap the previously brazed section as you continue around the fitting (*Figure 34*).

Step 10 After the filler metal has hardened and while it is still hot, wash the joint with a wet rag to clean any excess brazing flux from the joint. This also helps the copper oxides to crack and break away in flakes.

Step 11 Allow the joints to finish cooling naturally. Maintain the nitrogen flow until the joints are cool.

Figure 33 Multi-flame, or rosebud, heating tip.

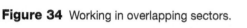

Figure 34 Working in overlapping sectors.

The technique for brazing varies somewhat, depending on the position in which the brazing is being done. The method just described is good for horizontal runs of pipe, but some alterations are needed for vertical-up and vertical-down brazing.

- *Vertical-up joints* — Heat the tube first, and then transfer the heat to the fitting. Move the heat back and forth between the fitting and the tube but be sure not to overheat the tube below the fitting. Doing so could cause the filler metal to flow out of the joint. When the brazing temperature is reached, apply the filler metal to the joint while applying heat to the wall of the fitting. This should cause the filler metal to run up into the fitting. Heat and capillary action must be used to defeat gravity.

- *Vertical-down joints* — Heat the tube before heating the fitting. When the brazing temperature is reached, apply more heat to the fitting while applying the filler metal to the joint. The filler metal should run down into the fitting.

2.4.0 Brazing Dissimilar Metals

Some accessories used in HVACR equipment have fittings that are made of steel, copper-clad steel, or brass. Accumulators and receivers, for example often have steel fittings, while some filter driers have copper-clad steel fittings. Sight glasses are sometimes made of brass. Special attention must be given to these situations. A brazing alloy with 45 percent silver and meeting BAg-5 specifications (*Figure 35*) is usually a better choice.

For example, when brazing dissimilar metals such as copper to brass or copper to steel, flux and a silver-bearing brazing alloy such as a BAg-5 should be used. A non-phosphorous alloy is generally required because the phosphorous could cause the joint to become brittle and fail. BAg-5 products do not contain phosphorous. Remember that, for copper-to-copper joints, a phosphorous-bearing filler metal acts as a flux, so no additional flux is required. When a non-phosphorous filler metal is used, a flux must be used, even on copper. As always, the pieces being joined must be very clean and free from any oil or grease.

Aside from choosing the correct filler metal and flux, the other important aspect of brazing dissimilar metals is the application of heat. The fittings on accessories that use steel or copper-clad steel fittings reach their melting point faster than copper, so there is a potential to overheat and distort the fittings during brazing. To avoid overheating, it is necessary to move the heat rapidly and uniformly around the joint and not let it linger too long in one place. Thin steel can turn cherry-red very quickly, since it transfers heat away much more slowly than copper, allowing its own temperature to rise quickly. When joining copper to steel, more heat needs to be applied to the copper, since it is the better conductor and thus will dissipate heat more quickly than the steel.

Also keep in mind that metals expand at different rates. Copper and brass will expand more than steel when heated. This may cause the joint to tighten up so much that the filler metal will not flow into the joint. For that reason, a joint clearance of 0.010" (0.25 mm) may be more suitable than the usual 0.001" to 0.003" (0.03 mm to 0.08 mm) clearance when brazing copper or brass to steel. However, if the heat is likely to loosen the fit between the two metals, start with a tighter fit.

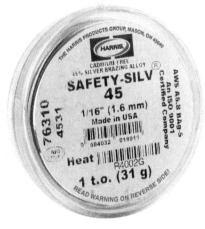

Figure 35 A 45 percent silver brazing alloy, in roll form (BAg-5).
Source: Harris Products Group

2.0.0 Section Review

1. An acetylene tank must *always* be used in an upright position.
   a. True
   b. False

2. Oxygen tanks are typically pressurized to about _____.
   a. 15 psig
   b. 250 psig
   c. 500 psig
   d. 2,000 psig

3. A rosebud tip is used for _____.
   a. soldering
   b. brazing larger tubing
   c. cutting
   d. brazing small tubing

4. Which of the following materials would be used for brazing dissimilar metals?
   a. A BCuP-3 filler metal
   b. A BAg-5 filler metal
   c. A BCuP-5 filler metal
   d. Any phosphorous-bearing alloy

Module 03104   Review Questions

1. The filler metals commonly used for soldering usually melt in a temperature range of _____.
   a. 200°F to 375°F
   b. 375°F to 500°F
   c. 750°F to 850°F
   d. 1,000°F to 1,200°F

2. The volume capacity of an acetylene B tank is _____.
   a. 10 ft^3
   b. 20 ft^3
   c. 30 ft^3
   d. 40 ft^3

3. Which of the following is a *false* statement regarding the filler metal used for soldering?
   a. The filler is a nonferrous metal.
   b. The filler is a metal alloy.
   c. The filler has a melting point below 842°F (450°C).
   d. The filler always contains some percentage of lead.

4. The measured length of a pipe should include the distance between the fitting faces plus _____.
   a. the cup depth of the two fittings
   b. half the total length of the two fittings
   c. the distance to the centers of the two fittings
   d. approximately $1/2$" on either end of the pipe

5. Acetylene becomes unstable when the pressure setting is above _____.
   a. 5 psig
   b. 7 psig
   c. 10 psig
   d. 15 psig

6. The desired joint clearance for a brazed copper-to-copper joint is _____.
   a. 0.001" to 0.003" (0.03 mm to 0.08 mm)
   b. 0.003" to 0.005" (0.08 mm to 0.13 mm)
   c. approximately 0.015" (3.81 mm)
   d. 5 percent of the pipe diameter

7. All brazing operations require the use of oxygen and acetylene.
   a. True
   b. False

8. The type of brazing filler metal that contains cadmium is _____.
   a. BCuP-1
   b. BCuP-2
   c. BAg-2
   d. BAg-5

9. Most brazed joints are made at a temperature range of _____.
   a. 375°F to 500°F (191°C to 260°C)
   b. 750°F to 850°F (399°C to 454°C)
   c. 850°F to 1,200°F (454°C to 649°C)
   d. 1,200°F to 1,550°F (649°C to 843°C)

10. The type of flame used for almost all heavy brazing applications is a(n) _____.
    a. oxidizing flame
    b. carburizing flame
    c. neutral flame
    d. feathering flame

Answers to odd-numbered Module Review Questions are found in *Appendix A*.

Answers to Section Review Questions

Answer	Section Reference	Objective
Section 1.0.0		
1. b	1.1.1	1a
2. c	1.2.0	1b
3. d	1.3.0	1c
Section 2.0.0		
1. a	2.1.1	2a
2. d	2.2.1	2b
3. b	2.3.0	2c
4. b	2.4.0	2d

User Update

SCAN ME

Did you find an error? Submit a correction by visiting **https://www.nccer.org/olf** or by scanning the QR code using your mobile device.

MODULE 03105

Basic Carbon Steel Piping Practices

Source: iStock@zozzzzo

Objectives

Successful completion of this module prepares you to do the following:

1. Identify and describe various types of steel pipe and fittings.
    a. Identify the characteristics and uses of steel pipe.
    b. Describe the characteristics of pipe threads and how they are measured.
    c. Identify various types of pipe fittings and describe how they are used.
    d. Describe how to measure pipe and determine cut lengths.
2. Describe the tools and methods used to cut and thread steel pipe.
    a. Identify pipe cutting and reaming tools.
    b. Identify pipe threading equipment and explain how to thread pipe.
3. Describe how to mechanically join and install steel pipe.
    a. Describe the tools and techniques used to connect threaded pipe.
    b. Describe pipe grooving methods and techniques.
    c. Explain how to assemble flanged steel pipe.
    d. Explain how to correctly install and support steel pipe.

Performance Tasks

Under supervision, you should be able to do the following:

1. Cut, ream, and thread steel pipe.
2. Join threaded pipe or pipe nipples using various fittings.

Overview

Iron and steel pipe are used in many HVACR systems. Steel pipe is used to carry water in hydronic systems and supply gas to furnaces. The ability to properly cut, thread, and join steel pipe is an important skill for an HVACR installer as well as service technicians.

Digital Resources for HVACR

SCAN ME

Scan this code using the camera on your phone or mobile device to view the digital resources related to this craft.

1.0.0 Steel Pipe and Fittings

Performance Tasks

1. Cut, ream, and thread steel pipe.
2. Join threaded pipe or pipe nipples using various fittings.

Objective

Identify and describe various types of steel pipe and fittings.

a. Identify the characteristics and uses of steel pipe.
b. Describe the characteristics of pipe threads and how they are measured.
c. Identify various types of pipe fittings and describe how they are used.
d. Describe how to measure pipe and determine cut lengths.

NOTE

Throughout this module, references to *black iron pipe* and *black pipe* refer to ungalvanized steel pipe.

Black iron pipe: A term used to describe carbon steel pipe that gets its black coloring from the carbon in the steel.

Galvanized: Describes steel with a zinc-based coating to prevent rust.

Wrought: Formed or shaped by hammering.

Cast iron: A brittle, hard metal containing between 2 and 4 percent of carbon and small amounts of other substances such as manganese and silicon. Some impurities, such as phosphorus and sulfur, are also present.

Carbon steel is part of a family of ferrous metals. The term *ferrous* simply means that it contains iron. Therefore, ferrous metal pipe is pipe that either contains, or is made from, iron.

Steel is made by adding carbon and other substances, such as manganese, to iron. The ferrous metal pipe used in HVACR work is steel pipe. Iron pipe is used in some plumbing applications such as drains and sewer lines, but it is not used in HVACR work. However, **black iron pipe**, and the shortened version, *black pipe*, are terms often used in the HVACR trade. When the terms are used by craftworkers in reference to piping, they are referring to steel pipe that has not been **galvanized**. Technically, black iron is iron that has been **wrought** and tempered, and the name differentiates it from **cast iron**.

Hydronic heating systems use hot water to deliver heat to the conditioned space. In these applications, the water may be carried in galvanized or black pipe. Black pipe is often used to supply gas to a furnace. *Figure 1* shows a gas manifold constructed of black pipe with copper branch lines feeding the equipment.

Steel pipe has the following advantages over other piping materials:

- It is very durable.
- It has exceptional structural strength.
- Its cost is low in comparison to copper. However, joining steel pipe is a more labor-intensive process, whether it is joined by threading or welding. In some applications, it can be joined using special press-to-connect fittings that save time.
- Once warm, it holds heat well.
- It does not expand and contract excessively. However, expansion and contraction in all piping systems must be considered when there are long runs of pipe or significant temperature changes are possible.

1.1.0 Characteristics of Steel Pipe

The two most common types of steel pipe are black pipe and galvanized pipe (*Figure 2*). Black and galvanized pipe are manufactured the same way. The only difference between the two is that black pipe does not have a zinc coating.

Figure 1 Gas piping manifold with copper branch lines.
Source: iStock@BanksPhotos

Figure 2 Galvanized and black pipe.

Carbon gives black pipe its characteristic color. Black pipe is most often used in the HVACR craft for natural gas piping, chilled and hot water, steam, and compressed air applications. It is used where corrosion will not adversely affect its uncoated surfaces, and where internal chemical treatment helps prevent corrosion.

Galvanized pipe is black pipe that has been dipped in molten zinc. The zinc protects the surfaces from both abrasion and corrosion. It gives the pipe a mottled, silvery color when it is new, and a dull, grayish color after it has aged. The outer coating is particularly useful for outdoor applications. However, the galvanizing material on the inside of the pipe is concerning in some situations. Galvanized pipe is most often used when specifications require it.

Both galvanized steel and black pipe are joined using threaded, grooved, or **flange** fittings. Flanges can be welded or threaded onto pipe. Flanges, as well as threaded and grooved pipe fittings, are manufactured (*Figure 3*). However, the pipe itself must be cut to size, then joined on the job. For that reason, the ability to measure, cut, thread, groove, and join galvanized and black iron pipe is an important skill for anyone installing HVACR systems.

Flange: A flat plate-like fitting attached to equipment or the end of a pipe, assembled with a gasket between the plates and held together with nuts and bolts.

1.1.1 Pipe Sizes and Wall Thickness

In the United States, pipe is sized in inches by its nominal size. For pipe sizes up to and including 12", the nominal size approximates the inside diameter (ID). For pipe 14" and larger, the nominal size reflects the outside diameter (OD) of the pipe.

There are times when the *nominal* size of a pipe and its actual inside or outside diameter differ significantly. But nominal size, not actual size, is always used to select and discuss piping. *Figure 4* shows the actual inside and outside diameters for a 1" nominal-sized steel pipe.

There are two ways to describe the wall thickness of a pipe. The first is by *schedule*. As schedule numbers get larger, pipe walls get thicker. The schedule numbers used for pipe are Schedule 5, 10, 20, 30, 40, 60, 80, 100, 120, 140, and 160.

It is important to remember that a schedule number only describes the wall thickness of a pipe for a given nominal size. Thus, $\frac{3}{4}$" Schedule 40 does not have the same wall thickness as 1" Schedule 40. All schedules or weights of a specific pipe size have the same outside diameter, so the same **threading dies** fit all of them. Threads are always formed on the outside of the pipe.

Threading dies: Cutting tools used to cut external threads by hand or using a machine.

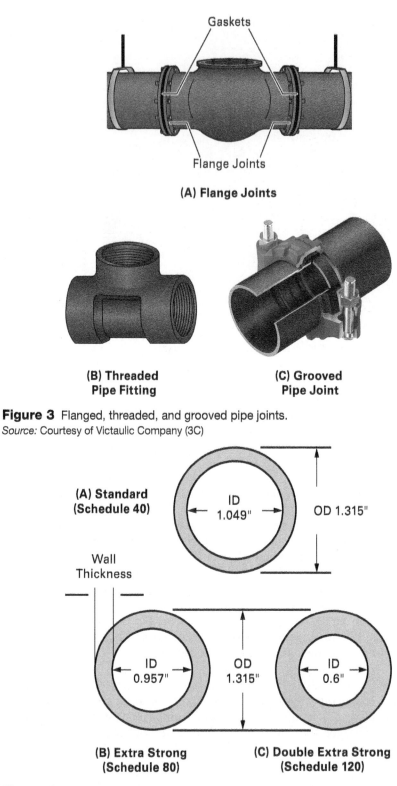

Figure 3 Flanged, threaded, and grooved pipe joints.
Source: Courtesy of Victaulic Company (3C)

(A) Flange Joints

(B) Threaded Pipe Fitting

(C) Grooved Pipe Joint

Gaskets

Flange Joints

(A) Standard (Schedule 40)

ID 1.049"

OD 1.315"

Wall Thickness

(B) Extra Strong (Schedule 80)

ID 0.957"

OD 1.315"

(C) Double Extra Strong (Schedule 120)

ID 0.6"

Figure 4 Steel pipe dimensions.

The second way to describe pipe wall thickness is by manufactured weight. There are three classifications in common use. In ascending order of wall thickness, they are:

- STD — Standard
- XS — Extra strong
- XXS — Double extra strong

The wall thickness and the inside diameter differ with each weight. The thicker the wall, the smaller the inside diameter, but the more pressure the pipe

How Long Have Iron and Steel Been Around?

The first use of iron has been traced back to nearly 5,000 years ago. A lot of early iron was extracted from meteorites that landed on Earth. The extensive use of iron for weapons and other purposes dates to the so-called Iron Age, which began around 3,200 years ago, or 1200 BC. The Iron Age was preceded by the Bronze Age, a period of about 800 years, which was preceded by the Stone Age.

Bronze is a blend of copper and tin. The tin was added to help overcome the brittle nature of copper. Steel is an iron-based alloy, made by adding material such as carbon, which acts as a hardening agent and gives the steel its exceptional strength. Steel has been around for thousands of years, but the manufacture of steel as we know it today dates back about 400 years to the invention of the Bessemer furnace (shown here), considered to be the first true blast furnace.

Source: iStock@Binnerstam

will withstand. Standard weight will prove adequate in most piping situations. In the HVACR craft, you will work with Schedule 40 pipe most of the time, with only occasional use of Schedule 80 pipe. The stronger and thicker pipe types are needed for applications such as high-pressure steam systems. XS and XXS pipe may also be used to contain highly toxic or hazardous materials, even at low pressures, to maximize safety.

Specifications, drawings, or data sheets sometimes specify pipe sizes in metric dimensions. You may need to convert these dimensions to US standard dimensions. The table in *Appendix 03105A* gives the dimensions of commonly used steel pipe in both US standard and metric sizes.

1.2.0   Pipe Threads

Sections of steel pipe are commonly joined by threading the end of the pipe and applying threaded fittings. The portion of a threaded pipe that is screwed into a fitting is referred to as the **makeup**, or *thread engagement*.

American National Standard pipe threads can be either straight or tapered. Tapered threads are identified with the acronym NPT, while straight threads are identified by NPS. Straight threads are rare, but they are sometimes used with pipe when fabricating a structure and no fluids are involved. Only tapered pipe threads (NPT) are used for HVACR work because they produce leak-tight and pressure-tight connections. When tight, they also produce a mechanically rigid assembly. Tapered threads can be cut by hand with a **stock** and threading die, or with an electric pipe-threading machine.

The tapered thread used on pipe is V-shaped with a thread angle of 60 degrees, and slightly rounded at the top (*Figure 5*). The rate of taper is $\frac{1}{16}$" per inch of length. (Only one side is visible in the figure, so the rate is shown as $\frac{1}{32}$".) There are about 7 to 12 perfect threads and several imperfect threads shown in *Figure 5*. *Imperfect threads* are those that are not cut to their full depth because of the taper, as opposed to full-depth perfect threads.

The actual number of perfect threads used depends on the size of the pipe being threaded. As shown in *Figure 5*, the threads closest to the end of the pipe are perfect threads; they are sharp at the top and bottom. The remaining threads are imperfect because they are not completely cut, resulting in rounded or imperfect edges. Sealing occurs as the female perfect threads begin to meet

Makeup: The length of the threads that engage and penetrate a pipe fitting.

Stock: A tool used to hold and turn threading dies to manually cut pipe threads.

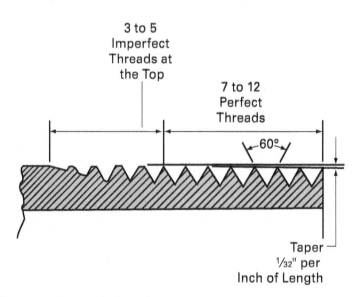

Figure 5 American National Standard taper (NPT) pipe threads.

the tapered imperfect threads. If the perfect threads are chipped or distorted, leaks will likely occur.

Threads are designated by specifying, in sequence, the nominal pipe size, the number of threads per inch, and the thread series symbols. The number of threads per inch is referred to as the **pitch** of the thread. For example, the thread specification $\frac{3}{4}$–14 NPT means the following:

Pitch: The number of threads per inch on threaded pipe and other threaded components.

- $\frac{3}{4} = \frac{3}{4}$" nominal pipe size (ID)
- 14 = 14 threads per inch (pitch)
- NPT = American National Standard tapered pipe thread

Male and female tapered pipe threads are fully engaged in two phases: hand-tight engagement followed by wrench tightening. *Table 1* shows the dimensions for hand-tight engagement as well as other NPT specifications for commonly used pipe sizes. In practice, about three turns are done by hand, followed by several turns with a wrench. When a pipe is properly threaded and tightened, about three threads (all imperfect) should remain visible.

TABLE 1 American National Standard Taper Pipe Thread (NPT) Dimensions

Nominal Pipe Size	Threads per Inch	No. of Usable Threads	Hand-Tight Engagement	Total Thread Makeup	Total Thread Length
$\frac{1}{8}$"	27	7	$\frac{3}{16}$"	$\frac{1}{4}$"	$\frac{3}{8}$"
$\frac{1}{4}$"	18	7	$\frac{1}{4}$"	$\frac{3}{8}$"	$\frac{5}{8}$"
$\frac{3}{8}$"	18	7	$\frac{1}{4}$"	$\frac{3}{8}$"	$\frac{5}{8}$"
$\frac{1}{2}$"	14	7	$\frac{5}{16}$"	$\frac{7}{16}$"	$\frac{3}{4}$"
$\frac{3}{4}$"	14	8	$\frac{5}{16}$"	$\frac{1}{2}$"	$\frac{3}{4}$"
1"	11.5	8	$\frac{3}{8}$"	$\frac{9}{16}$"	$\frac{7}{8}$"
$1\frac{1}{4}$"	11.5	8	$\frac{7}{16}$"	$\frac{9}{16}$"	1"
$1\frac{1}{2}$"	11.5	8	$\frac{7}{16}$"	$\frac{9}{16}$"	1"
2"	11.5	9	$\frac{7}{16}$"	$\frac{5}{8}$"	1"
$2\frac{1}{2}$"	8	9	$\frac{11}{16}$"	$\frac{7}{8}$"	$1\frac{1}{2}$"
3"	8	10	$\frac{3}{4}$"	1"	$1\frac{1}{2}$"
$3\frac{1}{2}$"	8	10	$\frac{13}{16}$"	$1\frac{1}{16}$"	$1\frac{5}{8}$"
4"	8	10	$\frac{13}{16}$"	$1\frac{1}{16}$"	$1\frac{5}{8}$"
5"	8	11	$\frac{15}{16}$"	$1\frac{3}{16}$"	$1\frac{3}{4}$"
6"	8	12	$\frac{15}{16}$"	$1\frac{3}{16}$"	$1\frac{3}{4}$"

Pipe threads may also be referred to with several other acronyms by manufacturers and others, both in print and in conversation. The acronyms include the following:

- MIP (Male Iron Pipe)
- MPT (Male Pipe Thread)
- FIP (Female Iron Pipe)
- FPT (Female Pipe Thread)
- IPT (Iron Pipe Thread)

The MIP, MPT, FIP, and FPT acronyms are useful when selecting pipe fittings and system components. They differentiate between male and female so that installers know what gender the mating portion must be.

1.3.0 Steel Pipe Fittings

Pipe fittings for steel pipe are generally made of cast iron, malleable iron, or galvanized (zinc-coated) iron. Malleable iron is produced by prolonged annealing of ordinary cast iron. This process makes the iron tough. Also, it can be bent or pounded to some extent without breaking.

Malleable iron fittings are typically used for gas piping. Cast iron fittings are used for steam and hydronic system piping. Galvanized fittings should be used with galvanized pipe.

Fittings are used in piping as follows:

- Joining pipe in a straight line, using unions and couplings
- Joining three sections at the same location, using tees
- Changing direction, using elbows
- Changing the size of connections or closing openings, using plugs, caps, and bushings

1.3.1 Tees and Crosses

Tees can be purchased in many sizes and patterns. They are typically used to create a branch at a right angle to the main flow. If all the outlets of a tee or cross are the same size, the fitting is called a regular tee or cross (*Figure 6*). If the outlet sizes vary, the fitting is called a reducing tee or cross.

Tees are specified by giving the straight-through, or *run*, dimensions first, followed by the branch dimensions. For example, a tee with one run outlet of 2", a second run outlet of 1", and a branch outlet of ¾", is called a 2" × 1" × ¾" tee. Always state the larger run size first, the smaller run size next, and the branch size last. It's worth mentioning that a tee with three different pipe sizes is rather rare and very difficult to find in local stock. The reality is that bushings or reducing couplings are usually required to change the outlet size(s) of a tee, simply due to availability at the time of need. Tees are also available with male threads on a run or branch outlet, but these are very rare. Unusual pipe fittings are not only difficult to find, but they are also more expensive due to the laws of supply and demand. If the need is known well in advance, unique fittings can be special ordered.

Whenever the flow of fluid through a pipe is being split into two directions, proper orientation of the tee is important to avoid *bull-heading*. Bull-heading refers to a stream of fluid advancing directly into the far wall of the fitting, as shown in *Figure 7*. This causes a lot of turbulence and increases pressure drop.

If, however, the connection is only being used to share fluid pressure with little or no fluid flow, any orientation is acceptable. For example, a tee may be used to provide a pressure input to two separate pressure switches. In these connections, there is no significant fluid flow.

A cross is a four-way distribution fitting. Instead of three connections, a cross has four connections. They are rarely needed in the HVACR craft.

Malleable iron: Iron that has been toughened by gradual heating or slow cooling.

Annealing: A process in which a material is heated to form it and then cooled to toughen it.

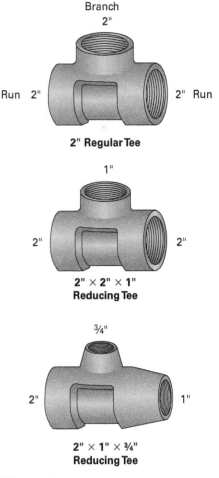

Figure 6 Tees.

Keep Pipe Fittings to a Minimum

Pipe fittings add resistance and reduce flow in piping systems. All piping systems should be planned in advance of the installation to minimize the number of fittings.

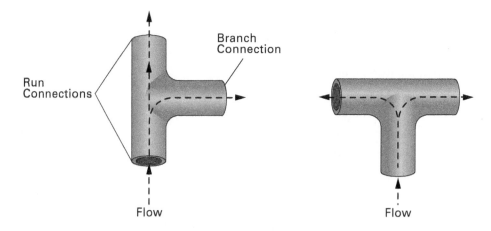

(A) Proper Tee Installation   **(B) Improper Tee Installation**

Figure 7 Orienting a tee for proper flow.

1.3.2 Elbows

Elbows (*Figure 8*), most often called *ells*, are used to change the direction of pipe. The most common ells are as follows:

- The 90-degree ell
- The 45-degree ell
- The *reducing ell*, which has two different connection sizes
- The *street ell*, which has male threads on one end and female threads on the other

(A) 90° Ell

(B) 45° Ell

(C) Reducing Ell

(D) Street Ell

Figure 8 Threaded elbows.
Source: ASC Engineered Solutions

1.3.3 Unions and Flanges

Unions make it possible to disassemble a threaded piping system. After loosening the union, the length of pipe on either end may then be turned. The two most common types of unions are the *ground joint union* (*Figure 9*) and the *flange union* (*Figure 10*).

The ground joint union connects two pipes by screwing the thread and shoulder pieces onto the pipes. Then, both the shoulder and thread parts are drawn together by the collar. This union creates a gas- and liquid-tight joint that can be disassembled quickly. Unions are often needed where equipment and piping component connections are made so that final assembly as well as disassembly are possible.

The flange union also connects two separate pipe runs or connects piping to equipment, valves, and similar components. The flanges screw onto the pipes to be joined and are then pulled together with nuts and bolts. A gasket between the flanges makes this connection gas- and liquid-tight. Flange unions are most often used for larger sizes of pipe, while ground joint unions are used on smaller sizes.

1.3.4 Couplings

Couplings (*Figure 11*) are fittings with female threads on each end. They are used to connect two sections of pipe in a straight line. The pipes may be the same size or different sizes. *Reducing couplings* are used to join pipes of different sizes. Couplings aren't used in place of unions because they cannot be disassembled in the same manner.

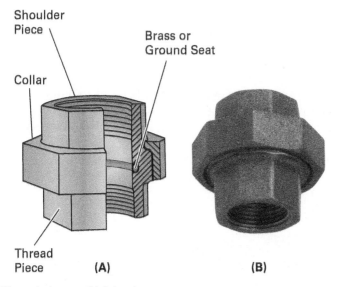

Figure 9 Threaded ground joint union.
Source: ASC Engineered Solutions (9B)

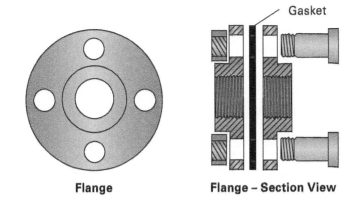

Figure 10 Flange union.

Ordinary Standard, Full,
and Half Couplings

Concentric Reducing
Coupling

Figure 11 Couplings.
Source: ASC Engineered Solutions

Full lengths of Schedule 40 pipe are typically threaded on each end by the manufacturer. Thread protectors are added to prevent damage. They may be plastic caps that simply slip over the end, or they may be threaded pieces that look like couplings. Although they may *look* like couplings, they should not be used to couple pipe. They are not usually made to the same standards as couplings.

1.3.5 Nipples

Nipples (*Figure 12*) are pieces of pipe 12" or less in length and threaded on both ends. They are used to extend the pipe to or from a fitting, or to join two fittings. Nipples are manufactured in many lengths, typically in $\frac{1}{2}$" increments through the 6" length. Nipples over 6" in length are sold in 1" increments. *Close nipples* (*Figure 13*) are short in length and have little or no unthreaded surface area between the threaded ends. Those that do have a short, unthreaded section may be referred to as *shoulder nipples*.

Nipples: Short lengths of pipe that are used to join fittings. Nipples are less than 12" long and have male threads on each end.

Black and Galvanized Pipe Nipples

Figure 12 Pipe nipples.
Source: ASC Engineered Solutions

Figure 13 Close nipple.
Source: Shutterstock.com/BW Folsom

Pipe Nipples

Pipe nipples are available from plumbing and heating suppliers in a variety of lengths. Sizes up to 6" in length are very common, but longer nipples are also available. It generally makes more sense to buy commonly used nipples in bulk, especially close nipples, rather than to cut and thread them in the field.

Most field threading equipment cannot be used to make close nipples. However, longer nipples can be fabricated using pieces of scrap pipe. This type of task can be accomplished in the shop during slow periods.

1.3.6 Plugs, Caps, and Bushings

Plugs are male threaded fittings used to close female openings in fittings. Plugs have a variety of drive sockets, such as square recessed, hexagon, and square as shown in *Figure 14*.

A cap is a fitting with a female thread. It is used for the same purpose as a plug, except that the cap fits on the male end of a pipe or nipple.

Bushings are fittings with a male thread on the outside and a female thread on the inside. They are usually used to connect the male end of a pipe to a fitting of a larger size. The ordinary bushing has a hexagon nut at the female end. Bushings can be used in place of a reducing fitting to accommodate a smaller pipe. However, project specifications may prohibit the use of bushings for some applications.

Cap   Bushing

Square and Hexagonal Socket
Countersunk Pipe Plugs

Square Socket Pipe Plug

Figure 14 Plugs, caps, and bushings.
Source: ASC Engineered Solutions

Fitting allowance: The distance from the end of the pipe to the center of the fitting. Also called *takeoff.*

| 1.4.0 | **Measuring Pipe** |

To accurately measure a section of pipe, it is important to understand the relationship between the pipe and the fittings. That in turn requires an understanding of two key terms.

Recall that makeup is the distance that the end of the pipe penetrates into the fitting. Refer to *Table 1* in *Section 1.2.0* to see the required makeup length for various sizes of pipe, along with other common characteristics. **Fitting allowance**, or *takeoff*, is the distance from the end of the pipe to the center of the fitting. The combination of makeup and fitting allowance equals the distance from the outer shoulder of the fitting to its center (*Figure 15*).

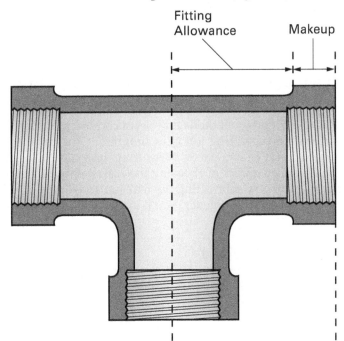

Figure 15 Makeup and fitting allowance.

There are several methods used to state the length of a threaded pipe, as shown in *Figure 16*. Several are described here in detail:

- *End to end* — Measuring the full length of the pipe, including the threads at both ends.

- *End to center* — Used for a piece of pipe with a fitting screwed onto one end only. The pipe length is equal to the total end-to-center measurement, minus the center-to-face dimension of the fitting, plus the length of the thread engagement.

- *Center to center* — Used to measure pipe with fittings screwed onto both ends. The pipe length is equal to the total center-to-center measurement between the fittings, minus the sum of the two center-to-face dimensions of the fittings, plus twice the length of the thread engagement.

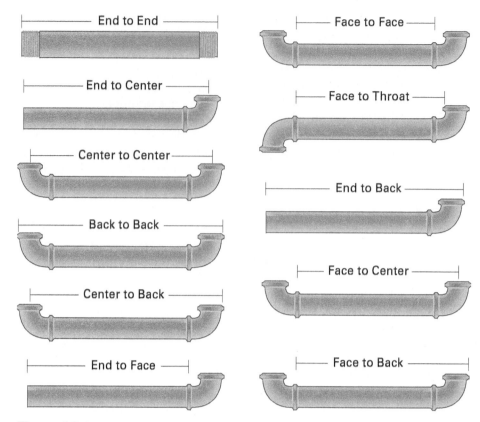

Figure 16 Approaches to measuring pipe.

- *Face to face* — Used under the same conditions as the center-to-center method. It is calculated by measuring the length of pipe between fittings (face to face), plus twice the length of the thread engagement.

Although these different methods may seem confusing at first, they quickly become clear once a pipefitting job is in progress. Other methods shown in *Figure 16* are also handy in some cases. Measuring the cut length of pipe between two fittings is simply a matter of choosing the easiest physical measurement to make, and then adding or subtracting the makeup and/or fitting allowance on each end as necessary. With practice, you'll make accurate measurements that result in a perfect fit.

1.0.0 Section Review

1. Which of these dimensions remains the same for all schedules of pipe of the same nominal size?
   a. Inside diameter
   b. Outside diameter
   c. Wall thickness
   d. Length

2. The number of perfect threads found on a proper pipe thread is _____.
   a. at least 10 but not more than 12
   b. dependent upon the size of the pipe
   c. dependent upon the schedule of the pipe
   d. dependent upon the type of fitting to be used

3. A close nipple is one that _____.
   a. is shorter than 3"
   b. is threaded on one end
   c. has no threads at all
   d. has little or no unthreaded surface

4. A pipe length measurement made by measuring the length of pipe between fittings and adding twice the length of the thread engagement is called a(n) _____.
   a. face-to-end measurement
   b. end-to-end measurement
   c. center-to-center measurement
   d. face-to-face measurement

2.0.0 Tools and Methods Used to Cut and Thread Steel Pipe	
Objective	**Performance Task**
Describe the tools and methods used to cut and thread steel pipe. a. Identify pipe cutting and reaming tools. b. Identify pipe threading equipment and explain how to thread pipe.	1. Cut, ream, and thread steel pipe.

Before threaded pipe can be assembled, the pipe must be cut to the proper length and threaded. Cutting and threading must be done properly to avoid problems during assembly and leaks.

2.1.0 Cutting Pipe

Standard pipe cutters (*Figure 17*) have from one to four cutting wheels. The cutter must be rotated around the pipe, so the more cutting wheels a cutter has, the less space required to rotate it around the pipe. In other words, a four-wheel cutter does not have to make a complete revolution around the pipe to make a cut. However, a single-wheel, or conventional, cutter must make complete revolutions. This is a disadvantage if there is limited clearance around the pipe, but that's not common.

A pipe cutter is rotated around the pipe to make a cut. The adjusting screw is tightened about one-quarter revolution after each full turn, keeping the wheel tight against the pipe. Avoid overtightening the cutting wheel, however.

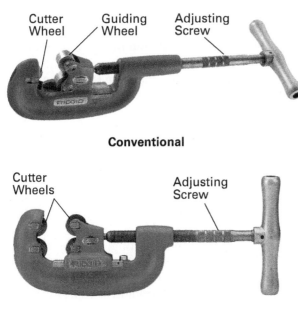

Figure 17 Pipe cutters.
Source: Ridge Tool Company

Doing so causes a larger burr to form inside the pipe, and it quickly dulls the cutting wheel. It can also distort the pipe (especially the lighter schedules, such as Schedule 40 and lower), or result in a shattered cutting wheel. Pipe that is out of round cannot be threaded and should be discarded. It's a good idea to save any usable pipe scraps to make nipples. True scrap should be recycled.

A tripod-mounted pipe vise (*Figure 18*) is used to secure the pipe for cutting and/or threading. The yoke vise is limited to smaller pipe sizes, but HVACR technicians typically work with the smaller sizes. Its jaws hold the pipe firmly and prevent it from turning while the pipe is cut or threaded. A yoke vise can typically handle pipe up to $2\frac{1}{2}$" in diameter, depending on the yoke size.

(A) Yoke Vise Tripod   (B) Chain Vise Tripod

Figure 18 Tripod-mounted pipe vises.
Source: Reed Manufacturing

The chain vise is used in the same way as the yoke vise. A chain vise can hold larger pipe than a yoke vise, but most can also accommodate pipe sizes down to $\frac{1}{8}$". Occasional lubrication of the chain is required, but the oil should be applied lightly. A stiff chain operates poorly, but too much oil will collect dirt and grit. The tightening mechanism on both vise types also needs to be lubricated occasionally.

2.1.1 Reaming the Pipe

Once the pipe is cut, a reamer (*Figure 19*) is used to remove the burr that forms on the inside of the pipe. If the burr is not removed, it collects deposits and restricts the flow of liquid. Because reamers are tapered, one reamer can be used on many pipe sizes. A typical tapered pipe reamer can handle pipe sizes from $\frac{1}{4}$" through 2". The larger reamer shown in *Figure 19* can ream pipe sizes from $2\frac{1}{2}$" through 4".

Pipe reamers for steel pipe, like the two shown in *Figure 19*, are also ratcheting. The reamer does not have to be rotated in a complete circle.

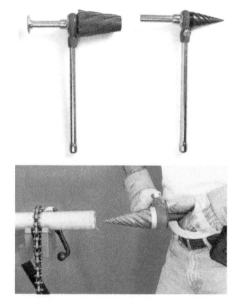

Figure 19 Pipe reamers.

2.2.0   Pipe Threading

Pipe can be threaded manually or using powered equipment. Hand threaders (*Figure 20*) are made up of two major parts: the die and the stock (handle). Dies are used to cut the threads, and the stock holds the die and provides turning leverage. The stock has a ratcheting head, allowing threads to be cut without making full rotations around the pipe. Although the style of dies shown in *Figure 20* are used with a stock for manual cutting, the same die can be inserted into a handheld power drive (*Figure 21*). Manual threading works well on small pipe and when there are only a few pieces to thread. Pipe sizes larger than 1", though, require a considerable amount of strength as well as weight to overcome the friction of the cutting teeth.

A handheld power drive is portable, yet heavy and surprisingly powerful. The pipe is secured on a tripod vise, just as it would be for manual threading. The rotating power head turns at about 30 rpm and accepts the same dies as the manual threader. They are typically limited in capacity to 2" pipe. When sizes

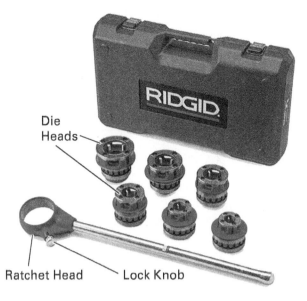

Figure 20 Manual threading equipment.
Source: Ridge Tool Company

Figure 21 Handheld power drive.
Source: Courtesy of Milwaukee Electric Tool Corporation

over 1" are being threaded, a support arm (not shown) is clamped to the pipe and the threader is rested against it to counteract the torque of the threader. The power drive is easily powerful enough to lift you from the ground when the friction of the cutting teeth is extreme.

A pipe threading machine (*Figure 22*) is used when large quantities of pipe must be threaded. These are multipurpose machines designed to rotate, cut, ream, and thread pipe. The built-in reamer and pipe cutter are nearly identical to handheld models. They are permanently mounted on most power threaders for quick access and use.

The pipe is loaded through the machine's speed chuck. The speed chuck has a wheel that spins easily with the hand to quickly lock the pipe into the teeth of the vise. Power threaders are usually operated using a foot pedal switch. Like pipe vises, threading machines can be mounted on a tripod or a shop bench.

CAUTION

Do not use a pipe threading machine to tighten fittings onto threaded pipe, as this could cause the fitting to be overtightened or damaged.

> **WARNING!**
>
> Pipe threading equipment is heavy and powerful. There are many sharp edges exposed, including the burr that results from cutting the pipe with a wheeled cutter. Piping materials may also be heavy and awkward to handle. Whether fabricating or installing pipe, always wear the proper PPE, including safety glasses, safety shoes or boots, and well-fitted heavy work gloves.

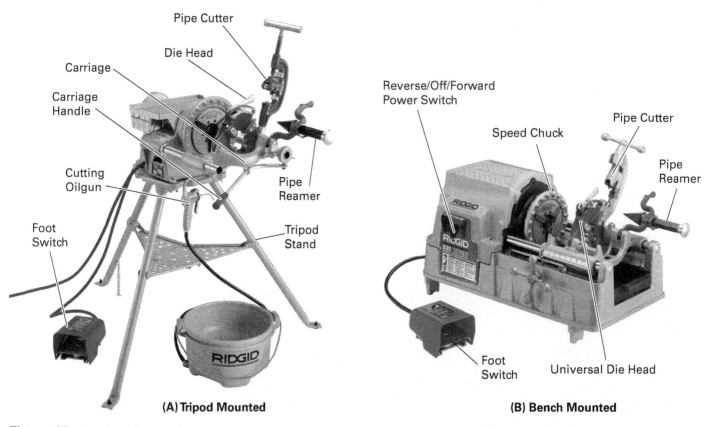

(A) Tripod Mounted

(B) Bench Mounted

Figure 22 Pipe threading machines.
Source: Ridge Tool Company

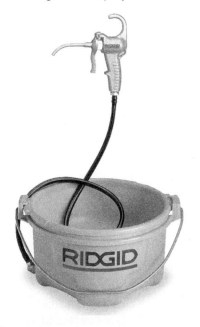

Figure 23 Threading oil basin and hand pump.
Source: Ridge Tool Company

Pipe threading machines often have built-in and automated oil pumps to lubricate the cutting teeth. For manual threading, you'll need an oil basin and pump (*Figure 23*). The basin is placed directly beneath the threading operation where it can capture oil as it drains away, as well as the threading chips.

> **WARNING!**
>
> Threading chips are extremely sharp, and very hot.

2.2.1 Using a Hand Threader and Vise

Pipe is often threaded by hand on small jobs where only a few joints need to be done. It is not an effective way to thread larger pipe or thread a lot of joints. Although hand threaders and dies are available to thread up to 2" pipe, threading pipe this size requires a tremendous amount of power. An average-sized human is not likely to succeed. Most manual threading is reserved for pipe sizes 1" or less due to the physical effort and time required.

When using a hand threader and vise, the pipe is first placed into the vise and secured. Prepare yourself by donning the appropriate PPE, including safety glasses and gloves. Threading produces extremely sharp and stiff metal debris and sharp edges, so the gloves should be chosen accordingly. To cut threads using a hand threader and vise, proceed as follows:

Step 1   Select the correct size die for the pipe being threaded.

Step 2   Inspect the die to make sure the cutters are free of nicks, excessive wear, and debris.

Step 3   Lock the pipe securely in a vise.

Step 4   Slide the die over the end of the pipe, pipe guide end first. The guide end is the side of the threader with the round opening, with no teeth.

Step 5   Push the die against the pipe with the heel of one hand. Most threader stocks have a ratcheting function to allow it to turn backwards freely. Use the ratcheting feature to make three or four short, slow clockwise turns. Keep the die pressed firmly against the pipe.

Step 6   When enough thread is cut to hold the die firmly to the pipe, apply some thread-cutting oil (*Figure 24*). The oil prevents the pipe from over-heating due to friction and lubricates the die. Oil the threads every two or three downward strokes.

Step 7   Back off one quarter turn after each full turn forward to clear out the metal chips. Continue until the pipe projects two threads from the die end of the stock. The threaded area cannot be too long, nor can it be too short.

Step 8   To remove the die from the pipe, rotate it counterclockwise.

Step 9   Carefully wipe off excess oil and any chips.

> **WARNING!**

Use a rag or a brush, not your bare hands, when wiping the pipe clear. The chips are sharp and can cause severe cuts.

Figure 24 Oiling the thread teeth.
Source: Ridge Tool Company

2.2.2 Using a Bench or Tripod Threading Machine

Each threading machine is slightly different. Become familiar with the manufacturer's operating procedures before attempting to operate any threading machine. Also, become thoroughly familiar with the maintenance and safety instructions for the machine. A poorly maintained machine is a safety hazard.

Did You Know?

How Much Does That Thing Weigh?

Power threaders are heavy and awkward to handle, although most are considered portable. They develop a lot of torque through their gearboxes, so their considerable mass works in their favor—until they need to be moved. In *Figure 22*, the threader mounted on a tripod—a Ridgid Model 300—weighs about 135 pounds (61 kg). The benchtop model shown—a Ridgid Model 535—weighs in at 285 pounds (130 kg). You're going to need some help with that!

When field pipe threading is on the schedule, make sure you are prepared to manage and mobilize the equipment safely.

NOTE

This section is based on the Ridgid Model 535 threading machine. While other machines are similar, ensure you are familiar with the machine in use and the terminology of the manufacturer.

Using a pipe threader includes the following procedures:

- Installing dies
- Checking cutting oil level
- Loading pipe into the threader
- Cutting and reaming pipe
- Threading the pipe
- Threader care and maintenance

Pipe Dies

Before cutting threads, the proper dies must be installed. Power threaders are equipped with a universal die head. The universal head can accommodate any die set needed for pipe sizes up through 2". There are four sets of dies that are used to thread pipe from $\frac{1}{8}$" through 2", and one set that is used to thread pipe $2\frac{1}{2}$" and larger. In other words, five sets of pipe dies cover all common pipe sizes.

Die sets are classified according to the number of threads per inch (tpi). *Table 2* provides the die information for common pipe sizes through 4", as well as the proper length of the threaded area. There are five dies, designed to cut 27, 18, 14, 11.5, or 8 threads per inch. The 8 tpi dies are used for pipe sizes over $2\frac{1}{2}$".

Die teeth for standard threaders are maintained in sets of four (*Figure 25*). Sets of five are needed only for a geared threader (*Figure 26*). Geared threaders are used for pipe sizes ranging from $2\frac{1}{2}$" to 6" and require five thread dies instead of four. The dies should have the numbers 1 through 4 on them, with geared threaders requiring one with a 5 on it as well. The numbers determine their position for insertion into the die head.

Once the proper set of dies has been identified, they must be installed in the threader die head. The die heads snap directly into a threading stock or into the bore of a power drive.

When using a power threader with a universal die head, the die teeth are inserted into their numbered positions. Then the die head must be adjusted to the pipe size being threaded. For example, to thread 1" pipe, a die set for 11.5 tpi must first be installed in positions 1 through 4, and then the die head is set to the 1" pipe size at the index mark (*Figure 27*).

To adjust the die head for different pipe sizes, loosen the clamp lever and move the size bar until the line underneath the desired pipe size lines up with the index line. Tighten the clamp lever after the adjustment has been made.

CAUTION

Threading stainless steel pipe requires special dies and lubricants. Standard dies will be damaged if used on stainless steel.

TABLE 2 Pipe Die Selection

Nominal Pipe Size	Die Required (Threads Per Inch)	Length of Threaded Area	Number of Threads Resulting
$\frac{1}{8}$"	27	$\frac{3}{8}$"	10
$\frac{1}{4}$"	18	$\frac{5}{8}$"	11
$\frac{3}{8}$"	18	$\frac{5}{8}$"	11
$\frac{1}{2}$"	14	$\frac{3}{4}$"	10
$\frac{3}{4}$"	14	$\frac{3}{4}$"	10
1"	11.5	$\frac{7}{8}$"	10
$1\frac{1}{4}$"	11.5	1"	11
$1\frac{1}{2}$"	11.5	1"	11
2"	11.5	1"	11
$2\frac{1}{2}$"	8	$1\frac{1}{2}$"	12
3"	8	$1\frac{1}{2}$"	12
4"	8	$1\frac{5}{8}$"	13

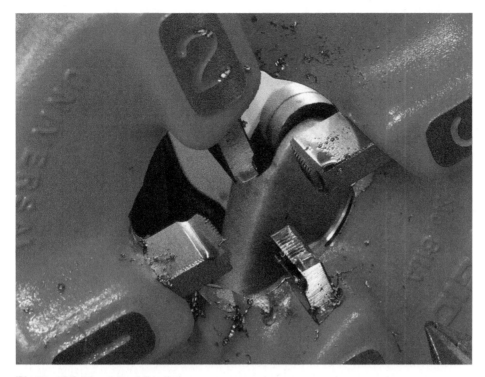

Figure 25 Threading die teeth.

Figure 26 Geared threader assembly.
Source: Ridge Tool Company

The throw-out lever is a quick-release handle used to close the dies into the threading position and then retract them when the task is complete. The lever is marked Open and Close. Lift this handle to the Open position to quickly move the dies away from the pipe. Newer machines with more automation can automatically open the dies when the threader has completed a thread.

Follow these steps to replace the dies in a universal die head:

Step 1   Disconnect the power source. Remove the die head from the machine, and place it on a workbench, with the numbers facing up.

Step 2   Open the throw-out lever.

Step 3   Loosen the clamp lever about three turns.

Step 4   Lift the tongue of the clamp lever washer up and out of the slot under the size bar.

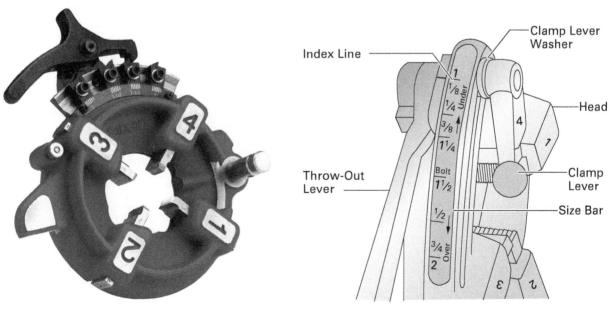

Figure 27 Universal die head, set at 1".
Source: Ridge Tool Company

Step 5 Slide the throw-out lever all the way to the end of the slot in the Over direction.

Step 6 Remove the old dies from the die head.

Step 7 Select a matched set of replacement dies. Clean and inspect them for damage.

Step 8 Insert the #1 die into the 1 position of the die head. Continue installing the dies in their matching positions.

Step 9 Slide the throw-out lever back so that the tongue of the clamp lever washer drops in the slot under the size bar.

Step 10 Adjust the die head size bar until the index line is lined up with the proper pipe size mark.

Step 11 Tighten the clamp lever.

Step 12 Reinstall the die head into the machine.

Checking the Cutting Oil Level

Before any threading operations are performed, check the cutting oil supply. While many machines have an integral lubricant sump and automatically apply cutting oil, oil can also be delivered to the threads manually.

Regardless of how it is applied, cutting oil must be applied liberally while threading to both lubricate and cool the workpiece and dies. This is more important when using a power threader because threading is continuous. The pipe and cutting teeth develop a great deal more heat as a result.

Follow these steps to check the cutting oil level in a threader with a built-in lubrication system:

Step 1 Slide the chip pan out from the base of the threading machine (*Figure 28*). Empty it if necessary.

Step 2 Check to see if the oil is up to the fill-level line in the reservoir. If there is debris on the strainer, remove it.

Step 3 Fill the reservoir with cutting oil as necessary.

Step 4 Lower the lubrication arm over the open reservoir.

Step 5 Turn the power selector switch to the Forward position.

Step 6 Step on the foot switch. Oil should flow from the lubrication arm.

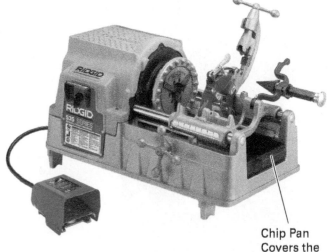

Chip Pan Covers the Oil Sump and Strainer

Figure 28 Threader chip pan.
Source: Ridge Tool Company

Step 7 Turn off the threader.

Step 8 Raise the lubrication arm and reinstall the chip pan.

If the oil does not flow from the lubrication arm when the pump is energized, you may need to prime the pump. Follow the manufacturer's instructions for priming and servicing the pump.

Loading Pipe into the Threader

On most threaders, the pipe can be inserted from either end of the chuck. Some models, however, can only be loaded from the rear. Either way, the operator must position the pipe so that the end to be threaded extends forward from the chuck. Follow these steps to load pipe into a pipe threader:

WARNING!

Before operating a pipe threader, don the required PPE, including safety glasses and work gloves. Pipe threaders have sharp cutting teeth and produce numerous sharp chips. Avoid loose clothing that can be caught in the rotating parts.

Step 1 Measure and mark the length of pipe to be worked if necessary.

Step 2 Insert the pipe into the threader. If the pipe is long enough to be held by the rear centering chuck, insert the pipe through the front or rear of the machine. If the pipe is short, insert the pipe through the front of the machine. For a long section of pipe, support the end behind the threader with a pipe roller or similar adjustable support.

Step 3 Position the end of the pipe so that it is roughly 4" (~10 cm) away from the speed chuck jaws.

Step 4 Make sure that the pipe is centered in the rear centering chuck (*Figure 29*) and turn it clockwise (as viewed from the operator's position) to close the jaws.

Rear Centering Chuck

Figure 29 Rear centering chuck on a threader.
Source: Ridge Tool Company

Step 5   Turn the wheel of the speed chuck clockwise to close the jaws. The wheel is designed to be repeatedly spun against its stop to tighten the jaws of the chuck on the pipe.

Cutting and Reaming Pipe

Follow these steps to cut and ream pipe using a pipe threader:

Step 1   Measure the pipe to determine where the cut should be and use a soapstone to mark the pipe all the way around to indicate the point where the pipe is to be cut. A wraparound (*Figure 30*), or pipe wrap, provides an excellent straightedge.

Step 2   Swing the cutter, thread die, and reamer up and out of the way.

Step 3   Load the pipe into the threader, placing a pipe stand as needed for a long section of pipe.

Step 4   Turn the rear centering chuck to center the pipe in the machine.

Step 5   Turn the speed chuck handwheel to lock the pipe in the chuck jaws.

Step 6   Open the cutter until it fits over the pipe.

Step 7   Lower the cutter onto the pipe and position the cutter wheel to the mark on the pipe.

Step 8   Rotate the cutter handle to tighten the cutting wheel against the pipe (*Figure 31*).

Step 9   Turn the power selector switch to the Forward position.

Step 10 Step on the foot switch to energize the drive and rotate the pipe.

Step 11 Rotate the cutter handle in small increments to apply pressure until the cut is complete.

Step 12 Release the foot switch.

Step 13 Open and raise the pipe cutter out of the way, then lower the reamer into place.

Figure 30 Wraparound, or pipe wrap.   **Figure 31** Pipe cutter lowered to the pipe.

Step 14 Press the safety latch on the reamer and slide the reamer assembly toward the pipe until the latch catches.

Step 15 Step on the foot switch to rotate the pipe.

Step 16 Rotate the reamer handwheel to move the reamer into the end of the pipe, removing any burrs and sharp edges on the inside of the pipe opening (*Figure 32*).

Step 17 Rotate the handwheel to move the reamer out of the pipe and release the foot switch. Raise the reamer assembly out of the way.

Step 18 Turn the power selector switch to the Off position.

Figure 32 Reaming the pipe.

Threading the Pipe

With the pipe properly prepared, it's time to thread the pipe. Follow these steps to thread the pipe:

Step 1 Load the pipe into the threading machine, placing pipe stands or rollers under the pipe as needed.

Step 2 Cut and ream the pipe to the required length.

Step 3 Make sure that the proper dies are in the die head for the pipe being threaded.

Step 4 Loosen the clamp lever and move the size bar to the proper setting for the pipe being threaded, then lock the clamp lever.

Step 5 Swing the die head down to the working position and close the throw-out lever.

Step 6 Lower the lubrication arm and direct the oil supply onto the cutting die.

Step 7 Turn the power selector switch to the Forward position. If the threader has multiple speeds, set the threader to a low speed.

Step 8 Step on the foot switch to start the machine.

Step 9 Turn the carriage to bring the die against the end of the pipe.

Figure 33 Threading the pipe.

Step 10 Apply light pressure on the handwheel to start the die. Release the handwheel once the dies have started to thread the pipe, as they will feed themselves into the pipe once started (*Figure 33*).

Step 11 Open the throw-out lever as soon as one full thread extends from the back of the dies; newer machines may disengage the dies automatically. Then release the foot switch to stop the machine.

Step 12 Turn the carriage handwheel to back the die away from the pipe.

Step 13 Swing the die and the oil discharge line up and out of the way.

Step 14 Check the threads by screwing a fitting onto the end of the pipe. If the threads are correct, you should be able to screw a fitting about 3 to $3\frac{1}{2}$ revolutions onto the new threads by hand.

Step 15 Turn the power selector switch Off.

Step 16 Open the two chucks and remove the pipe.

Threading Machine Care and Maintenance

By following a few simple guidelines, you can expect a long service life from a threader. Use the following guidelines to keep the threading machine in proper working condition:

- Never reverse the rotation of the threader while it is running. Always turn off the power and wait until the drive has stopped rotating before restarting in the opposite direction.

- At the start of each day, clean the chuck jaws using a stiff brush to remove rust, scale, chips, pipe coating, and other foreign matter. Apply lubricating oil to the machine bed rails, the cutter rollers, and the feed screw as necessary.

- Always use sharp dies that will produce smooth threads and require less torque, reducing machine stress.

- Read and follow the manufacturer's instructions on lubrication schedules for bearings and gears.

CAUTION

Make sure that the die remains flooded with cutting oil to prevent overheating both the die and the pipe.

NOTE

If the fitting cannot be threaded on as far as it should, or it threads on too far, adjust the die position setting slightly using the clamp lever on the die head. Cut off the first thread, discard the threaded portion, and repeat the threading operation. Several tests may be required to ensure the threads have the correct depth.

- The cutting oil should be periodically cleaned to remove accumulated sludge, chips, and other foreign matter. Replace the cutting oil when it becomes dirty or contaminated. Use oils designed specifically for threading operations.

- Remove the oil strainer in the sump periodically and clean it with solvent. Blow it clean with compressed air in a protected area. Do not operate the threader without the strainer in place.

2.0.0 Section Review

1. The tool shown in *Figure SR01* is used to _____.
   a. cut pipe
   b. thread pipe
   c. ream pipe
   d. deburr pipe

2. A manual threading operation stops when _____.
   a. five threads have been cut
   b. the threader stops automatically
   c. two threads are visible beyond the die
   d. the threads extend 1' from the end of the pipe

Figure SR01 Pipe cutter.
Source: Ridge Tool Company

3.0.0 Pipe Joining and Installation Procedures

Objective

Describe how to mechanically join and install steel pipe.

a. Describe the tools and techniques used to connect threaded pipe.

b. Describe pipe grooving methods and techniques.

c. Explain how to assemble flanged steel pipe.

d. Explain how to correctly install and support steel pipe.

Performance Task

2. Join threaded pipe or pipe nipples using various fittings.

Once the pipe has been threaded, it can be assembled using various fittings or connected directly to components. In most cases, unions are used to install important components into the piping system. Since the pipe can only be tightened in one direction, unions provide a point where the final joints can be assembled. They also provide a means of removing components for service and replacement without disassembling a great deal of pipe.

3.1.0 Assembling Threaded Pipe

Thread sealants are required for various applications depending on the material to be contained in the piping. Always check to make sure the sealant is compatible with the application. On an engineered project, the project specifications will usually identify accepted sealants.

Apply a joint compound or polytetrafluoroethylene (PTFE) tape to the male threads before assembling a pipe connection (*Figure 34*). PTFE tape, also referred to by the material's trade name of *Teflon®*, is often used to help seal pipe threads. Joint compound, commonly called **pipe dope**, is a putty-like material applied to seal a joint and provide lubrication for assembly.

The dope or tape should be applied only to the male threads. The tape must not hang over the threaded end of the pipe. Pipe dope should also be kept out of the inside of the pipe and fittings.

When using PTFE tape, no more than three wraps should be made. Apply the tape in a clockwise direction looking from the end of the pipe, the same direction as the fitting turns (*Figure 35*). Apply two to three layers, starting one thread from the end of the pipe. Remember to check the local codes to see if the use of tape is permitted.

Pipe dope: A thick but brushable sealant and lubricant applied to pipe threads.

NOTE

A special PTFE tape with a yellow tint is used on natural gas piping. Ensure that you use the correct product.

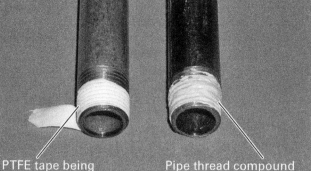

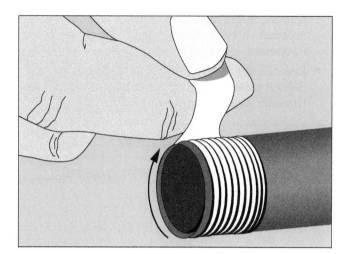

Figure 34 PTFE tape and pipe dope applied to pipe.   **Figure 35** Applying PTFE tape.

3.1.1 Pipe Wrenches

Pipe wrenches are usually sold in the following sizes, based on the length of the handle: 6", 8", 10", 14", 18", 24", 36", 48", and 60". Pipe wrenches are usually made of either cast iron or cast aluminum. Aluminum pipe wrenches are lighter and easier to handle, but they are also more expensive. Teeth and jaw kits can be purchased to repair worn wrenches that have not been seriously damaged.

Straight and offset pipe wrenches (*Figure 36*) are used to grip and turn round stock. They have teeth that are set at an angle. This angle allows the teeth of the wrench to gain grip in one direction only.

The chain wrench is used on pipe that is over 2" in diameter. The chain must be lubricated to prevent it from becoming stiff or rusty. The strap wrench is used to hold chrome-plated or other types of finished or polished pipe. It does not leave jaw marks or scratches on the pipe. Resin applied to the strap reduces slippage and adds to the holding power of the wrench.

Selecting the correct wrench size is important to ensure that the joint is properly tightened. A wrench that is too small does not provide sufficient leverage. A wrench that is too large allows even a small, lightweight worker to overtighten and damage a joint. Refer to *Table 3* to assist in selecting the proper wrench for a given pipe size.

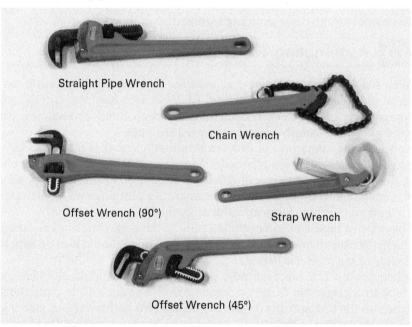

Figure 36 Pipe wrenches.

Using an extension on the wrench handle, such as a length of pipe often referred to as a *cheater bar*, is never recommended. The extension can slip off or the handle of the wrench can fail, causing an injury. Never use a pipe wrench that has a bent handle. Wrenches that are bent should be immediately taken out of service—they can fail in use and cause an injury.

A pipe wrench must be adjusted to fit the pipe properly. Maintain a gap of approximately $1/2$" between the pipe wall and the shank of the hook jaw on the wrench. See *Figure 37*. The adjusting nut should never be at the bottom of its travel.

Start the fitting onto the threaded pipe by hand. Turn the fitting clockwise. Finish tightening the fitting using pipe wrenches. Note the position and rotation of the two wrenches in *Figure 37*. The use of two wrenches may be necessary even when the pipe is firmly held by a pipe vise.

Refer to *Table 1* for the total thread makeup expected for various sizes of threaded pipe. Remember that the dimensions shown in the table are approximate. However, they do provide some guidance in determining how far to tighten a threaded fitting. Overtightening a fitting may create stress that later results in a leak or failure. A loose joint typically causes a leak and weakens the assembly structurally.

TABLE 3 Suggested Wrench-to-Pipe Sizes

Wrench Size	Nominal Pipe Size
6"	$1/8$" to $1/2$"
8"	$1/4$" to $3/4$"
10"	$1/4$" to 1"
12"	$1/2$" to $1\frac{1}{2}$"
14"	$1/2$" to $1\frac{1}{2}$"
18"	1" to 2"
24"	$1\frac{1}{2}$" to $2\frac{1}{2}$"
36"	2" to $3\frac{1}{2}$"
48"	3" to 5"
60"	3" to 8"

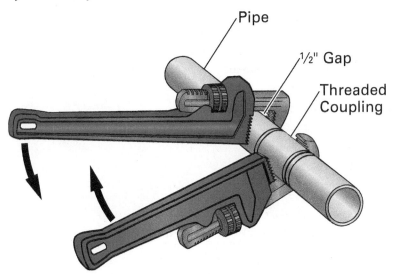

Figure 37 Tightening a joint using two pipe wrenches.

3.2.0 Grooved Pipe

Grooved pipe is so named because grooves are used to join the pieces together instead of welded, flanged, or threaded joints. Each joint in a grooved piping system serves as a union, allowing easy access to any part of the piping system for cleaning or servicing.

Grooved piping systems have a wide range of applications and can be used with a variety of piping materials. The following types of pipe can be joined by grooved couplings:

- Carbon steel
- Stainless steel
- Aluminum
- PVC plastic
- High-density polyethylene
- Ductile iron

A standard grooved pipe coupling consists of a rubber gasket and two housing halves that are bolted together. The housing halves are tightened together until they touch, so no special torquing of the housing bolts is required.

Risers: Pipes that travel vertically. Risers must be supported at each level of the structure as it passes through.

Grooved piping systems offer the option of rigid or flexible couplings. Rigid couplings (*Figure 38*) create a rigid joint useful for **risers**, mechanical rooms, and other areas where positive clamping with no flexibility within the joints is desired. Flexible couplings provide allowance for controlled pipe movement that occurs with expansion, contraction, and deflection. *Figure 39* shows the components and fit of a flexible coupling.

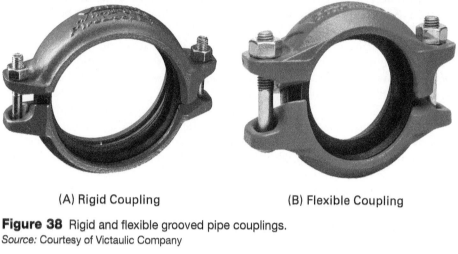

| (A) Rigid Coupling | (B) Flexible Coupling |

Figure 38 Rigid and flexible grooved pipe couplings.
Source: Courtesy of Victaulic Company

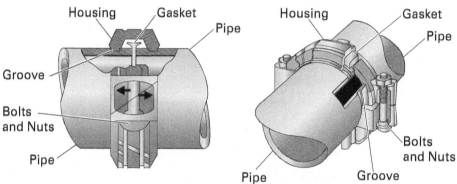

Figure 39 Grooved pipe coupling detail.

There is very little difference in the two types visually. The primary difference is in the way that the fittings fit the grooves. Rigid fittings grip and fit the grooves very tightly. Flexible couplings offer some freedom of movement in the pipe grooves. Flexible couplings help provide a virtually stress-free piping system.

Grooved pipe can be delivered to the jobsite precut to length and grooved, or the pipe can be cut and grooved on the job. Full lengths of pipe often arrive at the site with grooves. Pieces that must be cut to fit at the site are then grooved after the cut is made.

The groove in the pipe must be properly formed to effectively receive the *coupling housing key*. The groove must have enough depth to secure the coupling housing yet leave enough wall thickness for a full pressure rating.

Groove preparation varies with different pipe materials and wall thicknesses. The two methods of forming a groove in pipe are *roll grooving* and *cut grooving*.

Figure 40 shows the details of three different pipe grooves. Note that the cut groove removes material, while the roll groove does not. The dimensions indicated by the letters in the figure must comply with fitting vendor specifications.

- The A dimension is the distance from the pipe end to the first groove wall. This area is where the gasket seals against the wall of the pipe; it must be free from indentations, projections, or roll marks to provide a leakproof surface for the gasket.

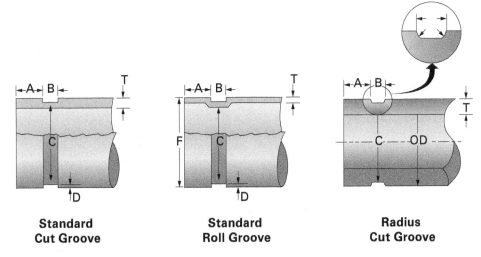

Figure 40 Details of pipe grooves.

- The B dimension is the groove width; it controls expansion and angular deflection based on its distance from the pipe end and its width, relative to the width of the coupling housing key.

- The C dimension is the proper inside diameter tolerance and is concentric with the outside diameter of the pipe. The D dimension, which indicates groove depth, may be changed if necessary to keep the C dimension within the stated tolerances.

- The F dimension is used with the standard roll groove only and represents the maximum allowable pipe end flare. As the groove is rolled in, the end of the pipe tends to flare out from the stress. If it flares too much, the joint cannot be used.

- The T dimension is the lightest grade or minimum thickness of pipe suitable for roll or cut grooving.

- The radius cut groove is often used for cast iron and PVC piping materials, where a sharp angle at the base of the groove creates a point of stress that leads to cracks and joint failure.

3.2.1 Roll Grooving

Power roll-grooving machines (*Figure 41*) are used to roll grooves in pipe. Grooving machines are available to groove 2" to 16" standard and lightweight steel, aluminum, stainless steel, and PVC pipe. The model shown in *Figure 41* is an accessory that can be installed on one of several Ridgid threading machines, increasing the machine's field versatility.

Roll grooving does not remove any metal from the pipe. The process forms a groove by displacing the metal, pushing it into the pipe. Since the groove is cold formed, it has slightly rounded edges that reduce pipe movement after the joint is assembled.

3.2.2 Cut Grooving

Cut grooving differs from roll grooving in that a groove is cut into the pipe, removing material in the process. Cut grooving is used for standard weight or heavier pipe. Pipe weights less than Schedule 40 are not thick enough for cut grooving. The cut removes slightly less than one-half of the pipe wall.

Cut-grooving machines (*Figure 42*) are designed to be rotated around a stationary pipe. This creates a groove that is of uniform depth and is concentric with the outside diameter of the pipe. Grooves can be cut with power equipment or with a stock and cutting head.

Figure 41 Power roll-grooving machine.
Source: Ridge Tool Company

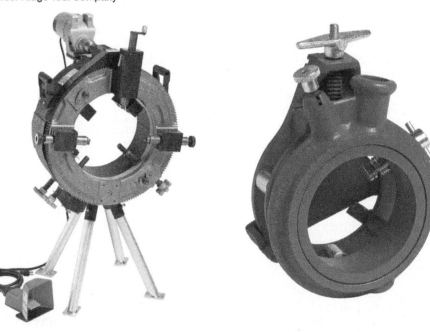

(A) Orbital Power Groover   (B) Manual Grooving Tool

Figure 42 Cut-grooving equipment.
Source: Courtesy of Victaulic Company

3.2.3 Selecting Gaskets

There are a variety of synthetic rubber gaskets available to cover a wide range of applications. To provide the maximum service life, the proper gasket selection is essential.

Several factors must be considered in determining the right gasket for a specific service. The first consideration is the temperature of the product flowing through the pipe. Temperatures beyond the recommended limits decrease gasket life. Also, the concentration of the product, duration of service, and continuity of service must be considered. There is a direct relationship between

In-Air Groover

Another tool that is used to roll grooves is the portable in-air groover. The portable in-air groover is a manually powered machine used to roll-groove piping that is already installed. Portable in-air grooving machines are available to groove a wide variety of pipe, up to Schedule 40 in most materials. In-air groovers require a different roll groove and drive roll to maintain the correct gasket seat and groove width dimension when grooving different types and sizes of pipe. Note the ratcheting drive handle for work in tight spaces.

Source: Ridge Tool Company

these factors and gasket life. It should also be noted that there are services for which specific gaskets are not recommended.

Always refer to the manufacturer's recommendations for selecting a joint gasket. Project specifications will usually identify the specific gasket material to be used for grooved joints.

3.2.4 Installing Grooved Pipe Couplings

Figure 43 and *Figure 44* show the joining sequence for grooved pipe. The procedure used to join grooved pipe is as follows:

Step 1   Check the pipe ends. To make a leakproof seal, the ends must be free from indentations, projections, or roll marks.

Step 2   Check to make sure that the gasket is suitable for the intended use. Some manufacturers color-code their gaskets. Apply a thin coat of lubricant to the lips and the outside of the gasket. Lubricants are typically silicone-based and available from the gasket and/or fitting manufacturer.

Step 3   Install the gasket over the pipe end. Be sure the gasket lip does not hang over the pipe end.

Step 4   Align and bring the two pipe ends together. Slide the gasket into position and center it between the grooves on each pipe. Be sure that no part of the gasket extends into the groove on either pipe.

Step 5   Assemble the housing segments loosely, leaving one nut and bolt off to allow the housing to swing over the joint.

Step 6   Install the housing, swinging it over the gasket and into position in the grooves on both pipes.

Step 7   Insert the remaining bolt and nut. Be sure that the bolt track head engages into the recess in the coupling housing (*Figure 45*).

Step 8   Tighten the nuts alternately and equally to achieve metal-to-metal contact at the angle bolt pads.

Note in *Figure 45* that the two coupling halves meet at an angle. This is a characteristic of one type of rigid (non-flexible) grooved coupling. The angular mating surfaces causes the coupling to pull harder against the walls of the groove, making it more rigid and resistant to axial movement.

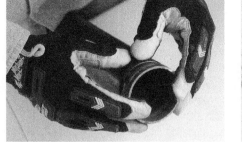

(A)

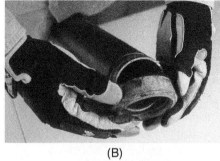

(B)

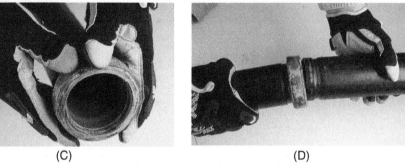

(C)

(D)

(E)

Figure 43 Joining grooved pipe (1 of 2).

(F)

(G)

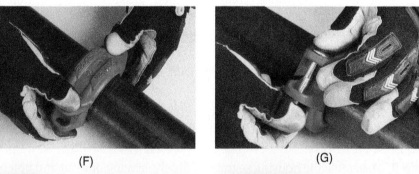

(H)

(I)

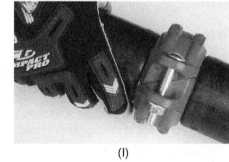

Figure 44 Joining grooved pipe (2 of 2).

Angular Mating Surfaces

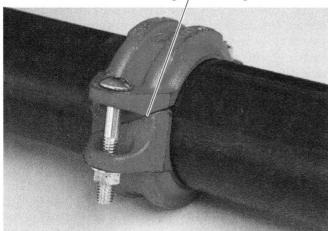

Figure 45 Assembled grooved joint.

Figure 46 Flanged joint.
Source: iStock@lcs813

3.3.0 Flanged Pipe

Larger pipes and valves are sometimes joined with flange fittings (*Figure 46*). Flange fittings are assembled using gaskets and bolts. Flange fittings require a gasket, and the selection is important here to establish a tight and reliable seal. As is the case with grooved pipe, the gasket material must be compatible with the transported material and its temperature range. Flanges can be welded or threaded onto pipe.

Piping systems are rarely assembled entirely with flanges. Most of the time, flanged fittings are used to connect valves and other components, or to provide access points in the piping. Most of the other pipe joints are then threaded, grooved, or welded.

3.3.1 Flange Faces

Flanges are divided into pressure classes assigned by the American Society of Mechanical Engineers (ASME) in *ASME Standard 16.5, Pipe Flanges and Flanged Fittings*. Typical pressure classes for forged carbon steel and cast iron flanges are shown in *Table 4*.

Note that the pressure class, with the number representing pressure in pounds per square inch, does not indicate the maximum pressure the flange can withstand in service under every set of conditions. Within a given pressure class, the maximum pressure rating for a flange is determined by two primary factors: the temperature of the process material and the material from which the flange is made. The design pressure rating for flange classes at various temperatures can be found in the tables of *ASME Standard 16.5*. The ratings are established based on the **hydrostatic testing** of flanges to their bursting point, and then by dividing the burst pressure by a factor of 3.0 to provide a significant safety margin.

Gasket selection is, in turn, affected by the pressure class of the joint because gaskets are also rated for pressure. In addition, the gasket dimensions and bolt patterns vary among flange classes. Differences in physical size and bolt diameters from one pressure class to another help prevent mismatches.

The surface finish of a flange face is an important characteristic. Flange faces are governed by the ASME standard to ensure that all flanges of each type and rating have the same characteristics. The *stock finish* is the standard of the industry (*Figure 47*). It creates a spiral originating from the center, like the groove of a vinyl music record. For this reason, it is described as *phonographic*. The base of the groove is rounded, as shown in *Figure 47*.

TABLE 4 Flange Pressure Classes

Forged Carbon Steel	Cast Iron
Class 150	Class 25
Class 300	Class 125
Class 400	Class 250
Class 600	Class 800
Class 900	
Class 1,500	
Class 2,500	

Hydrostatic testing: A method of testing pressure vessels, piping systems, and similar components for leaks by filling them with a liquid (typically water) and then applying pressure using air or an inert gas to aid in leak detection or failure analysis.

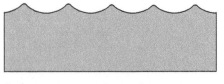

Figure 47 Stock flange-face profile.

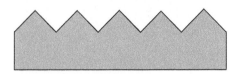

Figure 48 Spiral serrated flange-face profile.

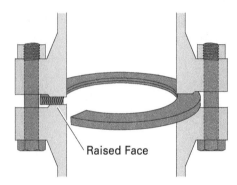

Raised Face

Figure 49 Raised-face flange and gasket assembly.

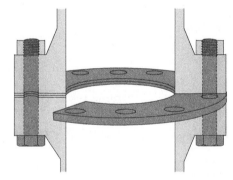

Figure 50 Flat-face flange and gasket assembly.

Another common finish is the spiral serrated finish (*Figure 48*). This results in a phonographic groove as well. However, it is formed with a pointed tool, leaving a 90-degree, V-shaped groove. Stock and spiral serrated finishes allow soft gasket material to compress into the grooves while the peaks of the spirals push into the material, effectively sealing the joint.

Some special applications require a flange with a perfectly smooth face, referred to as a *cold-water finish*. They are typically used without a gasket. However, this is a rare application requiring precise machine work on the two faces.

Other important characteristics of flanges are the shape and dimensions of the face. Raised-face and flat-face flanges are the most common.

Raised-Face Flanges

A raised-face flange has a wide, raised rim around the flange opening. It is the most common face for steel flanges. The raised face on Class 150 and Class 300 flanges is $1/16$" (1.6 mm) high, while the raised face on all other pressure ratings is $1/4$" (6.4 mm) high.

A flat ring gasket (*Figure 49*) is squeezed tightly between the raised faces of the flange. The face typically has a stock finish. Note that the bolts holding the two flange-halves together do not pass through the gasket. The raised-face flange gasket is not as large in diameter as the flange and is only in contact with the raised face. To summarize, raised-face flanges use a simple ring gasket with no bolt holes.

Flat-Face Flanges

The flat-face flange is flat across its entire face, as the name implies. This flange requires a gasket that covers the entire face; therefore, the gasket needs bolt holes (*Figure 50*). Flat-face flanges are commonly used to join Class 150 and Class 300 forged steel flanges with Class 125 and Class 250 cast iron flanges. The brittle nature of cast iron flanges requires that flat-face flanges be used so that full face contact is achieved between the flanges, spreading the stress across a larger area.

Assembling a Flange Joint

A torque wrench should be used to tighten flange bolts. The bolts must be tightened in a specific pattern. A typical pattern is shown in *Figure 51*. Always make sure that the gasket is properly positioned between the faces before tightening the bolts. Inspect the gasket for flaws before inserting it. Then install the bolts and nuts and tighten them finger tight, with the flange faces square to each other. Significant misalignment should not be corrected with the flange bolts. Misalignment must be corrected using proper pipefitting techniques.

Rather than tightening the bolts to the full torque the first time around, torque should be applied in several increments to prevent the joint from becoming distorted. For example, you can tighten the flange to one-third of the final torque on each pass, then tighten to the final torque on the third pass. The appropriate torque can be found in the flange manufacturer's documentation.

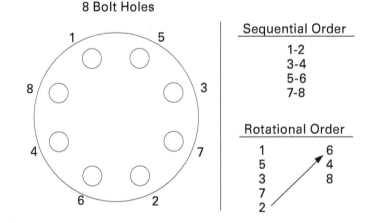

Figure 51 Typical flange bolt tightening pattern.

3.4.0   Installing Steel Piping

When installing steel piping, one important thing to remember is that the methods used are generally defined by the project specifications. Air conditioning and heating installations planned by architects and engineers include plans and specifications that describe the proposed installation in detail. After the job has been awarded to a contractor, the engineering drawings are supplemented by the working drawings of the installation contractor.

Although detailed piping drawings should be available for every piping system that is installed, this is not always the case. Therefore, the HVACR installer must often select the best route and install the piping. This is known as *field-fabricating* or *field-routing* a pipe run. One significant challenge is working around the electrical conduit, ductwork, and other piping system installations.

Some important points to consider when installing steel piping systems are as follows:

- Many specifications require that a pipe hanger or support be provided within 1' of any change in direction. For example, when an elbow is installed at the top of a riser to change it to a horizontal run, a hanger should be placed within 1' of the elbow. Another common requirement is to install a hanger or support within 1' of where the piping connects to equipment. The piping system must support itself and not rely on the equipment for significant support.

- To the extent possible, keep piping runs straight and avoid the excessive use of elbows, tees, and other fittings.

- In hydronic and steam piping systems, make sure piping is installed in accordance with the specifications to reduce noise and avoid damage from **water hammer**, thermal expansion, and vibration.

- Install isolating valves and unions in the piping system so selected portions of the piping can be isolated for repairs. Heavy, serviceable components in a piping system should always have hangers installed adjacent to them on each side.

- Install gas piping systems in accordance with the latest editions of NFPA 54, *National Fuel Gas Code*, as well as local codes and the related equipment manufacturer's installation instructions.

- If the equipment is not yet in place, rough-in the pipe to the approximate location and then plug or cap the pipe. The final piping can be accomplished when the unit is in place. If a material specification sheet exists for the job, check the sheet for the unit rough-in location information.

Water hammer: The noise and shaking in a piping system that occurs when a volume of liquid moving in a pipe suddenly stops or loses momentum.

CAUTION

Always make sure that any sealants used on piping are compatible with the substance being carried in the piping. The solvents in certain substances may dissolve a noncompatible sealant, resulting in dangerous leaks. A certain type of pipe dope is typically required for natural gas, while another type is required for LP (liquefied petroleum) gas. The same is true for PTFE tape. A special, yellow PTFE tape must be used for natural gas.

3.4.1 Pipe Hangers and Supports

Pipe must be properly supported to maintain the integrity of joints. Hangers must be correctly and securely installed and spaced in accordance with the job specifications and local codes.

Steel pipe uses many of the same hanger and support styles used for copper and plastic. However, steel pipe is much heavier and often larger, so anchoring and spacing requirements are much more stringent.

Hangers and supports provide horizontal and vertical support of pipes and piping. Hangers are generally supports that are anchored above the piping. Supports carry the piping load from below or from walls and other structural components. The principal purpose of hangers and supports is to keep the piping in alignment and to prevent it from bending or distorting.

For a hanger system to do its job, it must support the pipe consistently. Evenly spacing the hangers prevents any individual hanger from being overloaded. Proper piping support also relieves components such as pumps from carrying any significant piping load.

Table 5 lists the recommended maximum hanger spacing intervals for carbon steel pipe installed horizontally. Risers should always be supported using a riser clamp at every floor level at a minimum.

TABLE 5 Recommended Maximum Hanger Spacing Intervals for Carbon Steel Pipe

Pipe Size	Rod Diameter	Maximum Spacing
Up to 1¼"	⅜"	8'
1½" and 2"	⅜"	10'
2½" to 3½"	½"	12'
4" and 5"	⅝"	15'
6"	¾"	17'
8" to 12"	⅞"	22'

Table 5 is intended only to serve as an example. Always check the project specifications when determining pipe hanger spacing. Note that larger pipe requires fewer hangers, since the pipe is stronger. Piping systems must run straight without sagging.

It also should be noted that some piping systems must be installed with a pitch in one direction or another. In these cases, hanger rods suspended from a level surface must increase or decrease in length at every hanger to provide the proper pitch.

In areas with the potential for earthquakes, special care is required in selecting and installing pipe supports. Since pipes are attached directly or indirectly to the building structure, seismic action that applies stress to the building structure will also stress pipe joints. In an earthquake, a great deal of the property damage results from broken pipes that release flammable gas, water, or high-pressure steam. Special seismic restraints and methods are used to counteract this effect. The purpose of the restraints is to ensure that the pipe is securely fastened to the structure in the event of excessive vibration. For example, some codes require hangers and supports to be used at closer intervals than in non-seismically active areas.

Seismic action against a structure and its pipes can be lateral (up and down), as well as longitudinal (side to side). Therefore, seismic protection must account for both conditions. There are numerous methods and devices used to protect pipes from seismic action. *Figure 52* provides one example. The two cables limit the side-to-side movement of the hanger. The hanger itself is equipped with a spring to isolate vibration. Springs and other vibration isolation components are primarily selected based on the weight of the object being suspended or supported. Remember that local codes, as well as the project drawings and specifications, specify the type of seismic protection to be used. Failure to follow these requirements will create liability for the contractor and could lead to personal injury.

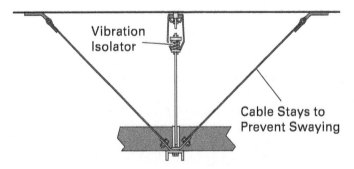

Figure 52 Example of seismic protection for suspended pipe.

There are many ways to support pipe runs. A number of these are covered in NCCER Module 03103, *Basic Copper and Plastic Piping Practices*. Single pipes can be supported using a clevis hanger or pipe clamp as shown in *Figure 53*. *Figure 54* shows examples of methods used to secure piping runs to walls and ceilings. Piping runs are often supported on trapeze hangers suspended from a ceiling or beam. *Figure 55* shows several examples of this approach.

The thermal expansion of pipes must be considered in hot water and steam heating systems. To compensate for thermal expansion, rollers and spring hangers such as those shown in *Figure 56* are used.

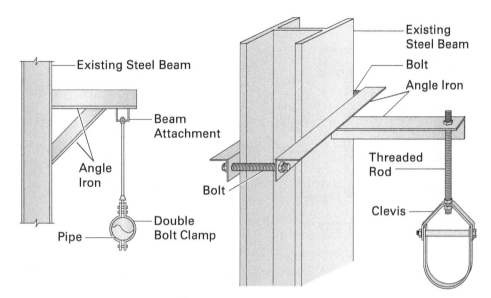

Figure 53 Individual pipe supports.

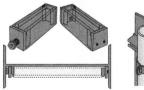

Surface Mounted to Ceiling   **Surface Mounted to Wall**

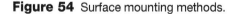

Surface Mounted to Steel Columns

Figure 54 Surface mounting methods.

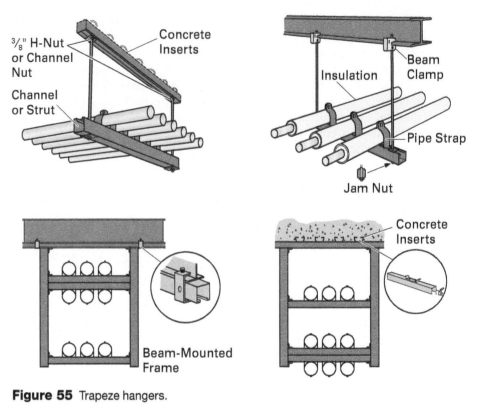

Figure 55 Trapeze hangers.

Single Pipe Roll

Roller Chair

Spring
Cushion Hanger

Adjustable Pipe
Roll Support

Variable
Short Spring Hanger

Light-Duty
Spring Hanger

Figure 56 Rollers and spring hangers.
Source: ASC Engineered Solutions

3.0.0 Section Review

1. The type of wrench that is *best* used on polished stainless steel or chrome-plated pipe is the _____.
   a. chain wrench
   b. offset wrench
   c. straight wrench
   d. strap wrench

2. The two main categories of grooved couplings are _____.
   a. large and small
   b. rigid and flexible
   c. flanged and standard
   d. left-handed and right-handed

3. Flanged fittings should be tightened with a torque wrench.
   a. True
   b. False

4. PTFE tape typically approved for use on natural gas piping is colored _____.
   a. white
   b. yellow
   c. green
   d. blue

Module 03105   Review Questions

1. The type of pipe used for the natural gas supply to a furnace is _____.
   a. galvanized
   b. wrought iron
   c. cast iron
   d. black iron

2. The inside diameter of a given nominal size of Schedule 80 pipe is _____.
   a. greater than that of Schedule 40
   b. less than that of Schedule 40
   c. less than that of Schedule 120
   d. the same as that of Schedule 40 and 120

3. Which of the following schedules of pipe has the *thickest* walls?
   a. Schedule 5
   b. Schedule 20
   c. Schedule 40
   d. Schedule 120

4. All weights of steel pipe have the same _____.
   a. inside diameter
   b. outside diameter
   c. thread diameter
   d. nominal size

5. The standard thread pitch for a $\frac{1}{2}$" pipe size is _____.
   a. $11\frac{1}{2}$ threads per inch
   b. 14 threads per inch
   c. 17 threads per inch
   d. 18 threads per inch

6. A tee fitting described as $2 \times 1\frac{1}{2} \times 1$ has a _____.
   a. large run size of 2", small run size of $1\frac{1}{2}$", and branch size of 1"
   b. large run size of 1", small run size of $1\frac{1}{2}$", and branch size of 2"
   c. large run size of $1\frac{1}{2}$", small run size of 2", and branch size of 1"
   d. large run size of 2", small run size of 1", and branch size of $\frac{1}{2}$"

7. Pipe nipples are used to _____.
   a. close openings in other fittings
   b. connect two lengths of pipe when making straight runs
   c. change the direction of the pipe run
   d. make extensions from a fitting or join two fittings

8. A measurement of the length of a pipe between fittings that includes the threads is a(n) _____.
   a. end-to-center measurement
   b. end-to-end measurement
   c. face-to-end measurement
   d. face-to-face measurement

9. What should be done to the wheels of the cutting tool if the pipe end is flattened while it is being cut?
   a. Sharpen them
   b. Replace them
   c. Polish them
   d. Adjust them

10. Overtightening the cutting wheel on a pipe cutter _____.
    a. causes a larger burr to form on the inside of the pipe
    b. will make the threading process go faster
    c. is better for cutting through galvanized pipe
    d. will make it easier to thread the pipe

11. Before using a threading machine to thread pipe, you should *always* _____.
    a. apply a coat of heavy grease
    b. become familiar with its operating and safety instructions
    c. wipe the unthreaded end of the pipe with cutting oil
    d. apply pipe dope to the unthreaded end of the pipe

12. PTFE tape is installed by wrapping it counterclockwise as viewed from the end of the pipe.
    a. True
    b. False

13. What type of joint uses two housing halves bolted together over the ends of properly prepared pipe ends?
    a. Flanged
    b. Threaded
    c. Grooved
    d. Brazed

14. Generally, 2" steel pipe should be supported with metal hangers at intervals of _____.
    a. 4'
    b. 6'
    c. 8'
    d. 10'

15. A clevis hanger would be used to _____.
    a. secure a pipe to a wall
    b. support multiple pipes suspended from a ceiling
    c. support an individual pipe suspended from a ceiling
    d. provide seismic protection for piping runs

Answers to odd-numbered Module Review Questions are found in *Appendix A*.

Cornerstone of Craftsmanship

Kathleen Egan
Project Manager
NCCER

How did you choose a career in the industry?

I grew up with family and friends that worked in various trades, including HVACR, pipefitting, and equipment maintenance. After completing college and receiving a degree that would likely place me in larger cities, my father and a few friends suggested I learn a skilled trade as well. Then, no matter where I moved, I would always have a skill and a job available to me. I contacted several companies and asked to do ride-a-longs to see what a day of work would be like. In the end, I chose the HVACR trade.

What types of training have you been through?

To learn the HVACR trade, I attended the Advanced Training Institute (ATI) in Las Vegas. Relative to HVACR, I also completed factory training on large chillers in Tampa, FL and in Holland. I've been certified to operate a lot of material handling equipment, including hoist motors, forklifts, boom and scissor lifts, telehandlers, and skid steer loaders. I also have a universal EPA certification.

Beyond my HVACR training, I have earned my MBA in Project Management and hold a Project Management Professional (PMP®) certification.

How important is education and training in construction?

I think it is important in life overall, but in the HVACR trade, it is imperative if you wish to grow. I think confidence is very important; continuing education and training builds more confidence which leads to safe judgment in the field. As Albert Einstein once said, "Once you stop learning, you start dying."

How has training and the construction industry impacted your life?

Working in the industry has taught me to always be objective and look at where other people are coming from. This has helped me bridge communication gaps and excel as a project manager. Craftworkers are largely a brotherhood/sisterhood of people that understand nonverbal communication and respect the hard work of those within the trades.

It has also given me both mental and physical strength that is significantly different from that developed during four years of university. I am now confident that, whatever I face, I can figure out the mechanics of the situation and resolve the problem.

What kinds of work have you done in your career?

I've had several HVACR jobs. I worked for two different firms in Las Vegas, a residential company, and a commercial HVACR contractor. I also worked for Disney on Ice, touring the world for seven years and installing ice floors along with my husband.

Tell us about your current job.

I am currently a project manager at NCCER. I dove headfirst into two major IT projects, so my education continues!

What do you enjoy most about your job?

I enjoy helping people communicate and collaborate on a project. As a project manager, I am continuing to troubleshoot and resolve problems, just as I did as an HVACR technician.

What factors have contributed most to your success?

Hard work. There is no replacement for this. I was always told, "If it was easy, everyone would be doing it." Hard work always pays off.

Remaining humble has always been a factor too. There is always someone who knows more than you. There is always something to learn from every person that crosses your path!

Would you suggest construction as a career to others? Why?

Yes! It can be a career or a chance to skill up. It's an opportunity to learn many things and increase your skills. Once you have learned a skill, no one can take it away from you. It's in your toolbox for life. How you wish to build your future determines the combination, timing of education, and skill development required to get you there.

What advice would you give to those new to the field?

- Being green is not easy, but there is a reason for the process. If you have the grit to stick it out, it gets fun!
- If you aren't sure about something, ask!
- Stay hydrated and take good care of your feet!

Interesting career-related fact or accomplishment?

- I was the Associate MVP at Home Depot for 4 years. I hustled!
- I was the first woman to serve as the Ice Lead for Feld Entertainment. I worked hard for three years toward that promotion. I ran crews of at least thirty people all over the world, working with people who spoke many different languages. I also prompted safety policy changes for the entire European region of Feld Entertainment, improving the workplace.

- I had to lead what we call a *double load in and load out* at The O2 Arena in London over the Christmas holidays. We unloaded and built a 140' ice floor for a week of Disney shows, then disassembled it and moved it out for a one-night New Year's Eve concert. We were back in at 2 a.m. to recreate the ice floor for another week of shows following the concert.
- Another very interesting task was to produce an ice floor in 10 hours for Ice in Brussels (Belgium) with a 12 a.m. start time. That is a very small window of opportunity to create an ice floor!

How do you define craftsmanship?

Taking pride in the job you are doing. The devil is in the details—pay attention to the details!

Answers to Section Review Questions

Answer	Section Reference	Objective
Section 1.0.0		
1. b	1.1.1	1a
2. b	1.2.0	1b
3. d	1.3.5	1c
4. d	1.4.0	1d
Section 2.0.0		
1. a	2.1.0; Figure 17	2a
2. c	2.2.1	2b
Section 3.0.0		
1. d	3.1.1	3a
2. b	3.2.0	3b
3. a	3.3.1	3c
4. b	3.4.0	3d

User Update

SCAN ME

Did you find an error? Submit a correction by visiting **https://www.nccer.org/olf** or by scanning the QR code using your mobile device.

APPENDIX A REVIEW QUESTION ANSWER KEYS

MODULE 03101

Answer	Section Reference
1. a	1.1.0
3. c	1.1.0; 1.3.0
5. c	1.2.0
7. b	1.3.0
9. a	2.2.0
11. a	2.3.1
13. d	3.2.1
15. c	3.3.1

MODULE 03102

Answer	Section Reference
1. c	1.2.1; *Table 4*
3. a	1.2.2
5. d	1.2.4; *Table 5*
7. d	1.4.1
9. a	1.4.5
11. d	2.1.6
13. b	3.1.0
15. d	3.4.2

MODULE 03106

Answer	Section Reference
1. c	1.1.0
3. d	1.2.0; *Table 1*
5. b	1.3.1; *Table 2*
7. d	2.1.2; *Figure 17*
9. a	2.2.0; *Figure 19*
11. c	3.1.1
13. b	3.2.1
15. d	4.1.2
17. b	4.2.2
19. c	4.2.3

MODULE 03108

Answer	Section Reference
1. c	1.1.1
3. b	1.1.4
5. a	1.2.1
7. b	1.2.4
9. c	2.2.1
11. a	2.3.1
13. b	3.1.1
15. a	3.2.1

MODULE 03107

Answer	Section Reference
1. d	1.1.2
3. c	1.1.3
5. a	1.1.3
7. b	1.2.2
9. a	1.3.1
11. d	2.1.1
13. c	2.3.3; *Figure 23*
15. b	3.2.0
17. b	3.4.2
19. b	3.6.7

MODULE 03109

Answer	Section Reference
1. d	1.1.0
3. c	1.2.0
5. a	1.3.1
7. b	2.1.2
9. a	2.4.1
11. c	2.5.1
13. b	2.5.2
15. b	3.3.2

MODULE 03103

Answer	Section Reference
1. c	1.1.0
3. a	1.1.1
5. b	2.1.2
7. c	2.3.2; *Figure 41*
9. a	2.5.0
11. d	3.1.4
13. a	3.2.1
15. c	3.2.2

MODULE 03104

Answer	Section Reference
1. b	1.0.0
3. d	1.1.3
5. d	1.3.0
7. b	2.1.2
9. d	2.2.1

MODULE 03105

Answer	Section Reference
1. d	1.1.0
3. d	1.1.1
5. b	1.2.0; *Table 1*
7. d	1.3.5
9. b	2.1.0
11. b	2.2.2
13. c	3.2.0
15. c	3.4.1; *Figure 53*

MODULE 03106 CALCULATIONS

Question 7

Select the correct formula from the power formulas, based on the result needed and the information available. Since you have the current (*I*) and voltage (*E*) and need the power (*P*), use the following formula:

$$I \times E = P$$
$$2.7A \times 120V = \textbf{324W}$$

APPENDIX 03101A

Samples of NCCER Training Recognition Documents

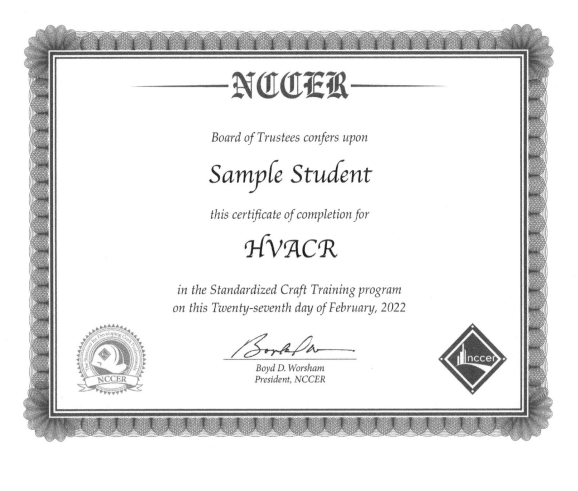

NCCER

Board of Trustees confers upon

Sample Student

this certificate of completion for

HVACR

*in the Standardized Craft Training program
on this Twenty-seventh day of February, 2022*

Boyd D. Worsham
President, NCCER

APPENDIX 03102A

Conversion Factors

US Standard to Metric Conversions		
Weight		
1 ounce	=	28.35 grams
1 pound	=	435.6 grams, or 0.454 kilograms
1 (short) ton	=	907.2 kilograms
Length		
1 inch	=	2.54 centimeters
1 foot	=	30.48 centimeters
1 yard	=	91.44 centimeters or 0.9144 meters
1 mile	=	1.609 kilometers
Area		
1 square inch	=	6.452 square centimeters
1 square foot	=	929 square centimeters or 0.0929 square meters
1 square yard	=	0.836 square meters
Volume		
1 cubic inch	=	16.39 cubic centimeters
1 cubic foot	=	0.02832 cubic meters
1 cubic yard	=	0.765 cubic meters
Liquid Measure		
1 fluid ounce	=	0.095 liters
1 pint	=	473.2 cubic centimeters
1 quart	=	0.926 liters
1 gallon	=	3,785 cubic centimeters or 3.785 liters
Temperature		
°F	=	Use the formula: $°C = \frac{5}{9}(°F - 32)$

Metric to US Standard Conversions

Weight		
1 gram	=	0.0353 ounces
1 kilogram	=	2.205 pounds
1 metric ton	=	2,205 pounds
Length		
1 millimeter	=	0.039 inches
1 centimeter	=	0.394 inches
1 meter	=	3.281 feet or 1.094 yards
1 kilometer	=	0.621 miles
Area		
1 square millimeter	=	0.0016 square inches
1 square centimeter	=	0.155 square inches
1 square meter	=	10.76 square feet or 1.196 square yards
Volume		
1 cubic centimeter	=	0.061 cubic inches
1 cubic meter	=	35.31 cubic feet or 1.308 cubic yards
Liquid Measure		
1 cubic centimeter	=	0.061 cubic inches
1 liter	=	1.057 quarts, 2.113 pints, or 61.02 cubic inches
Temperature		
°C	=	Use the formula: $°F = (\frac{9}{5} \times °C) + 32$

APPENDIX 03106A

Schematic Symbols

Switches

Disconnect	Magnetic Circuit Breaker	Thermal Circuit Breaker	Limit			Maintained Position
			Spring Return			
			Normally Open	Normally Closed	Neutral	

Held Closed   Held Open

Liquid Level		Vacuum & Pressure		Temp. Actuated		Air or Water Flow	
Low	High	Low	High	Normally Open (1)	Normally Closed (2)	Normally Open (1)	Normally Closed (2)

Conductors		Fuses	Coils			
Not Connected	Connected		Relays, Timers, Etc.	Overload Thermal	Solenoid	Transformer

or

Pushbuttons

Single Circuit		Double Circuit	Mushroom Circuit	Maintained Contact
Normally Open	Normally Closed			

Timer Contacts
Contact Action is Retarded When Coil Is:

General Contacts
Starters, Relays, Etc.

Energized		De-energized		Overload Thermal	Normally Open	Normally Closed
Normally Open	Normally Closed	Normally Open	Normally Closed			

(1) Make on Rise
(2) Make on Fall

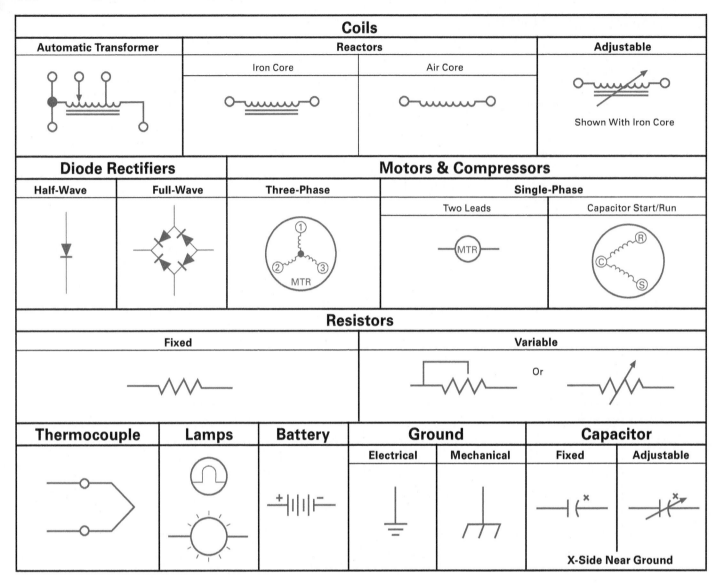

APPENDIX 03107A

Pressure Conversion Chart

Use the following conversion chart to convert common pressure units. First, find the unit of pressure you wish to convert from in the left-hand column. Then, locate the pressure unit you wish to convert to in the top row. The box where they meet shows the conversion factor. Multiply your original unit by the conversion factor to get the converted unit.

For example, if you want to convert 15 psi to atmospheres, multiply 15 by 0.068 to get 1.02 atmospheres.

	Atmospheres	Pounds per sq in (psi)	Kilopascals (kPa)	Kg per sq cm	Inches of Mercury at +32°F (+0°C)	Feet of Water at +60°F (+16°C)	Inches of Water at +60°F (+16°C)	Bar
Atmospheres	1.000	14.700	0.010	1.033	29.921	33.910	406.900	1.013
Pounds per sq in (psi)	0.068	1.000	0.145	0.070	2.036	2.307	27.680	0.069
Kilopascals (kPa)	101.325	6.895	1.000	0.010	0.295	0.335	4.019	0.010
Kgs per sq cm	0.968	14.220	98.067	1.000	28.960	32.840	394.100	0.981
Inches of Mercury	0.033	0.491	3.386	0.035	1.000	1.132	13.590	0.034
Feet of Water	0.029	0.433	2.989	0.030	0.882	1.000	12.000	0.030
Inches of Water	0.002	0.036	0.249	0.003	0.074	0.083	1.000	0.002
Bar	0.987	14.500	100	1.020	29.530	33.490	401.800	1.000

APPENDIX 03105A

Comparison of Imperial and Metric Pipe Sizes (1 of 2)

Nominal Pipe Size (in)	Nominal Pipe Size (mm)	OD (in)	OD (mm)	Schedule Designations (ANSI/ASME)	Wall Thickness (in)	Wall Thickness (mm)	Lb/ft	Kg/m
$\frac{1}{8}$	6	0.405	10.30	10/10S	0.049	1.24	0.1863	0.28
$\frac{1}{8}$	6	0.405	10.30	STD/40/40S	0.068	1.73	0.2447	0.36
$\frac{1}{8}$	6	0.405	10.30	XS/80/80S	0.095	2.41	0.3145	0.47
$\frac{1}{4}$	8	0.540	13.70	10/10S	0.065	1.65	0.3297	0.49
$\frac{1}{4}$	8	0.540	13.70	STD/40/40S	0.088	2.24	0.4248	0.63
$\frac{1}{4}$	8	0.540	13.70	XS/80/80S	0.119	3.02	0.5351	0.80
$\frac{3}{8}$	10	0.675	17.10	10/10S	0.065	1.65	0.4235	0.63
$\frac{3}{8}$	10	0.675	17.10	STD 40/40S	0.091	2.31	0.5676	0.84
$\frac{3}{8}$	10	0.675	17.10	XS 80/80S	0.126	3.20	0.7388	1.10
$\frac{1}{2}$	15	0.840	21.30	5/5S	0.065	1.65	0.5383	0.80
$\frac{1}{2}$	15	0.840	21.30	10/10S	0.083	2.11	0.671	1.00
$\frac{1}{2}$	15	0.840	21.30	STD 40/40S	0.109	2.77	0.851	1.27
$\frac{1}{2}$	15	0.840	21.30	XS 80/80S	0.147	3.73	1.088	1.62
$\frac{1}{2}$	15	0.840	21.30	160	0.188	4.78	1.309	1.95
$\frac{1}{2}$	15	0.840	21.30	XX	0.294	7.47	1.714	2.55
$\frac{3}{4}$	20	1.050	26.7	5/5S	0.065	1.65	0.6838	1.02
$\frac{3}{4}$	20	1.050	26.7	10/10S	0.083	2.11	0.8572	1.28
$\frac{3}{4}$	20	1.050	26.7	STD/40/40S	0.113	2.87	1.131	1.68
$\frac{3}{4}$	20	1.050	26.7	XS/80/80S	0.154	3.91	1.474	2.19
$\frac{3}{4}$	20	1.050	26.7	160	0.219	5.56	1.944	2.89
$\frac{3}{4}$	20	1.050	26.7	XX	0.308	7.82	2.441	3.63
1	25	1.315	33.4	5/5S	0.065	1.65	0.8678	1.29
1	25	1.315	33.4	10/10S	0.109	2.77	1.404	2.09
1	25	1.315	33.4	STD/40/40S	0.133	3.38	1.679	2.50
1	25	1.315	33.4	XS/80/80S	0.179	4.55	2.172	3.23
1	25	1.315	33.4	160	0.25	6.35	2.844	4.23
1	25	1.315	33.4	XX	0.358	9.09	3.659	5.45
$1\frac{1}{4}$	32	1.660	42.2	5/5S	0.065	1.65	1.107	1.65
$1\frac{1}{4}$	32	1.660	42.2	10/10S	0.109	2.77	1.806	2.69
$1\frac{1}{4}$	32	1.660	42.2	STD/40/40S	0.140	3.56	2.273	3.38
$1\frac{1}{4}$	32	1.660	42.2	XS/80/80S	0.191	4.85	2.997	4.46
$1\frac{1}{4}$	32	1.660	42.2	160	0.250	6.35	3.765	5.60
$1\frac{1}{4}$	32	1.660	42.2	XX	0.382	9.70	5.214	7.76

Comparison of Imperial and Metric Pipe Sizes (2 of 2)

Nominal Pipe Size (in)	Nominal Pipe Size (mm)	OD (in)	OD (mm)	Schedule Designations (ANSI/ASME)	Wall Thickness (in)	Wall Thickness (mm)	Lb/ft	Kg/m
1½	40	1.900	48.3	5/5S	0.065	1.65	1.274	1.90
1½	40	1.900	48.3	10/10S	0.109	2.77	2.085	3.10
1½	40	1.900	48.3	STD/40/40S	0.145	3.68	2.718	4.05
1½	40	1.900	48.3	XS/80/80S	0.200	5.08	3.631	5.40
1½	40	1.900	48.3	160	0.281	7.14	4.859	7.23
1½	40	1.900	48.3	XX	0.400	10.16	6.408	9.54
2	50	2.375	60.3	5/5S	0.065	1.65	1.604	2.39
2	50	2.375	60.3	10/10S	0.109	2.77	2.638	3.93
2	50	2.375	60.3	STD/40/40S	0.154	3.91	3.653	5.44
2	50	2.375	60.3	XS/80/80S	0.218	5.54	5.022	7.47
2	50	2.375	60.3	160	0.344	8.74	7.462	11.11
2	50	2.375	60.3	XX	0.436	11.07	9.029	13.44
2½	65	2.875	73	5/5S	0.083	2.11	2.475	3.68
2½	65	2.875	73	10/10S	0.120	3.05	3.531	5.26
2½	65	2.875	73	STD/40/40S	0.203	5.16	5.793	8.62
2½	65	2.875	73	XS/80/80S	0.276	7.01	7.661	11.4
2½	65	2.875	73	160	0.375	9.53	10.01	14.9
2½	65	2.875	73	XX	0.552	14.02	13.69	20.37
3	80	3.500	88.9	5/5S	0.083	2.11	3.029	4.51
3	80	3.500	88.9	10/10S	0.120	3.05	4.332	6.45
3	80	3.500	88.9	STD/40/40S	0.216	5.49	7.576	11.27
3	80	3.500	88.9	XS/80/80S	0.300	7.62	10.25	15.25
3	80	3.500	88.9	160	0.438	11.13	14.32	21.31
3	80	3.500	88.9	XX	0.600	15.24	18.58	27.65
3½	90	4.000	101.6	5/5S	0.083	2.11	3.472	5.17
3½	90	4.000	101.6	10/10S	0.120	3.05	4.973	7.40
3½	90	4.000	101.6	STD 40/40S	0.226	5.74	9.109	13.56
3½	90	4.000	101.6	XS 80/80S	0.318	8.08	12.500	18.6
3½	90	4.000	101.6	XX	0.636	16.15	22.850	34.01

ADDITIONAL RESOURCES

03101 Introduction to HVACR

Apprenticeship.gov. **https://www.apprenticeship.gov/**.

Fundamentals of HVAC/R. Carter Stanfield and David Skaves. Latest Edition. Hoboken, NJ: Pearson Prentice Hall.

NCCER's Build Your Future initiative (**www.byf.org**) aims to be the catalyst for recruiting the next generation of craft professionals. The initiative works to equip industry and education with the resources needed to effectively reach students, educate them about the industry, and promote careers in construction. The site's offerings include a craft labor demand map, interactive construction career path, and a blog targeted to industry and students.

NCCER Module 70101, *Your Role in the Green Environment*. Latest Edition. Alachua, FL: NCCER.

The O*NET Resource Center, the nation's primary source of occupational information. **www.onetcenter.org/**.

Refrigeration and Air Conditioning: An Introduction to HVAC/R. Latest Edition. Larry Jeffus and David Fearnow. Hoboken, NJ: Pearson Prentice Hall.

US Department of Labor, Employment and Training Administration. **https://www.dol.gov/agencies/eta/apprenticeship**.

03102 Trade Mathematics

Math for the Technician. Leo A. Meyer. Haywood CA: Lama Books.

Mathematics for Carpentry and the Construction Trades. Latest Edition. Alfred P. Webster. Upper Saddle River NJ: Prentice Hall.

Metric Conversion Charts and Calculators. **metric-conversions.org**.

03106 Basic Electricity

Electronics Fundamentals: Circuits, Devices, and Applications. Thomas L. Floyd, David M. Buchla, and Gary D. Snyder. 9th Edition. New York, NY: Pearson.

NCCER Module 26102, *Safety for Electricians*.

NCCER Module 26103, *Introduction to Electrical Circuits*.

NCCER Module 26104, *Electrical Theory*.

NCCER Module 26112, *Electrical Test Equipment*.

NCCER Module 26210, *Circuit Breakers and Fuses*.

NCCER Module 26501, *Managing Electrical Hazards*.

Principles of Electric Circuits: Conventional Current. Thomas L. Floyd and David M. Buchla. 10th Edition. New York, NY: Pearson.

03108 Introduction to Heating

National Fuel Gas Code, NFPA 54. Boston, MA: National Fire Protection Association. Available at **http://www.nfpa.org**.

NCCER Module 03202, *Chimneys, Vents, and Flues*.

NCCER Module 03203, *Introduction to Hydronic Heating*.

NCCER Module 03209, *Troubleshooting Gas Heating*.

NCCER Module 03211, *Heat Pumps*.

Refrigeration and Air Conditioning: An Introduction to HVAC/R, Larry Jeffus. Latest Edition. New York, NY: Pearson Education Inc.

03107 Introduction to Cooling

International Institute of Ammonia Refrigeration (IIAR). **www.iiar.org**.

Refrigeration and Air Conditioning: An Introduction to HVAC/R. Latest Edition. New York, NY. Pearson.

The US Clean Air Act. **http://www.epa.gov/air/caa/**.

03109 Air Distribution Systems

Air Distribution Systems: Introduction to Thermo-Fluids Systems Design. 2012. A. G. McDonald and H. L. Magande. Chichester, UK: John Wiley & Sons, Ltd.

Mechanical Insulating Level One, NCCER. Latest Edition. 2018. New York, NY: Pearson.

Mechanical Insulating Level Two, NCCER. Latest Edition. 2018. New York, NY: Pearson.

Mechanical Insulating Level Three, NCCER. Latest Edition. 2018. New York, NY: Pearson.

NFPA 90, Installation of Air Conditioning and Ventilation Systems. Latest Edition. Quincy, MA: National Fire Protection Association.

Weatherization Technician. Latest Edition. NCCER. New York, NY: Pearson.

03103 Basic Copper and Plastic Piping Practices

The Copper Tube Handbook. A publication of the Copper Development Association. **https://www.copper.org/applications/plumbing/cth/**.

The Plastic Pipe and Fittings Association. **https://www.ppfahome.org/**.

Plastics Pipe Institute, Inc. **https://plasticpipe.org//**.

03104 Soldering and Brazing

Filler Metal Selection Guide. The Harris Products Group. Available at **www.harrisproductsgroup.com**.

Gas Welding, Cutting, Brazing, & Heating Torch Instruction Manual. The Harris Products Group. Available at **www.harrisproductsgroup.com**.

A Guide to Brazing and Soldering. The Harris Products Group. Available at **www.harrisproductsgroup.com**.

www.brazingbook.com. Lucas-Milhaupt, Inc. – An interactive resource on the subject of brazing.

03105 Basic Carbon Steel Piping Practices

The Pipefitter's Blue Book. Latest Edition. W.V. Graves. Webster, TX: Grave Publishing Company.

Pipefitter's Handbook. Latest Edition. Forrest Lindsey. New York, NY: Industrial Press, Inc.

GLOSSARY

Absolute pressure: The pressure measured relative to a perfect vacuum, where there is no measurable pressure. Absolute pressure is expressed in pounds per square inch absolute (psia). Absolute pressure = gauge pressure + atmospheric pressure.

Acetone: A colorless organic solvent that is volatile and extremely flammable. In the HVACR trade, it is used as a carrier for acetylene gas in cylinders.

A_k factor: An area factor used with registers, grilles, and diffusers that reflects the free area, relative to a square foot, for airflow.

Alloy: Any substance made up of two or more metals.

Alternating current (AC): An electrical current that changes direction on a cyclical basis.

Ampere (A): The basic unit of measurement for electrical current, represented by the letter *A*; typically spoken and written as *amp*.

Analog meter: A meter that uses a needle to indicate a value on a scale.

Anhydrous: Containing no water.

Annealed: Describes a metal that has been heat-treated, making it more formable yet tougher after it cool.

Annealing: A process in which a material is heated to form it and then cooled to toughen it.

Annual Fuel Utilization Efficiency (AFUE): Industry standard for defining furnace efficiency, expressed as a percentage of the total heat available from the relevant fuel. The AFUE considers the thermal efficiency of a unit as well as the losses of efficiency that occur during startup, warmup, and shutdown.

Arc: A visible flash of light and release of heat that occurs when electrical currents cross an air gap.

Arc flash: The intense light and heat produced by an electrical energy release such as an *arc fault* that occurs when an electrical connection occurs through the air to ground or by a connection between different phases of power.

Area: The amount of space contained in a two-dimentional object such as a rectangle, circle, or square.

Atmospheric pressure: The pressure exerted on Earth's surface by the weight of the atmosphere, equal to 14.7 psi, or 29.92 inHg, for 70°F air at sea level.

Atomized: Broken into tiny pieces or fragments, such as a liquid being broken into tiny droplets to create a fog-like spray.

Balance point: An outdoor temperature value that represents the matching of a structure's heat loss with the capacity of a given heat pump to produce heat. At the balance point, a heat pump must run continuously to maintain the indoor setpoint, without falling below or rising above it.

Barometer: An instrument used to measure atmospheric pressure, typically in inches of mercury (inHg).

Barometric pressure: The actual atmospheric pressure at a given place and time.

Black iron pipe: A term used to describe carbon steel pipe that gets its black coloring from the carbon in the steel.

Blower door: An assembly that fits a door frame or other opening and contains a fan used to depressurize a building by drawing the air out.

Boots: Sheet metal fittings designed to transition from the branch duct to the receptacle for the register, grille, or diffuser to be installed.

Brazing: A heat-based method of joining metals using another alloy with a melting point lower than the metal(s) being joined. The alloy serves as a filler metal to bond and fill any gaps between the pieces by capillary action. Brazing uses filler metals that have a melting point above 842°F (450°C).

Brine: Water containing large amounts of salt. In the HVACR industry, the term typically describes any water-based mixture that contains salt or glycols to lower its freezing point.

British thermal unit (Btu): The amount of heat needed to raise the temperature of one pound of water by one degree Fahrenheit.

Capillary action: The movement of a liquid along the surface of a solid in a kind of spreading action.

Carbon monoxide (CO): A colorless, tasteless, odorless, and toxic gas that is lighter than air and a common byproduct of combustion processes. CO reacts with blood hemoglobin to form a substance that significantly reduces the flow of oxygen to all parts of the body.

Cast iron: A brittle, hard metal containing between 2 and 4 percent of carbon and small amounts of other substances such as manganese and silicon. Some impurities, such as phosphorus and sulfur, are also present.

Chamfer: Breaking what would usually be a 90-degree angle to create a symmetrical angled surface on an edge. Chamfer is also used as a noun, as a name for the angled edge.

Chiller: In the HVACR industry, refers to a machine that uses the mechanical refrigeration cycle to cool water for industrial processes or comfort cooling.

Chlorofluorocarbon (CFC): A compound used as the basis for a class of refrigerants that contain chlorine, fluorine, and carbon.

Clamp-on ammeter: An ammeter with operable jaws that are placed around a conductor to sense the magnitude of the current flow.

Coefficients: Numerals that multiply a variable, e.g., the numeral 2 in the math expression $2b$.

Combustion: The process by which a fuel is burned in the presence of oxygen.

Comfort cooling: Cooling related to the comfort of building occupants. The temperature range for comfort cooling is considered to be 60°F (~16°C) to 80°F (~27°C).

Compounds: As related to refrigerants, substances formed by a union of two or more elements in definite proportions by weight; only one molecule is present.

Compressor: In a refrigerant circuit, the mechanical device that converts low-pressure, low-temperature refrigerant vapor into a high-temperature, high-pressure refrigerant vapor.

Condensate: Water produced by the condensation of water vapor.

Condenser: A heat exchanger that transfers heat from a refrigerant to air or water flowing over it, enabling the change of state from a vapor to a liquid.

Condensing furnace: A furnace equipped with a secondary heat exchanger that extracts additional heat from flue gases that would otherwise be vented to the atmosphere.

Condensing units: Cooling equipment that typically contains a compressor, condenser coil, condenser fan, and other devices required for compressor control and operation.

Conduction: The process by which heat is directly transferred between materials in contact when there is a difference in temperature.

Conductor: Relevant to electricity, a material through which an electrical current easily passes; Relevant to heat transfer, a material that readily accepts or transfers heat by conduction.

Constants: Elements in an equation with fixed values.

Contactor: Control device consisting of a magnetic coil and one or more sets of contacts used as switching devices in line-voltage, high-current circuits.

Continuity: Having a continuous path for current to follow. The absence of continuity indicates an open circuit.

Convection: The air movement caused by the tendency of hotter air to rise and colder, denser air to fall, resulting in heat being transferred.

Cooling tower: A heat-rejection unit that moves air through falling water to cool it.

Copper plating: The bonding of a thin layer of copper to another metal surface. When plating is done intentionally, it is an electro-chemical process using an electric current to encourage and strengthen the bond. In a refrigerant circuit, it is a problem that occurs when copper molecules are in circulation, primarily at locations where there is a friction fit between two other metal surfaces.

Copper-clad: Components that have been coated or covered with a thin copper layer.

Cup depth: The distance that a tube inserts into a fitting, usually determined by a stop inside the fitting.

Current: The rate at which electrons flow through an electrical circuit.

Desiccant: A material or substance, such as silica gel or calcium chloride (CaCl), that seeks to absorb and hold water from any adjacent source, including the atmosphere.

Dew point: The temperature at which air becomes saturated with water vapor (100 percent relative humidity).

Digital meter: An instrument that provides a direct numerical reading of a measured value.

Direct current (DC): An electric current that flows in only one direction, such as the current provided by a common battery.

Dry-bulb temperature: The temperature measured using a standard thermometer, representing the sensible heat present.

Dual-fuel system: A heating system typically comprised of a heat pump and a fossil-fuel furnace. Heat pumps supplemented by electric heat are not considered dual-fuel systems.

Electromagnet: A coil of wire wrapped around a soft iron core that, when energized, creates a powerful magnetic field.

Electronically commutated motor (ECM): A DC-powered motor that operates at varying speeds based on the programming of its AC-powered electronic control module that also converts AC to three-phase DC. ECMs operate more efficiently than standard motors.

Enthalpy: The total heat content (sensible and latent) of a refrigerant or other substance.

Evaporator: A heat exchanger that transfers heat from the air or liquid flowing over it to the cooler refrigerant flowing through it, typically causing the refrigerant to vaporize.

Expansion device: A device that provides the pressure drop to reduce high-pressure liquid refrigerant to a lower pressure and significantly reducing the temperature at which it boils. Also known as a *metering device*.

Exponent: A small figure or symbol placed above and to the right of another figure or symbol to show how many times the latter is to be multiplied by itself (e.g., $b^3 = b \times b \times b$).

External static pressure (ESP): The cumulative resistance of all obstructions and ductwork in an air distribution system beyond the blower assembly itself.

Ferrule: A ring or bushing placed around a tube that squeezes or bites into the tube when compressed, forming a seal.

Fillet: A rounded internal corner or shoulder of filler metal often appearing at the meeting point of a piece of tubing and a fitting when the joint is soldered or brazed.

Fitting allowance: The distance from the end of the pipe to the center of the fitting. Also called *takeoff*.

Flame rectification: A method of proving the existence of a pilot flame by applying an AC current to a flame rod, which is then rectified to a DC current as it flows back to a ground source. Monitoring for a DC current flow provides the means of proving a pilot flame has been established.

Flange: A flat plate-like fitting attached to equipment or the end of a pipe, assembled with a gasket between the plates and held together with nuts and bolts.

Flashback arrestors: Valves that prevent a flame from traveling back from the tip and into the hoses.

Fluorocarbon: A compound formed by replacing one or more hydrogen atoms in a hydrocarbon with fluorine atoms.

Flux: A chemical substance that prevents oxides from forming on the surface of metals as they are heated for soldering, brazing, or welding.

Force: A push or pull on a surface.

Fusible link: Mechanically, a device that holds a fire damper in the open position and is designed to melt at a specific temperature. Once melted, the damper opens by spring pressure. Fusible links are not reusable and must be replaced; An electrical safety device that melts to open a circuit when exposed to excessive heat or a current that causes excessive heat.

Galvanized: Describes steel with a zinc-based coating to prevent rust.

Gauge pressure: Pressure measured on a standard gauge, expressed as pounds per square inch gauge (psig), that is calibrated to read zero with standard atmospheric pressure applied.

Geothermal heat pump: A heat pump that transfers heat to or from the ground, using the earth as a source of heat in the winter and a heat sink in the summer. The system uses water as the heat-transfer medium between the earth and the refrigerant circuit.

Greenhouse effect: Describes the process through which the sun's warmth and the heat generated from Earth becomes trapped in the lower atmosphere, primarily due to gases in the atmosphere that prevent the heat from being radiated to space.

Ground fault: An unintentional, electrically conducting connection between an energized conductor and another conductor connected to earth, such as an equipment frame, or directly to earth. The fault current flows to the earth ground, as well as through a person or other conductive object in the path.

Ground fault circuit interrupters (GFCIs): Devices that interrupt and de-energize an electrical circuit to protect a person from electrocution in the event of a ground fault.

Halocarbon: Any compound containing carbon and one or more halogens (chlorine, fluorine, bromine, astatine, and iodine). For example, HCFC-22 contains chlorine, making it a halocarbon. Refrigerants bearing chlorine have been or are being phased out for environmental reasons.

Halogens: A term used to identify substances containing one or more of the elemental halogens chlorine, fluorine, bromine, astatine, or iodine.

Hard-drawn: Describes a material, such as hard-drawn copper tubing, that has been drawn through dies to form its shape and dimensions. Each die is progressively smaller than the previous until the desired size is reached.

Hazard Risk Assessment: A process, often conducted by safety personnel, used to identify and evaluate the potential hazards associated with a task and develop ways to control or eliminate them.

Heat exchanger: A component used to transfer heat between fluids while keeping the fluids physically separated.

Heat of compression: Heat energy added to a vapor during the compression cycle. For compressors that place the motor in the vapor stream, motor heat is also added to the vapor.

Heat pumps: Comfort systems or water heaters that can produce heat by reversing the standard mechanical refrigeration cycle.

Heat transfer: The movement of heat from a warmer substance to a cooler substance.

Hot surface igniter (HSI): A ceramic device that becomes extremely hot when current is passed through it, used to ignite a fuel/air mixture.

Hydrocarbons: Compounds containing only hydrogen and carbon atoms in various combinations.

Hydrochlorofluorocarbon (HCFC): A compound used as the basis for a class of refrigerants that contain hydrogen, chlorine, fluorine, and carbon.

Hydronic: Describes a system that uses water as a heat transfer medium.

Hydrostatic testing: A method of testing pressure vessels, piping systems, and similar components for leaks by filling them with a liquid (typically water) and then applying pressure using air or an inert gas to aid in leak detection or failure analysis.

Induced-draft furnaces: Gas furnaces in which a motor-driven fan creates (induces) a draft through the heat exchanger, assisting the passage of combustion byproducts and enabling more effective heat exchanger designs.

Inductive loads: Electrical loads that operate by converting electrical current into magnetism, which is often used to do mechanical work. An electric motor is a good example of an inductive load.

Infiltration: The undesirable and uncontrolled movement of air into a building through cracks and crevices.

Inrush current: A significant rise in electrical current associated with energizing inductive loads such as motors.

Insulator: A substance that inhibits the flow of electrical current; the opposite of conductor. It can resist heat transfer.

Ladder diagram: A simplified schematic diagram in which the load lines are arranged like the rungs of a ladder between vertical lines that represent the voltage source.

Latent heat: The heat energy absorbed or rejected when a substance is in the process of changing state (e.g., solid to liquid, liquid to gas, or the reverse of either change) without a change in temperature.

Latent heat of condensation: The heat released from a vapor when changing to its liquid state, e.g., steam condensing to water.

Latent heat of fusion: The heat released from or absorbed by a liquid when changing to or from a solid, e.g., ice to water or water to ice.

Latent heat of vaporization: The heat absorbed by a liquid when changing to a vapor state, e.g., water to steam.

Line duty: Describes a control that is capable of carrying the same current that a load consumes, installed in the load's power circuit.

Loads: Devices that convert electrical energy into another form of energy (heat, mechanical motion, light, etc.). Motors are one of the most common electrical loads in HVACR systems.

Makeup: The length of the threads that engage and penetrate a pipe fitting.

Malleable: A characteristic of metal that allows it to be pressed or formed to some degree without breaking or cracking. For example, copper is malleable, while cast iron is not.

Malleable iron: Iron that has been toughened by gradual heating or slow cooling.

Mass: The quantity of matter present.

Mastic: A flexible sealant that is painted or sprayed to seal duct joints and insulation seams.

Mechanical refrigeration cycle: A process that depends on machinery and phase-changing, heat-transfer mediums (refrigerants) to provide a cooling effect, typically using the refrigerant to absorb heat from one location and transfer it to another. May also be referred to as the *vapor compression cycle*.

Mixture: As related to refrigerants, a blend of two or more components that do not have a fixed proportion to one another and that, however thoroughly blended, are conceived of as retaining a separate existence; more than one type of molecule remains present.

Motor starters: Magnetic switching devices used to control heavy-duty motors.

Multimeter: A test instrument capable of reading voltage, current, and resistance. Also known as a *volt-ohm-milliammeter* (*VOM*).

Multipoise furnace: A furnace that can be configured for upflow, counterflow, or horizontal installation, relative to airflow.

Natural-draft furnaces: Gas furnaces that rely on the natural tendency of hot air to rise to support combustion and assist the passage of combustion byproducts through the heat exchanger.

Newton (N): The amount of force required to accelerate one kilogram at a rate of one meter per second.

Nipples: Short lengths of pipe that are used to join fittings. Nipples are less than 12" long and have male threads on each end.

Nonferrous: A term used to describe metals or metal alloys that do not contain iron.

Noxious: Harmful to health; potentially poisonous or fatal.

Ohms (Ω): The basic unit of measurement for electrical resistance, represented by the symbol Ω (omega).

Orifices: Small, precisely drilled holes.

Oxidation: The process of combining with oxygen at a molecular level. Copper and oxygen join to form copper oxides, appearing as darkened deposits on the copper surface. Common rust is essentially iron oxide.

Pilot duty: Describes a control device that cannot carry the current of a significant load, and is therefore installed in a control circuit, typically controlling the action of a line duty device.

Pipe dope: A thick but brushable sealant and lubricant applied to pipe threads.

Pitch: The number of threads per inch on threaded pipe and other threaded components.

Pitot tube: A special tube housing a second tube used to capture pressure measurements from a moving air stream.

Plenum: A chamber at the inlet or outlet of an air handling unit. Ducts typically attach to a plenum.

Polygons: Shapes formed when three or more straight lines are joined in a regular pattern.

Power: The rate of doing work, or the rate at which energy is dissipated. Electrical power is measured in *watts*.

Pressure drop: The reduction in pressure between two points in a pipe or tube. Pressure drop results from friction in piping systems, which robs energy from the flowing fluid or vapor.

Primary air: Air that is introduced into a burner and mixes with the gas before combustion occurs.

Psychrometric chart: A comprehensive chart that presents the important properties of air and water vapor as they relate to each other.

Pump-down control: A control scheme that includes a refrigerant solenoid valve that closes to initiate the system off cycle. Once the valve closes, the compressor continues to operate, pumping most or all the remaining refrigerant out of the evaporator coil. Pump-down control eliminates the possibility of excessive refrigerant condensing in the evaporator coil during the off cycle. Primarily used in refrigeration applications, a thermostat typically controls the solenoid valve, while a low-side pressure switch and/or a timer controls the compressor.

R-value: A number, such as R-19, that indicates the ability of a material to resist the flow of heat. The higher the R-value, the better the insulating ability.

Radiation: In the context of heat transfer, the transmission of heat by electromagnetic waves passing through air; also referred to as *thermal radiation*.

Reaming: Using a tool to remove the sharp lip and burrs left inside a pipe or tube after cutting.

Reclaim: To process a used or contaminated refrigerant until it is returned to the standards of purity required of new refrigerant.

Recover: To remove refrigerant from a system and temporarily store it in containers approved for that purpose.

Rectifier: A device that converts AC voltage to DC voltage.

Recycle: To circulate recovered refrigerant through filtering devices in the shop or field that remove moisture, acid, and other contaminants.

Redundant gas valves: Gas-flow control valves that contain two gas valves in series.

Refrigerant floodback: A significant amount of liquid refrigerant that returns to the compressor through the suction line during operation. Refrigerant floodback can have several causes, such as a metering device that overfeeds refrigerant or a failed evaporator fan motor.

Relative humidity (RH): The ratio of moisture present in a given sample of air compared to the volume of moisture it can hold when saturated (at dew point), expressed as a percentage.

Relay: A magnetically operated control device consisting of a coil and one or more sets of contacts.

Resistance: An electrical property that opposes the flow of current through a circuit. Resistance is measured in *ohms* and represented by the letter *R* in formulas.

Resistive loads: Electrical loads that consume only active power, with no magnetic field involved. Heating elements found in electric furnaces, stoves, and toasters are examples of resistive loads.

Revolutions per minute (rpm): The common unit of measure for rotational speed.

Risers: Pipes that travel vertically. Risers must be supported at each level of the structure as it passes through.

Rupture disk: A pressure-relief device that protects a vessel or other container from damage if pressures exceed a safe level. A rupture disk typically consists of a specific material at a precise thickness that will break or fracture when the pressure limit is reached, creating a controlled weakness, thereby protecting the rest of the container from damage.

Saturation temperature: The temperature at which liquid changes to a vapor at a given pressure, which is also its boiling point. In a saturated condition, both liquid and vapor are likely present in the same space, as is the state of refrigerant stored in a cylinder.

Secondary air: Air that is introduced into a burner above or around the combustion process to further support combustion and venting of the combustion byproducts.

Sectional boiler: A boiler consisting of two or more similar sections that contain water, with each section usually having an equal internal volume and surface area. Boiler sections are generally made from cast iron and are shipped in pieces to be assembled on-site.

Sensible heat: Heat energy that can be measured by a thermometer or sensed by touch.

Short circuit: The bypassing of a load by a conductor, allowing two separate energized conductors to make direct contact; causes instantaneous and extreme current flow due to the loss of resistance.

Slugging: Traditionally refers to a significant volume of oil returning to the compressor at once, primarily at startup. For example, a trap in a suction line may fill with oil during the off cycle, and then leave the trap as a *slug* at start-up. The term is incorrectly used by some to describe refrigerant floodback.

Solder: A fusible alloy used to join metals, with melting points below 842°F (450°C).

Soldering: A heat-based method of joining metals using another alloy with a melting point lower than the metal(s) being joined. The alloy is used as a filler metal to bond and fill any gaps between the pieces by capillary action. Soldering uses filler metals that have a melting point below 842°F (450°C).

Solenoid coil: An electromagnetic coil of wire used to control a mechanical device such as a valve or a relay.

Specific heat: The amount of heat required to raise the temperature of one pound of a substance by one degree Fahrenheit. Expressed as Btu/lb/°F. The specific heat of water (H_2O) is the standard at a specific gravity of 1.00.

Spuds: Threaded metal fittings that screw into the gas manifold, into which a gas orifice is installed.

Standing pilot: A gas pilot flame that remains continuously lit.

Static pressure (SP): The pressure exerted uniformly in all directions by air within a duct system, usually measured in inches of water column (w.c.) or centimeters of water column (cm H_2O).

Stepper motor: A small DC-powered motor that moves in discrete steps and can be commanded to any position, with each step representing a small fraction of a complete 360-degree rotation.

Stock: A tool used to hold and turn threading dies to manually cut pipe threads.

Stratify: To form or arrange into layers. In air distribution, layers of air at different temperatures will tend to stratify unless they are forced to move and mix. Warmer air stratifies above cooler air.

Subcooling: The reduction in temperature of a refrigerant liquid after it has completed the change in state from a vapor. Only after the phase change is complete can the liquid begin to decrease in temperature as additional heat is removed.

Superheat: The increase in temperature of a refrigerant vapor after it has completed the change of state from a liquid to a vapor. Only after the phase change is complete can the vapor begin to increase in temperature as additional heat is absorbed.

Surety bond: A funded guarantee that a contractor will perform as agreed.

Surge chamber: A vessel or container designed to hold both liquid and vapor refrigerant. The liquid is generally fed out of the bottom into an evaporator, while vapor is drawn from the top of the container by the compressor to maintain refrigerant temperature through pressure control.

Sustainable construction: Construction that has a minimum effect on the land, natural resources, and raw materials throughout a building's life cycle, and generally results in reduced energy consumption as well.

Swaging: The process of using a tool to shape metal. In context, it describes the process of forming a socket at the end of a copper tube that is the correct size to accept another piece of tubing, acting as a coupling.

Takeoffs: Connection points installed on a trunk duct that allow the connection of branch ducts.

Thermal conductivity: The measure of a material's capacity to conduct heat, expressed as the volume of heat movement over some unit of time. Metals typically offer high levels of thermal conductivity, while materials like spun fiberglass and foam insulation do not conduct heat well.

Thermistor: A semiconductor device that changes resistance with a change in temperature, typically used as a temperature sensor.

Thermocouple: A device comprised of wire made from two different metals joined at the end that generates a scalable voltage based on the temperature of the junction.

Thermoplastic: Describes plastic materials that become more plastic (elastic) when heated and harden again when cooled and can generally do so repeatedly.

Thermoset: Describes plastic materials that are irreversibly hardened when manufactured and cannot be melted and reformed.

Threading dies: Cutting tools used to cut external threads by hand or using a machine.

Time-delay fuses: Fuses with a built-in time delay to accommodate the inrush current of inductive loads.

Ton of refrigeration: Unit for measuring the rate of heat transfer. One ton is defined as 12,000 Btus per hour, or 12,000 Btuh.

Total pressure: The sum of the static pressure and the velocity pressure in an air duct.

Transfer grille: A grille usually installed on walls or doors, with one mounted on each side of an opening, that allows air to pass freely into or out of an enclosed space.

Transformer: Two or more coils of wire wrapped around a common core. Used to raise and lower voltages.

Transitions: Joints that accommodate a change in duct size.

Ultraviolet (UV) light: A form of electromagnetic radiation naturally generated by the sun that has a shorter wavelength than visible light. UV light can also be generated artificially and used to destroy bacteria, viruses, and similar contaminants.

U-tube manometer: A U-shaped instrument that measures small pressures by displacing a column of liquid.

Vacuum: Any pressure that is less than the prevailing atmospheric pressure; the absence of positive pressure.

Vapor retarder: A barrier placed over insulation to stop water vapor from passing through the insulation and condensing on a cold surface.

Variables: Elements of an equation that may change in value.

Velocity: The speed at which air is moving, usually measured in feet per minute.

Velocity pressure: The pressure developed in a duct due to the linear movement of the air.

Venturi: A ring or panel surrounding the blades on a propeller fan to improve fan performance.

Vestibule: The compartment of a gas furnace that houses the gas valve, manifold, and pilot assembly.

Voltage: The driving force that makes current flow in a circuit. Voltage, often represented by the letter *E* in formulas, is measured in *volts*. Also known as *electromotive force* (*emf*), voltage represents the difference of electrical potential.

Volts (V): The unit of measurement for voltage, represented by the letter *V*. One volt is equivalent to the force required to produce a current of one ampere through a resistance of one ohm.

Volume: The amount of space contained within a three-dimensional shape.

Water hammer: The noise and shaking in a piping system that occurs when a volume of liquid moving in a pipe suddenly stops or loses momentum.

Watts (W): The unit of measurement for power consumed by a load.

Wet-bulb temperature: Temperature measured with a thermometer fitted with a wetted wick wrapped around the sensing bulb. The bulb is exposed to a moving airstream and is cooled by evaporation, the rate of which depends on the amount of moisture in the airstream.

Wetting: A process that reduces the surface tension so that molten (liquid) solder flows evenly throughout the joint.

Witness mark: A mark made as a means of determining the proper positioning of two pieces of pipe or tubing when joining. A witness mark is typically in addition to a first mark, made in a location that will not be obscured by the joining process; the first mark is often obscured during the process.

Work hardening: A permanent hardening caused by a change in the crystalline structure of metal, caused by repetitive bending or forming.

Wrought: Formed or shaped by hammering.